AF352880

Recent Advances in Nanogenerators

Recent Advances in Nanogenerators

Editors

Ya Yang
Zhong Lin Wang

MDPI • Basel • Beijing • Wuhan • Barcelona • Belgrade • Manchester • Tokyo • Cluj • Tianjin

Editors
Ya Yang
Beijing Institute of
Nanoenergy and
Nanosystems, Chinese
Academy of Sciences
China

Zhong Lin Wang
School of Materials Science
and Engineering, Georgia
Institute of Technology
USA

Editorial Office
MDPI
St. Alban-Anlage 66
4052 Basel, Switzerland

This is a reprint of articles from the Special Issue published online in the open access journal *Nanoenergy Advances* (ISSN 2673-706X) (available at: https://www.mdpi.com/journal/nanoenergyadv/special_issues/recent_advances_in_nanogenerators).

For citation purposes, cite each article independently as indicated on the article page online and as indicated below:

LastName, A.A.; LastName, B.B.; LastName, C.C. Article Title. *Journal Name* **Year**, *Volume Number*, Page Range.

ISBN 978-3-0365-5019-0 (Hbk)
ISBN 978-3-0365-5020-6 (PDF)

© 2022 by the authors. Articles in this book are Open Access and distributed under the Creative Commons Attribution (CC BY) license, which allows users to download, copy and build upon published articles, as long as the author and publisher are properly credited, which ensures maximum dissemination and a wider impact of our publications.
The book as a whole is distributed by MDPI under the terms and conditions of the Creative Commons license CC BY-NC-ND.

Contents

About the Editors

Prof. Dr. Ya Yang

Ya Yang has developed coupled nanoenergy materials and devices for next-generation energy-scavenging technology. He has published a book and over 200 SCI papers, which have been cited over 17,000 times with an H-index of 76 (Web of Science). He obtained the runner-up prize of the National Natural Science Foundation of China as a fourth author in 2018. Moreover, he has held five international conferences and delivered over 50 invited talks. He is the Editor-in-Chief of Nanoenergy Advances. His research interests include ferroelectric materials and devices, hybridized and coupled nanogenerators, ferro-pyro-phototronic devices and thermo-phototronic devices.

Prof. Dr. Zhong Lin Wang

Dr. Zhong Lin (Z.L.) Wang (US citizen) received his Ph.D. in Physics from Arizona State University in 1987. He served as a Visiting Lecturer in SUNY (1987–1988), Stony Brook, as a research fellow at the Cavendish Laboratory in Cambridge (England) (1988–1989), Oak Ridge National Laboratory (1989–1993) and at the National Institute of Standards and Technology (1993–1995) before joining Georgia Tech as a faculty member in 1995. He is the Hightower Chair in Materials Science and Engineering and Regents' Professor at Georgia Tech. His research interests include nanogenerators and self-powered nanosystems, piezotronics for smart systems, hybrid cells for energy harvesting, piezo-phototronics for energy science and optoelectronics.

nanoenergy advances 

Review

Advances of High-Performance Triboelectric Nanogenerators for Blue Energy Harvesting

Huamei Wang [1,2], Liang Xu [1,2,*] and Zhonglin Wang [1,2,3,4,*]

1   Beijing Institute of Nanoenergy and Nanosystems, Chinese Academy of Sciences, Beijing 101400, China; wanghuamei@binn.cas.cn
2   School of Nanoscience and Technology, University of Chinese Academy of Sciences, Beijing 100049, China
3   CUSTech Institute, Wenzhou 325024, China
4   School of Materials Science and Engineering, Georgia Institute of Technology, Atlanta, GA 30332-0245, USA
*   Correspondence: xuliang@binn.cas.cn (L.X.); zlwang@gatech.edu (Z.W.)

Abstract: The ocean is an enormous source of blue energy, whose exploitation is greatly beneficial for dealing with energy challenges for human beings. As a new approach for harvesting ocean blue energy, triboelectric nanogenerators (TENGs) show superiorities in many aspects over traditional technologies. Here, recent advances of TENGs for harvesting blue energy are reviewed, mainly focusing on advanced designs of TENG units for enhancing the performance, through which the response of the TENG unit to slow water agitations and the output power of the device are largely improved. Networking strategy and power management are also briefly discussed. As a promising clean energy technology, blue energy harvesting based on TENGs is expected to make great contributions for achieving carbon neutrality and developing self-powered marine systems.

Keywords: triboelectric nanogenerator; network; blue energy; wave energy; energy harvesting

Citation: Wang, H.; Xu, L.; Wang, Z. Advances of High-Performance Triboelectric Nanogenerators for Blue Energy Harvesting. *Nanoenergy Adv.* **2021**, *1*, 32–57. https://doi.org/10.3390/nanoenergyadv1010003

Academic Editor: Joao Oliveira Ventura

Received: 8 July 2021
Accepted: 22 August 2021
Published: 26 August 2021

Publisher's Note: MDPI stays neutral with regard to jurisdictional claims in published maps and institutional affiliations.

Copyright: © 2021 by the authors. Licensee MDPI, Basel, Switzerland. This article is an open access article distributed under the terms and conditions of the Creative Commons Attribution (CC BY) license (https://creativecommons.org/licenses/by/4.0/).

1. Introduction

Covering over 70% of the earth's surface, ocean plays a crucial role for lives on the planet and can be regarded as an enormous source of blue energy, whose exploitation is greatly beneficial for dealing with energy challenges for human beings [1–3]. With extreme climate conditions taking place more frequently nowadays, the world feels the urge to take immediate action to alleviate climate deterioration caused by global warming [4–6]. Carbon neutrality is thus put forward as a goal to reach balance between emitting and absorbing carbon in the atmosphere [7]. One of the most effective methods is to develop and expand the use of clean energy that generates power without carbon emission, such as the enormous blue energy [8]. Meanwhile, with the increasing activities in ocean, equipment deployed in the far ocean is facing problems regarding an in situ and sustainable power supply, where blue energy is an ideal source for developing new power solutions for such applications, allowing self-powered marine systems and platforms, though the harvesting scale can be much smaller [9–11].

The ocean blue energy is typically in five forms: wave energy, tidal energy, current energy, thermal energy, and osmotic energy, among which the wave energy is promising for its wide distribution, easy accessibility, and large reserves. The wave energy around the coastline is estimated to be more than 2 TW (1 TW = 10^{12} W) globally [12]. However, the present development of wave energy harvesting is challenged by its feature as a type of high-entropy energy, which refers to the chaotic, irregular waves with multiple amplitudes and constantly changing directions that are randomly distributed in the sea [13,14]. Most significantly, wave energy is typically distributed in a low-frequency regime, yet the most common and classic method of blue energy harvesting at status quo, the electromagnetic generator (EMG), performs rather poorly in low-frequency energy harvesting, which relies on propellers or other complex mechanical structures to drive bulky and heavy magnets and metal coils in order to transform mechanical energy into electricity [8]. Thus, it usually has high cost and low reliability.

First proposed by Wang in 2012 [15], the triboelectric nanogenerator (TENG, also called Wang generator) derived from Maxwell's displacement current shows great prospect as a new technology to convert mechanical energy into electricity [16], based on the triboelectrification effect and electrostatic induction [17–20]. TENGs present superiorities including light weight, cost-effectiveness, easy fabrication, and versatile material choices [21–25]. The concept of harvesting blue energy using the TENG and its network was first brought out in 2014 [20]. As a new form of blue energy harvester, the TENG surpasses the EMG in that it intrinsically displays higher effectiveness under low frequency, owing to the unique feature of its output characteristics [22,26–28]. Moreover, adopting the distributed architecture of light-weighted TENG networks can make it more suitable for collecting wave energy of high entropy compared with EMGs, which are oversized in volume and mass [29–33].

Gaining attention, the material selection of TENGs, aimed at fulfilling their requirements and maximizing output, is governed by the mechanism. In triboelectrification process, surface charge density, a fundamental parameter regarding the performance of TENGs, relates to the electron affinity difference of triboelectric material pairs. As a result, polymers prevail as the dielectric layer since their various functional groups enhance charge transfer and capturing, as manifested in the triboelectric series, and they also possess the merits of excellent flexibility, scalability, and low cost [34] (Figure 1a). On top of that, as deduced from the Maxwell's displacement current, other key factors contributing to output include dielectric permittivity, surface morphology, and robustness [35]. In addition, the rapid development of TENGs requires additional material properties that suit devices of various modes under conditioned application scenarios, such as a low friction coefficient for sliding mode TENGs, and high electrostatic breakdown strength and chemical stability for high-voltage TENG devices.

In this paper, recent advances of TENGs for harvesting wave energy are reviewed. The review mainly focuses on advanced designs of TENG units for enhancing the performance of wave energy harvesting, including rolling ball structure, multilayer structure, grating structure, pendulum structure, mass-spring structure, spacing structure, water-solid contact structure, and charge pumping strategy. Through such designs, the response of the TENG unit to slow wave agitations and the output power of the device are largely improved. Networking strategy and power management are also briefly discussed. Finally, further development challenges and the prospect of blue energy harvesting by the TENG are discussed.

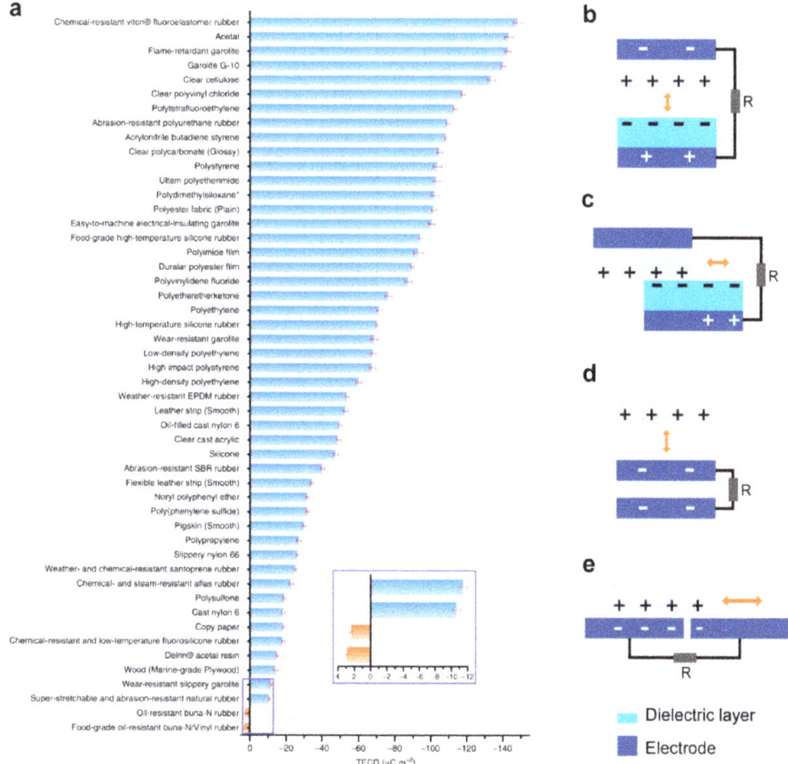

Figure 1. Triboelectric series and four fundamental working modes of TENGs. (**a**) Quantified triboelectric series. Reprinted with permission from ref. [34], Copyright 2019, Springer Nature. (**b**) Vertical contact-separation mode. (**c**) Lateral sliding mode. (**d**) Single-electrode mode. (**e**) Freestanding triboelectric-layer mode.

2. TENG Systems for Blue Energy Harvesting

2.1. Fundamental Working Modes of TENGs

TENGs generate electricity by the coupling of triboelectrification and electrostatic induction, which are classified into the four fundamental working modes [19]: vertical contact-separation mode, lateral sliding mode, single-electrode mode, and freestanding triboelectric-layer mode (Figure 1b–e).

As shown in Figure 1b, in vertical contact-separation mode, the two dielectric surfaces are oppositely charged after physical contact due to triboelectrification. When the two surfaces are vertically separated with a gap in between, a potential drop is produced between the two electrodes attached to the backsides of the two dielectric layers due to the separation of positive and negative static charges, which drives current to flow between the electrically connected electrodes to balance the electrostatic field. When the gap vanishes, the potential drop due to static charges exists no more and so the induced free charges flow backwards. In this way, an alternate current (AC) output is generated in the external circuit under periodic contact and separation movement of the TENG [36].

For the lateral sliding mode shown in Figure 1c, the two dielectric surfaces are charged through triboelectrification during initial sliding motion. Under the full alignment of the two surfaces, no potential difference by static charges is created across the electrodes, since surface static charges of opposite signs completely compensate each other. When a relative displacement paralleled to the interface is introduced under lateral sliding, the mismatched area of the dielectric layers leads to bare static charges and so a potential difference appears between electrodes. The sliding back and forth of the TENG hence results in periodical changes of potential that drive free electrons to flow between the electrodes [37].

In single-electrode mode TENG (Figure 1d), the moving dielectric layer no longer has to be bound to electrodes or electrically connected by wires. The only electrode essential to the mode is one electrically connected to the ground, which can be regarded as another electrode. The surface of the dielectric layer is first charged under full contact with the electrode. When they start to move apart from each other, the induced charges in the electrode decrease in order to balance the electric potential, through charge exchange with the ground. Then, when the dielectric layer moves back to contact again, electrons flow in the opposite direction to re-establish an electrostatic equilibrium until the two surfaces fully overlap. Although charge transfer is not effective as a result of electrostatic screening effect, the triboelectric layer can move freely without any restrains [38].

The freestanding triboelectric-layer mode TENG can also work with the moving dielectric layer disconnected to electrodes, yet without any screening effect (Figure 1e). When the dielectric layer charged by triboelectrification approaches a pair of electrodes asymmetrically, a potential difference is induced across the two electrodes, causing electrons to flow between them to balance the local potential distribution. Under the back-and-forth movement of the dielectric layer, electrons oscillate across the paired electrodes, generating an AC current output [39].

2.2. TENG Systems for Harvesting Blue Energy

For effectively harvesting distributed wave energy, the TENG is conceived to be organized in networks, which can have hierarchical structure of modules [29]. The network structure also enables the device to be applied in different scales of harvesting, ranging from self-powered systems to large-scale clean energy (Figure 2a). In the development of TENGs for blue energy, efforts are mainly focusing on four aspects: TENG unit design, networking strategy, power management, and application system (Figure 2b). The primary and most significant part is the fundamental design of the TENG unit, which relies on continuous improvements focusing on the structure, principle, and material to enhance its energy harvesting performance and to meet demands raised by various ocean environments, both on the water surface and beneath it. Networking strategy is then adopted to add outputs of single TENG units and expand them in a reliable way. It decides the connection pattern of massive TENG units and the coupling effect between TENG units, which could further enhance the performance. Before finally supplying electricity power to the application system, power management is required to manipulate the TENG output for a better match with appliances and improve the power efficiency with circuit approaches. This review mainly emphasizes methods to advance the design of TENG units, which is the most challenging part, and networking strategy and power management are also discussed.

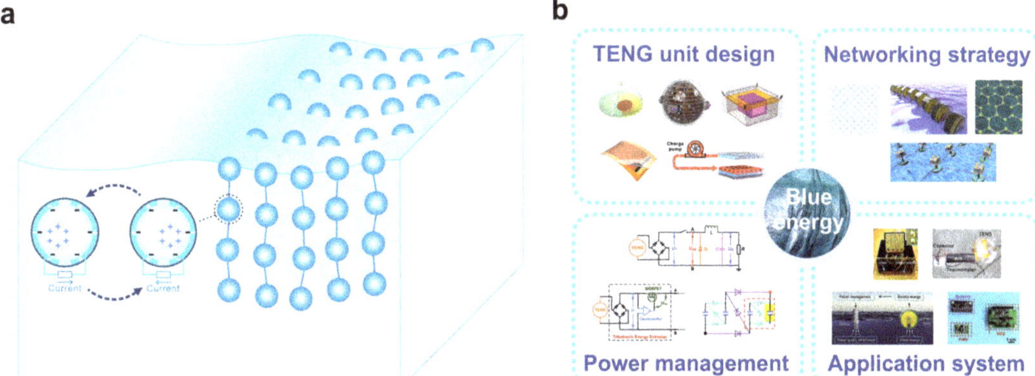

Figure 2. Schematics of blue energy harvesting based on TENGs. (**a**) Schematic diagram of the TENG network for harvesting wave energy. Reprinted with permission from ref. [40], Copyright 2021, IOP publishing, Ltd. (**b**) Schematic diagram of major aspects for blue energy harvesting based on TENGs, including TENG unit design, networking strategy, power management, and application system. Reprinted with permission from ref. [41], Copyright 2019, Elsevier. Reprinted with permission from ref. [42], Copyright 2019, Elsevier. Reprinted with permission from ref. [43], Copyright 2017, Elsevier. Reprinted with permission from ref. [44], Copyright 2015, American Chemical Society. Reprinted with permission from ref. [45], Copyright 2018, Elsevier. Reprinted with permission from ref. [46], Copyright 2019, Elsevier. Reprinted with permission from ref. [47], Copyright 2018, Elsevier. Reprinted with permission from ref. [48], Copyright 2020, Elsevier. Reprinted with permission from ref. [49], Copyright 2017, Elsevier. Reprinted with permission from ref. [50], Copyright 2020, Springer Nature. Reprinted with permission from ref. [51], Copyright 2014, American Chemical Society. Reprinted with permission from ref. [52], Copyright 2020, Springer Nature. Reprinted with permission from ref. [11], Copyright 2019, Elsevier. Reprinted with permission from ref. [29], Copyright 2017, Springer Nature.

3. Advanced Designs of TENG Units for Blue Energy

3.1. Rolling Ball Structure

The rolling ball structure is the most fundamental and typical structure for blue energy harvesting based on TENGs [29,53,54]. The structure basically features a rolling ball inside a rocking spherical shell, which is simple and capable of rocking in water. Due to the small rolling friction, it can be sensitive to tiny wave agitations. Moreover, the structure naturally has a low frequency related to the size of the ball and shell, which can be tuned to reach resonance with water wave motion. Since the rolling ball normally functions as a freestanding triboelectric layer and no wire is connected to the ball, TENG units based on such a structure are durable. Wang et al. reported the rolling-ball-structured TENG based on the freestanding triboelectric-layer mode for the first time, which consists of one rolling nylon ball and two stationary electrodes covered by Kapton films [53] (Figure 3a). Wave vibration drives the ball to roll back and forth between electrodes, producing alternate current in the external circuit connecting the electrodes. However, due to the limited contact area and charge density, the output charge and current are relatively low, reaching only 24 nC.

Advancement on the structure includes the evolution of material and structure of the ball and shell components. As proposed by Xu et al., the UV-treated silicon rubber is adopted as the material for the rolling ball, since its softness can contribute to enhancing the real contact area and durability of the TENG units [54] (Figure 3b). The UV treatment enables Si−O−Si chain scission and oxidative conversion of radical groups in rubber, altering the electron affinity and strengthening triboelectrification with silicone rubber dielectric layer. In addition, microstructures are created on the dielectric surface by mixing polyformaldehyde particles into the rubber to make the surface less sticky, lowering the damping force, thus resulting in a high charge output of the TENG unit even under small agitations (Figure 3c). The charge output of single unit is improved to 72.6 nC.

For further enhancing the contact of the ball, Cheng et al. designed a TENG consisting of a soft liquid/silicone rolling ball and a hollow acrylic sphere attaching two Cu electrodes [55]. The silicone rubber is shaped in a hollow sphere for sufficient softness while containing water inside to ensure full contact (Figure 3d). In this way, the structure of triboelectric rubber can expand the contact area and generate charges as high as 500 nC, more than 10 times that of the conventional structure of polytetrafluoroethylene (PTFE) dielectric layer and hard core (Figure 3e). Systematically speaking, the energy harvesting performance of the soft-contact spherical TENG is largely affected by the contact area, which relates to many factors, including sphere hardness, fill rate, acceleration speed, and bulk density. A theoretical model based on contact electrification and mainly concerning the sphere hardness was quantitatively established by Guan et al. [56]. The result indicates that the dynamic contact of sphere is affected by core mass and fill rate, which is optimized at 83%.

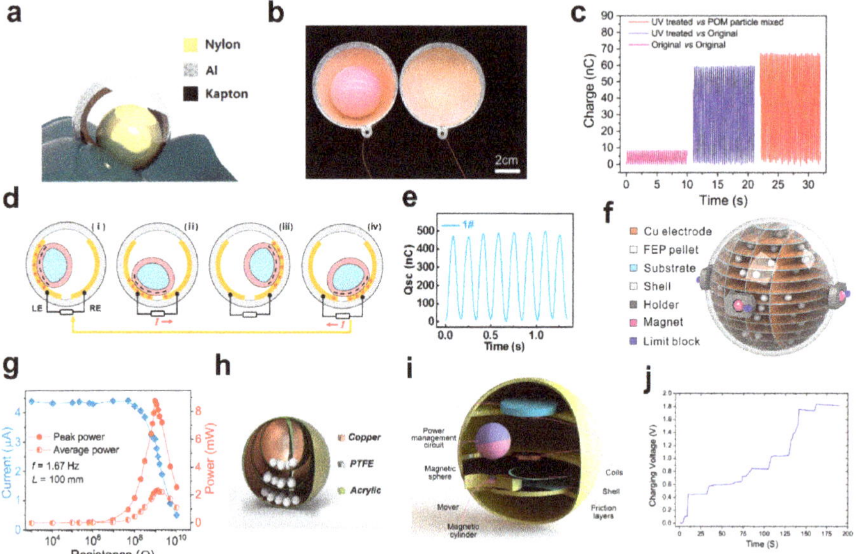

Figure 3. Design and output of TENGs based on rolling ball structure. (**a**) Schematic diagram of TENG with the first rolling ball structure based on freestanding mode. Reprinted with permission from ref. [53], Copyright 2015, Wiley. (**b**) Photo of TENG adopting UV-treated rubber rolling ball and shell. (**c**) Charge output of a single TENG unit before and under UV treatment and POM doping. (**b,c**) Reprinted with permission from ref. [54], Copyright 2018, American Chemical Society. (**d**) Working principle of TENG based on soft liquid/silicone rolling ball structure. (**e**) Charge output of TENG with soft liquid/silicone rolling ball structure. (**d,e**) Reprinted with permission from ref. [55], Copyright 2019, Elsevier. (**f**) Schematic diagram of TENG with 3D electrode structure intercalated by FEP pellets. (**g**) Current and power output of 3D-electrode-structured TENG in water waves. (**f,g**) Reprinted with permission from ref. [42], Copyright 2019, Elsevier. (**h**) Schematic diagram of TENG structure featuring nest-assembling multiple shells. Reprinted with permission from ref. [57], Copyright 2019, Elsevier. (**i**) Schematic diagram of hybrid TENG containing power management, EMG module, and TENG module. (**j**) Charging curve to a supercapacitor by the hybrid TENG in water. (**i,j**) Reprinted with permission from ref. [58], Copyright 2019, American Chemical Society.

Another improvement approach concerning the ball structure is adopting multiple balls. Yang et al. featured TENG units with a 3D electrode structure intercalated by fluorinated ethylene propylene (FEP) pellets in internal channels as the tribo-material [42] (Figure 3f). A pair of 3D-structured electrodes are made by electrode plates of connected copper layers on the substrates, largely enhancing the surface area of the electrodes and improving the utilization of the internal space of the shell. The device can output under different modes of external excitation as controlled by inertial forces and gravity. For harmonic translational mode, the increase in displacement extent and frequency leads to rapid rising of the transferred charges and current, obtaining a high charge output of 520 nC. Under impact translational mode, the structure of multiple rolling balls widens the current peak, boosting the peak power density and average power density to 32.6 W m^{-3} and 8.69 W m^{-3}, respectively. Meanwhile, an average power density of 2.05 W m^{-3} can be achieved in real water waves agitated at 1.45 Hz (Figure 3g). Moreover, the amount of FEP pellets influences charge output due to the cancel-out of electrostatic induction by FEP pellets in the two 3D electrodes when more than half of the device volume is filled. Considering the evolution of the shell structure, a similar TENG structure reported by Pang et al. is characterized by nest-assembling multiple shells with decreasing sizes, with PTFE balls filled into the gap space between neighboring shells [57] (Figure 3h). On the interior of each shell, gaped Cu electrodes are painted with good adhesion, conductivity, and durability. The PTFE balls can move freely on electrode surfaces. Such unique structure design also improves the utilization of limited device volume and elevates the contact area between balls and electrodes.

Combining the TENG with the EMG by making the rolling ball magnetic is another enhancement method, which was achieved by Wu et al. The packaging hollow shell of the hybridized device is divided into four zones by acrylic disks, which are set for power management, EMG module, and TENG module, respectively [58] (Figure 3i). On one hand, the magnetic sphere drives the EMG module via electromagnetic induction, as two coils are embedded on both sides of the magnet; on the other hand, it motivates the mover by magnetic attractive force, since the mover is centered by a magnetic cylinder. Then, the mover can drive the TENG module through its interaction with PTFE friction layers adhered to Tai Chi-shaped electrodes. Eventually, such design can collect wave energy from any orientation. It was also placed on a buoy to test in lake Lanier. During 162 s, the electricity stored in a supercapacitor reached 1.64 mJ (Figure 3j).

3.2. Multilayer Structure

Considering that triboelectrification is a phenomenon happening at the surface, to improve the output, enhancing the contact area in a certain space with a multilayer structure should be an effective method. The strategy is realized in many different types of TENGs, and the output charge and power are greatly improved. The major challenge lies in the structure design to drive the complex structure by external agitations, which requires delicate structure designs.

An integrated multilayered TENG based on soft membrane was proposed by Xu et al. The structure includes the inner oscillator and the outer shell, which are connected by elastic bands to form a spring-levitated oscillator structure [43] (Figure 4a). The oscillator is based on air-driven soft membranes, which shape into air-pocket-like structures to split the air chamber into upper and lower parts. Accordingly, the electrodes of TENGs attached to the membranes form upper and lower multilayered TENG arrays. When the outer shell moves up and down as pushed by waves, the inner oscillator goes down and up relative to the shell and compresses the air chambers alternately. Differences in air pressure reshape membranes, making the upper and lower TENGs contact and separate periodically. Thus, the agitations from water waves can effectively drive the multilayered TENGs via the air-driven mechanism. The device with a series of integrated TENGs working simultaneously and effectively can produce high transferred charges of 15 µC per cycle (Figure 4b).

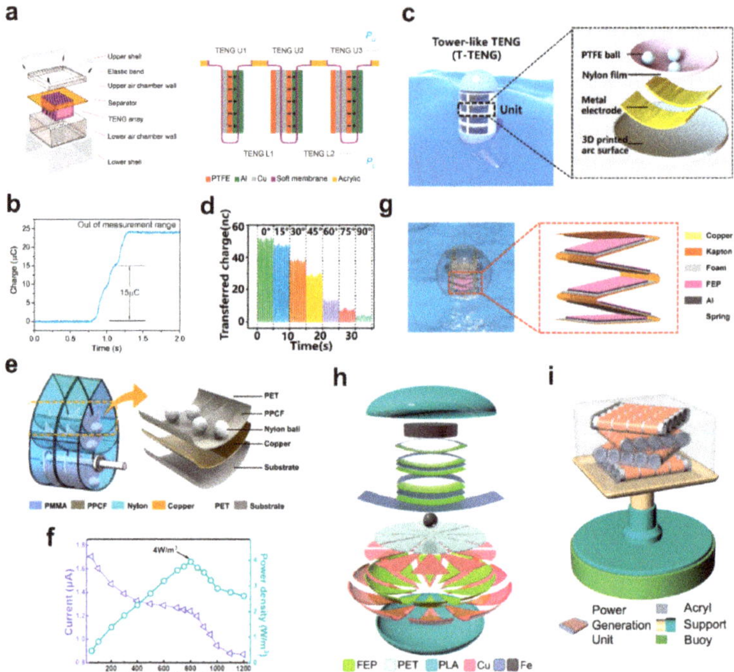

Figure 4. Design and output of TENGs based on multilayer structure. (**a**) Schematic diagrams of integrated multilayered TENG based on soft membrane and spring levitation. (**b**) Charge output of integrated soft membrane TENG per cycle. (**a,b**) Reprinted with permission from ref. [43], Copyright 2017, Elsevier. (**c**) Schematic diagram of TENG with tower-like structure containing multiple units in one tubular block. (**d**) Charge output of tower-like TENG collecting wave energy in different directions. (**c,d**) Reprinted with permission from ref. [59], Copyright 2019, American Chemical Society. (**e**) Schematic diagram of TENG with nodding duck structure. (**f**) Current and power density of nodding duck-structured TENG under water waves. (**e,f**) Reprinted with permission from ref. [60], Copyright 2021, American Chemical Society. (**g**) Photo and schematic figure of TENG with zigzag multilayered structure. Reprinted with permission from ref. [61], Copyright 2018, Wiley. (**h**) Schematic diagram of oblate TENG. Reprinted with permission from ref. [62], Copyright 2019, Wiley. (**i**) Schematic diagram of floating buoy-based TENG structure. Reprinted with permission from ref. [47], Copyright 2018, Elsevier.

Similarly, Xu et al. reported a tower-like TENG structure consisting of multiple parallel connected units in one tubular block [59] (Figure 4c). In the unit, PTFE balls roll on 3D-printed arc surface shell, which is attached with two stationary electrodes coated by melt adhesive reticulation nylon film, forming a freestanding mode TENG. Only one rectifier is needed in a block since all PTFE balls move in the same phase under agitations. The structure is also capable of collecting water wave energy in arbitrary directions (Figure 4d).

Liu et al. reported a nodding duck-structured TENG that consists of external nodding duck blocks installed with shafts and bearings, and internal multitrack TENG units [60] (Figure 4e). Each unit is made up of a pair of Cu electrodes, free-rolling nylon balls, and dielectric layers placed on arc-shaped tracks. Together, they form a multilayer-structured TENGs of three layers that share electrodes and dielectric films, maximizing the utilization of internal space. A multiple-track structure was also introduced to avoid arbitrary movements of balls in different TENG layers. Upon triggering of external waves, nodding duck shell rotates around the shaft, leading balls to roll back and forth on the tracks. Under the combination of gravity, supporting force, and damping motion, an AC output is generated with a maximum peak power density of 4 W m^{-3} by one block at 800 MΩ under waves excited at 0.21 Hz (Figure 4f).

A spherical TENG that consists of a zigzag multilayered structure with spring assistance was proposed by Xiao [61] (Figure 4g). The multilayered structure is fabricated by deforming a thin Kapton film at evenly distanced intervals to serve as the substrate for five TENG units on both sides.

Liu et al. introduced an oblate spherical TENG structure, which is divided into the upper part for rough sea energy harvesting and the lower part for tranquil ocean energy collection [62] (Figure 4h). The upper part adopts a multilayer structure, which is composed of a pie iron and three arched units. The units are made up of two joint spring steel plates, one coated with FEP film. The upper part operates as the pie iron presses FEP film to contact the plates. The lower part is composed of a radial patterned FEP film with copper electrodes underneath, while another same-structured PET film is suspended above it. The lower part works as the external excitation triggers the iron shot to push FEP films into contact (shot on them) or separation (shot off them). Such structure also enhances self-stabilization and costs less, since the utilization of spring plates with good elasticity makes extra supporting structures for the units unnecessary.

Kim et al. reported a floating buoy-based TENG that contains a power generation unit and a height-adjustable support, which connects the unit to a floating buoy [47] (Figure 4i). There are four arrays of cylinder TENGs in the unit, each rotationally stacked upward for every 45°, forming a multilayered structure for energy harvesting in arbitrary directions. For each freestanding mode TENG, a solid PTFE bar oscillates inside a tube, with the outer surface of the tube being attached by four Cu electrodes. The height-adjustable support changes the force applied on the unit according to the alteration of the water wave amplitude. The acrylic protection layer can tolerate harsh environmental conditions.

3.3. Grating Structure

The grating structure is regarded as a high-performance TENG design that has superior output. It can convert low-frequency agitations to high-frequency electrical output through a periodic grating electrode pattern. However, large friction at the interface makes it difficult to be agitated by slow water waves.

Bai et al. proposed a radial-grating-disk-structured TENG, where every TENG unit is made of facing sides of a stator disk and a rotator disk, patterned with radial-grating-structured electrodes [46] (Figure 5a). The acrylic surface as one tribo-surface is frosted to reduce the friction force for easier agitation and better durability of the rotation device. In this way, the device can be successfully driven by the slow water waves, and the average power density of the device reaches 7.3 W m^{-3} at a wave frequency of 0.58 Hz (Figure 5b). Its capability of driving a self-powered total dissolved solids testing system is demonstrated (Figure 5c), indicating that the high-power wave energy harvester can be applied for in situ, real-time mapping of water quality.

3.4. Pendulum Structure

The pendulum structure features swinging mass blocks, which temporarily transform and store kinetic energy as potential energy to drive TENGs to output electricity. The structure inherently features with high sensitivity under mechanical excitations, and improves the energy conversion efficiency since pendulums can convert impact agitations to continuous swinging in its inherent frequency and can elongate the operation time of TENGs if the friction is small enough.

In this previously discussed work, Bai et al. set the TENG units of rotators and stators in a tandem with two swing mass blocks on both sides, together encapsulated in a waterproof acrylic shell [46] (Figure 5a). While stators are fixed on the shell, rotators are linked with swinging mass blocks by shafts; therefore, when the eccentric blocks are agitated by waves, rotators can rotate synchronously. The tandem structure allows unit expansion, which scales up the total output and reduces the matched resistance through simple parallel connection of units. As the number of units expends, the peak current increases from 8 µA to 120 µA, and the charge output raises from 0.2 µC to 3.4 µC, under a harmonic reciprocating rotation. When interacting with waves, the nature of the TENG in frequency response resulted by various configurations was revealed. An additional mass block is capable of adjusting frequency characteristics of the device, since changing block mass alters the resonant frequency of the TENG.

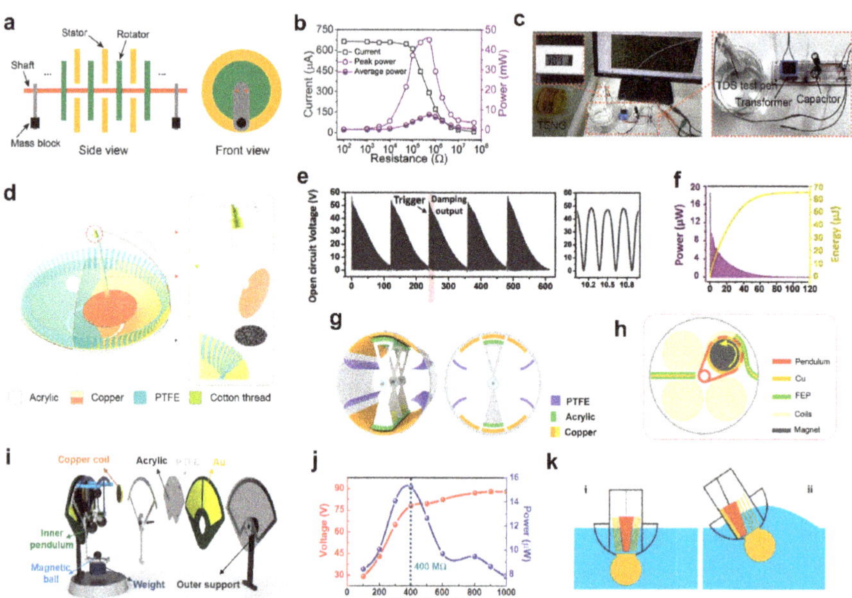

Figure 5. Design and output of TENGs based on grating structure and pendulum structure. (**a**) Schematic diagrams of radial-grating-disk-structured TENG based on pendulum. (**b**) Current and power output of grating-structured TENG agitated by water waves. (**c**) Demonstration of grating-structured TENG driving a self-powered total dissolved solids testing system. (**a–c**) Reprinted with permission from ref. [46], Copyright 2019, Elsevier. (**d**) Schematic diagram of TENG adopting a pendulum triboelectric layer and thin stripes. (**e**) Damping output of the pendulum TENG. (**f**) Power output curve of the pendulum TENG. (**d–f**) Reprinted with permission from ref. [41], Copyright 2019, Elsevier. (**g**) Schematic diagram of cylindrical TENG with a bearing-based swing. Reprinted with permission from ref. [63], Copyright 2020, Wiley. (**h**) Schematic diagram of hybrid TENG adopting rotational pendulum structure. Reprinted with permission from ref. [64], Copyright 2019, Elsevier. (**i**) Schematic diagram of hybrid TENG with chaotic pendulum structure. (**j**) Voltage and power output of TENG module in the hybrid device. (**i,j**) Reprinted with permission from ref. [65], Copyright 2020, Elsevier. (**k**) Working principle of active resonance TENG device. Reprinted with permission from ref. [66], Copyright 2021, Elsevier.

Other different types of pendulum structures are utilized as well to suit diverse designs of TENGs. Lin et al. reported a freestanding mode TENG that is mainly composed of an electrode layer, a pendulum triboelectric layer, and thin stripes [41] (Figure 5d). The pendulum is a copper layer capable of free oscillation with high sensitivity to excitations, since it is connected to the outer shell by a cotton thread. The electrode layers are an internal circular one and an external ring one, covered by PTFE films. Thin strips on the outer edge of acrylic shell act by electrification with the pendulum layer. Such design can be easily agitated to swinging and output electricity continuously (Figure 5e,f).

Jiang et al. proposed a TENG structure that consists of a bearing-based swing component and a cylindrical acrylic shell, whose internal surface is attached by equally sized Cu electrodes and PTFE stripes [63] (Figure 5g). The swing component is made of acrylic blocks with upper hollow and lower solid structures, the top and bottom of which are both adhered by arc-shaped acrylic strip. Swing motion of the block is realized by adding copper blocks to the bottom to lower the gravity center.

Apart from swinging pendulum devices, Hou et al. demonstrated a hybrid TENG adopting the rotational pendulum structure, which consists of a pendulum rotor, TENG blades, coils, and a cylindrical-tube-shaped frame [64] (Figure 5h). The pendulum rotor is made up of magnets twined by a copper ring, which are installed on the center shaft with ceramic bearings. The TENG blades connected in parallel are sandwiched flexible structure of FEP/Cu/FEP, with one side freestanding and the other side fixed on the frame. The blades and the Cu ring form the TENG part. The EMG is composed of coils that curl around the shaft and are embedded in the bottom, and the magnets of pendulum rotor, which are just above the coils. As the pendulum rotor rotates clockwise and anticlockwise around the shaft, a contact-separation mode TENG forms as the copper ring touches blades, and the EMG works under the periodic movement of magnets passing coils. The bearing support on the magnet rotor enables rotation even under tiny external excitations.

Another hybridized device reported by Chen et al. adopts a chaotic pendulum structure for the virtue of low working frequency [65] (Figure 5i). The chaotic pendulum consists of a major pendulum that swings back and forth under oscillation, and an inner pendulum that moves in a different chaotic manner following the major one. Both the TENG and the EMG are driven by the chaotic inner pendulum. The major pendulum of a weighted magnetic ball is used to enhance the oscillation frequency of the inner pendulum. The TENG is made of PTFE films and Au electrodes fixed on the inner swing. The EMG consists of the inner pendulum, which is composed of three magnetic balls evenly spaced on a rotating shaft, and three coils attached to the inside. When the main swing is pushed by water excitations, the freestanding mode TENG works based on the chaotic movement of the inner swing to send PTFE films moving back and forth on the electrodes, converting mechanical energy into electricity. The EMG part generates magnetic flux change of copper coils, as the magnetic balls begin moving under gravity and external magnetic incentive condition. The maximum output of the TENG reaches 15.21 μW and the EMG is up to 1.23 mW under an external excitation frequency of 2.5 Hz (Figure 5j).

To collect wave energy in all directions, Zhang et al. designed an active resonance TENG device that mainly integrates a simple pendulum, a floating tumbler, and flexible ring TENG [66] (Figure 5k). The tumbler intercepts wave energy, which is then captured by the pendulum, and finally transformed into electricity through TENGs. In the initial floating state, TENGs with flexible ring multilayered structure remain separate. The wave propagation makes the device lean aside; the pendulum thus brings the TENGs into full contact. A similar process proceeds as waves leave the system, generating two rounds of contact–separation output in one wave agitation. The pendulum with a circle-table-shaped core and the tumbler composed of a hemispherical float and a steel ball can create resonance effect by themselves, regardless of the external excitation frequency. Together with their damped motion, the energy harvesting efficiency is also enhanced. Moreover, the omnidirectional structural design ensures effective collection of omnidirectional water wave energy by the device.

3.5. Mass-Spring Structure

The mass-spring structure can produce oscillation under external agitations, which can be tuned to the frequency of the water waves to achieve resonance. It normally vibrates along the vertical direction with the effect of gravity. The structure shares the energy transfer mechanism of the pendulum structure, along with its merits.

Hu et al. reported a TENG built on a suspended 3D structure within a cube, which is a spiral with a seismic mass at the bottom that behaves like a spring [67] (Figure 6a). The top of the spiral is fixed to the inner top surface of the cube. One of TENG plates is attached to the bottom of the spiral, and the other plate is at bottom of the cube. Initially, the TENG plates slightly contact because the spiral is released. When the external disturbance pushes the cube's bottom to displace, the seismic mass vibrates with it and so the spiral begins oscillation. TENG plates thus separate as a result of spiral contraction, forming a vertical contact-separation mode TENG. A maximum output power density of 2.76 W m^{-3} on a load of 6 MΩ can be achieved.

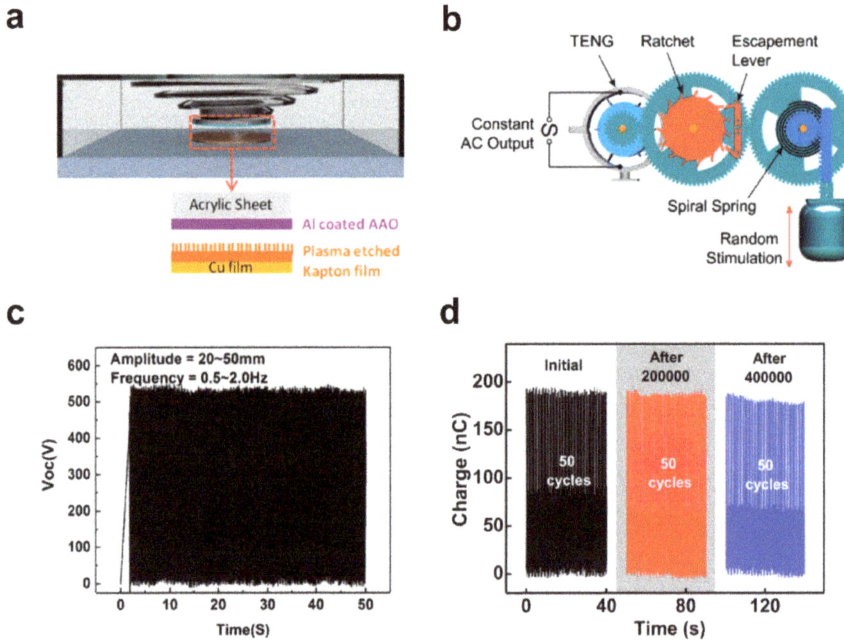

Figure 6. Design and output of TENGs based on mass-spring structure and spacing structure. (**a**) Schematic diagram of TENG based on suspended 3D structure. Reprinted with permission from ref. [67], Copyright 2013, American Chemical Society. (**b**) Schematic diagram of TENG device adopting the principle of auto-winding mechanical watch. (**c**) Constant output of the TENG device under low frequencies and varying working conditions. (**b**,**c**) Reprinted with permission from ref. [68], Copyright 2021, Elsevier. (**d**) Charge output of cylindrical TENG with a bearing-based swing during cycling test. Reprinted with permission from ref. [63], Copyright 2020, Wiley.

The previously discussed multilayered TENG structure raised by Xu et al. contains a spring-levitated oscillator, where the inner oscillator and the outer shell are connected by elastic bands [43] (Figure 4a). The shell acts as a mounting base that also protects the oscillator. Energy transfer is achieved by the oscillator, whose elastic bands provide elastic support for the oscillator both horizontally and perpendicularly, making it stable under movement. The device can oscillate several times under single agitation.

The work by Xiao et al. mentioned beforehand also contains multilayered TENG units that rely on the structure for energy conversion [61] (Figure 4g). Four flexible springs are to support the mass block, and four rigid springs are attached to the mass's top to exert resilience force to press the TENG units upon wave excitation. Four shafts are also fixed for the mass and springs to move along with.

He et al. developed a mechanical regulator to drive TENGs with steady AC output, using the principle of auto-winding mechanical watch, which is to store random energy by transformation into potential energy and then release it at a designed pace [68] (Figure 6b). The structure involves more complex mechanisms by springs. The device consists of three parts: energy harvest and storage module, energy controllable release module, and energy conversion module. The first module contains input shaft, input gear train, and flat spiral spring, which transforms environmental energy into elastic potential energy. The second module is composed of escapement-spring-leaf and transmission gear train. The escapement-spring-leaf is an integration of hairspring system and escapement, in order to reserve their kinetic merits and simplify the transmission chain to reduce energy loss. The last module is made up of output gear train, one-way shaft system, flywheel, and TENG units. When the system works, the spiral spring is first compressed by wave movements, driving the rotor of TENGs to rotate under the control of the escapement-spring-leaf. With such design, constant output is realized under even a low frequency of 0.5 Hz and random working stimulations (Figure 6c).

3.6. Spacing Structure

The spacing structure introduces an air gap between the triboelectric layer and electrodes, which reduces friction and surface wear, thus enhancing the robustness and durability of the TENG device, and the triboelectric charges can be replenished by soft structures and intermittent contact.

The feature was first realized in the work by Lin et al. that has been discussed in the pendulum structure section. In the device, the pendulum triboelectric layer and electrode layer are not allowed to contact intimately [41] (Figure 5d). A freestanding gap is reserved to reduce material abrasion and enhance oscillation degree. Hence, the TENG device boosts superior durability for long time operation, with one trigger lasting for more than 120 s. PTFE soft structures around the electrodes allow a supplement of charges on the pendulum triboelectric layer.

In the work by Jiang et al., an air gap also exists between the interfaces of triboelectrification, namely, the surfaces of arc-shaped acrylic strips and Cu electrodes, as have been introduced in the pendulum structure section [63] (Figure 5g). Although the electrodes contact with PTFE brushes in the swing process for charge-replenishing purpose, the friction resistance is low. Moreover, the swing motion of disks relative to the central steel shaft is smoothed by bearings with lubrication oil. Apart from that, animal furs can also be adopted as low-friction materials to further reduce wearing and enhance charge output. As a result, the performance remains steady after 400,000 cycles (Figure 6d).

3.7. Water-Solid Contact Structure

The water-solid contact (L-S contact) structure involves two components: electrodes covered by dielectric films and liquid. The working mechanism includes two steps: firstly, charges are generated on the dielectric surface through L-S contact electrification; secondly, electrostatic induction works through the asymmetric screening of triboelectric charges by liquid. Here, liquid can be regarded as the freestanding layer, and so TENGs based on L-S contact are mainly in freestanding mode. Essentially, L-S contact TENGs can only operate under the condition that they are constantly shifting between the states of being submerged in liquid and emerging out of liquid. Surface charge density is enhanced since contact area is increased at the liquid-solid interface. Moreover, the structure includes no extra mechanical component. The friction at the interface is also suppressed.

The first work for wave energy harvesting based on the mechanism was proposed by Zhu et al. [51]. The specific working process of the TENG is described in Figure 7a. After L-S contact electrification, the FEP film is distributed with negative triboelectric charges. The partial submerge of electrode A by water wave induces an interfacial electrical double layer, formed by positive charges in water like hydroxonium to screen the negative surface charges on FEP. Accordingly, the potential difference between electrode A and B rises until B starts to dive into water too. When the whole device is submerged, triboelectric charges are completely screened by ions in water and no electrostatic induction exists any more (Figure 7b). In addition, the FEP film surface is intentionally patterned with hydrophobic nanowires to create nanoscale roughness. The modified surface can not only realize water infiltration into aligned nanowires to enlarge contact area, but also ensures immediate water repellency as water leaves the surface.

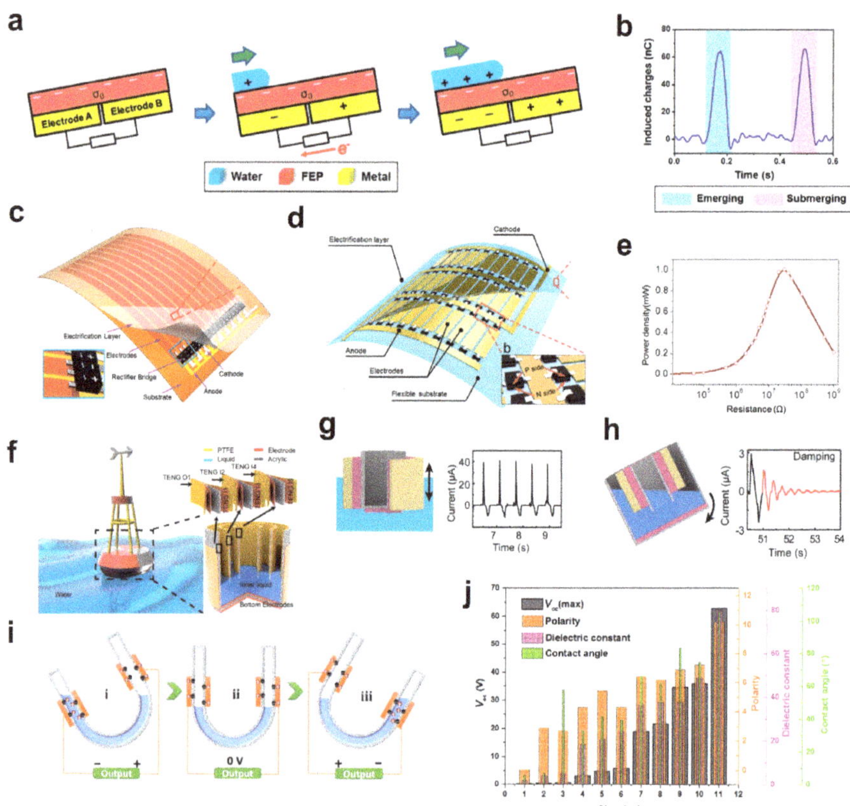

Figure 7. Design and output of TENGs based on L-S contact structure. (**a**) Working principle of the first reported L-S contact TENG. (**b**) Charge output of the L-S contact TENG. (**a**,**b**) Reprinted with permission from ref. [51], Copyright 2014, American Chemical Society. (**c**) Schematic diagram of TENG with surface-mounted bridge rectifier arrays. Reprinted with permission from ref. [44], Copyright 2015, American Chemical Society. (**d**) Schematic diagram of TENG with two-dimensional networked structure. (**e**) Power output of TENG with two-dimensional networked structure. (**d**,**e**) Reprinted with permission from ref. [69], Copyright 2018, American Chemical Society. (**f**) Schematic diagram of buoy-like TENG. (**g**) Working principle and output of the outer TENG of buoy-like device under up-down movement. (**h**) Working principle and output of the inner TENG of buoy-like device under shaking or rotation movement. (**f**–**h**) Reprinted with permission from ref. [70], Copyright 2018, Wiley. (**i**) Working principle of U-tube TENG. (**j**) Output of eleven types of liquids with different polarities, dielectric constants, and contact angles. (**i**,**j**) Reprinted with permission from ref. [71], Copyright 2018, Springer Nature.

Further optimization focuses on the electrode structure, the film nanostructure, and liquid type. An integration approach was proposed by Zhao et al. that mounts bridge rectifier arrays on the TENG substrate surface to connect parallel electrodes together [44] (Figure 7c). Paralleled arrays of strip-shaped Cu electrodes are sputtered on Kapton substrate. The bridge rectifiers rectify currents between any pair of neighboring electrodes, by connecting pairs of adjacent electrodes through input pins, and linking all the negative output pins and positive output pins respectively to form comb-shaped joint cathode and insulated anode. In this way, the output can be added up to form pulsed direct current (DC) between the anode and cathode. The outermost electrification PTFE layer also protects and insulates all the conductive parts from water, with PTFE nanoparticles of 200 nm evenly distributed on its surface.

Zhao et al. further reported a TENG featuring a two-dimensional networked structure that yields high and stable output regardless of the wave type [69] (Figure 7d). The characteristic is realized through a rectifying chip based on p-n junction that is connected to both ends of electrode units, where the chip acts as a gate that only allows induced current from the anode to cathode. The electrode units in strip shape form a two-dimensional array on the Kapton substrate, and the two chips at the opposite sides of every electrode unit are in serial connection. A stable short circuit current of 13.5 µA and electric power of 1.03 mW are produced (Figure 7e).

As an intriguing structure design, the buoy-like TENG device proposed by Li et al. involves an outer TENG whose bottom electrode is on the outside, and an inner TENG whose bottom electrode is inside the buoy [70] (Figure 7f). As PTFE film rubs water, charges are induced on the surface, leading to a potential difference between the bottom electrode and the electrode on the backside of PTFE. The outer TENG works under up-down wave movements, while the parallel-connected inner TENGs function under shaking or rotation movements (Figure 7g,h). Consequently, the device can collect energy from different types of wave excitations.

The work constructed by Pan et al. aimed to study the influence of liquid properties on TENG output, by a U-tube TENG that includes a FEP U-tube, Cu electrodes, and liquid solution [71] (Figure 7i). The liquid flows on the coarsely nanostructured inner surface of the U-tube. As the U-tube rolls rightward or leftward and drives the liquid to overlap the FEP U-tube, currents are generated on the Cu electrodes wrapped outside the tube. It is concluded that liquid with higher polarity, dielectric constant, and contact angle yields higher output, and pure water, which is the eleventh specimen, showed the best experimental result in eleven types of liquids (Figure 7j).

3.8. Charge Pumping Strategy

As is known, the TENG is based on triboelectrification to generate static charges, and electrostatic induction to drive free charge transfer. However, the limited surface charge density by triboelectrification greatly hinders the output enhancement of TENGs [21]. Compared to other methods, developing new mechanisms on charge accumulation can address the issue fundamentally under ambient conditions.

The charge pumping strategy is shown to effectively enhance the charge density [45]. The utilization of the strategy requires a floating layer that can accumulate and bind large amounts of charges for electrostatic induction, and a charge pump that can pump charges into the floating layer simultaneously (Figure 8a). The realization of such mechanism involves a pump TENG, which is a normal TENG with a rectifier to make charge flow unidirectionally, and a main TENG (Figure 8b). The floating layer of M2 is achieved owing to the insulation from electrodes M1 and M3 of the main TENG by two dielectric layers, D1 and D2. Therefore, M2 cannot exchange charges with the main TENG electrodes and can only accept electron flow of one direction from the rectifier. The charge pumping process from the pump TENG to the main TENG is similar to charging a capacitor, and can last until the dielectric layers break down electrically. The pumped charges inside the floating layer are like static charges in normal TENGs, thus achieving large output through electrostatic induction. The energy collected by pump TENG adds up through charge accumulation in the main TENG, achieving an ultrahigh effective surface charge density of 1020 µC m^{-2} in ambient conditions, which does not rely on intensive rubbing or contact, improving energy efficiency (Figure 8c).

Another similar method reported by Liu et al. utilizes only one TENG combined with a self-voltage-multiplying circuit [72] (Figure 8d). Charge amplification is realized through altering the connection pattern of external capacitors accompanying the motion of the TENG. A voltage-multiplying circuit (VMC) module is adopted that can automatically switch the capacitors between parallel and serial connection during the contact and separation process of the TENG, owing to the unidirectional property of diodes. The doubled charge output from the VMC can flow back to the TENG in each cycle, thus greatly enhancing the charge density. A high output is achieved with a Zener diode for voltage stabilization (Figure 8e).

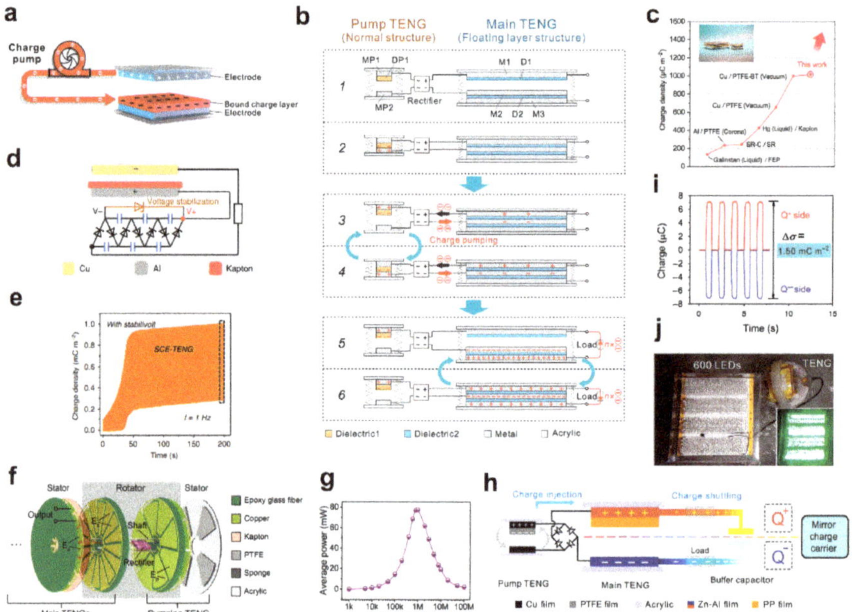

Figure 8. Design and output of TENGs based on charge pumping strategy. (**a**) Schematic diagram on mechanism of charge pumping. (**b**) Working principle of TENG based on charge pumping. (**c**) Ultrahigh surface charge density of charge-pumping TENG compared with other works. (**a–c**) Reprinted with permission from ref. [45], Copyright 2018, Elsevier. (**d**) Schematic diagram of TENG combining VMC. (**e**) Output of TENG with VMC. (**d,e**) Reprinted with permission from ref. [72], Copyright 2019, Springer Nature. (**f**) Schematic diagram of rotation and sliding TENG combined with charge pumping. (**g**) Average power output of rotation and sliding TENG based on charge pumping. (**f,g**) Reprinted with permission from ref. [73], Copyright 2020, Wiley. (**h**) Schematic diagram of TENG based on charge shuttling. (**i**) High effective charge density of TENG based on charge shuttling. (**j**) Demonstration of TENG device based on charge shuttling for wave energy harvesting. (**h–j**) Reprinted with permission from ref. [52], Copyright 2020, Springer Nature.

The work reported by Bai et al. develops a successful combination of charge pumping mechanism with rotation and sliding TENGs [73] (Figure 8f). The device is composed of a pump TENG and a main TENG, each with two disks as the rotator and the stator. A novel synchronous rotation structure is designed to allow direct injection of bound charges from the pump TENG to the main TENG. Specifically, an inversion structure allows the electrode disk of pump TENG to rotate, which is assembled with the main TENG by shafts to form the rotator, while the output electrodes of the main TENG are part of the stator. In this way, the electrodes of the pump TENG and storing bound charge layer can rotate in a synchronous mode. After the pump TENG injects charges into the storing electrodes of the main TENG, opposite charges are induced in the underlying output electrodes. Then, with the rotational sliding of storing electrodes, induced charges move following the motion of the bound charges, generating ultrahigh charge density and an average power density of 1.66 kW m^{-3} under a low drive frequency of 2 Hz (Figure 8g). Moreover, the decoupling of surface charge generation and friction enables lubricant to be applied to the interfaces to enhance durability. Ultrafast accumulation of bound charges is achieved in the rotation TENG that also possesses good expandability to charge multiple main TENGs by one pump TENG.

Wang et al. proposed a fundamentally different mechanism of charge shuttling to generate output on TENGs, which is not based on the electrostatic induction of static charges but on the shuttle of corralled charges in a TENG and a buffer capacitor [52] (Figure 8h). The corralled mirror charge carriers are accumulated through charge pumping strategy, and shuttle due to the interaction of two quasi-symmetrical conduction domains, which doubles the output. The device based on charge shuttling is composed of a pump TENG, a main TENG, and a buffer capacitor. The electrodes of the main TENG and the buffer capacitor form two conduction domains respectively presenting a quasi-symmetrical structure with Q^+ side and Q^- side. The pump TENG injects charges into domains through a rectifier. Upon contact and separation of the main TENG, its capacitance changes while that of the buffer capacitor remains, inducing a voltage difference between them to drive charges to shuttle in a quasi-symmetrical way, generating electricity on two loads. Consequently, the charge output is doubled by the two shuttling mirror charge carriers, achieving a total effective charge density of 1.5 mC m^{-2} with a Zener diode for stabilization (Figure 8i). The mechanism was successfully applied in an integrated device for wave energy harvesting. The peak current of the device reached about 1.3 mA under wave agitations of 0.625 Hz. The maximum peak power was 126.67 mW at 300 kΩ, corresponding to a volume power density of 30.24 W m^{-3} (Figure 8j).

4. Networking Strategy and Power Management

4.1. Networking Designs

The networking design of TENG units for blue energy harvesting involves mechanical connection and electrical networking. The ideal mechanical connecting method should be highly resilient, anti-fatigue, mechanically robust, and resistant to corrosion, while the electrical networking demands maximized output, avoidance of the disturbing electrostatic induction, and sufficient protection from the interference and corrosion by sea water. Meanwhile, the interconnection of TENG units is expected to introduce a coupling effect to enhance the total harvesting efficiency.

The coupled TENG network for wave energy harvesting was first realized by Xu et al. [54]. The work investigated the coupling design in details. Three connection methods were proposed and tested, of which the flexible connection shows higher efficiency and better performance (Figure 9a,b). The output of the linked unit is 10 times greater than that without linkage due to the coupling between TENG units. The network of 16 units is rectified and electrically connected for self-powered sensing by wave agitations (Figure 9c).

In order to enhance the robustness of mechanical connection, another intriguing way is to construct a network structure that can heal by itself after being broken by strong waves, as put forward by Yang et al. [42]. The idea was accomplished by the design of a self-adaptive magnetic joint (SAM-joint), which is composed of a rotatable spherical magnet and a limit block mounted around the TENG device shell (Figure 9d). The SAM-joint functions by self-adaptive mechanism of the pole and anisotropic restriction on the degree of freedom, enabling the network structure with characteristics of self-assembly, self-healing, and facile reconfiguration, which greatly improve the autonomy and robustness of the TENG network (Figure 9e).

Liu et al. put forward a study concerning four types of electrical networking topology [74]. The effect of wire resistance and output phase asynchrony of different units on the network output were analyzed and concluded to be crucial (Figure 9f).

Liu et al. designed a special type of plane-like power cable for TENGs to meet the demands raised by both electrical networking and mechanical linkage, which consists of spring steel tapes and three polymer films on the outside [48] (Figure 9g). Firstly, the plane shape that limits the entanglement among interconnected units, along with the highly resilient steel tape as the structural skeleton and the hydrophobic PTFE film, fulfills requirements for mechanical connection. Secondly, the conducting wires of TENGs are sandwiched in nearby steel tapes to avoid water screening effect for higher electrical output. Moreover, the power cable structure has an additional merit of generating electricity by the cable itself as a TENG device. Because as the cable contacts and separates with water, the steel tapes act as electrodes, which are covered by PTFE films, forming a L-S contact mode TENG.

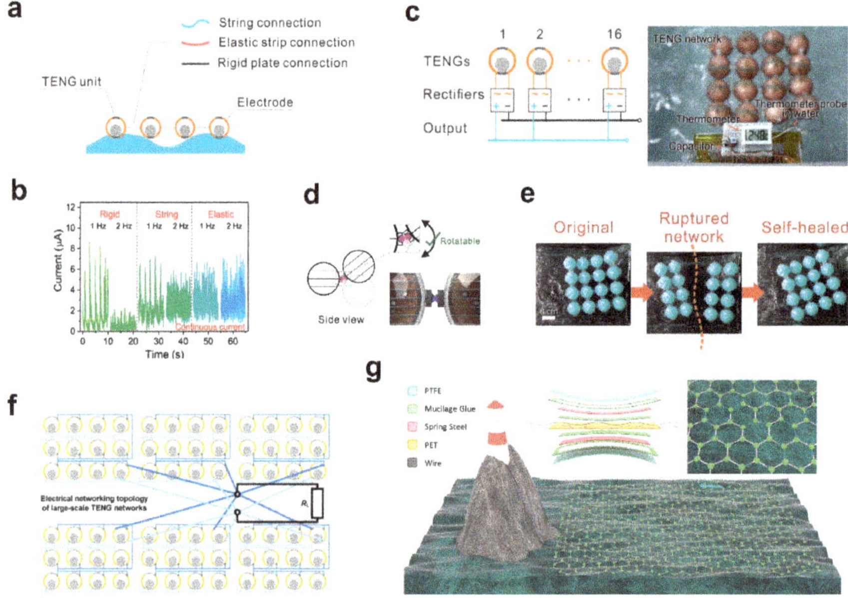

Figure 9. Networking design and output of TENGs. (**a**) Schematic diagram of three connection methods of coupled TENG network for wave energy harvesting. (**b**) Current output of coupled TENG network by three connection methods. (**c**) Schematic diagram of TENG network rectification and demonstration of coupled TENG network for self-powered sensing by wave agitations. (**a–c**) Reprinted with permission from ref. [54], Copyright 2018, American Chemical Society. (**d**) Schematic diagram of SAM-joint mechanism. (**e**) Demonstration of self-healing of TENG network by SAM-joint. (**d,e**) Reprinted with permission from ref. [42], Copyright 2019, Elsevier. (**f**) Schematic diagram of electrical networking topology of TENG network. Reprinted with permission from ref. [74], Copyright 2020, Elsevier. (**g**) Schematic diagram of plane-like power cable and perspective. Reprinted with permission from ref. [48], Copyright 2020, Elsevier.

4.2. Power Management

TENGs are generally characterized to have high internal impedance, which does not match the impedance of most general electronics. Direct powering the electronics by TENGs will result in low efficiency. Meanwhile, a switch strategy is required to maximize the power generated in a single cycle [21]. Moreover, for the network with multiple TENG units, there are new challenges to maximize the total output of modules or the whole network by power management.

The power management module (PMM) proposed by Xi et al. consists of two components: a tribotronic energy extractor to enhance the energy transfer efficiency from the TENG device to the circuit system, and a DC-DC buck converter to generate DC output on the load [49] (Figure 10a). The core goal of power management is to optimize and maximize the TENG output power on the loads through step-down flow, which is achieved by the PMM in a universal, efficient, and autonomous way because it is compact in size and applicable for various forms of TENGs with a self-management mechanism.

Xi et al. then utilized the PMM to support an application system based on wave energy [11] (Figure 10b). The harvested energy by TENGs passing through the PMM can drive a microprogrammed control unit (MCU), several microsensors, and a transmitter due to the enhanced energy transfer efficiency. The MCU based on an intelligent monitoring mechanism can deploy electricity for each sensor by varied priority and data transmission cycle, accomplishing a sustainable and autonomous wireless sensing for acceleration, magnetic intensity, and temperature.

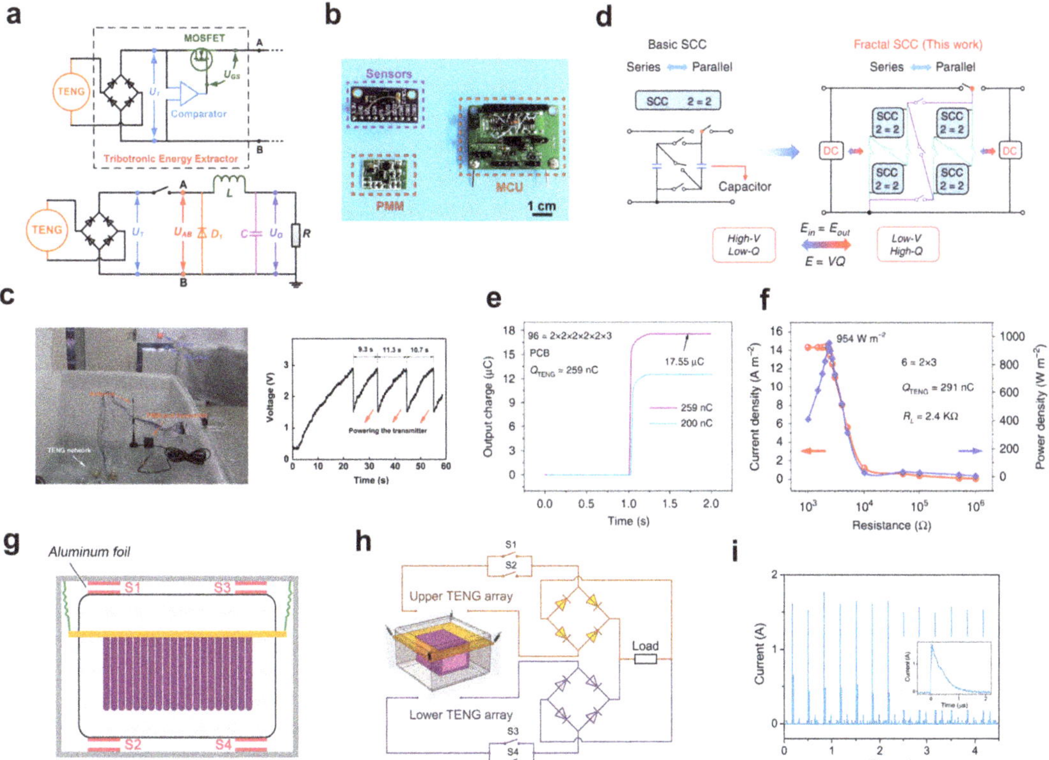

Figure 10. Power management and output of TENGs. (**a**) Circuit diagram of PMM including tribotronic energy extractor and DC-DC buck converter. Reprinted with permission from ref. [49], Copyright 2017, Elsevier. (**b**) Demonstration of utilizing PMM to support MCU and sensors. Reprinted with permission from ref. [11], Copyright 2019, Elsevier. (**c**) Demonstration of combining PMM with a hexagonal network of TENGs to power wireless transmitter. Reprinted with permission from ref. [75], Copyright 2019, Wiley. (**d**) Schematic diagram comparing SCC with FSCC. (**e**) Charge output of TENG adopting power management based on FSCC. (**f**) Current and power output of TENG adopting power management based on FSCC. (**d–f**) Reprinted with permission from ref. [50], Copyright 2020, Springer Nature. (**g**) Schematic structure of air-membrane TENG device with contact switches. (**h**) Schematic diagram of the switch circuit. (**i**) Pulse current of the device with switches. Inlet: single current pulse. (**g–i**) Reprinted with permission from ref. [43], Copyright 2017, Elsevier.

Liang et al. proposed a hexagonal network of TENGs combined with the above PMM, successfully powering a wireless transmitter to send signals every 10 s under wave agitations, which further verified the application of the PMM in wave energy harvesting [75] (Figure 10c).

In addition, the transformer of switched-capacitor-convertor (SCC) is also a good candidate for power management since it is magnet-free, light-weight, and easy for integration (Figure 10d). Liu et al. put forward an SCC that adopts fractal design (FSCC), which optimizes the topology to decrease the impedance and lower the switch loss of transistors by integration on printing circuit boards [50]. FSCC also shows high step-down ratio and electrostatic voltage applicability. By coupling TENGs with FSCC, the power management system can boost the charge output 67 times (Figure 10e), and reaches a high power density of 954 W m^{-2} in pulse mode at 1 Hz (Figure 10f).

Apart from the above module, a switch circuit can also intensify the current and power output, as put forward in the mentioned multilayered air-membrane TENG by Xu et al. [47]. The switches are formed by attaching Al foils on air chambers and their corresponding contact spots on the outer shell, and are closed when matching foils contact. As the inner oscillator reaches its highest or lowest position and creates a voltage owing to the charge separation, switches are closed and a pulse current driven by the voltage appears in the external circuit (Figure 10g,h). Such switch circuit results in swift charge transfer that no longer depends on mechanical contact-separation speed of electrodes, rising the peak current to 1.77 A and the instantaneous power to 313 W, at a load of 100 Ω (Figure 10i), which further enhances previous experimental results by Cheng et al. [74]. Apart from that, the possible energy output per cycle is also increased, the principle of which was thoroughly discussed by Zi et al. [21].

5. Summary and Perspectives

In this paper, major aspects in the construction of TENG networks for blue energy harvesting are reviewed. The design of TENG unit for performance improvement is mainly discussed, from structure advancement to mechanism alteration, including rolling ball structure, multilayer structure, grating structure, pendulum structure, mass-spring structure, spacing structure, water-solid contact structure, and charge pumping strategy. Different kinds of structure designs are born with advantages to cater to specific needs, including the naturally low frequency of rolling ball structure to match with slow wave agitations, the superior output density of multilayer and grating structures, the high sensitivity and elongated operation time under mechanical excitations to improve the energy conversion efficiency through pendulum and mass-spring structures, and the outstanding robustness and durability by spacing structure. Principle innovation such as charge pumping is significant as it can bring large promotion to the system. Networking strategy and power management are also briefly discussed. As a promising clean energy technology, blue energy harvesting based on TENGs is expected to make great contributions for achieving carbon neutrality and developing self-powered marine systems. Revealed as a type of mechanical energy harvester more suitable for low-frequency excitations, the key to the commercialization of TENGs lies in the combination of high power density and robustness. The following aspects are suggested to be focused upon in future investigations:

(1) The design of TENG units is still quite crucial for further enhancing the power density, especially in a real ocean environment, which has much more complex wave conditions than in the lab, and the design of the device can be further validated and optimized based on current devices [76]. A detailed comparison on typical devices is shown in Table 1. In general, devices based on multilayer structure and grating structure intrinsically output with greater power density. Making the components soft can expand contact area, which enhances triboelectrification. Mechanism innovations regarding charge pumping strategy achieve ultrahigh charge density. Rolling ball, pendulum, and mass-spring structures can make blue energy harvesting more adaptive to changing directions and broad frequency of waves, with better durability. To reach higher output of TENGs from the material aspect, polymers can be improved in dielectric permittivity, electrostatic breakdown strength, stability, contact status, and mechanical robustness, through surface morphology and molecular functionalization as well as bulk composition modification [77].

(2) The durability of the TENG with friction or contact interfaces should be further examined and optimized. For long-term operation at sea, the device should achieve high reliability.

(3) The networking strategy is important for organizing and coupling TENG units together to reach higher efficiency as an integrated system. It is less investigated in the past for its complexity, which should be emphasized with further development of blue energy.

Table 1. Summary of typical TENG units.

Device	Feature	Typical Output				Dimension Per Unit	Material	Mode	Year	Note
		Q_{sc}	I_{sc}	Power	Power Density					
rolling-structured TENG [53] (RF-TENG)	rolling	24 nC (wave, 1.43 Hz)	1.2 µA (wave, 1.43 Hz)			sphere diameter 6 cm	Nylon, Al, Kapton	freestanding	2015	low friction
ball-shell-structured TENG [54] (BS-TENG)	rolling	72.6 nC (motor)	1.8 µA (motor, 3 Hz)	**peak**: 1.28 mW (motor, 5 Hz) **average**: 0.31 mW (motor, 5 Hz)	**peak**: 7.13 W m^{-3} (motor, 5 Hz) **average**: 1.73 W m^{-3} (motor, 5 Hz)	sphere diameter 7 cm	silicon rubber, POM, Ag-Cu	freestanding	2018	low damp-ing force
3D electrode TENG [42]	rolling, multi-layer	0.52 µC (motor)	5 µA (motor, 2 Hz)	**peak**: 8.75mW (motor, 1.67 Hz) **average**: 2.33 mW (motor, 1.67 Hz)	**peak**: 32.6 W m^{-3} (motor, 1.67 Hz) **average**: 8.69 W m^{-3} (motor, 1.67 Hz) 2.05 W m^{-3} (wave)	sphere diameter 8 cm	FEP, Cu	freestanding	2019	enhanced contact area
air-driven membrane structure TENG [43]	multilayer, mass-spring	15 µC (rectified, motor)	187 µA (motor) 1.77 A (contact switch)	**peak**: 10 mW (motor) 313 W (contact switch)	**peak**: 13.23 W m^{-3} (motor, core device)	rectangular inner part: 12 cm × 9 cm	PTFE, soft membrane, Al, Cu	contact separation	2017	high output
spring-assisted spherical TENG [61]	multilayer, mass-spring	0.67 µC	120 µA	**peak**: 7.96 mW	**peak**: 15.2 W m^{-3}	sphere diameter 10 cm	Kapton, FEP, spring, Cu, Al	contact separation	2018	
nodding duck structure multi-track TENG [60] (NDM-FTENG)	multilayer, rolling		~1.1 µA (two devices, wave)		**peak**: 4 W m^{-3} (motor, 0.21 Hz)	10 cm × 20 cm (width by height)	PPCF (PVDF/PDMS composite films), nylon, Cu, PET, PMMA	freestanding	2021	
tandem disk TENG [46] (TD-TENG)	grating, pendu-lum, multi-layer	3.3 µC (wave, 0.58 Hz)		**peak**: 45.0 mW (wave, 0.58 Hz) **average**: 7.5 mW (wave, 0.58 Hz)	**peak**: 7.89 W m^{-3} (wave, 0.58 Hz) **average**: 1.3 W m^{-3} (wave, 0.58 Hz) 7.3 W m^{-3} (wave, 0.58 Hz, core device)	volume 0.0057 m^3	PTFE, acrylic, Cu	freestanding	2019	high power density
single pendulum inspired TENG (P-TENG) [41]	pendulum, spacing	18.2 nC (motor, 0.017 Hz)				sphere diameter 13 cm	PTFE, Cu, acrylic, cotton thread	freestanding	2019	durable

Table 1. *Cont.*

Device	Feature	Typical Output				Dimension Per Unit	Material	Mode	Year	Note
		Q_{sc}	I_{sc}	Power	Power Density					
robust swing-structured TENG [63] (SS TENG)	pendulum, spacing	256 nC (wave, 1.2 Hz)	5.9 μA (wave, 1.2 Hz)	**peak**: 4.56 mW (motor, 0.017 Hz)	**peak**: 1.29 W m^{-3} (motor, 0.017 Hz)	cylindrical shell: length 20 cm, outer diameter 15 cm	PTFE, Cu, acrylic	freestanding	2020	durable
active resonance TENG [66] (AR-TENG)	pendulum, multi-layer	0.55 μC (wave)	120 μA (wave)	**peak**: 12.3 mW (wave)	**peak**: 16.31 W m^{-3} (wave, core device)	volume 754 cm^3 (core device)	FEP, Kapton, Cu	contact separation	2021	omnidirecti-onal
spiral TENG [67]	mass-spring		15 μA (wave)		**peak**: 2.76 W m^{-2} (motor, 30 Hz)	sphere diameter 14 cm	Kapton, Cu, Al	contact separation	2013	
liquid solid electrification enabled generator [51] (LSEG)	L-S contact	75 nC (motor, 0.5 m/s)	3 μA (motor, 0.5 m/s)	**average**: 0.12 mW (motor, 0.5 m/s)	**average**: 0.067 W m^{-2} (motor, 0.5 m/s)	planar: 6 cm × 3 cm	water, FEP, Cu	freestanding	2014	
networked integrated TENG [69] (NI-TENG)	L-S contact		13.5 μA (motor, 0.5 m/s)	**peak**: 1.03 mW (motor, 0.5 m/s)	**peak**: 0.147 W m^{-2} (motor, 0.5 m/s)	planar: 10 cm × 7 cm	Kapton, PTFE, water	freestanding	2018	
TENG based on charge shuttling [52] (CS-TENG)	charge pumping, multi-layer	53 μC (rectified, wave, 0.625 Hz)	1.3 mA (wave, 0.625 Hz)	**peak**: 126.67 mW (wave, 0.625 Hz)	**peak**: 30.24 W m^{-3} (wave, 0.625 Hz)	sphere diameter 20 cm	PTFE, PP, Cu, Zn-Al	contact separation	2020	high charge output

(4) The interaction of devices with water motion is crucial for power-take-off (PTO) performance of the system, which should be theoretically investigated and optimized based on fluid-structure interaction dynamics.

(5) Adaptability to the ocean environment involves packaging of the device, antifouling, and anticorrosion, which can ensure that the function of the network is not damaged in the severe ocean environment. The complete encapsulation of the device by waterproof materials can protect the circuit and core device from water.

(6) Hybrid harvesting that can harness various other energy forms at sea, such as wind, rain drops, and sun light, can further improve the utilization efficiency of certain ocean area [78].

(7) Environmentally friendly designs and degradable materials are also highly required to reduce the environment risks of such systems in ocean [79].

(8) Power management that optimizes at the module level or network level is highly required, which extends the present work to larger systems. It is expected to greatly improve the total efficiency of the whole system.

Author Contributions: Conceptualization, L.X. and Z.W.; writing—original draft preparation, H.W.; writing—review and editing, L.X. and Z.W.; supervision, L.X. and Z.W.; project administration, L.X. and Z.W.; funding acquisition, Z.W. and L.X. All authors have read and agreed to the published version of the manuscript.

Funding: This research was funded by the National Key R & D Project from Minister of Science and Technology, China (No. 2016YFA0202704), the Key Research Program of Frontier Sciences, CAS (ZDBS-LY-DQC025), National Natural Science Foundation of China (No. 51605033, 51735001), and Youth Innovation Promotion Association, CAS (No. 2019170).

Conflicts of Interest: The authors declare no conflict of interest.

References

1. Isaacs, J.D.; Schmitt, W.R. Ocean Energy: Forms and Prospects. *Science* **1980**, *207*, 265. [CrossRef]
2. Salter, S.H. Wave power. *Nature* **1974**, *249*, 720–724. [CrossRef]
3. Ellabban, O.; Abu-Rub, H.; Blaabjerg, F. Renewable energy resources: Current status, future prospects and their enabling technology. *Renew. Sust. Energ. Rev.* **2014**, *39*, 748–764. [CrossRef]
4. Marshall, B.; Ezekiel, C.; Gichuki, J.; Mkumbo, O.; Sitoki, L.; Wanda, F. Global warming is reducing thermal stability and mitigating the effects of eutrophication in Lake Victoria (East Africa). *Nat. Preced.* **2009**. [CrossRef]
5. Barbarossa, V.; Bosmans, J.; Wanders, N.; King, H.; Bierkens, M.F.P.; Huijbregts, M.A.J.; Schipper, A.M. Threats of global warming to the world's freshwater fishes. *Nat. Commun.* **2021**, *12*, 1701. [CrossRef]
6. Yamaguchi, M.; Chan, J.C.L.; Moon, I.-J.; Yoshida, K.; Mizuta, R. Global warming changes tropical cyclone translation speed. *Nat. Commun.* **2020**, *11*, 47. [CrossRef] [PubMed]
7. van Soest, H.L.; den Elzen, M.G.J.; van Vuuren, D.P. Net-zero emission targets for major emitting countries consistent with the Paris Agreement. *Nat. Commun.* **2021**, *12*, 2140. [CrossRef]
8. Tollefson, J. Power from the oceans: Blue energy. *Nature* **2014**, *508*, 302–304. [CrossRef]
9. Qin, Y.; Alam, A.U.; Pan, S.; Howlader, M.M.R.; Ghosh, R.; Hu, N.-X.; Jin, H.; Dong, S.; Chen, C.-H.; Deen, M.J. Integrated water quality monitoring system with pH, free chlorine, and temperature sensors. *Sens. Actuators B Chem.* **2018**, *255*, 781–790. [CrossRef]
10. Callaway, E. Energy: To catch a wave. *Nature* **2007**, *450*, 156–159. [CrossRef]
11. Xi, F.; Pang, Y.; Liu, G.; Wang, S.; Li, W.; Zhang, C.; Wang, Z.L. Self-powered intelligent buoy system by water wave energy for sustainable and autonomous wireless sensing and data transmission. *Nano Energy* **2019**, *61*, 1–9. [CrossRef]
12. Khaligh, A.; Onar, O.C. *Energy Harvesting—Solar, Wind, and Ocean Energy Conversion Systems*; CRC: Boca Raton, FL, USA, 2009; Volume 89.
13. Chu, S.; Majumdar, A. Opportunities and challenges for a sustainable energy future. *Nature* **2012**, *488*, 294–303. [CrossRef] [PubMed]
14. Wang, Z.L. Entropy theory of distributed energy for internet of things. *Nano Energy* **2019**, *58*, 669–672. [CrossRef]
15. Fan, F.-R.; Tian, Z.-Q.; Wang, Z.L. Flexible triboelectric generator. *Nano Energy* **2012**, *1*, 328–334. [CrossRef]
16. Wang, Z.L. On Maxwell's displacement current for energy and sensors: The origin of nanogenerators. *Mater. Today* **2017**, *20*, 74–82. [CrossRef]
17. Wang, Z.L.; Chen, J.; Lin, L. Progress in triboelectric nanogenerators as a new energy technology and self-powered sensors. *Energy Environ. Sci.* **2015**, *8*, 2250–2282. [CrossRef]
18. Wu, C.; Wang, A.C.; Ding, W.; Guo, H.; Wang, Z.L. Triboelectric Nanogenerator: A Foundation of the Energy for the New Era. *Adv. Energy Mater.* **2019**, *9*, 1802906. [CrossRef]

19. Niu, S.; Wang, Z.L. Theoretical systems of triboelectric nanogenerators. *Nano Energy* **2015**, *14*, 161–192. [CrossRef]
20. Wang, Z.L. Triboelectric nanogenerators as new energy technology and self-powered sensors—Principles, problems and perspectives. *Faraday Discuss.* **2014**, *176*, 447–458. [CrossRef] [PubMed]
21. Zi, Y.; Niu, S.; Wang, J.; Wen, Z.; Tang, W.; Wang, Z.L. Standards and figure-of-merits for quantifying the performance of triboelectric nanogenerators. *Nat. Commun.* **2015**, *6*, 8376. [CrossRef]
22. Zi, Y.; Guo, H.; Wen, Z.; Yeh, M.-H.; Hu, C.; Wang, Z.L. Harvesting Low-Frequency (<5 Hz) Irregular Mechanical Energy: A Possible Killer Application of Triboelectric Nanogenerator. *ACS Nano* **2016**, *10*, 4797–4805. [CrossRef]
23. Wang, J.; Li, S.; Yi, F.; Zi, Y.; Lin, J.; Wang, X.; Xu, Y.; Wang, Z.L. Sustainably powering wearable electronics solely by biomechanical energy. *Nat. Commun.* **2016**, *7*, 12744. [CrossRef] [PubMed]
24. Bae, J.; Lee, J.; Kim, S.; Ha, J.; Lee, B.-S.; Park, Y.; Choong, C.; Kim, J.-B.; Wang, Z.L.; Kim, H.-Y.; et al. Flutter-driven triboelectrification for harvesting wind energy. *Nat. Commun.* **2014**, *5*, 4929. [CrossRef]
25. Xiong, J.; Cui, P.; Chen, X.; Wang, J.; Parida, K.; Lin, M.-F.; Lee, P.S. Skin-touch-actuated textile-based triboelectric nanogenerator with black phosphorus for durable biomechanical energy harvesting. *Nat. Commun.* **2018**, *9*, 4280. [CrossRef]
26. Kim, W.J.; Vivekananthan, V.; Khandelwal, G.; Chandrasekhar, A.; Kim, S.-J. Encapsulated Triboelectric–Electromagnetic Hybrid Generator for a Sustainable Blue Energy Harvesting and Self-Powered Oil Spill Detection. *ACS Appl. Electron. Mater.* **2020**, *2*, 3100–3108. [CrossRef]
27. Chen, H.; Wang, J.; Ning, A. Optimization of a Rolling Triboelectric Nanogenerator Based on the Nano–Micro Structure for Ocean Environmental Monitoring. *ACS Omega* **2021**, *6*, 21059–21065. [CrossRef]
28. Yuan, Z.; Wang, C.; Xi, J.; Han, X.; Li, J.; Han, S.-T.; Gao, W.; Pan, C. Spherical Triboelectric Nanogenerator with Dense Point Contacts for Harvesting Multidirectional Water Wave and Vibration Energy. *ACS Energy Lett.* **2021**, *6*, 2809–2816. [CrossRef]
29. Wang, Z.L. Catch wave power in floating nets. *Nature* **2017**, *542*, 159–160. [CrossRef]
30. Wang, Z.L.; Jiang, T.; Xu, L. Toward the blue energy dream by triboelectric nanogenerator networks. *Nano Energy* **2017**, *39*, 9–23. [CrossRef]
31. Chen, H.; Xing, C.; Li, Y.; Wang, J.; Xu, Y. Triboelectric nanogenerators for a macro-scale blue energy harvesting and self-powered marine environmental monitoring system. *Sustain. Energy Fuels* **2020**, *4*, 1063–1077. [CrossRef]
32. Jurado, U.T.; Pu, S.H.; White, N.M. Grid of hybrid nanogenerators for improving ocean wave impact energy harvesting self-powered applications. *Nano Energy* **2020**, *72*, 104701. [CrossRef]
33. Rodrigues, C.; Ramos, M.; Esteves, R.; Correia, J.; Clemente, D.; Gonçalves, F.; Mathias, N.; Gomes, M.; Silva, J.; Duarte, C.; et al. Integrated study of triboelectric nanogenerator for ocean wave energy harvesting: Performance assessment in realistic sea conditions. *Nano Energy* **2021**, *84*, 105890. [CrossRef]
34. Zou, H.; Zhang, Y.; Guo, L.; Wang, P.; He, X.; Dai, G.; Zheng, H.; Chen, C.; Wang, A.C.; Xu, C.; et al. Quantifying the triboelectric series. *Nat. Commun.* **2019**, *10*, 1427. [CrossRef]
35. Wang, Z.L. On the first principle theory of nanogenerators from Maxwell's equations. *Nano Energy* **2020**, *68*, 104272. [CrossRef]
36. Niu, S.; Wang, S.; Lin, L.; Liu, Y.; Zhou, Y.S.; Hu, Y.; Wang, Z.L. Theoretical study of contact-mode triboelectric nanogenerators as an effective power source. *Energy Environ. Sci.* **2013**, *6*, 3576–3583. [CrossRef]
37. Niu, S.; Liu, Y.; Wang, S.; Lin, L.; Zhou, Y.S.; Hu, Y.; Wang, Z.L. Theory of Sliding-Mode Triboelectric Nanogenerators. *Adv. Mater.* **2013**, *25*, 6184–6193. [CrossRef]
38. Niu, S.; Liu, Y.; Wang, S.; Lin, L.; Zhou, Y.S.; Hu, Y.; Wang, Z.L. Theoretical Investigation and Structural Optimization of Single-Electrode Triboelectric Nanogenerators. *Adv. Funct. Mater.* **2014**, *24*, 3332–3340. [CrossRef]
39. Niu, S.; Liu, Y.; Chen, X.; Wang, S.; Zhou, Y.S.; Lin, L.; Xie, Y.; Wang, Z.L. Theory of freestanding triboelectric-layer-based nanogenerators. *Nano Energy* **2015**, *12*, 760–774. [CrossRef]
40. Wang, Z.L. From contact-electrification to triboelectric nanogenerators. *Rep. Prog. Phys.* **2021**. [CrossRef]
41. Lin, Z.; Zhang, B.; Guo, H.; Wu, Z.; Zou, H.; Yang, J.; Wang, Z.L. Super-robust and frequency-multiplied triboelectric nanogenerator for efficient harvesting water and wind energy. *Nano Energy* **2019**, *64*, 103908. [CrossRef]
42. Yang, X.; Xu, L.; Lin, P.; Zhong, W.; Bai, Y.; Luo, J.; Chen, J.; Wang, Z.L. Macroscopic self-assembly network of encapsulated high-performance triboelectric nanogenerators for water wave energy harvesting. *Nano Energy* **2019**, *60*, 404–412. [CrossRef]
43. Xu, L.; Pang, Y.; Zhang, C.; Jiang, T.; Chen, X.; Luo, J.; Tang, W.; Cao, X.; Wang, Z.L. Integrated triboelectric nanogenerator array based on air-driven membrane structures for water wave energy harvesting. *Nano Energy* **2017**, *31*, 351–358. [CrossRef]
44. Zhao, X.J.; Zhu, G.; Fan, Y.J.; Li, H.Y.; Wang, Z.L. Triboelectric Charging at the Nanostructured Solid/Liquid Interface for Area-Scalable Wave Energy Conversion and Its Use in Corrosion Protection. *ACS Nano* **2015**, *9*, 7671–7677. [CrossRef]
45. Xu, L.; Bu, T.Z.; Yang, X.D.; Zhang, C.; Wang, Z.L. Ultrahigh charge density realized by charge pumping at ambient conditions for triboelectric nanogenerators. *Nano Energy* **2018**, *49*, 625–633. [CrossRef]
46. Bai, Y.; Xu, L.; He, C.; Zhu, L.; Yang, X.; Jiang, T.; Nie, J.; Zhong, W.; Wang, Z.L. High-performance triboelectric nanogenerators for self-powered, in-situ and real-time water quality mapping. *Nano Energy* **2019**, *66*, 104117. [CrossRef]
47. Kim, D.Y.; Kim, H.S.; Kong, D.S.; Choi, M.; Kim, H.B.; Lee, J.-H.; Murillo, G.; Lee, M.; Kim, S.S.; Jung, J.H. Floating buoy-based triboelectric nanogenerator for an effective vibrational energy harvesting from irregular and random water waves in wild sea. *Nano Energy* **2018**, *45*, 247–254. [CrossRef]
48. Liu, G.; Xiao, L.; Chen, C.; Liu, W.; Pu, X.; Wu, Z.; Hu, C.; Wang, Z.L. Power cables for triboelectric nanogenerator networks for large-scale blue energy harvesting. *Nano Energy* **2020**, *75*, 104975. [CrossRef]

49. Xi, F.; Pang, Y.; Li, W.; Jiang, T.; Zhang, L.; Guo, T.; Liu, G.; Zhang, C.; Wang, Z.L. Universal power management strategy for triboelectric nanogenerator. *Nano Energy* **2017**, *37*, 168–176. [CrossRef]
50. Liu, W.; Wang, Z.; Wang, G.; Zeng, Q.; He, W.; Liu, L.; Wang, X.; Xi, Y.; Guo, H.; Hu, C.; et al. Switched-capacitor-convertors based on fractal design for output power management of triboelectric nanogenerator. *Nat. Commun.* **2020**, *11*, 1883. [CrossRef]
51. Zhu, G.; Su, Y.; Bai, P.; Chen, J.; Jing, Q.; Yang, W.; Wang, Z.L. Harvesting Water Wave Energy by Asymmetric Screening of Electrostatic Charges on a Nanostructured Hydrophobic Thin-Film Surface. *ACS Nano* **2014**, *8*, 6031–6037. [CrossRef]
52. Wang, H.; Xu, L.; Bai, Y.; Wang, Z.L. Pumping up the charge density of a triboelectric nanogenerator by charge-shuttling. *Nat. Commun.* **2020**, *11*, 4203. [CrossRef]
53. Wang, X.; Niu, S.; Yin, Y.; Yi, F.; You, Z.; Wang, Z.L. Triboelectric Nanogenerator Based on Fully Enclosed Rolling Spherical Structure for Harvesting Low-Frequency Water Wave Energy. *Adv. Energy Mater.* **2015**, *5*, 1501467. [CrossRef]
54. Xu, L.; Jiang, T.; Lin, P.; Shao, J.J.; He, C.; Zhong, W.; Chen, X.Y.; Wang, Z.L. Coupled Triboelectric Nanogenerator Networks for Efficient Water Wave Energy Harvesting. *ACS Nano* **2018**, *12*, 1849–1858. [CrossRef]
55. Cheng, P.; Guo, H.; Wen, Z.; Zhang, C.; Yin, X.; Li, X.; Liu, D.; Song, W.; Sun, X.; Wang, J.; et al. Largely enhanced triboelectric nanogenerator for efficient harvesting of water wave energy by soft contacted structure. *Nano Energy* **2019**, *57*, 432–439. [CrossRef]
56. Guan, D.; Cong, X.; Li, J.; Shen, H.; Zhang, C.; Gong, J. Quantitative characterization of the energy harvesting performance of soft-contact sphere triboelectric nanogenerator. *Nano Energy* **2021**, *87*, 106186. [CrossRef]
57. Pang, Y.; Chen, S.; Chu, Y.; Wang, Z.L.; Cao, C. Matryoshka-inspired hierarchically structured triboelectric nanogenerators for wave energy harvesting. *Nano Energy* **2019**, *66*, 104131. [CrossRef]
58. Wu, Z.; Guo, H.; Ding, W.; Wang, Y.-C.; Zhang, L.; Wang, Z.L. A Hybridized Triboelectric–Electromagnetic Water Wave Energy Harvester Based on a Magnetic Sphere. *ACS Nano* **2019**, *13*, 2349–2356. [CrossRef]
59. Xu, M.; Zhao, T.; Wang, C.; Zhang, S.L.; Li, Z.; Pan, X.; Wang, Z.L. High Power Density Tower-like Triboelectric Nanogenerator for Harvesting Arbitrary Directional Water Wave Energy. *ACS Nano* **2019**, *13*, 1932–1939. [CrossRef]
60. Liu, L.; Yang, X.; Zhao, L.; Hong, H.; Cui, H.; Duan, J.; Yang, Q.; Tang, Q. Nodding Duck Structure Multi-track Directional Freestanding Triboelectric Nanogenerator toward Low-Frequency Ocean Wave Energy Harvesting. *ACS Nano* **2021**, *15*, 9412–9421. [CrossRef]
61. Xiao, T.X.; Liang, X.; Jiang, T.; Xu, L.; Shao, J.J.; Nie, J.H.; Bai, Y.; Zhong, W.; Wang, Z.L. Spherical Triboelectric Nanogenerators Based on Spring-Assisted Multilayered Structure for Efficient Water Wave Energy Harvesting. *Adv. Funct. Mater.* **2018**, *28*, 1802634. [CrossRef]
62. Liu, G.; Guo, H.; Xu, S.; Hu, C.; Wang, Z.L. Oblate Spheroidal Triboelectric Nanogenerator for All-Weather Blue Energy Harvesting. *Adv. Energy Mater.* **2019**, *9*, 1900801. [CrossRef]
63. Jiang, T.; Pang, H.; An, J.; Lu, P.; Feng, Y.; Liang, X.; Zhong, W.; Wang, Z.L. Robust Swing-Structured Triboelectric Nanogenerator for Efficient Blue Energy Harvesting. *Adv. Energy Mater.* **2020**, *10*, 2000064. [CrossRef]
64. Hou, C.; Chen, T.; Li, Y.; Huang, M.; Shi, Q.; Liu, H.; Sun, L.; Lee, C. A rotational pendulum based electromagnetic/triboelectric hybrid-generator for ultra-low-frequency vibrations aiming at human motion and blue energy applications. *Nano Energy* **2019**, *63*, 103871. [CrossRef]
65. Chen, X.; Gao, L.; Chen, J.; Lu, S.; Zhou, H.; Wang, T.; Wang, A.; Zhang, Z.; Guo, S.; Mu, X.; et al. A chaotic pendulum triboelectric-electromagnetic hybridized nanogenerator for wave energy scavenging and self-powered wireless sensing system. *Nano Energy* **2020**, *69*, 104440. [CrossRef]
66. Zhang, C.; He, L.; Zhou, L.; Yang, O.; Yuan, W.; Wei, X.; Liu, Y.; Lu, L.; Wang, J.; Wang, Z.L. Active resonance triboelectric nanogenerator for harvesting omnidirectional water-wave energy. *Joule* **2021**, *5*, 1613–1623. [CrossRef]
67. Hu, Y.; Yang, J.; Jing, Q.; Niu, S.; Wu, W.; Wang, Z.L. Triboelectric Nanogenerator Built on Suspended 3D Spiral Structure as Vibration and Positioning Sensor and Wave Energy Harvester. *ACS Nano* **2013**, *7*, 10424–10432. [CrossRef]
68. He, G.; Luo, Y.; Zhai, Y.; Wu, Y.; You, J.; Lu, R.; Zeng, S.; Wang, Z.L. Regulating random mechanical motion using the principle of auto-winding mechanical watch for driving TENG with constant AC output—An approach for efficient usage of high entropy energy. *Nano Energy* **2021**, *87*, 106195. [CrossRef]
69. Zhao, X.J.; Kuang, S.Y.; Wang, Z.L.; Zhu, G. Highly Adaptive Solid–Liquid Interfacing Triboelectric Nanogenerator for Harvesting Diverse Water Wave Energy. *ACS Nano* **2018**, *12*, 4280–4285. [CrossRef]
70. Li, X.; Tao, J.; Wang, X.; Zhu, J.; Pan, C.; Wang, Z.L. Networks of High Performance Triboelectric Nanogenerators Based on Liquid–Solid Interface Contact Electrification for Harvesting Low-Frequency Blue Energy. *Adv. Energy Mater.* **2018**, *8*, 1800705. [CrossRef]
71. Pan, L.; Wang, J.; Wang, P.; Gao, R.; Wang, Y.-C.; Zhang, X.; Zou, J.-J.; Wang, Z.L. Liquid-FEP-based U-tube triboelectric nanogenerator for harvesting water-wave energy. *Nano Res.* **2018**, *11*, 4062–4073. [CrossRef]
72. Liu, W.; Wang, Z.; Wang, G.; Liu, G.; Chen, J.; Pu, X.; Xi, Y.; Wang, X.; Guo, H.; Hu, C.; et al. Integrated charge excitation triboelectric nanogenerator. *Nat. Commun.* **2019**, *10*, 1426. [CrossRef] [PubMed]
73. Bai, Y.; Xu, L.; Lin, S.; Luo, J.; Qin, H.; Han, K.; Wang, Z.L. Charge Pumping Strategy for Rotation and Sliding Type Triboelectric Nanogenerators. *Adv. Energy Mater.* **2020**, *10*, 2000605. [CrossRef]
74. Liu, W.; Xu, L.; Liu, G.; Yang, H.; Bu, T.; Fu, X.; Xu, S.; Fang, C.; Zhang, C. Network Topology Optimization of Triboelectric Nanogenerators for Effectively Harvesting Ocean Wave Energy. *iScience* **2020**, *23*, 101848. [CrossRef] [PubMed]

75. Liang, X.; Jiang, T.; Liu, G.; Xiao, T.; Xu, L.; Li, W.; Xi, F.; Zhang, C.; Wang, Z.L. Triboelectric Nanogenerator Networks Integrated with Power Management Module for Water Wave Energy Harvesting. *Adv. Funct. Mater.* **2019**, *29*, 1807241. [CrossRef]
76. Liu, Y.; Liu, W.; Wang, Z.; He, W.; Tang, Q.; Xi, Y.; Wang, X.; Guo, H.; Hu, C. Quantifying contact status and the air-breakdown model of charge-excitation triboelectric nanogenerators to maximize charge density. *Nat. Commun.* **2020**, *11*, 1599. [CrossRef]
77. Yu, Y.; Li, Z.; Wang, Y.; Gong, S.; Wang, X. Sequential Infiltration Synthesis of Doped Polymer Films with Tunable Electrical Properties for Efficient Triboelectric Nanogenerator Development. *Adv. Mater.* **2015**, *27*, 4938–4944. [CrossRef] [PubMed]
78. Xu, L.; Xu, L.; Luo, J.; Yan, Y.; Jia, B.-E.; Yang, X.; Gao, Y.; Wang, Z.L. Hybrid All-in-One Power Source Based on High-Performance Spherical Triboelectric Nanogenerators for Harvesting Environmental Energy. *Adv. Energy Mater.* **2020**, *10*, 2001669. [CrossRef]
79. Chen, G.; Xu, L.; Zhang, P.; Chen, B.; Wang, G.; Ji, J.; Pu, X.; Wang, Z.L. Seawater Degradable Triboelectric Nanogenerators for Blue Energy. *Adv. Mater. Technol.* **2020**, *5*, 2000455. [CrossRef]

nanoenergy advances

Review

Ferroelectric Materials Based Coupled Nanogenerators

Jabir Zamir Minhas [1,2], Md Al Mahadi Hasan [1,2] and Ya Yang [1,2,3,*]

1   CAS Center for Excellence in Nanoscience, Beijing Key Laboratory of Micro-Nano Energy and Sensor, Beijing Institute of Nanoenergy and Nanosystems, Chinese Academy of Sciences, Beijing 101400, China; jabir.ucas@gmail.com (J.Z.M.); mahadihasan@binn.cas.cn (M.A.M.H.)

2   School of Nanoscience and Technology, University of Chinese Academy of Sciences, Beijing 100049, China

3   Center on Nanoenergy Research, School of Physical Science and Technology, Guangxi University, Nanning 530004, China

*   Correspondence: yayang@binn.cas.cn

Abstract: Innovations in nanogenerator technology foster pervading self-power devices for human use, environmental surveillance, energy transfiguration, intelligent energy storage systems, and wireless networks. Energy harvesting from ubiquitous ambient mechanical, thermal, and solar energies by nanogenerators is the hotspot of the modern electronics research era. Ferroelectric materials, which show spontaneous polarization, are reversible when exposed to the external electric field, and are responsive to external stimuli of strain, heat, and light are promising for modeling nanogenerators. This review demonstrates ferroelectric material-based nanogenerators, practicing the discrete and coupled pyroelectric, piezoelectric, triboelectric, and ferroelectric photovoltaic effects. Their working mechanisms and way of optimizing their performances, exercising the conjunction of effects in a standalone device, and multi-effects coupled nanogenerators are greatly versatile and reliable and encourage resolution in the energy crisis. Additionally, the expectancy of productive lines of future ensuing and propitious application domains are listed.

Keywords: ferroelectric materials; nanogenerators; piezoelectricity; triboelectricity; pyroelectricity; bulk ferroelectric photovoltaic effect (BPVE); harvesting; coupled effects

Citation: Minhas, J.Z.; Hasan, M.A.M.; Yang, Y. Ferroelectric Materials Based Coupled Nanogenerators. *Nanoenergy Adv.* **2021**, *1*, 131–180. https://doi.org/10.3390/nanoenergyadv1020007

Academic Editor: Joao Oliveira Ventura

Received: 9 October 2021
Accepted: 19 November 2021
Published: 25 November 2021

Publisher's Note: MDPI stays neutral with regard to jurisdictional claims in published maps and institutional affiliations.

Copyright: © 2021 by the authors. Licensee MDPI, Basel, Switzerland. This article is an open access article distributed under the terms and conditions of the Creative Commons Attribution (CC BY) license (https://creativecommons.org/licenses/by/4.0/).

1. Introduction

With the advancement in modern mobile electronics, wireless communication systems, and the internet of things (IoT), researchers became more concerned with the miniaturization and multi-functionality of devices, e.g., low power, flexibility, etc. Mobile devices are handy and carry information that is immediately accessible and transmittable wirelessly. The devices are small enough to be worn or fastened with objects of everyday use, such as goggles, clothes, wristwatches, and many others. These minuscule devices need small operational powers, usually in micro-watt or milli-watt, for which the use of the traditional chemical batteries is a miser and are the cause of environmental threaten, too; hence, small-scale sustainable power solutions are in need. The make-up to this innovates the maturing of energy harvesting units, which can efficiently harvest energy, such as mechanical, heat, and light energy, to electrical energy and reuse sources of energies from the ambient atmosphere [1–4]. In 2006, Z. L. Wang and J. Song developed nanogenerators (NGs) by transforming nanoscale mechanical energy into electricity using arrays of piezoelectric ZnO nanowire and marked energy harvesters to jump into a new span [5]. The basic theory of nanogenerators was established based on the concept of the use of the Maxwell displacement current as a driving force to convert environmental energies into electrical energy signals. Therefore, NGs work by current generations and employ the concept of internal polarization of the material [6]. Out of thirty-two crystal structures, twenty-one crystal classes do not possess central symmetry and are non-centrosymmetric, twenty of which show polarization as the reaction of mechanical stress forces are piezoelectric and, among these, twenty (1, 2, m, 222, mm2, 4, −4, 422, 4 mm, −42 m, 3, 32, 3 m, 6, −6, 622 m

6 mm, −62 m, 23, −43 m). Among these twenty piezoelectrics, ten show spontaneous polarization variations, induced by temperature changes are pyroelectric (1, 2, m, mm2, 3, 3 m, 4, 4 mm, 6, 6 mm). Their polarization turned out to be reversed with the implementation of the electric field; additionally, the particular materials are ferroelectrics [7,8]. Ferroelectrics are functional dielectric materials that have shown spectacular properties of spontaneous polarization at zero electric fields. The spontaneous polarization (P_s) can be switched to reversed direction when a large poling electric field (in units of kV/cm to MV/cm) is applied. Figure 1a(i) demonstrates the typical ferroelectric P-E curve [9], showing both the states are steady and thermodynamically stable. Thus, ferroelectric materials have been of great research interest for almost over a century [10,11]. In conventional ferroelectric (FE) materials, also known as proper ferroelectrics, the spontaneous polarization (P_s) is responsible for the change in structure (i.e., structural phase transition) breaking the crystal symmetry at curie temperature (Tc). So, the order parameter of phase transition in proper ferroelectrics is spontaneous polarization. Ferroelectrics have posted a wide range of applications contributing to frequency filters, pressure and temperature sensors, actuators, hydrophones, oscillators, and many others [12–18]. Since all the ferroelectrics are piezoelectric and pyroelectric, they speculate the phenomenon of piezoelectricity, pyroelectricity, and photovoltaic ferroelectric effects. These effects are the result of responses from external stimuli, such as mechanical, thermal, and solar energies, respectively [19–22]. The piezoelectric coefficient of semiconductor piezoelectric ZnO is ~12 pC/N, which is quite low, in contrast to the ferroelectric materials with perovskite structure, e.g., $BaTiO_3$ and $Pb(Zr, Ti)O_3$, which exhibit high piezoelectric coefficients of 100 pC/N and 200 pC/N, respectively [23]. These properties of ferroelectrics promote their use in the fabrication of nanogenerators for energy scavenging and harvesting from low frequency natural and artificial energy sources from the environment [24,25]. Along with energy storage units, self-powered systems can also be integrated by nanogenerators for powering functional devices. Many nanogenerators have been fashioned after, including piezoelectric NGs (PENG), pyroelectric NGs (PyENG), triboelectric NGs (TENG), and photovoltaics (PVC) [26–29]. Piezoelectric and triboelectric NGs transform mechanical stress forces and/or energies from wind/airflow, water waves, human motions, biomechanical energies, etc., to electrical energy, accompanied by changing polarization degrees in ferroelectrics [30–35]. Pyroelectric NGs works by temperature variations from industrial heat wastes, radiations from the sun, etc. [36–41]. The photovoltaic effect in ferroelectrics demonstrated that the photocurrent is not limited by the bandgap of material; rather, it is associated with material polarity and separation of light-induced photo carriers [42–45]. At present, hybrid nanogenerators and multi-effect coupled NGs are of practical interest, developed by integrating the above mentioned effects into a single structure and, hence, contributing to the maximization of energies for high electrical power outputs, depending on the strength of external stimuli [6,46–50]. The ever first one-structured coupled nanogenerator, based on piezo–tribo–pyro–photovoltaic effects, was proposed by professor Ya Yang and his co-workers in 2015 [51]. Ferroelectric materials possess permanent dipole moments in the electric field, which increases their polarization density and facilitates their use in wearable, flexible electronics. The conventional semiconductor materials do not own this property and so ferroelectrics gather more attention [52]. The most active ferroelectric materials encapsulated in nanogenerator application are inorganic ferroelectric ceramics and organic polymers; the best of them include $BaTiO_3$ (BTO) [53], $PbZr_{1-x}Ti_xO_3$ (PZT) [54], $Na_{0.5}Bi_{0.5}TiO_3$ (NBT) [55], $KNaNbO_3$ (KNN) [56], $BiFeO_3$ (BFO) [57], $Pb(Mg_{1/3}Nb_{2/3})O_3$-$PbTiO_3$ (PMN-PT) [58], PVDF [59–61], and P(VDF-TrFE) [62]. In comparison with inorganic ferroelectrics, very few organic ferroelectrics exist. For example, single-component polar organic molecules, such as CDA, DNP, CT complexes, and polymers, such as PVDF, nylon-11, and organic-inorganic composites, such as $HdabecoReO_4$, TGS, TSCC, are excellent for applications in piezoelectric and triboelectric nanogenerators [63]. Other reports on semiconductor ferroelectric have also been found. Ferroelectric polymers are highly flexible and have found applications in wearable and foldable devices. Ferroelectric ceramics

are hard and are often complexed with other ferroelectric/non-ferroelectric material and stacked to multilayered architecture, in order to improve their output features with various structural morphologies and, e.g., $(Ba_{0.85}Ca_{0.15})(Ti_{0.90}Zr_{0.10})O_3$-x $BiHoO_3$/PDMS [64], PDMS/PZT [65], and CNTs-PMNT/PDMS [66]. In this peer review paper, we briefly talk about the basic properties of ferroelectric materials, together with related phenomena for energy conversion, i.e., piezoelectricity, pyroelectricity, triboelectricity, and ferroelectric photovoltaic effects, as well as operating conditions of various types of ferroelectric material-based NGs. Mechanism of ferroelectrics-based hybrid and multi-effect coupled NGs, with their structure-related performances and power conversion efficiencies, are also discussed. Some recent applications, including self-powered micro/nano-systems, multifunctional sensors, and wearable flexible devices, are discussed. Furthermore, the advantages of ferroelectric-based NGs and future prospects are devised.

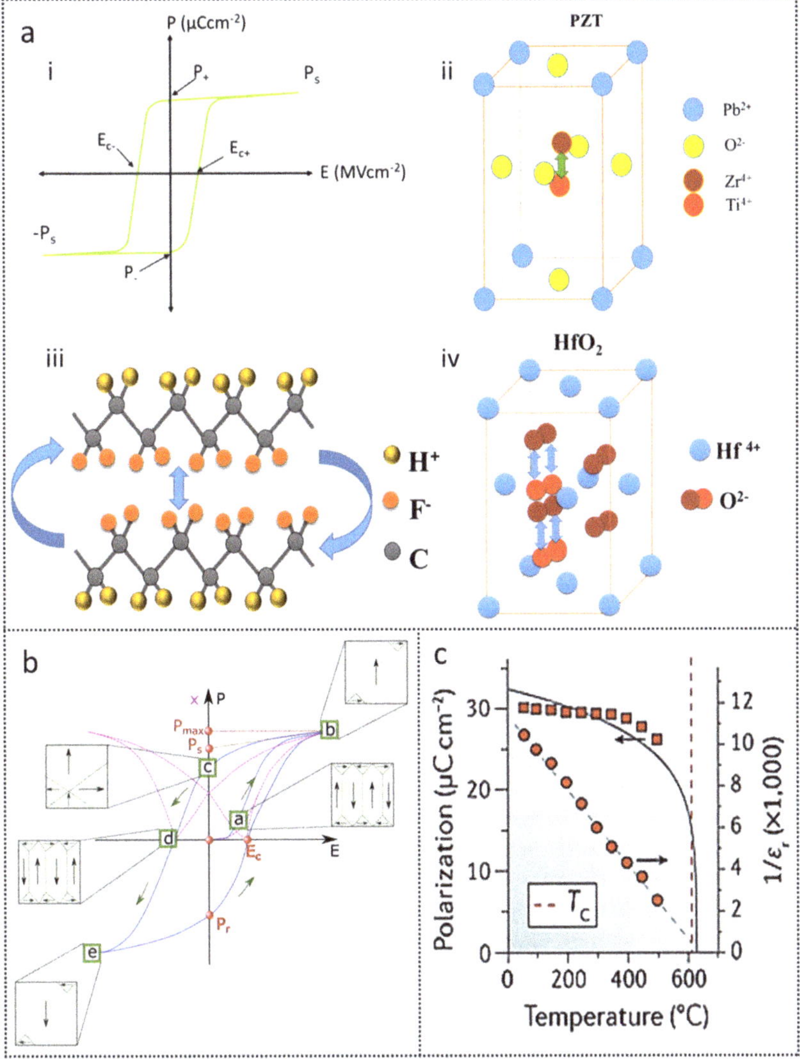

Figure 1. Hysteresis curve and typical ferroelectric crystals structures. (**a**) Hysteresis curve for ferroelectric materials with important factors of coercive field (E_c), remanent (P_R), and spontaneous polarization (P_s); (**i**) PZT crystal, representing the

two stable positions of Zr^{4+} or Ti^{4+} ions; (**ii**) PVDF polymer chain with the two orientations for polarization; (**iii**) hafnium oxide with switching O^{2-} oxygen ions; (**iv**) O^{2-} ion switching in fluoride structure of orthorhombic hafnium oxide (HfO_2). (**b**) Domain wall movement. Reprinted with permission from reference [67], Copyright 2017, AIP Publishing. (**c**) Transition temperature dependence of ferroelectrics. Reprinted with permission from reference [68], Copyright 2016, Springer Nature.

2. Ferroelectric Materials

2.1. Fortunes of Ferroelectric Materials

Ferroelectric materials in their ferroelectric phase exhibit net dipole moment, i.e., the centers of charges (positive and negative) do not coincide, resulting in robust polarization lasting permanently below a particular temperature, called the curie temperature (T_c), in the non-poled state. Before the time of the second world war, from 1920–1943, ferroelectric materials remained of academic and theoretical interest for small applications zones. In the 1950s, $BaTiO_3$ (BTO) discovered that strong ferroelectric behavior evoked the electronic ceramic industry and became an interesting candidate for piezoelectric transducers. By the 1960s, researchers undertook ferroelectric thin films and realized practicing non-volatile memories, but quality maintenance, restricted their practical applications, up until the 1980s; later in the 1990s, ferroelectrics, alongside microsensors, found widespread applications in radio frequency and microwave devices. In 1994, the bypass ferroelectric capacitors installed in digital mobile phones provided a breakthrough in the leading micro-digital industry [69–71]. For the past two decades, ferroelectric materials subsisted a great application tempt in energy harvesting.

2.2. Crystal Structures

Joesph Valasek has discovered the phenomenon of ferroelectricity in Rochelle salt ($NaKC_4H_4O_6.4H_2O$) orthorhombic crystal structure first. Up until now, more than seven hundred ferroelectric materials have been demonstrated, inclusive of oxides, polymers, ceramics, and liquid crystals, as well. Typical ferroelectric materials have either ABO_3 perovskite or hydrogen-bonded potassium dihydrate phosphate KH_2PO_4 (KDP) structures, but others also exist, such as GeTe, $SrAlF_5$, SbSI, SbSeI, etc. [72,73]. Perovskite ferroelectric materials are with crystal structure isomorphous to the mineral perovskite calcium titanium oxide $CaTiO_3$, general formula ABO_3, whereupon A and B stand as cation, and O is the oxygen ion. The ionic radius of cation A and B is particularly around 1.2 Å to 1.6 Å and 0.6 Å to 0.7 Å, respectively. Oxygen atoms/ions are positioned at the face center and A ion at the cube corner, forming octahedral surrounding B ions and staying at the body center. Structure stability requires that the valencies of ions be balanced. Moreover, the structural distortions, such as octahedral distortion or tilting, associated with low space group symmetries, cause relief of B-site electronic instabilities, leading to high values of ferroelectric polarization, piezoelectric coefficients, unusual photovoltaic behaviors, and dielectric constants [7]. Figure 1a(ii) represent perovskite structure-based inorganic ferroelectric materials, such as lead zirconate titanate PZT ($Pb(Ti_xZr_{1-x})O_3$), barium titanate BTO ($BaTiO_3$), and layered strontium bismuth tantalite SBT ($SrBi_2Ta_2O_9$); in all of them, the central cation switches its position between the stable states [9]. The literature revealed the existence of morphotropic phase boundary (MBP) in PZT, with compositional variations that benefited piezo ceramics a lot [74–77]. Furthermore, the nature of lead (Pb) is weighty and detrimental to humans and the environment. Therefore, lead-free materials (e.g., $BiFeO_3$ (BFO), $KNaNbO_3$ (KNN), etc.) which are non-toxic, environmentally friendly, and biocompatible, are of more research interest [78,79]. Some ferroelectrics, with more complex perovskite structures, have also been observed with cations of different valencies but fixed molar concentrations. For example, $PbMg_{1/3}Nb_2O_3$ (PMN), $PbSc_{1/2}Ta_{1/2}O_3$ (PST), and $Bi_{1/2}Na_{1/2}TiO_3$ (BNT) [80,81]. Ferroelectric hafnium oxide (HfO_2) is also of great interest and becoming a spotlight for the memories industry [9].

Ferroelectrics with ilmenite structure are similar to perovskites, with distinction in the A cation, which is small enough to fill ABO_3 structure coordinated site. Oxygen ions are packed closely in a hexagonal layer, with cations A and B located between the layers at octahedral sites $LiNbO_3$ and $LiTaO_3$, which are two important uniaxial ferroelectric materials of this class [82–84].

Polyvinylidene fluoride (PVDF) with CH_2CF_2 monomer was observed to be the first ferroelectric polymer, without any particular curie temperature, as melting occurs first. PVDF exists in four different polymorphs (i.e., the $\alpha-$, $\beta-$, $\gamma-$, and $\delta-$phases), depending on the configuration of carbon–carbon links. Fluorine atoms are electronegative and make the C–F bond polar, resulting in the molecule having a net dipole moment orthogonal to molecular length carbon backbone chain. Though the molecules of polymer arrange in the unit cell and, hence, the dipoles balance each other in the α-phase; so, the PVDF in α-phase is not ferroelectric, but its field application produces a net dipole moment on unit cell and put it in δ-phase. Annealing of these systems at high temperatures crystalizes them in the γ-phase, which also has a net dipole moment perpendicular to the carbon backbone and is polar. Orthorhombic β-phase can be achieved by stretch or draw in previously derived phases and is an all-trans configuration with high spontaneous polarization and a dielectric constant [85]. So, electric poling makes β-PVDF-phase a strong piezoelectric and pyroelectric phase, when compared with other polymorphs. β-PVDF-phase can be directly obtained by the copolymerization of vinylidene fluoride with trifluroethylene TrFE (~10 to 46% by wt.); the resultant P(VDF-TrFE) material has the same activity rate as pure PVDF but exhibits high remnant polarization [82,86,87]. Figure 1a(iii,iv) show the switching behavior of polar polymer chains in organic ferroelectric polyvinylidene fluoride (PVDF) polymerized, with tetrafluoroethylene (TrFE) and O^{-2} ion switching in fluoride structure of orthorhombic hafnium oxide (HfO_2), respectively [9]. Hexafluoropropylene HFP is another important copolymer material for the PVDF matrix. Other polymers that exhibit ferroelectric behavior include odd-numbered nylons, but these are only weakly piezoelectric. These polymeric ferroelectrics are useful in flexible electronics, especially in wearable devices.

Another class of ferroelectrics comprises of improper ferroelectrics, for those in which the order parameter is not the polarization; rather, the spontaneous polarization arises as to the by-product of phase transition, as a secondary effect. The temperature dependence of the permittivity does not follow Curie–Wiess law; additionally, the phase transition is not suppressed by the electric fields in improper ferroelectrics [88]. The dielectric constant of improper ferroelectric remains low, usually near the phase transition temperature, and is conducive to the large pyroelectric figure of merits. Examples of improper ferroelectric involve iron-iodine-boracite [$FeB_7O_{13}I$ (TMO)], dicalcium-lead-propionate [$Ca_2Pb(CH_3CH_2COO)_6$ (DLP)], [Hdabeco]ClO_4, etc., [89]. A subset of ferroelectric materials forms ferroelastic materials, which may show spontaneous strain and possibly exhibit two or more stable orientation states under zero electric field or mechanical stress. $Pb_3(PO)_4$ is a particular example of ferroelastic materials [68].

Anti-ferroelectric (AFE) structures are mainly characterized by the phase transition from the low symmetry state, usually from the low-temperature phase to the high symmetry state (usually high-temperature phase). Contrary to ferroelectrics, the anti-ferroelectrics do not have permanent electric polarization. Therefore, an anti-ferroelectric crystal lattice can be considered to be made up of two interpenetrating sublattices with equal but opposite polarizations. However, high spontaneous polarization (P_s) appears as the transition from the anti-ferroelectric phase to the ferroelectric phase. They exhibit low dielectrics losses and coercive fields. A typical anti-ferroelectric material with a perovskite structure is $PbZrO_3$, with Sr doping energy density of $14.5\,J/cm^3$, which had been observed at $900\,kV/cm$ by Hao et al. [90]. Anti-ferroelectric Zr doped HfO_2 had shown the largest energy density of $46\,J/cm^3$ at $4.5\,MV/cm$ [91]. Ferroelectric relaxors have also been demonstrated in the past few years; relaxors belong to disordered crystals when nonequilibrium ions are added into normal ferroelectric materials or heated above transition temperatures and de-poled under critical electric fields. They exhibit broad and suppressed dielectric peaks but higher susceptibilities. The ferroelectric $(Sr_xBa_{1-x})Nb_2O_6$ (SBN) contains five Sr or Ba ions; so, one of the A-site remains unfilled, therefore, varying the Sr/Ba concentration to an increased ratio, they transform from normal ferroelectrics to relaxor ferroelectric, as reported for x = 0.75 in reference [92]. The 0.67 Pb(Mg$_{1/3}$Nb$_{2/3}$)O$_3$-0.33PbTiO$_3$ composite ferroelectric relaxor shows two phase transitions at 34 °C (from rhombohedral to tetragonal) and at 144 °C (from tetragonal to cubic) [93]. PVDF shows the relaxor behavior with improved dielectric properties and piezoelectric responses by carrying temperature treatment and further copolymerization to TrFE-CFE [94,95].

2.3. Ferroelectricity and Hysteresis

Ferroelectricity is the material property of exhibiting spontaneous polarization (P$_s$). Ferroelectric materials have high dielectric constants, and field removal does not bring polarization back to its original direction spontaneously, which signifies that the polarization is persistent. This characteristic behavior of ferroelectric materials is defined by a schematic, non-linear curve observed between the electric field (E) and polarization (P), known typically as hysteresis, displayed in Figure 1a(i) [9]. Field escalations align polarizations in distinct dipolar regions of a ferroelectric crystal, reaching a saturation value of spontaneous polarization (P$_s$) and setting off the field to zero gives remanent polarization (P$_+$), which is slightly smaller than P$_s$ and the result of charge displacement. Negative field value reduces the polarization to zero at the coercive field (E$_{c-}$), and further increases cause reverse saturation polarization ($-$P$_s$) and remanent polarization (P$_-$) on returning to zero field value. Polarization, again, reaches zero at E$_{c+}$, towards a positive applied field and back to P$_s$. An important feature of ferroelectrics is that remanent polarization (2P$_R$ = |P$_+$| + |P$_-$|) persists, even after externally applied E is abolished; this makes them suitable for application in non-volatile random-access memories. BaTiO$_3$ shows a remanent polarization and coercive field of 2.55 µC/cm^2 and 3.14 kV/cm, respectively [96]. NBT (Na$_{0.5}$Bi$_{0.5}$TiO$_3$) has a remanent polarization of 38 µC/cm^2 and coercive field of 70 kV/cm only [97]. However, doping of ST (SrTiO$_3$) to NBT reduces the value of remanent polarization and coercive field to 0.007 µC/cm^2 and 0.15 kV/mm [98]. During the process of poling for polarization reorientations in the same direction, the bound charges restrain at the opposite faces of polarization moment, i.e., positive at the negative end and vice versa. The bound charges establish a depolarization field and side of the material, in order to maintain charge neutrality, while the screening for free charge accumulates on the surface, compensating the bound charge keeping the surface neutral [97,98].

2.4. Ferroelectric Domains and Phase Transition

In the ferroelectric materials, certain regions, called domains, exist and are separated by interfaces known as domain walls. Each region containing polarization mixtures has a different polarization direction for other regions, inferring that the virgin ferroelectric material does not show a particular net polarization direction. When ferroelectric materials are poled, their domains begin to set in the field direction and show a net polarization. This signifies that polarization switching is quantum-mechanically functionalized by domain wall motion. Figure 1b shows the domain wall motions and corresponding polarization switching in BTO ceramics [67]. Two unique domains are labeled as 180° domains, with polarizations aligned antiparallel to each other and 90° domains with polarizations aligned perpendicular to each other. The motion of the 180° domain walls gives rise to dielectric and piezoelectric response, whilst for 90° to the only dielectric response. The fact is important for ferroelectric material's use as energy harvesters and has been observed in many perovskite structure ferroelectrics. The symmetry of the crystal determines the favorable domain formation; hence, ferroelectric materials need to be non-centrosymmetric. The polar ferroelectric behavior exists below a particular temperature, called the critical temperature or curie point (Tc), above which nonpolar highly symmetric para-electric phase transition occurs, reflecting the optical mode softening at the BZ center [99]. BaTiO$_3$ is cubic in its para-electric phase and tetragonal in its ferroelectric phase at T$_c$ 120 °C. At 5 °C, a second-order phase transition from tetragonal to orthorhombic occurs for BTO [7], the transition/curie temperature for KDP (~$-150\ °C$) and Rochelle salt (~$24\ °C$) for PVDF ~$150\ °C$ [100–102]. In PbTiO$_3$ tetragonal ferroelectric phase occurs below $490\ °C$, assisted by cation displacement in <100>. LiNbO$_3$ and LiTaO$_3$ are two important uniaxial ferroelectric materials, comprising of high curie temperatures at $1200\ °C$ and $620\ °C$, respectively, and containing only $180°$ domains [7]. Studies show that for PZT ferroelectric, the piezoelectric response is related to the volume snippet of $180°$ domain walls. Pyroelectric coefficients can be tuned at phase transitions ascribed with temperature-controlled domain wall motions, as observed in PZT films subjected to tensile strains [103–106]. Huan et al. reported the grain size effect of BTO and found that by reducing grain size, the width of the $90°$ domain diminished, leading to enhanced piezoelectric properties of BTO [107]. However, the materials with complexed domain wall structures show different properties from single domain wall ferroelectrics. Multiferroic BiFeO$_3$ (BFO) is ferroelectric at Tc = 1103 K with a rhombohedral structure and possesses $71°$, $109°$, and $180°$ domain walls. Seidel et al. reported the observation of room-temperature electronic conductivity at $109°$ and $180°$ ferroelectric domain walls in insulating BeFiO$_3$, proving domain walls as discrete functional entities with potential in nanoelectronic applications [108]. Ghara et al. demonstrated the formation of highly conductive domain walls in multiferroic GaV$_4$S$_8$, which can be annihilated magnetically by driving the system to a single domain state [109]. Werner et al. reported the stable, metal-like conductivity of charged domain walls in lithium niobate crystals, with orders 13 higher than the bulk, high stability for the temperatures $\leq 70\ °C$, and promoted advanced LN-semiconductor optoelectronic devices [110].

Lattice defects can also pin the motion of domain walls, as observed oxygen vacancies and dipole defects lower the polarization values in KBNNO pyro and piezo response. The defects can couple with polarization and beget anisotropic lattice deformations, complementing the curie point, as shown in Figure 1c. Hence, ferroelectric material properties can be tailored by controlled poling, defects, and engineering at a small scale, with non-ferroelectric materials producing composite multilayer structures [68].

3. Energy Harvesting Ferroelectric Materials for Nanogenerators (NGs)

Ferroelectric materials can efficiently harvest energy, with variations of the internal dipole moments or potentials, induced by surrounding ambient energies. These available ambient energies are being utilized individually and by coupling with physical phenomena related to mechanical, thermal, and solar energy to generate electricity. In this section, we will discuss the basic energy conversion phenomenon related to ferroelectric materials and the corresponding energy harvesting nanogenerators (NGs).

3.1. Piezoelectric Effect and Piezoelectric Nanogenerators

The piezoelectric effect is interwined to the causation of electric charges in a material; when subjected to mechanical stress or strain forces, the center of cations and anions move apart in an asymmetric manner, the material becomes electrically polarized and builds piezoelectric potential. The polarity of the developed charge depends on the stress direction, i.e., either the stress is compressive or tensile. Whilst in converse piezoelectric effect, the applied electric field develops mechanical strain in the material, and field direction determines either the material expands or contracts. P. Cure and J. Curie were the first to observe the phenomenon of piezoelectricity (in 1880) in quartz and Rochelle salt [111]. The dielectric polarization (P) is defined by a linear relation for direct piezoelectric effect:

$$P_i = d_{ij}\, X_j \tag{1}$$

where in d and X represent the piezoelectric coefficient or constant and the applied stress, respectively. The piezoelectric coefficient d evinces piezoelectric material performance and is anisotropic. The subscripts i and j indicate the directions of dielectric displacement and applied stress, respectively [112]. Ferroelectric materials (such as PZT, KNN, and PMN) with morphotropic (MPB) or polymorphic phase boundary (PPB) relate to the rotation of polarization and indicate the presence of multiple phases within the material, show exceptional piezoelectric response, and piezoelectric coefficient [113,114] BTO; PVDF has also been studied widely in piezoelectric harvesters of mechanical energies, known as piezoelectric nanogenerators (PENG). The operational mechanism of ferroelectric PENG is based on the volume density model; that is, compressing the ferroelectric dipole density increases over the reduced material thickness and stretching material dipole density declines over increased material thickness [85,115]. Consequently, ferroelectric material polarization changes as dipole density vary and results in the generation of the piezoelectric signal. Figure 2a(I) portrays a poled piezoelectric generator (PENG) in a stress-free situation; polarized to value P_s, the charge will sit on the surface to establish charge balance. Subjecting the material to compressive stress, as in Figure 2a(II), the polarization level decreases, and the surface charge becomes free to flow generating electric current signal across the load R. Figure 2a(III) depicts that, when removing stress or applying tensile stress, the material polarization level increases, generating current in the opposite direction to keep charge balance. Typically, the current signal, generated by PENGs, is AC; therefore, the rectification of the output signal is needed. For energy harvesting by PENGs, the selection of load resistance R must lead to an optimum value of piezo potential and current for power generation, which is possible by impedance matching of R and energy material. Furthermore, the infinite value of R leads to an open-circuit device condition, which is suitable to develop sensitive voltage sensors, and the zero value of R leads to a short-circuiting device [116]. Perovskite ceramic ferroelectrics, e.g., BTO and KNN, have usually high piezoelectric coefficients; hence, ferroelectric ceramics-based PENGs show high-output power performances. Guo et al. estimated the piezoelectric coefficient (d_{33}) of ~30 pC/V and dielectric constant of (ε) ~340 for 140 nm thick BTO film, deposited on $Pt/TiO_x/SiO_2/Si$, with the $LiNbO_3$ buffer layer [117]. Huan et al. accounted for the grain size effects on the piezoelectric coefficient of BTO, they found that reducing the grain size to 1 μm gives a maximum d_{33} of 519 pC/V, the dielectric constant of (ε) ~6079, and electromechanical coupling factor (K_p) of 39.5% [107]. Du et al. derived the piezoelectric features of $(K_{0.5}Na_{0.5})NbO_3$ (KNN) with the perfect perovskite phase, orthorhombic symmetry, piezoelectric coefficient (d_{33}) of 120 pC/N, electromechanical coupling factor (K_p) of 0.40, and curie temperature of 400 °C [118]. Matsubara et al. showed that, for KNN, there was an estimated ~180 pC/N piezoelectric coefficient (d_{33}), with an electromechanical coupling factor (K_p) ~0.39 and curie temperature of 420 °C [119]. As the ceramic materials make it difficult to achieve flexibility, they should be deposited with smaller dimensions, i.e., as thin films or nanoparticles. Chen et al. studied lead-free $Ba_{0.9}Ca_{0.1}Ti_{0.90}Sn_{0.10}O_3$-x$La_2O_3$ for x = 0.03 mol%, and they got an optimized d_{33} = 496 pC/N and K_p = 41.7, with coexisting orthorhombic and tetragonal phase [120]. Ferroelectric polymer composites (e.g., $ZnSnO_3$:PDMS) can easily be employed in PENGs and has the advantage of low-cost, large-area manufacturing and large mechanical durability, even under high-stress conditions. Ferroelectric polymers, such as PVDF and P(VDF-TrFE), with piezoelectric coefficients 18 pC/N and −21 pC/N with dielectric constants of 8.4 and 10 are promising for total flexible PENGs, they are even foldable, stretchable, and twistable [94,121].

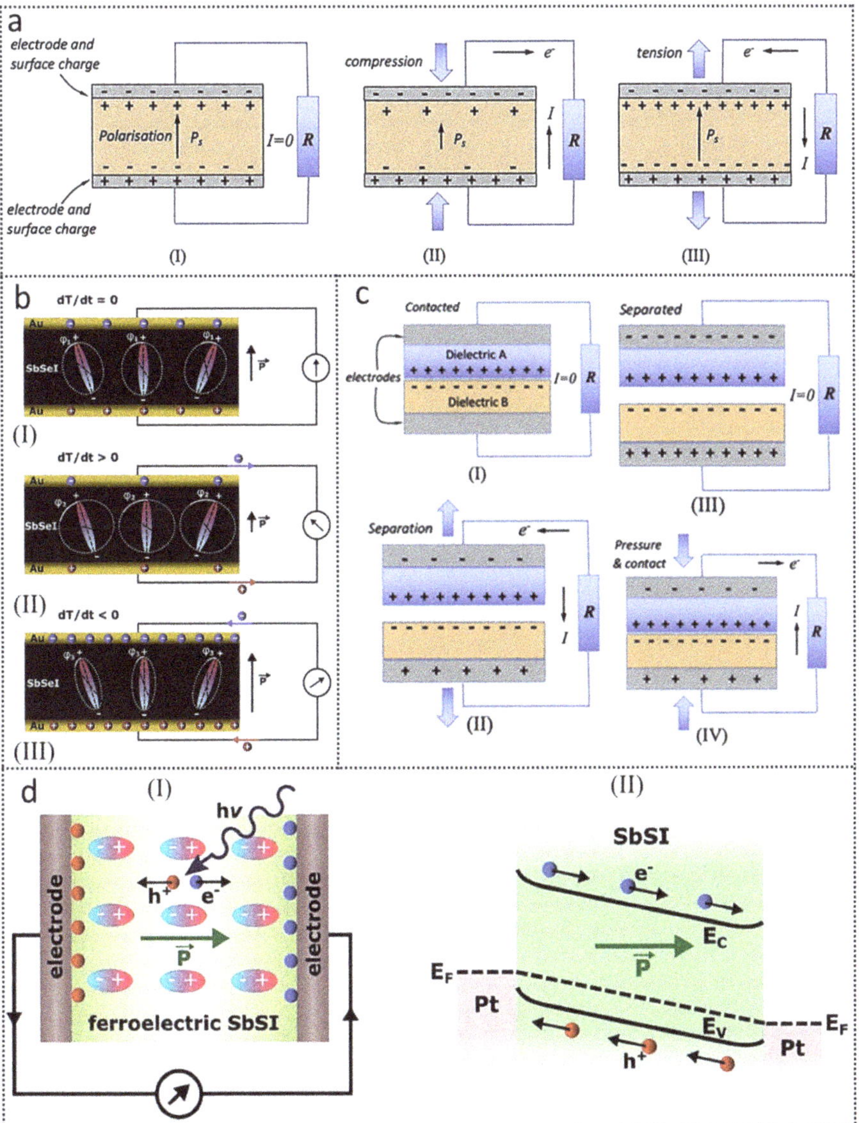

Figure 2. Various effects in ferroelectrics. (**a**) Piezoelectric ferroelectric nanogenerators. Reprinted with permission from reference [116], Copyright 2020, IScience. (**b**) The pyroelectric effect in SbSeI nanowires φ_1, φ_2, and φ_3 represent the degrees of dipole oscillations. Reprinted with permission from reference [122], Copyright 2019, Elsevier. (**c**) Triboelectric effect in contact-separation mode. Reprinted with permission from reference [116], Copyright 2020, Elsevier. (**d**) Ferroelectric photovoltaic effect in SbSI. Reprinted with permission from reference [29], Copyright 2019, MDPI.

3.2. Pyroelectric Effect and Pyroelectric Nanogenerators

The pyroelectric effect is associated with the variations in spontaneous polarization with temporal temperature changes and, hence, the generation of electric current. Thermal energy, generated by nature, mechanical frictions, and machines, can efficiently be altered into electric energy, owing to the pyroelectric effect. Pyroelectric coefficient p is defined as:

$$p = \frac{\partial P_s}{\partial T} \tag{2}$$

where p represents the efficiency of pyroelectric material, and pyroelectrics have high pyroelectric coefficients. P_s is the spontaneous polarization of pyroelectric, and T is the temperature [123]. Pyroelectric nanogenerators include three main layers: the metallic top layer, acting as the top electrode, is designed to collect heat efficiently; the middle layer is the pyroelectric material layer, facilitating heat conversion into electricity by variations in internal polarization; and the third metal layer is the bottom electrode [124,125]. The pyroelectric materials have a unique polar axis. Figure 2b shows SbSeI ferroelectric-based PyENG [122]; when the material temperature is fixed, the electric dipole oscillates randomly around aligning axes, φ_1, reaching their equilibrium, as shown in Figure 2b(I). No current was observed since P_s remained the same. Thermal fluctuations produce polarization change, which heads to the separation of bound charges and their accretion at the electrodes. Increasing temperature creates net dipole moments, by enhanced dipole oscillation ($\varphi_2 > \varphi_1$) and spontaneous polarization, decreases by increased thermal agitations. This leads to the depletion of bound charge, so free charge redistributes itself to compensate for the bound charge and, consequently, pyroelectric current flows (Figure 2b(II)). Cooling the sample reverts the current direction and increases spontaneous polarization as the dipole oscillation reduces ($\varphi_3 < \varphi_1 < \varphi_2$) (Figure 2b(III)). The pyroelectric effect lasts only until temperature fluctuation remains. The output power of PyENGs can be significantly improved by enhancing the absorption of thermal energy with structural modifications, increasing the pyroelectric coefficient by material alteration, and strengthening spontaneous polarization carried out by thermal expansion [122]. Pyroelectric materials exhibit piezoelectricity, too; so, accordingly, the temperature variations encourage stress production and piezo polarization, i.e., secondary pyroelectric effect. Therefore, the bulk pyroelectric coefficient is the additive of both the primary and secondary coefficients [126]. However, in the hybrid systems, with polar inclusions embedded into polar matrixes, it is possible to control the individual component polarization, so the pyroelectric and piezoelectric responses can be compensated. For example, Ploss et al. reported that 27% volume of lead titanate (PT) embedded in PVDF-TrFE matrix, followed by parallel or antiparallel poling inclusions, can cause piezoelectric-compensated pyroelectric material or pyroelectric-compensated piezoelectric material to be constructed, respectively [127]. Similar trends were observed by Meirzadeh et al. for α-glycine crystals doped with minute amounts of different L-amino acids [128].

Inorganic ferroelectric PyNGs, e.g., PZT, show low-output power, due to a small pyroelectric constant, ~-80 nC/cm^{2}K; polymer PVDF ferroelectric is found to be most favorable and shows a high coefficient, $\sim$200 µC/cm^{2}K [126,129]. For BTO, Song et al. found the pyroelectric coefficient to be dependent on temperature, from 16 nC/cm^{2}K to 57 nC/cm^{2}K for 299–310 K and 45.2 nC/cm^{2}K for 310–324 K for light-induced pyroelectric effect [130].

3.3. Triboelectric Effect and Triboelectric Nanogenerators

The triboelectric effect is delineated as the exchange of charge among two different materials in contact with each other. Triboelectric nanogenerators TENGs work by coupling triboelectrification (static polarized charge production on the material surface) and electrostatic induction [131–133]. Electrostatic induction drives the harvesting of ambient mechanical energy to electrical energy by variations in potential, which are incited by impulsively agitated material layer separation (particularly by displacement). Periodic mechanical triggering of the triboelectric layer generates AC output. TENGs works by four different mechanisms, taking account of the change in the polarization direction of two triboelectric material surfaces/layers and the configuration of electrodes, which are insulated carefully from each other. The four distinct modes are (i) vertical contact-separation (CS) mode, prompted by series of contacting and separating the triboelectric layers attached to the electrodes. The mode shows high-output peak current and figure of merit (FOM), which can further be enhanced by a stacked multilayer structure [134]. Figure 2c is a schematic of CS-mode TENG [116]. In Figure 2c(I), two layers are pressed by external force, in order to create contact and generation of charges at the interface of dielectric triboelectric layers A and B. Releasing outside force (Figure 2c(II)) leads to charge separation among the layers, establishing the potential difference that drives free charges in the electrode for potential balance setting current across the load R. When the charge is balanced, the current disappears, see Figure 2c(III). Again, pressing layers to make contact results in the flow of free charge, accumulated on the surface of the electrode, back into the circuit, resulting in current, but reversed, direction. The electric potential (V) of TENG is defined as:

$$V = \frac{\sigma_{tr}}{\varepsilon_0 \varepsilon_r} d \tag{3}$$

In the above equation, σ_{tr} is the triboelectric charge density, d is the gap distance betwixt two triboelectric material layers, ε_0 is the permittivity of free space, and ε_r is the relative permittivity of triboelectric material. Clearly, the potential produced is a function of layer gap and depends on layer movement, so, the current. (ii) Lateral-sliding (LS) mode [116] is triggered by sliding the tribo-layers, studies revealed that longer sliding distances lead to lower FOM, but charge densities are high. The output performance of LS mode TENGs can be improved by structure grating, which can also be altered to either rotational or cylindrical gratings [135]. (iii) In the single-electrode (SL) mode, the back electrode is removed, which offers a very small amount of charge transfer and voltage [136]. (iv) In the freestanding triboelectric layer (FL) mode, the triboelectric layers are free from electrodes and can locomote with or without contacting with static electrodes, resulting in high-output power and energy conversion efficiency records [137]. Introducing the ferroelectric material layer to TENG speculates the increased surface charge density, e.g., the barium titanate ferroelectric ceramic layer, on polytetrafluoroethylene (PTFE) TENG [138]. Liu et al. obtained a high piezoelectric and pyroelectric coefficient of 150 pC/N and 29.7 nC/cm^{2}K for BTO polarized disc by wind-driven TENG at speeds of 14 m/s, which resulted in the high-output voltage of 1000 V in <10 ms [40]. The literature revealed the improved output performance of PVDF-TrFE, based TENG, due to well-oriented polarization. TENGs offer broad material availability and high efficiency, yet low operation frequency [139].

3.4. Ferroelectric Photovoltaic Effect and Photovoltaic Cells

The ferroelectric photovoltaic effect refers to the photovoltaic response, i.e., photovoltage and/or photocurrent, observed in ferroelectric materials when exposed to light. The photovoltaic phenomenon in ferroelectrics is significantly dissimilar to the conventional photovoltaic effect (PVE) observed in semiconductor pn-junction. The built-in electric field, across the depletion region of the pn-junction, separates the generated photo-carriers, thereby defining that the open-circuit voltage for particular photovoltaic effects in semiconductors has a bandgap constraint. However, in the ferroelectric photovoltaic effect, the built-in field is because of remnant polarization (P_R) and goes throughout the ferroelectric material, and there is no bandgap restriction for charge separation. Therefore, the photo-potential generated in ferroelectrics is much greater to abnormal value than their band gaps. This phenomenon is the so-called abnormal photovoltaic effect or bulk photovoltaic effect (BPVE) and had been known in barium titanate and lithium niobate for decades. However, its clear explanation is still a mystery and the photovoltage integrated can be visualized as the photovoltage of domains [140,141]. The bulk ferroelectric photovoltaic effect has been observed in bulk ferroelectric materials, such as BFO (BiFeO$_3$), with large photovoltage stipulating the internal bias field presence. The thin films of BTO (BaTiO$_3$) evidenced that the direction of photocurrent and photovoltage can be rolled over by reverting polarization, proving the polarization as a dominant feature in the bulk photovoltaic effect [140,142,143]. Figure 2d(I) represents the ferroelectric photovoltaic effect mechanism in poled SbSI nanowires with Pt metal contacts [29]. The SbSI nanowires proposed the bulk photovoltaic ferroelectric (BPVE) mechanism, shining light with energy greater than the bandgap of SbSI ferroelectric, resulting in the absorption of light photons, and promoting the photo-generation of electron and holes as excess carriers. Poling the SbSI ferroelectric caused band bending and, thus, the internal electric field, due to the spontaneous polarization of nanowires (Figure 2d(II)). The internal field determines spatial separation charge carriers in ferroelectric photovoltaic devices and these excess carriers contribute to the photovoltaic output current [29].

3.5. Coupled Effects NGs

Recently, hybridized nanogenerators NGs have also been reported by integrating individual piezoelectric, pyroelectric, triboelectric, and PVC harvesters, in series or parallel, for the realization of self-powering of devices with higher consumption powers by utilizing multi-energies. The hybrid stacked structures enhance the overall output power but restrict the interaction among various effects in the device, as different electrodes export electrical energy. Furthermore, it is discouraging for reducing device size and heavy productions [6,46,56,144]. Therefore, it is greatly aspired to deploy multifunctional energy harvesting materials for establishing the coupled nanogenerators, based on a single structure. Ferroelectric materials are multifunctional, thereby defining that the piezoelectric, pyroelectric, and photoelectric properties concur, also a common feature among piezoelectricity, pyroelectricity, triboelectricity, and the ferroelectric photovoltaic effect is that they all work by having an influence on the polarization of ferroelectric materials. This similarity indicates the potential to couple these effects, in order to magnify the charge quantity and electric power from various energy resources simultaneously and promote their use in multi-effect coupled NGs, whenever all the mechanical, thermal, and solar energies are either available individually or simultaneously. The multi-effect coupled NGs are composed of multilayers, with only one electrode pair that is highly versatile, reliable, simply smaller, and low cost. The multi-effects coexist and interact to produce electrical output; their performance hangs on the strength and type of external stimuli. Device design determines the performance of multi-effect coupled nanogenerators. By now, a sandwich layer layout with a function material embedded in two electrodes, planer structure consisting of two parts, functional material layer, coplanar electrodes realized by laser etching or masking, and heterojunction design (which commonly intervene in PVCs) have been contrived [6,47,145].

4. Device Structure and Performances of Ferroelectric NGs

Exceptional output device performances are always in search of better design and tailoring material properties.

4.1. Device Design and Output Power Optimization in Piezoelectric PENGs

The piezoelectric ceramic perovskites show high piezoelectric constants and electromechanical response because of polarization rotations. The ferroelectric, ceramic-based PENGs with active layers of lead zirconate titanate (PZT) [146,147], barium titanate (BTO) mboxciteB148-nanoenergyadv-1435434,B149-nanoenergyadv-1435434,B150-nanoenergyadv-1435434, lead magnesium niobate titanate (PMN-PT) [151–154], sodium bismuth titanate (NBT) [155–159], and sodium potassium niobate (KNN) [160,161] have been reported with high-output performances. Ferroelectric thin film PENGs have also been reported on flexible plastic substrates with metal electrodes and by transferring techniques [146,154]. However, ceramic PENGs have limitations of large-area devices fabrication; additionally, ceramics are frangible in nature and cannot withstand very strong strain-producing forces. To get out of these problems, reports on a polymer matrix supported by ferroelectric powder composites have been found, which fortified a large area of manufacturing devices, reduced cost, sustain high-stress forces, and enhanced mechanical resilience, but are not admissible in low magnitude and frequency input forces [146,153]. This time, highly durable and elastic polymer ferroelectric-based PENGs were proposed, with high mechanical permanence and optimistic output performances [62,162–164]. Semi-metallic/conductive layers or electrodes are deposited over the ferroelectric active layers to get charge separation without poling; graphene is a potential conductive layer, due to high thermal and electrical properties over the large surface area. Additionally, surface treatments or doping can make graphene either p-type or n-type, which can further align the polarized charges efficiently. Polyvinylidene fluoride (PVDF) and its copolymers are rosy for polymeric ferroelectric-based PENGs. PVDF exhibit a smaller dielectric constant but have extremely high parameters of voltage (g_{33} = 28.26) and piezoelectric constants (d_{33} = 16.2 pC/N) [63,163]. The d_{33} value can be further enhanced by introducing ceramic fillers with relatively larger permittivity, e.g., PZT, PLZT (PZT:La), PTO, BTO, and PMN-PT into the PVDF matrix. The quantity of ceramic powder fillers is most important, and it controls the harvesting properties of PENGs; the positive and negative piezocoefficients of ceramic and polymeric ferroelectrics may rule out each other's effects, thereby reducing the output performances, so tedious measurements must be taken into account [165,166]. Yaqoob et al. [167] fabricated a novel tri-layer PVDF-BTO/n-Gr/PVDF-BTO piezoelectric NG, with 60 μm thickness. The PVDF/BTO layers stacked on both sides of the N-graphene layer deposited over Si substrate. Figure 3a demonstrates the making process of tri-layer PENG. The working mechanism of tri-layer PENG is illustrated in Figure 3b, escribing that between n-Gr and PVDF-BTO, the layer charge barriers are created, due to the accumulation of opposite charges near the interfaces. The lower n-Gr/PVDF-BTO layer acts as a blocker and restricts recombination and sustains the charge on the upper layer, and dipole alignment happens on both sides. Extreme bending of PENG shows high flexibility. FESEM studies confirmed that BTO nanoparticles were dispersed uniformly in the PVDF matrix and graphene over PVD-BTO film in the β-tetragonal phase. The high peak-to-peak open-circuit voltage of 1.5 V and 10 V was observed in un-poled and poled PENG under the force of 2 N by human finger tapping. Figure 3c shows the maximum short-circuit current output of 2.5 μA and instantaneous powers of tri-layer PENG was found saturated at 10 MΩ load resistance, with a maximum value of 5.8 μW at 1 MΩ. PENG was found to be highly stable, even under continuous pressing and releasing at 2 N for 1000 s, and is attributed with N-graphene, suggesting graphene-enhanced PENG performance and realizing that the tri-layer structure is efficient for mechanical energy harvesting and perspective to use as a pressure sensor and power source of self-powered devices. Hanani et al. [33] investigated a bio-flexible PENG with potential in self-powered biomedical devices, as well as lead-free, bio-ceramic $Ba_{0.85}Ca_{0.15}Zr_{0.10}Ti_{0.90}O_3$ (BCZT), functionalized with polydopamine that is embedded with polylactic acid. BCZT@PDA/PLA is sandwiched between two Cu-foils and encapsulated with Kapton tape to keep it water- and dust-proof and secure, as of repeated mechanical excitations, see Figure 3d. PLA has a high piezoelectric coefficient d_{13} of 10 pC/N, without poling or being self-poled. The BCZT nanoparticle's surface was functionalized via core-shell saturation (Figure 3e), with a PDA layer to promote interaction among the two. Under gentle finger tapping, the mechanical performances, open-circuit voltage, and short-circuit current were quantified to be 14.40 V and 0.55 μA, respectively; however, the increased imparting pressures caused V_{oc} to rise. Young modulus of 2.1 GPa endures mechanical robustness, and the high-output power density of 7.54 mW/cm^3 is capable of driving the 1 μF capacitor, with an energy storage of 3.92 μJ, only in 115 s, under gentle finger tapping and longer stability at 23 Hz for 14,000 cycles by sewing machine, which showed no performance degradation for 4800 cycles, even after one year.

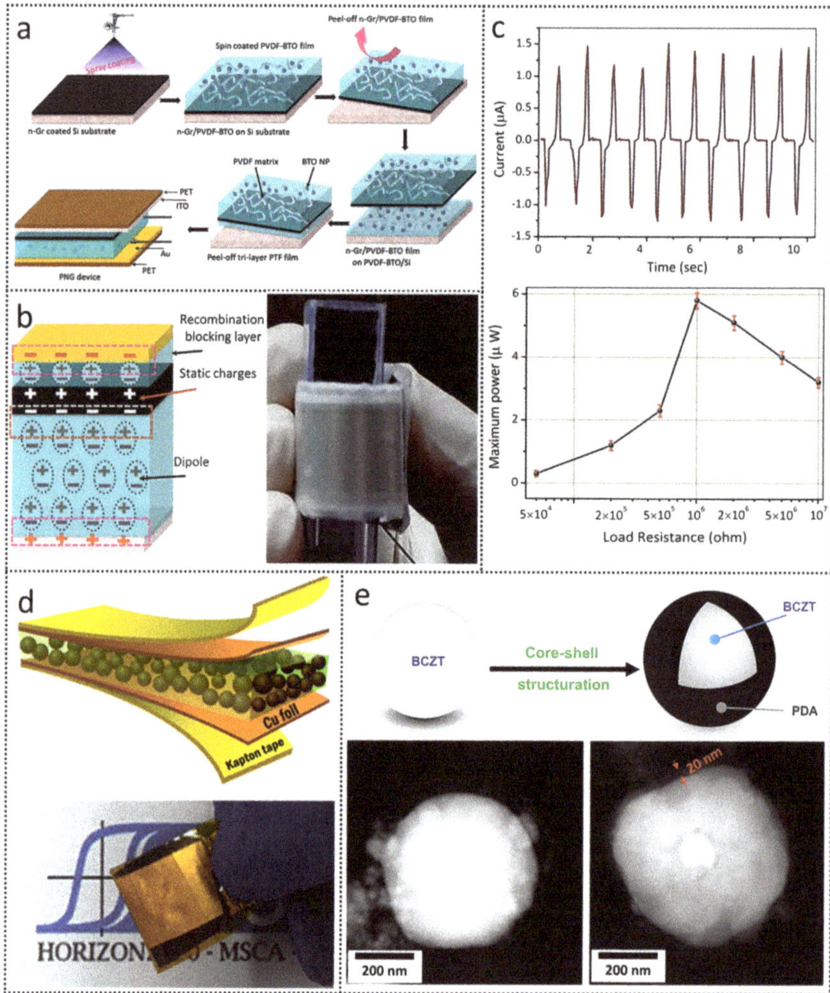

Figure 3. Device structure and performances of piezoelectric ferroelectric nanogenerators (PENGs). (**a**) Fabrication steps for tri-layer n-Gr/BTO-PVDF on BTO-PVDF/Si PENG. (**b**) The working mechanism of tri-layer PENG, along with an optical image of the extreme bending situation. (**c**) Short-circuit current and instantaneous power output of tri-layer PENG. (**a–c**) Reprinted with permission from reference [167], Copyrights 2017, Elsevier. (**d**) Schematic of Bio-flexible BCZT-based nanogenerator. (**e**) Illustration of the core-shell structuration of BCZT nanoparticles, by PDA, along with microimages of BCZT and BCZT@PDA. (**d**,**e**) Reprinted with permission from reference [33], Copyrights 2021, Elsevier.

Xiaohu et al. [168] exposed $BiFeO_3$-PDMS composite (~100 nm)-based flexible PENG on polyethylene terephthalate (PET) substrate and Al-foil electrodes (PET/Al/BFO-PDMS/Al/PET), see Figure 4a. $BiFeO_3$ nanoparticles (NPs) had a rhombohedral structure, with 180° domain switching of permanent polarization. The PENGs showed the maximum output, open-circuit voltage of 3 V for 40% BFO NPs in PVDS matrix and a short-circuit current of 250 nA, under hand pressing, can be seen in Figure 4b. Simulated piezo potential distribution is indicated on the right side of the voltage-time graph, as a response to the compressive stress of 10 KPa. The fabricated BFO PENG can light commercial LEDs. Lee et al. [169] reported a novel fiber-type piezoelectric NG, based on PTO ($PbTiO_3$) nanotubes, grown radially on Ti metal fiber; the PTO/Ti core-shell fiber was embedded in the PDMS cylindrical templates, with inter-wire spacing formed and thin, insulating PDMS. Poling (via convex and concave bending of the PENG) the output voltage and current of 623 mV and 1.0 nA/cm^2 was estimated. When a second core-shell was added in the back of the first (Figure 4c), poled in radial direction because of radial alignment of PTO domains, the output voltage estimated in- and out-of-plane bending were ±320 and ±390 mV (for 0° and 180°) and ±370 and ±450 mV (for 90° and 270°), respectively. Under natural wind, blowing the output voltage of 69.74 ± 10.97 mV with 0–360°, bending was observed. The idea captivates applications in portable and flexible devices. Yan et al. [170] reported a PDMS-based, flexible PENG, with BTO nanofibers in three different alignment modes (random, horizontal, and vertical) and tetragonal structure, as shown in Figure 4d, poled for 12 h at 120 °C and an electric field of 5 kV/mm. Random (BTNF-R), horizontal (BTNF-H), and vertical (BTNF-V) alignment of BTO nanofiber PENGs showed output voltages of 0.56 V, 1.48 V, and 2.67 V and currents 57.78 nA, 103.33 nA, and 261, 40 nA, respectively, under a poled situation and periodic mechanical compressive pressure of 0.002 MPa, with a harvesting power of ~0.1841 µW in vertical mode (BTNF-V). This pointed out that the vertically lined BTO nanofibers were finest in dielectric and piezoelectric response and used to light commercial LED and charge a 1 µF capacitor to 0.46 V in just 34 s (shown in Figure 4e), stating wonderful alignment for use as harvester units or storage entities for wireless networks. Jung et al. [171] presented a PENG based on NaNbO$_3$ nanowires, composited with the PDMS matrix and Au/Cr coated films (PS/Au-Cr-coated kapton/NaNbO$_3$-PDMS/Au/Cr-coated Kapton; PS is the supporting polyester film, the strain neutral line lies near the vicinity of PS film). Estimated output voltage V_{oc} of 3.2 V, short-circuit current I_{sc} of 72 nA, and power density of 0.6 mW/cm^3, under the compressive strain produced by vibrations of 0.33 Hz. Production of NaNbO$_3$, at relatively low temperature and domain control (by poling) make them perfect nanogenerator members. Park et al. [150] devised MIM structure-based Au/BTO/Pt PENG with a ribbon structure over a plastic substrate, using PDMS stamp, the Kapton film, and poled for 15 h at 140 °C, with a field application of 100 kV/cm. Poling caused d_{33} to increase from 40 pm/V to 105 pm/V; periodic bending (by finger) resulted in an output voltage and current values of ~1.0 V and ~26 nA and the power density of 7 mW/cm^3. In comparison to inorganic ferroelectrics, very few organic ferroelectrics exist. For example, single-component polar organic molecules, such as CDA, DNP, CT complexes, polymers (such as PVDF and nylon-11), and organic-inorganic composites (such as HdabecoReO4, TGS, and TSCC), are excellent for applications in piezoelectric and triboelectric nanogenerators. Isakov et al. [172] reported a piezoelectric nanogenerator, based on organic ferroelectric nanofibers of hybrid 1,4-diazabicyclo[2.2.2]octance perrhenate (dabacoHReO$_4$), with an output voltage of 100 mV under moderate strain values. Ferroelectric diphenylalanine peptides are also becoming promising for piezoelectric energy harvesting and are biocompatible. Zelenovskii et al. [173] reported the fabrication of the 2D layered biomolecular crystals of diphenylalanine, with a piezoelectric constant d_{33} of 20 pm/V and voltage coefficient g_{33} of 0.75 Vm/N, which is suitable for microfluidic device applications. Nguyen et al. [174] estimated strong piezoelectricity in diphenylalanine (FF) peptide nanotubes, with a piezoelectric constant d_{33} of 17.9 pm/V, open-circuit output voltage of 1.4 V, and power density of 3.3 nW/cm^2. In a report by Lee et al. [175], peptide-based piezoelectric energy harvesters, at a force of 42 N, can produce an output voltage of 2.8 V, current of 27.4 nA, and power of 8.2 nW, which is sufficient to power multiple liquid-crystal display panels. Table 1 gives a quick insight into various piezoelectric nanogenerators (PENG), along with their output characteristics and working conditions, made with various structural variations and morphologies, indicating the high-output characteristics.

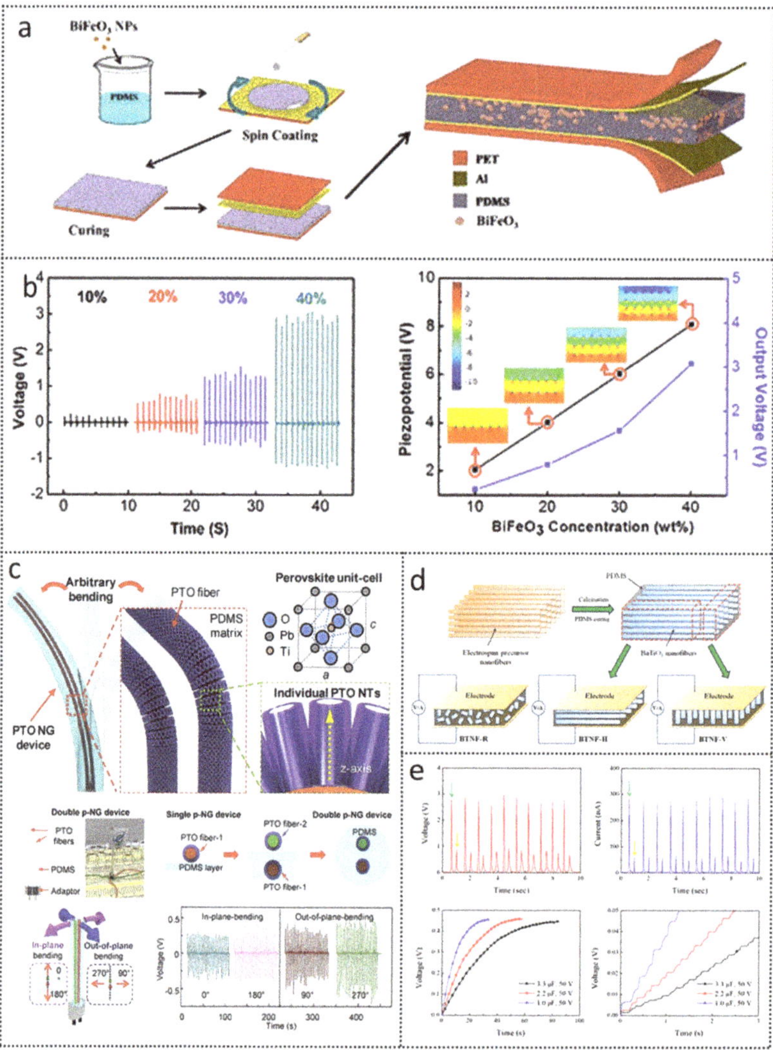

Figure 4. Device structure and performances of piezoelectric ferroelectric nanogenerators (PENGs). (**a**) Lead-free flexible BFO-PDMS PENG. (**b**) Time-dependent voltage and COSMOL simulated piezo potential for 0–40% BFO nanoparticle concentration PENGs. (**a**,**b**) Reprinted with permission from reference [168], Copyrights 2016, ACS Publications. (**c**) PTO/Ti fiber PENG in arbitrary bending direction with PDMS polymer matrix. The bottom line shows the double PTO fiber PENGs, along with two possible bending directions (in- and out-plane) and corresponding output voltages. Reprinted with permission from reference [169], Copyrights 2017, John Wiley and Sons. (**d**) Schematic of BTO-based PENG nanogenerator in three different alignment modes. (**e**) Rectified voltage-current signals and voltage-time graphs for different capacitors. (**d**,**e**) Reprinted with permission from reference [170], Copyrights 2016, ACS Publications.

Table 1. Piezoelectric nanogenerators, based on ferroelectric materials, for energy harvesting.

Material Name	Open-Circuit Voltage V_{oc}	Short-Circuit Current I_{sc}	Short-Circuit Current Density J_{sc}	Maximum Area Power Density	Maximum Power	Work Conditions	Reference
PVDF/SM-KNN (3%)	21 V	22 µA	5.5 µA/cm^2	115.5 µW/cm^2	-	1.1 kPa pressure	[34]
Ag/(K, Na)NbO$_3$	240 V	23 µA	-	-	1.13 mW	0.1 MPa mechanical stress	[176]
FAPbBr$_3$@PVDF	30 V	-	6.2 µA/cm^2	27.4 µW/cm^2	-	0.5 MPa mechanical stress	[177]
BaTiO$_3$/PDMS	5.5 V	350 nA	350 nA/cm^2	-	-	1 MPa mechanical stress	[23]
PVDF-TrFE/SnS nanosheets	17.28 V	-	0.94 µA/cm^2	10.69 µW/cm^2	-	0.5 MPa pressure	[178]
rGO/PVDF nanofiber mat	16 V	700 µA	-	-	7.4 µW	Finger press	[179]
BaTiO$_3$@PVDF	6 V	1.5 µA	-	8.7 mW/cm^2	-	Impact force of 700 N with frequency 1–3 Hz	[180]
Pb$_{0.67}$Zr$_{0.33}$TiO$_3$/PDMS	152 V	17.5 µA	-	-	1.1 mW	100 N stress force	[75]
P(VDF-TrFE)/GeSe nanosheets	17.58 V	-	1.14 µA/cm^2	9.76 µW/cm^2	-	0.5 MPa pressure	[181]
0.5Ba(Zr$_{0.2}$Ti$_{0.8}$)O$_3$-0.5(Ba$_{0.7}$Ca$_{0.3}$)TiO$_3$/PVDF-TrFE	13.01 V	-	-	-	1.46 µW	Cyclic tapping of 6 N and 10 Hz	[182]
0.94(Bi$_{0.5}$Na$_{0.5}$)TiO$_3$-0.06BaTiO$_3$ (NBT-BT)/PVDF	9 V	8–9 µA	-	-	80 µW	250 N impact force	[183]
Ba(Ti$_{0.88}$Sn$_{0.12}$)O$_3$-GFF/PVDF	26.9 V	597 nA	-	-	-	5 N impact force	[184]

4.2. Device Design and Output Power Optimization in Pyroelectric PyENGs

Low-grade heat wastes coming from the industry of which maximum portion is hard to recuperate, a way to rescue these heat wastes is harvesting them into electric energy by temperature fluctuations either by Seebeck effect or by pyroelectricity. Thermoelectric nanogenerators require a steady gradient temperature that is infrequent in nature, otherwise, by pyroelectric nanogenerators, PyENG, lined with temporal temperature sways. The material and structure designs are crucial for the improvement in the output of PyENGs because the pyroelectric coefficients have small values. Improving these various approaches has already been reported, including the control over crystallinity for im-proving pyroelectric coefficient [185], introducing strain coupling effects [186], or polymer modifications [187]. Kim et al. [185] used high dipole moment solvents for improving the crystallinity and pyroelectric coefficient of P(VDF-TrFE); the solvents they used were THF (tetahydrofuran), MEK (methylethylketone), DFM (dimethylformamide), and DMSO (dimethylsulfoxide). High crystallinity and improved pyroelectric constant were observed and figured to be associated with the long-chain lengths of P(VDF-TrFE); they exhibited high molecular weight and 1.4 times higher output voltage and current. Hence, pyroelectric coefficient enhancement, by utilizing high dipole moment solvents, is a promising way to improve output PENG performance. Ghosh et al. [187] presented improved pyroelectricity of PVDF PyENG by introducing the Er^{+3} via formation of self-polarized β-phase and porous flower-like structures, as well as enhanced thermal sensitivity, triggered by IR irradiations. Under a temperature change rate of 1 K/s, the estimated output current was 13 nA. The 4.7 µF capacitor was easily charged by Er-PVDF PyENG, which operates low-power electronic devices with great ease. Mistewicz et al. [122] demonstrated low-temperature heat waste by SbSeI-based PyENG, with a pyroelectric coefficient of 44(5) nCK^{-1}/cm^2 at 327 K. The output voltage, current, and power of 12 mV, 11 nA, and 0.59(4) µW/m^2 were generated by PyENG for temperature fluctuations of 324 K to 334 K. Yang et al. [188] reported KNbO$_3$ single-crystal nanowire (~150 nm), along a [011] direction composite with PDMS ($KNbO_3$-PDMS in 3:7 and pyroelectric constant 0.8 nC/cm^{2}K) PyENG, embedded with Ag and ITO electrodes, Figure 5a. Positive poling caused the dipoles in the $KNbO_3$ nanowires to align from top to bottom and reverse, i.e., bottom to top, for negative poling. Increased temperature from RT resulted in output voltage and current peaks to be positive, which go opposite for lowing the temperature (see Figure 5b). At a temperature change of about 40 K, the observed output voltage and current peaks were 10 mV and 120 pA, and the output current is directly related to temperature variations (dT/dt > 0). They also deployed the generator for harvesting the solar energy, as heat is induced by the sun, showing potential for self-powered widgets. Lee et al. [186] designed an innovatively stretchable pyroelectric nanogenerator (SPNG), by coupling thermally induced piezoelectric and pyroelectric effects, using P(VDF-TrFE) and PDMS; the piezoelectric effect was generated by thermal energy, only by accomplishing the different thermal expansions (122×10^{-6}/K for P(VDF-TrFE) and 310×10^{-6}/K for PDMS) and generating a compressive strain in P(VDF-TrFE) by PDMS. Figure 5c shows a schematic of stretchable PyENG, Ag/AgNWs/P(VDF-TrFE)/Au/PDMS, along with an optical micrograph of 500 μm × 700 μm P(VDF-TrFE) in β-phase and 200 μm × 200 μm PDMS. The temperature sensitivity of stretched nanogenerator was analyzed by comparing it with a normal P(VDF-TrFE)/Ni/SiO2/Si nanogenerator, and negative peak signals were reported for temperature elevation from RT (right part of Figure 5c). The sensitive, stretchable pyroelectric NG, under extremely low temperature variations (ΔT) of 0.64 K, to higher variations 18.5 K, showcased highly stable output voltage performance of 8 mV to 2.48 V under extreme stretch circumstances over a normal nanogenerator 2 mV to 0.54 V. Almost 15% stretching, caused only the PDMS layer to expand by 30 μm, and temperature variation of 22 K from RT caused an output voltage and current density of 2.5 V and 570 nA/cm^2, respectively. Yang et al. [37] reported a PZT film-based pyroelectric nanogenerator (PyENG) with Ni, a top and bottom electrode connected, and Cu-wires (fixed with Ag paste) for electrical measurements and encased with Kapton tape. For temperature changes of 45 K, at the rate of 0.2 K/s, the output open-circuit voltage, short-circuit current, and short-circuit current density of PENG reached 22 V, 430 nA, and 171 nA/cm^2, respectively; the corresponding maximum power density of 0.215 mW/cm^3 was estimated, and a single electrical pulse can power a liquid crystal display (LCD) for more than the 60 s. The proposed PyENG showed potential applications as wireless sensors, drivable by a rechargeable Li-ion battery with a voltage of 2.8 V, see Figure 6a. The pyropotential calculation of PZT film showed distributions from −200 V to 200 V for temperature variations from 295 K to 340 K. With the measured electric potential of 22 V, much smaller than the theoretical value, they showed increased output current, as the surface area of PyENG was doubled than original. Xue et al. [27] proposed the self-powered wearable pyroelectric nanogenerator, based on Al-coated PVDF film, exercised in an N95 respirator for the sustainable harvesting and monitoring of human breathing energy, see Figure 6b. They observed that the open-circuit output voltage V_{oc} of ~42 V is generated by the temperature oscillations rate, 13 °C/s, encouraged by humans breathing exhausted gas, at an ambient temperature of 5 °C with short-circuit current of 2.5 μA and the maximum power output of 8.31 μW. The power produced can drive a liquid crystal display (LCD) and/or LED directly. They employ the fabricated PyENG as self-power sensors for human health monitoring and temperature. Gao et al. [189] proposed a P(VDF-TrFE)-based pyroelectric nanogenerator, driven by temperature fluctuations, induced up to 23 °C/s by water vapors during the process of evaporation and condensation. As can be seen in Figure 6c, the oscillation airflow environment was created by a fan with an air speed of 1–2 m/s. The PyENG was capable of generating a high-output voltage and current density of 145 V and 0.12 μA, with a power density of 1.47 mW/cm^3 and 4.12 μW/cm^3 by volume and area. The water-operated PyENG could run a low-power digital watch, blue LED light, and charge a capacitor of 2.2 μF, giving a new map to harvest energy from the wastewater of industries. Yang et al. [36] prepared a self-powered temperature sensor, based on a pyroelectric nanogenerator, using a single micro/nanowire of PZT on a thin glass substrate, packaged by PDMS. Employing this device to touch the heat source the voltage was found to vary linearly with temperature increments, as well as the response and reset time of 0.9 s and 3 s. The minimum constraint on detecting temperature variations at RT was 4 K, and the device was as sensitive to detecting the temperature of the fingertip; large temperature fluctuations could light up the LCD. Some other pyroelectric nanogenerators, PyENG, have been listed in Table 2.

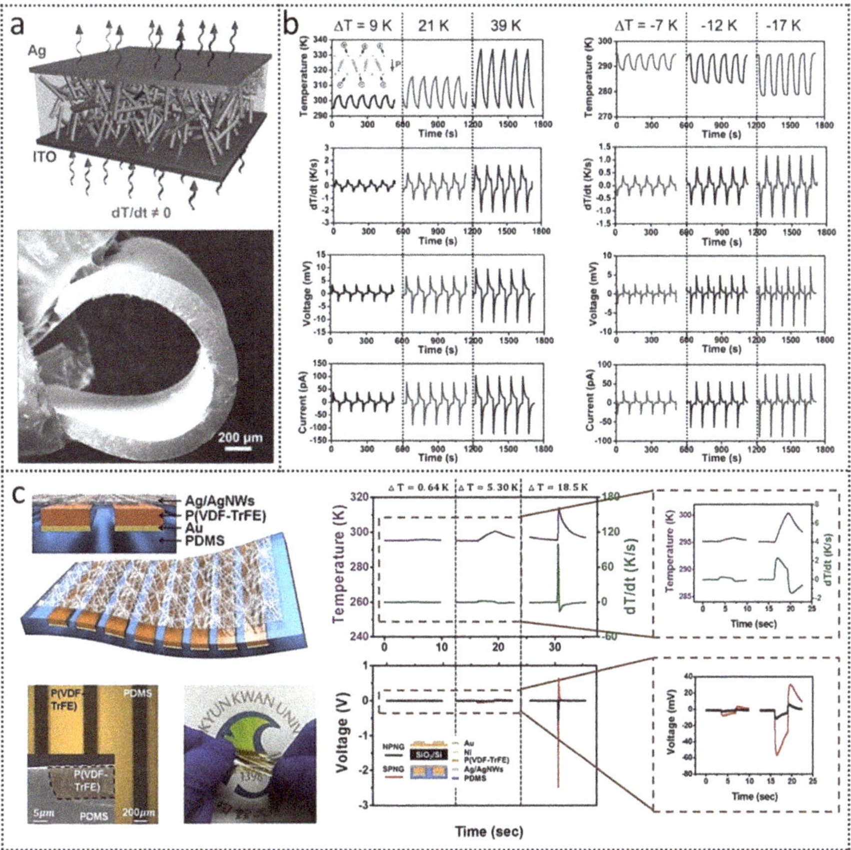

Figure 5. Device structure and performances of pyroelectric ferroelectric nanogenerators (PyENGs). (**a**) Schematic of KNbO$_3$-PDMS composite pyroelectric nanogenerator, with Ag ad ITO electrodes; the bottom line is the SEM image of KNbO$_3$ in the bent state. (**b**) Cyclic changes in temperature, with corresponding differential curves of open-circuit voltage V$_{oc}$ and short-circuit current I$_{sc}$. (**a,b**) Reprinted with permission from reference [188], Copyrights 2012, John Wiley and Sons. (**c**) Schematic of highly stretchable P(VDF-TrFE) pyroelectric nanogenerator (SPNG); the bottom line left image indicates the top view of SPNG, along with photo image. The right hand image indicates the measured output characteristics for both stretchable and normal pyroelectric nanogenerators. Reprinted with permission from reference [186], Copyrights 2015, John Wiley and Sons.

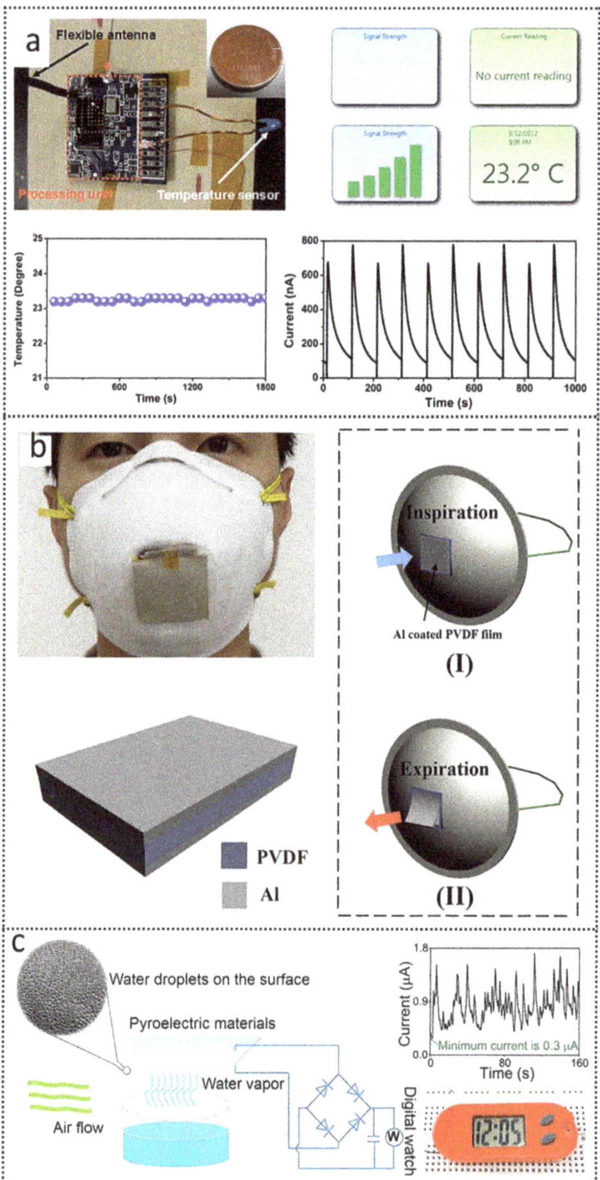

Figure 6. Device structure and performances of pyroelectric ferroelectric nanogenerators (PyENGs). (**a**) Photograph of wireless temperature sensor, based on PZT PyENG, along with temperature signals from a wireless sensor in off- and on-state, temperature signal data, as a function of time, was recorded for half an hour (for every one minute) and enhanced the output current profile of as-deposited PyENG for doubled surface area. Reprinted with reference from [37], Copyrights 2012, ACS Publishers. (**b**) Physical schematic image of the wearable PyENG with PVDF film, driven by human respiration, along with states of (I) inspiration and (II) expiration. Reprinted with permission from reference [27], Copyrights 2017, Elsevier. (**c**) Powering digital watch from P(VDF-TrFE)-based PyENG, powered by hot water vapors. Reprinted with permission from reference [189], Copyrights 2016, Elsevier.

Table 2. Pyroelectric nanogenerators, based on various ferroelectric materials.

Material Name	Output Characteristics	Pyroelectric Coefficient p	Operating Conditions ΔT, dT/dt	Maximum Power/Power Density	Energy Density	Reference
$0.7Pb(Mg_{1/2}Nb_{2/3})O_3$-$PbTiO_3$	V_{oc}: 1.1 V I_{sc}: 10 nA	$104\ nC/cm^2K$; 300 K $235\ nC/cm^2K$; 370 K	(300 to 370 K) 0.5 K/s	-	-	[190]
$0.94Na_{0.5}Bi_{0.5}TiO_3$-$0.06BaZr_{0.2Ti}0.8O_3$ (NBT-BZT): $xSiO_2$	V_{oc}: 3.5 V I_{sc}: 90 nA	$20 \times 10^{-4}\ C/m^2K$	25 °C to 50 °C	-	$110\ \mu J/cm^3$	[191]
$(1-x)(0.98Bi_{0.5}Na_{0.5}TiO_3$-$0.02BiAlO_3)$-$x(Na_{0.5}K_{0.5})NbO_3$	V_{oc}: 3.3 V I_{sc}: 81.5 nA	$8.04 \times 10^{-4}\ C/m^2K$	25 °C to 50 °C and 3 °C/s	-	$23.32\ \mu J/cm^3$	[192]
PVDF	V_{oc}:192 V I_{sc}: 12 μA	$2.7\ nC/cm^2K$	60 °C to 80 °C	14 μW	-	[129]
PVDF/GO	V_{oc}: 60 mV I_{sc}: 45 pA	$27\ nC/m^2K$	ΔT = 20 °C and 2.12 K/s	$1.2\ nW/m^2$	-	[193]
PMnN-PMS-PZT (Zr/Ti:95/5)	V_{oc}: 25.3 V	$5957\ \mu C/m^2K$	18 °C to 65 °C and 3.15 °C/s	25.7 μW	-	[194]
0.68PMN-0.32PTO	I_{sc}: 90 nA	$-550\ \mu C/m^2K$	Temperature oscillations of 10 K	$526\ W/cm^3$	$1.06\ \mu J/cm^3$	[195]

4.3. Device Design and Output Power Optimization in Triboelectric Nanogenerators TENG

The electrical outputs generated from TENG are proportional to the intensity of the mechanical input strengths. Triboelectric nanogenerators can be supported by ferroelectrics because of their switchable intrinsic polarization. It increases the triboelectric charging behavior to enhance their output characteristics by coupling residual dielectric and surface polarization. Organic ferroelectrics, such as P(VDF-TrFE), show controllable ferroelectric polarization, leading to elevated surface charge density and improved TENG performance. High dielectric, inorganic ferroelectric ceramics and polymer composites of TENG need precise control of the nanoparticles' dispersion into the dispersive media. The precision control avoids the agglomeration that leads to the degradation of device performances. Park et al. [196] reported TENG with multilayer nanocomposites, comprised of alternate layers of organic and inorganic ferroelectric materials P(VDF-TrFE) and BTO nanoparticles with Al and Cu electrodes (Figure 7a). The organic soft layer P(VDF-TrFE) significantly transfers the vertically applied stress to the BTO inorganic layer. It exhibits a dielectric constant of 17.06 at 10 kHz in a multilayer structure, which yields superior interfacial polarization. The multilayer structure facilitates less leakage current, instigated by the P(VDF-TrFE) barrier between BTO layers. Poling at 30 MV/m further induced the surface potential of -2.85 V. As a result, more increased output characteristics of current, voltage, and dielectric constants, compared to only P(VDF-TrFE) and a single layer of BTO dispersed in P(VDF-TrFE) (Figure 7b,c). Multilayered TENG had a pressure sensitivity of 0.94 V/kPa with an output power density of 29.4 μW/cm^2 and an output current density of 1.77 μAcm^{-2} and voltage of 44.5 V, with active application in healthcare monitoring devices. Fang et al. [197] reported TENG with self-assembled nanospheres of polystyrene with PVDF porous film. Its unique structure showed an outstanding high-output voltage of 220 V being able to directly pole single crystal of $0.65Pb(Mg_{1/3}Nb_{2/3})O_3$-$0.35PbTiO_3$ (PMN-PT). They realized the application of TENG in ferroelectric FET, with PMN-PT as a gate material, posting the potential of TENG in a self-powered memory system. Ferroelectric ceramics, polymerized with non-ferroelectric materials, have a great impact on the output performance enhancement of TENGs. Sahu et al. [64] demonstrated the increase in output performance of TENG by introducing a ferroelectric material layer of $(Ba_{0.85}Ca_{0.15})(Ti_{0.90}Zr_{0.10})O_{3x}\ BiHoO_3$ nanoparticles (BCZTBH for x = 0.02 to 0.1) into non-ferroelectric PDMS negative triboelectric layer. They correlated the piezoelectric properties of BCZTBH and integrated a multistacked piezo/tribo hybrid nanogenerator. They also concluded that mechanical excitation gives a higher output voltage of 300 V, a current of 6.6 μA, and a power density of 157 mW/m^2. Further, they integrated the multi-stacked device to charge a capacitor, in order to monitor the biomedical energy released from the body in various yoga poses. The module, as IR receivers and transmitters, were installed into a self-powered, wireless IR sensor communication system for high reliability and efficiency checks. The literature revealed that the annealing of poly(vinylidene fluoride-co-trifluroethylene) leads to crystal orientation instability and improper polarization axis match with the substrate. It prohibits its merits and obstructs its applications to TENGs. So, crystal orientation tailoring is in need. To treat the problem, researchers have worked to introduce a graphene layer [198], employed self-assembled monolayer-modified Au-substrates [198,199], and organo-silicate lamella [200], lattice matching with molecularly ordered PTFE substrates via epitaxy [201]. All these epitaxial fabrication and graphene layer insertions yielded fabulous results by controlling the orientation of the crystal polarization axis by recrystallization after melting. However, during TENG operation, insurance of adequate charge density on the surface is a big problem that leads to inferior performance of TENG. To get out of this problem, the right angle alignment of the b-axis of PVDF-TrFE with the substrate is proposed. Eom et al. [139] reported on the TENG that was based on epitaxially grown PVDF-TrFE over chitin film (Figure 7d), the crystallographic overlap between two enabled tailoring the PVDF-TrFE crystal orientation with polarization axis (b-axis) orthogonal to chitin (c-axis) substrate. The observed remanent polarization (~4.2 $\mu C/cm^2$) was quite high, corresponding to a sweep voltage of ±80 V and coercive field ~60 MV/m higher than its theoretical value of 50 Mv/m, attributed with the series connection of PVDF-TrFE and chitin film. Chitin film showed severe distortions in humid circumstances, which can cause the short-circuiting of TENG. To avoid penetration of water molecules into chitin they deposited PVDF-TrFE film on two sides of chitin substrate by spin coating. Figure 7e depicts PVDF-TrFE/chitin hybrid layer TENG in vertical contact separation mode with an Al electrode. The 200 °C annealed hybrid film structure showed 11.7 and 4.9 times higher output current when compared with freestanding chitin film and non-epitaxial PVDF-TrFE/chitin hybrid film with a maximum power density of ~418 nW/cm^2 at 100 MΩ (Figure 7f). They fabricated triboelectric sensors of epitaxial PVDF-TrFE/chitin films. The allowed monitoring of subtle pressures suggested that tailoring the crystal orientation of PVDF-TrFE is encouraging for developing high-performance TENGs.

Since ferroelectrics have large remanent polarization. They can trap the spatial charge on electrodes for longer times and limit the TENG application in energy harvesting and sensing. To avoid this, exploitation of PLZT, a flexible anti-ferroelectric ceramic with large polarization differences ($\Delta P = P_r - P_{max}$), was carried out by Zhang et al. [202]. They proposed that $Pb_{0.94}La_{0.04}Zr_{0.98}Ti_{0.02}O_3$, with zero remanent polarization (P_r) and large maximum polarization (P_{max}), can enable all the dielectric polarization-induced charges, when used as an inner layer in TENG. A conducting fabric was attached with a natural rubber mat and PVDF/PZLT on LNO/Ni, as electrodes behave as a tribo-positive and -negative layer, as shown in Figure 8a. PLZT–TENG showed a record high $\Delta P = 1.065 \times 10^5$ nC/cm^2 at 1300 kV/cm, much higher than PZT-TENG ($PbZr_{0.52}Ti_{0.48}O_3$), shown in Figure 8b,c. PLZT–TENG delivered a high-output voltage of 456 V, with a high current density of 11.6 $\mu A/cm^2$ and charge transfer density of 13.5 nC/cm^2 at the impulse force of 5 N. Their study revealed that PLZT-based TENG exhibits excellent mechanical and polarization stability and can serve as self-powered sensors for joint motion detection, along with biomedical energy harvesting. Their study has paved a path to upgraded and optimized TENGs. Singh et al. [203] determined improved triboelectrification, by incorporating ZnO nanorods to PVDF polymer and PTFE and fastening to a moveable wooden frame with double-sided foam tape. They reported that the fabricated ZnO/PDMS TENG showed 21% and 60% increased voltage (~119 V) and current (1.6 mA), as compared to PVDF/PTMS TENG, with output voltage and current 98 V and 1mA, as well as power output 10.6 mW/cm^2 and 6.4 mW/cm^2 at 150 MΩ. They proposed that the increase in triboelectric energy, observed by them in ferroelectric polymer TENGs, is because of β-phase of PVDF enhancement. The surface roughness of films, hydrophobicity, and decreased work function is due to ZnO nanorod incorporations.

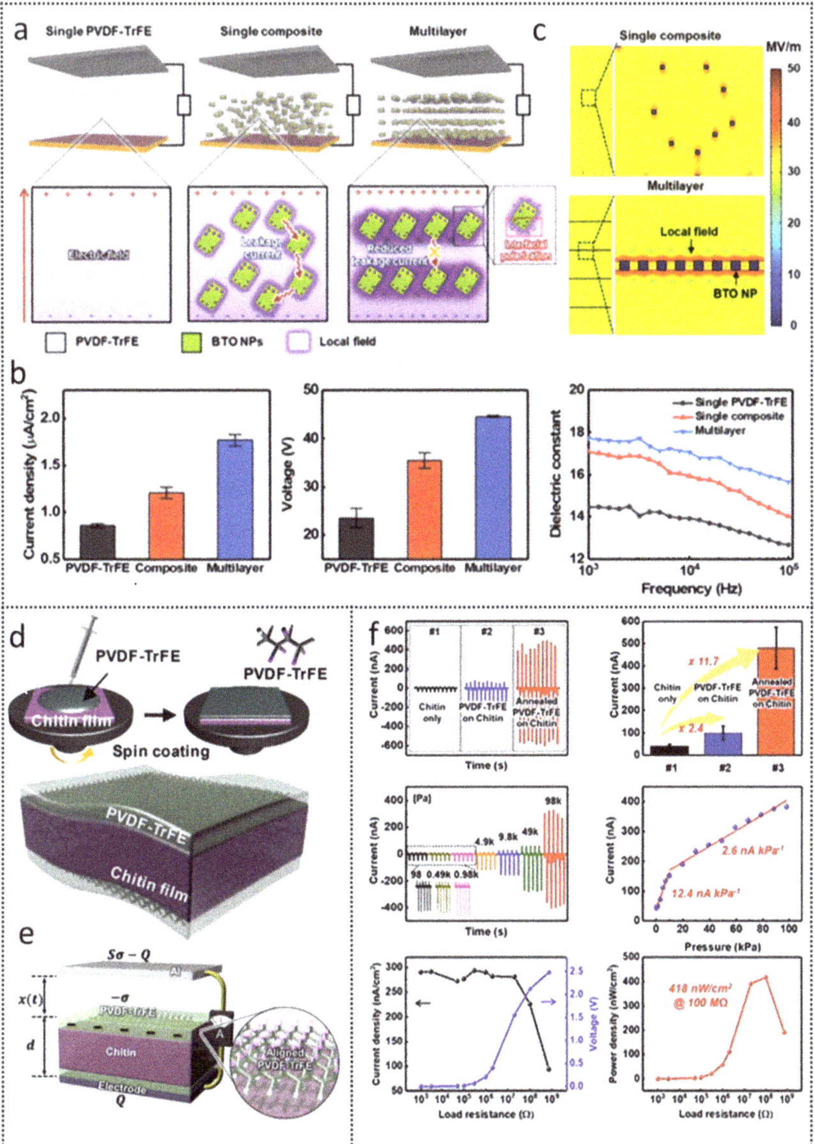

Figure 7. Device structure and performances of triboelectric ferroelectric nanogenerators (TENGs). (**a**) Schematic illustration of PVDF-TrFE/BTO composite TENG with single PVDF-TrFE, single PVDF-TrFE/BTO, and multilayer PVDF-TrFE/BTO layers. (**b**) Plots for output current density, voltage, and dielectric constant of three TENGs at vertical pressure of 98 kPa and 2 Hz. (**c**) Simulated potential in single and multilayered PVD F-TrFE/BTO composites. (**a–c**) Reprinted with permission from reference [196], Copyrights 2020, ACS Publisher. (**d,e**) Schematics of chitin-based PVDF-TrFE TENG. (**f**) Current and power plots of chitin /PVDF-TrFE hybrid film TENG. (**d–f**) Reprinted with permission from reference [139], Copyrights 2020, ACS Publications.

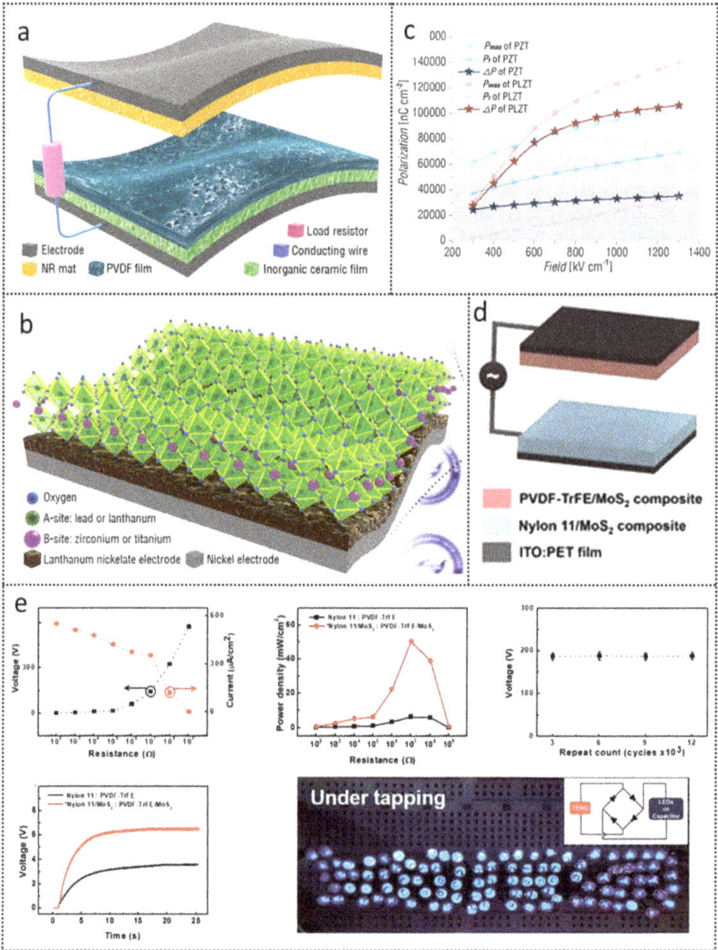

Figure 8. Device structure and performances of triboelectric ferroelectric nanogenerators (TENGs). (**a**,**b**) Schematic illustration of TENG with PLZT inner dielectric layer on flexible electrodes. (**c**) Maximum polarization, remanent polarization, and polarization difference for PLZT TENG and PZT TENG. (**a**–**c**) Reprinted with permission from reference [202], Copyrights 2021, Elsevier. (**d**) Structure of MoS_2, composited with PVDF-TrFE and nylon-11; right is an optical image of single PVDF-TrFE; PVDF-TrFE composited with MoS_2 and with nylon-11 and nylon-11. MoS_2, bottom line. (**e**); Voltage, current, and power conversion dependence, as for various load resistances and LED lighting. (**d**,**e**) Reprinted with permission from reference [204], Copyrights 2019, ACS Publications.

The continuous reiteration of TENG contact and separation is inevitable and leads to the life-shortening of the device. To overcome this wearing problem in TENGs, serval attempts had already been made, inclusive of which are the use of shape memory polymer as a contact material to heal damaged surfaces by heat application, proposed by Lee et al. [205]. The atomic layer deposition (ALD) technique for the in-filtering of inorganic materials from polymers was reported by Yu et al. [206]. Application of permanent magnet isolating outside vibrations was introduced by Huang et al. [207]. All the ways are innovative, but a much more facile way to avoid wearing issues, compatible with TENGs, is in need. Kim et al. [204] reported Nylon-11- and PVDF-TrFE-based TENGs, composited with bulk MoS_2, as shown in Figure 8d. Both composite ferroelectric layers, along with increased polarization, led to an increased surface charge density and produced high-output voltage, current, and power density of 145 V, 350 $\mu A/cm^2$, and 50 nW/cm^2. The highest power density was reported at 10 MΩ, and the repeated output voltage measures for 12,000 cycles at 6.5 Hz operating frequency gave same voltage results. Additionally, can light LEDs without charge storage (see Figure 8e). Kim et al. [208] have reported a much simpler way to enhance TENG durability, by employing the concept of reduced contact number cycles in ferroelectric, PVD-based TENG, with ITO as the electrode and PDMS as an elastomer. They showed that the up-sided polarization (58 pm/V) of PVDF led to extremely fast charge accumulation and larger TE-outputs, when compared with a downside (36 pm/V) and non-polarized (1.1 pm/V) PVDF under contact state. Up-side polarized TENG showed as Li-ion battery chargeable, even by 40% reduced contact cycles. Figure 9a,c depicts convention TENG vs. PVDF ferroelectric TENG; SEM and power curves are depicted in Figure 9b,d. Continuous contact caused surface deterioration at a pushing cycle of 1.7×10^6 for vibration of 1 Hz. The power is reduced to 0.18 μW from 85 μW at a load resistance of 10 MΩ and mechanical force of 35 N for conventional TENG. In PVDF-based TENG, random contacts were performed, which significantly lowered surface wear and mitigated power after longer usage. The surface of PVDF remained clean and flat, even after the pushing cycle of 1.7×10^6, with the power loss from 1.23 mW to 0.87 mW, only a bit smaller. The results lasted for many load resistances and forces, even in non-contact mode (Figure 9e), realizing that stable power can be generated from random contacts, with output peak-to-peak values of voltage, current, and charge (410 V, 25 μA, and 60 nC, respectively). The deposited TENG show potential in wireless temperature sensor from water waves. Park et al. [209] developed the triboelectric nanogenerator, which relied on a cyclo-diphenylaladine (cyclo-FF) nanowire array with high-output performance voltage and a current of 350 V and 10 μA, respectively, which is sufficient to light up 100 LEDs.

Some more triboelectric nanogenerators have been listed in Table 3.

Table 3. Triboelectric nanogenerators, based on various ferroelectric materials.

Material Pair	TENG Mode	Short-Circuit Current I_{sc}/Current Density J_{sc}	Open-Circuit Voltage V_{oc}	Maximum Power/Power Density	Work Function	Work Conditions	Reference
BTO/FEP	CS-mode	100 μA	1000 V	67 mW	-	14 m/s wind energy	[40]
ZnO-PVDF@PTFE	FL	1.6 μA	119 V	10.6 $\mu W/cm^2$	4.57 eV		[203]
Al-NR@PVDF-PI/(NBT-15ST(60))-Al	CS	32.5 $\mu A/cm^2$	1020 V	-	-	Instantaneous force of 50 N	[55]
PVDF-NaNbO$_3$	CS	3.98 μA	181 V	0.17 mW	4.0 eV		[210]
PZT/GFF	SL	59.05 mA/m^2	1640 V	10.8 W/m^2	-	10 N, 4 Hz	[211]
PDMS/PVDF	CS	22 μA	255 V	832.05 mW/m^2	-	0.5 Hz	[212]
BFO-GFF/PDMS	CS	3.67 $\mu A/cm^2$	115.22 V	151.42 $\mu W/cm^2$	-	1 Hz	[57]
BTO-PTFE	CS	10.4 $\mu A/cm^2$	45 V	-	-	50 N, 2 Hz	[46]
PVDF-TrFE/Mxene	CS	140 mA/m^2	270 V	4.02 W/m^2	-	7 N, 6 Hz	[121]
PI/PVDF-TrFE	CS	17.2 μA	364 V	2.56 W/m^2	-	2 Hz	[213]

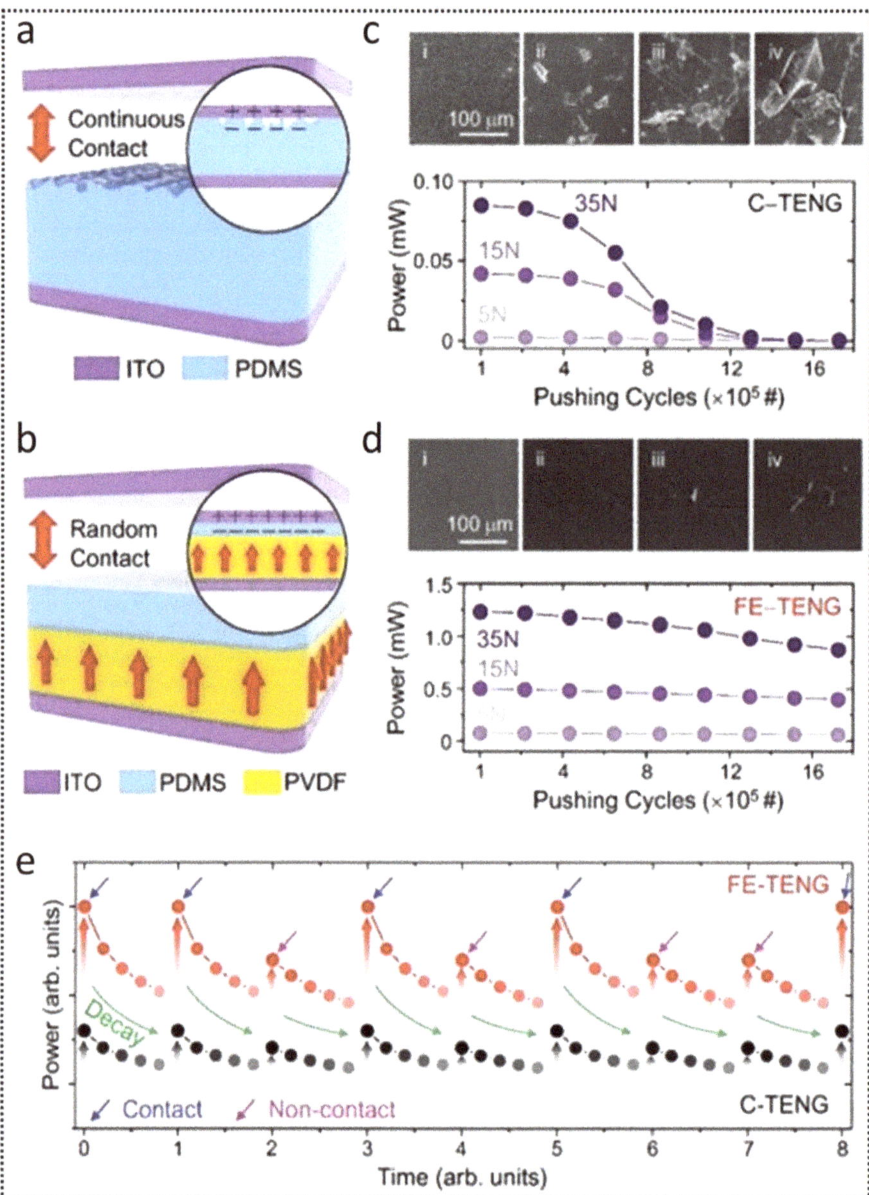

Figure 9. (**a**,**c**) Schematic of conventional (ITO/PDMS) and ferroelectric TENG (ITO/PDMS/PVDF) with magnified ITO, PDMS, and PVDF surfaces in contact (**b**,**d**). Pushing morphology-dependent power generation of conventional and ferroelectric TENG. (**e**) Power generation and decay for both types, with contact and noncontact conditions. (**a–e**) Reprinted with permission from reference [208], Copyrights 2019, John Wiley and Sons.

4.4. Device Design and Output Power Optimization in Ferroelectric Photovoltaics (PVC)

Ferroelectric photovoltaic devices are marvelous for harvesting solar energy. For this scientist's endeavor to tailor the properties of photovoltaic devices, a prime way is the implication of ferroelectric-based materials with narrow bandgap, designed with micro or nanostructure domains and unifying both conventional and ferroelectric photovoltaic effects. Inorganic ferroelectric oxides exhibit relatively wide band gaps, E_g, ranging from, e.g., 2.2–2.7 eV for $BiFeO_3$ [214,215], 2.48 eV for $LiNbO_3$ [216], 3.3 eV for $KNbO_3$ [217], and 1.5–3 eV for $(KNO)_x(BNNO)_{(1-x)}$ [218], etc., and are susceptible to the high absorption of UV-light. Maximum solar spectrum light absorption happens around 1.5 eV and gives small photocurrents. The bulk photovoltaic effect (BPVE) in ferroelectric materials assists in parting the photogenerated charge carriers (e–h pairs), without requisition of conventional pn-junction photovoltaics. The past studies were only restricted to theoretical studies, which showed low-power conversion efficiency PCE of perovskite ferroelectrics ($\sim 10^{-4}$, much smaller than conventional semiconductor PV materials), assisted by low conductivity ($\sim 10^{-9}$ S/m) and leading to a high recombination rate of photo-generated carriers [219–221]. However, the recent reports showed that the intrinsic electric field present in non-centrosymmetric ferroelectrics, due to spontaneous polarization, P_s, has supported the power conversion efficiency (PCE) to transcend beyond the Schokley–Queisser limit [222]. This has ruled into an emerged era of solar energy harvesting, assisted by optimization of bandgap engineering, carried by chemical ordering and structural modifications in thicknesses of the layer material and/or electrode and domain interfaces [223]. He et al. [224] have presented a first-principle study to investigate the effect of charge ordering in Bi-based ferroelectric perovskite oxides ($ABiO_3$) by the combination of A-site cations (Ca, Cd, Zn, and Mg) with Bi^{3+} or Bi^{5+} charge disproportions. They revealed that the cationic size, along with chemical featuring, shifted the valance band edge towards the high energy side, reducing band gap to ~2 eV, which ushered strong visible light absorption, smaller effective masses, and a large polarization field, alongside producing a dispersive Bi 6s state in the near vicinity of the Fermi level. Han et al. [225] determined a photocurrent effect in $(1-x)KNbO_3$-$xBaFeO_{3-\delta}$ (KN-BF, where x = 0.0 to 0.1) ferroelectric semiconductor ceramics with bandgap (E_g) ~1.82 eV at x = 0.1, ~43% less than pure KN, E_g = 3.22 eV, and which coexisted with cubic and tetragonal phase. The activation energies for KN-BF; x = 0.07 domains and domain walls were 0.26 eV and 0.52 eV, attributed to the ionization conduction mechanism of oxygen and remanent polarization of 3.50 $\mu C/cm^2$ for pure KN. They analyzed that the third states of Fe are important in KN-BF for band reduction, under light illumination of 1.5AM, for x = 0.07, demonstrated J_{sc} and V_{oc}, for positive voltage 1.92 nA/cm^2 and 0.31 V, while under dark 11.39 nA/cm^2 and 0.217 V, respectively. Their study promoted ferroelectric semiconductor ceramics as a potential member of photovoltaic devices. Alkathy et al. [226] reported bandgap tuning by doping of $Ba_{0.92}Bi_{0.04}Na_{0.04}Ti_{0.96}M_{0.04}O_3$ (BNBT-M); (M = Ni, Fe, Co), all the samples crystallized in tetragonal structure with the estimated E_g and corresponding power conversion efficiency PCE for BNBT, BNBT-Fe, BNBT-Ni, and BNBT-Co were 2.95, 0.468521; 2.38, 0.387803; 2.24, 0.368166; and 2.54, 0.409616. Their study revealed a guide to find perovskite ferroelectric oxide for visible light PV applications. Jimenez et al. [227] reported stability of bandgap of $BiFe_{1-x}Cr_xO_3$ (BFCO), with high Cr concentration over $La_{0.7}Sr_{0.3}MnO_3$ (LMSO) coated $SrTiO_3$ substrate by high-pressure processing (HP) technique. For PV response, the Pt/BFCO/LSMO/STO device structure was employed, with BFCO as an absorber layer. Additionally, HP 12.5-BFCO/SiTrO$_3$ showed an energy band gap (E_g) of 2.57 eV and remanent polarization P_r of 40 $\mu C/cm^2$, see Figure 10a. The non-polarized illuminated device showed V_{oc} and J_{sc} at RT to be 0.10 V and 6×10^{-3} mA/cm^2, with a power output of 0.6 $\mu W/cm^2$, unveiling their potential in solar energy harvesting and polarization, depicted with slightly higher values, 0.16 V, 40×10^{-3} mA/cm^2 and 6.4 $\mu W/cm^2$ (Figure 10b), attributed to reduced recombination rates and ohmic resistances, due to ferroelectric domains. The grain boundaries also play an important role in photocurrent and PV performance, as they can lead to the increased area of minority carrier collection, due to the development of built-in potential on the grain boundaries (GB) [228], which can serve as recombination centers and reduce the photocurrent densities [229]. Mengwei et al. [230] reported that hexagonal h-$YMnO_3$ (YMO) films over Si substrates with PV device structure Au/h-YMO/Au; h-YMO, as an absorber layer, poled to ± 700 V for 100 s at RT and led to the production of a high short-circuit current density of ~3.92 mA/cm^2 and open-circuit voltage -0.41 V, under the illumination of 1 sun (or 100 mW/cm^2). The J_{sc} and V_{oc} reached 3.16 mA/cm^2 under 65 mW/cm^2 intensity and 4.01 mA/cm^2 under 140 mW/cm^2 intensity, see Figure 10c. This is accomplished by the magnification of the impulsive force, relying on polarization and grain boundary effects for the separation of photogenerated charge carriers. Their study suggested downward band bending, induced by the potential difference and assisted by grain and interfaces between grain boundaries, behaving as a conductive path for effective transport of carriers. The bandgap, E_g, was found at 1.55 eV, and another absorption edge at 1.35 eV suggested the absorption in near IR region. Tu et al. [231] reported a $Bi_{0.93}Ni_{0.07}FeO_3$ (BNFO)-based PV device, with a field-modulated PV effect, carried by the pn-junction, in conjunction with a polarization-modulated Schottky barrier, under irradiation of 405 nm, with $E_g \sim 3.95$ eV and ITO/BNFO/Au heterostructure, see Figure 10d. The currents density and voltages are taken at an irradiation intensity of 1000 W/m^2, under 405 nm, with electric poling at 1 kV/cm ~8 A/m^2 and 0.9 V, with PCE ~1% and EQE ~11%. The photo-response of as-grown devices, under 532 nm and 633 nm irradiation, showed a weak response, mainly attributed to the transitions with intermediated defect levels. Nan et al. [43] investigated the temperature dependence of photocurrents in BTO ceramics and found it enhanced, by approximately 121.9% to 179.6% at 80 K to 240 K, assisted with ferro-pyro-phototronic effect, convincing the activation of shallow traps and polarization, as compared to RT, which was useful for detecting 405 nm light at low temperatures. Jia et al. [215] realized a temperature-modulated, self-powered UV photodetector, based on the ITO/BFO/Ag device, for spotting light of 365 nm with intensity 271.4 mW/cm^2 and observed increased photocurrent ~60% when the bottom of the device was subjected to temperature variations (heating) of around 42.5 °C. The amplified current is manifested with the change in carrier concentration, convinced by the temperature effect and electron shift in the BFO layer, instigated by the thermo-phototronic effect. Mai et al. [232] reported heterostructure, designed for the conversion of IR light into electricity, by employing BFO as a ferroelectric layer and upconversion layer of YOT (Y_2O_3:Yb). The structure revealed the convertible and stable PV-effect, under the illumination of 980 nm laser light, with the energy much smaller than the BFO E_g, to induce a photovoltaic effect in BFO itself. YOT mutates the incident to produce photovoltage $V_{oc} \sim 0.4$ V and photocurrent I_{sc} ~25 nA. This kind of heterojunction structure lined a pathway to bulk photovoltaic effects in ferroelectrics, by expanding the absorption of the solar spectrum. Table 4 discusses the output performances of some other ferroelectric photovoltaic (FPV) devices.

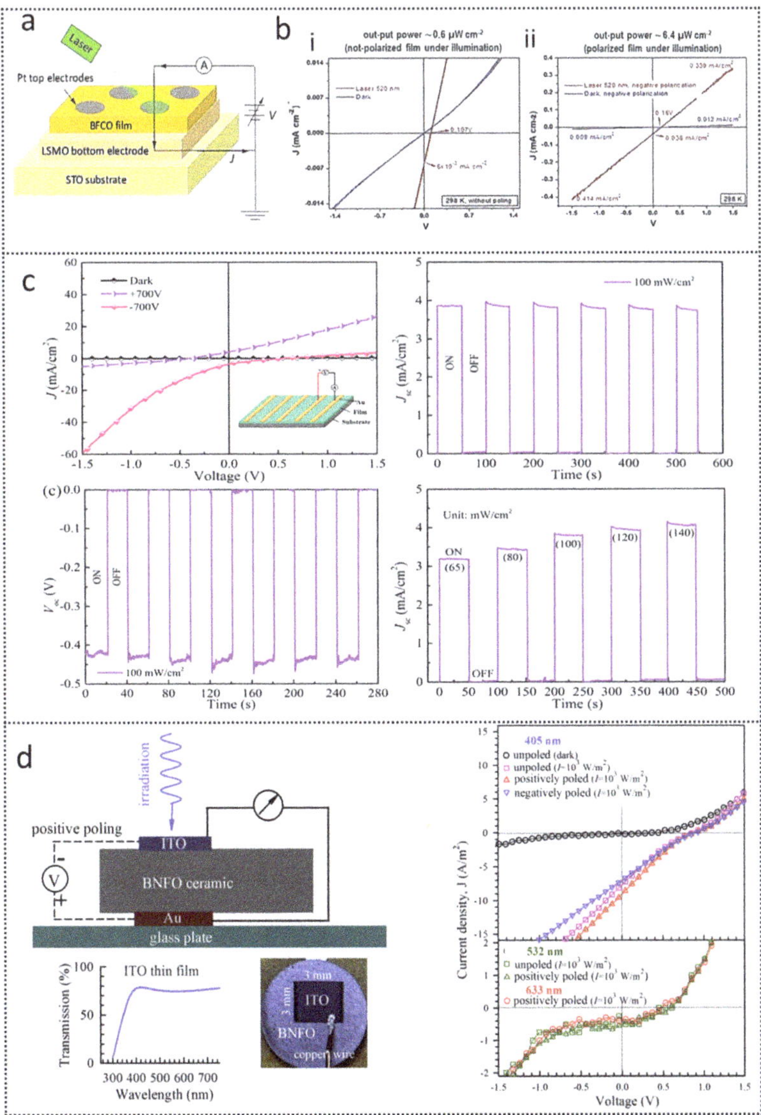

Figure 10. (**a**) The electrical configuration of the BFCO/LSMO/STO device for ferroelectric polarization and photoresponse, with and without illumination. (**b**) Short-circuit photocurrent density J-V curve at room temperature and non-polarized film (**i**); current density measurement J-V curve at room temperature and polarized film, with and without illumination (**ii**). (**a**,**b**) Reprinted with permission from reference [227], Copyrights 2021, Elsevier. (**c**) J-V characteristic, with inset showing a schematic of Au/h-YMO/Au device with photocurrent and photovoltage curve, as a function of light intensity. Reprinted with permission from reference [230], Copyrights 2021, Elsevier. (**d**) Schematics of a ITO/BFNO/Au photovoltaic device, optical transmission of ITO film, and top view of test disk. J-V curves of poled and unpoled under dark and illumination of 405 nm, 532 nm, and 633 nm. Reprinted with permission from reference [231], Copyrights 2019, Elsevier.

Table 4. Output performances of some ferroelectric photovoltaic devices.

Material	Short-Circuit Current I_{sc}, Current Density J_{sc}	Open-Circuit Voltage V_{oc}	Work Condition	Responsivity (R)/ Detectivity (D)/ Gain (G)	Output Power	PCE (%)	E_g (eV) Work Function (φ)	Reference
$Bi_{0.98}Ca_{0.02}Fe_{0.95}Mn_{0.05}O_3$ (BCMFO)	0.26 mA	−0.92 V	white light (280 mW/cm^2)	-	-	0.0075	E_g = 2.41 eV φ = 4.82 eV	[233]
$NiO/BaTiO_3/ZnO$	2.2×10^{-2} A		254 nm UVC, 5.57 mW 100 °C	R = 3.98 A/W D = 3 × 10^9 nHz$^{1/2}$/W G = 19.4	-	1	-	[234]
$BaTiO_3$ nanoparticles	393.1 nA	-	405 nm, 131.0 mW/cm^{2}80–240 K	R = 1.46 × 10^{-6} A/W D = 1.59 × 10^7 jones G = 4.46 × 10^{-6}	8.02 nW	-	φ = 5.3 eV	[43]
$CsPbBr_3$ nanowires	-	3 V	405 nm, 0.2 mW/cm^2	R = 4.4 × 10^3 A/WG = 1.3 × 10^4	-	-	-	[39]
$K_{0.49}Na_{0.49}Ba_{0.02}(Nb_{0.99}Ni_{0.01})O_{2.995}$	0.03 μA/cm^2	0.11 V	405 nm laser, 50 mW	-	60 μW	0.12	-	[223]
SbSeI nanowires	0.36 pA 0.0074 μA/cm^2	97 mV	488 nm Ar laser, 127 mW/cm^2	-	0.119 nW	-	E_g = 1.862 eV	[29]
$ZnO/Bi_5FeTi_3O_{15}$	2.2 mA/cm^2	0.15 V	405 nm, 200 mW/cm^2	-	0.09 mW/cm^2	-	E_g = 3.08 eV φ = 4.86 eV	[235]
$Bi_6Fe_{1.6}Co_{0.2}Ni_{0.2}Ti_3O_{15}/Bi_2FeCrO_6$	10.3 mA/cm^2	0.66 V	AM 1.5G illumination	-	-	3.40	E_g = 1.62 eV/1.74 eV	[236]
$NiO/BFCrO/WS_2$	1.80 mA/cm^2	0.78 V	1 sun illumination	-	-	7.07	-	[237]
$Sm:BiFeO_3$	0.58 nA/cm^2	0.90 V	Xenon lamp (100 mW/cm^2)	-	-	1.65	-	[238]

4.5. Device Design and Output Power Optimization of Coupled Effect Nanogenerators

The atmosphere is rich in energies, whether natural or artificial; therefore, the energy harvesters of light, heat, and mechanical energy can coexist as hybrid- and multi-effect-coupled nanogenerators, stemming into continuous and intensified production of electricity. The different ferroelectric, material-based coupled energy nanogenerators have been reported to date, including the meld of photovoltaic and mechanical energy, photovoltaic and thermal energy, mechanical and thermal energy, and some comprising of all of these collectively and synchronously. It is believed that the coupled nanogenerators can be proved highly nifty for harvesting multi-energies from the environment. Ferroelectric materials-based nanogenerators have been reported with high-output characteristics, but their coupling is still a daring challenge. Zhao et al. [239] have fabricated a flexible nanogenerator, employing the ferro-pyro-phototronic effect, cited by the coupling of photovoltaic and pyroelectric effects in ferroelectric $BaTiO_3$ (BTO), with device structure ITO/BTO/Ag and Kapton membrane, in order to achieve flexibility, Figure 11a. A ferro-photo-phototronic effect makes use of a temperature fluctuation-incited pyroelectric field, in order to attune the photoexcited charge carriers for outstripping the photocurrent in the wake of coupling between ferroelectric material, pyroelectricity, and photoexcitation. The device was poled at 2.3 kV/mm for 30 min and responded well to detecting light and temperature variations, applied simultaneously with the individual detection. The photocurrent and current plateaus were found to increase with increasing light intensity (7.78 mW/cm^2 to 127.6 mW/cm^2), from 5.4 nA to 43.2 nA and 5 nA to 40 nA at room temperature, owing to the ferro-pyro-phototronic effect. Heating at ΔT = 22.4 K, the lower coupled current plateaus were observed from 4.46 nA to 35 nA and illuminated under same light intensities; however, the coupled photocurrent showed similar increasing behavior from 29.8 nA to 69.7 nA. Cooling (ΔT = −14.5 K) the device led to a reduction of photocurrent (from −19.2 nA to −12 nA) but enhanced current plateaus (from 5.9 nA to 44.9 nA). Their results showed that the coupled photocurrent peak and photocurrent peak enhancement ratio, under the state of heating + light, were 451.9% and 61.3%, while the photocurrent plateaus and photocurrent plateau enhancement ratio, generated under cooling + light, were 17.2% and 11.1% sharper than light generated via the photovoltaic effect alone, at a light intensity of 127.6 mW/cm^2, see Figure 11b. For heating + light, the currents from heating and light are in one direction, superimpose, and are attributed with the higher temperature difference, while reduced current plateau might be associated with the reduced polarization of BTO, due to large thermal agitations. The reverse is true for cooling + light. Similar increasing trends for voltage peak/voltage, peak enhancement ratio values were 391.5%/125% for heating + light, and plateau/voltage plateau enhancement ratios were observed to be 50.9%/92.8% for cooling + light, respectively. Therefore, the fabricated device boost in current and voltage plateau was carried by the device under simultaneous light + cooling applications, whilst photocurrent and photovoltage were enhanced under the condition of light + heating. They developed 16 photo-pyro sensing matrices (4 × 4) (shown in Figure 11c) and achieved a photodetector system for light illumination detection of 405 nm and temperature variations, demonstrating the possible application in environmental and artificial intelligent sensors. Zhao et al. [240] reported photovoltaic pyroelectric-coupled NG with a planner structure, based on radially polarized BTO, as a functional layer and coplanar laser-engraved ITO electrode (see Figure 11d). The pyro-phototronic effect enabled BTO to exhibit a high pyroelectric constant of 16 nC/cm^{2}K and showed the dramatically increased photocurrent; it can perform fast sensing of 365 nm ultraviolet illuminated light with a 0.5 s response time. Heating temperature modulations led to the current enhancement of almost >30 times, under 0.6 mW/cm^2 intense light, as compared to only light illuminated current; however, the photocurrent enhancement ratios were found to decrease as the light intensity increased. Under the cooling + light condition, the currents found were larger than that of heating + light condition; this is because of elevated pyroelectric coefficient. The current plateau dependence on the ferro-pyro-phototronic effect can be visualized in Figure 11e. The currents were measured individually, under the illumination of 81.8 mW/cm^2, 1.25 K/s heating, and simultaneous heating-light. The heating + light showed that photocurrent is not the superposition of individual currents produced, and coupling light and heating caused a decreased current plateau, rather than individual light illuminated. The current plateau increased by ~23%, by simultaneous cooling and light shining on the sample, indicating the coupling enhancement "1 + 1 > 2" of photocurrent plateau, associated with the band bending, induced at the BTO/ITO interface. The estimated peak currents under illumination, pyroelectric heating, and cooling were 50 nA, 29.3 nA, and 36 nA, respectively, while the total peak currents of illumination, along with heating and cooling, estimated were 71 nA (<50 nA + 29.3 nA) and 0.2 nA (<50 nA − 36 nA), respectively. Hence, the coupled current's relation with individual currents is not additive. Zhang et al. [241] have reported the ferroelectric-metal-semiconductor BiFeO$_3$/20sAu/ZnO heterostructure as intriguing, coupled pyroelectric and photovoltaic effects. They analyzed that photocurrent density, induced by light illumination of 405 nm, was 1.4 times that of BFO/ZnO structure associated with the introduction of Au nanoparticles that tuned the energy band alignments and by light-induced temperature oscillations, across the heterojunction interface, resulting in improved photoelectric performances. They additionally responded deliberately to a wide range of solar spectrum wavelengths, 360 nm to 1060 nm, by coupled pyroelectric–photovoltaic effects. The BTO/20sAu/ZnO heterojunction produced short-circuit current densities of 191.9 µA/cm^2 and 90.8 µA/cm^2 under light illumination of 405 nm and 360 nm with 60 mW/Cm^2 and 26 mW/cm^2 intensities. For wavelengths of 532 nm and 635 nm, with intensities 150 mW/cm^2 and 650 mW/cm^2, the heterojunction current densities in short-circuiting were 3.7 µA/cm^2 and 2.8 µA/cm^2, while the photoelectric response of simple BTO/ZnO was negligible. So, Au nanoparticle induction led to the broad range of light detection by the device. Ji et al. [145] reported piezo–pyro–photoelectric coupled effect in BTO with coupling enhancement of 1 + 1 > 2 for charge quality, the maximum output power obtained for pyroelectric, photovoltaic, and pyroelectric–photovoltaic units were 117 nW, 11 nW, and 189 nW, respectively. The total photocurrent Ipyro + photo + piezo estimated was ~0.67 µA, with a charge transfer of 67.74 µC. Wang et al. [242] proposed a one structure-based coupled nanogenerator to scavenge mechanical and thermal energies, via the tribo–piezo–pyroelectric effect, using PVDF nanowires as pyro–piezoelectric active material and PVDF–PDMS composite as a triboelectric layer. Their hybrid nanogenerator showed improved charging performance, rather than individual TENG, PENG, and PyENG. The triboelectric–piezoelectric (TiPENG) unit showed the problem of lower voltage, while the pyroelectric unit had to face the low current performance; so, individual units were unable to power the light bulbs connected in series. However, the synergetic effect of hybridized nanogenerator with maximum output energy supply lightened the bulbs much more efficiently under the state of both units working simultaneously. Song et al. [243] developed Ag/BTO/Ag sensor arrays, based on a pyro–piezoelectric conjunction nanogenerator and realized that, with the advantage of 1 + 1 = 2, evidencing the simultaneous detection of both temperatures and pressure-assisted with zero loss and interference of overlying multi-functionality of device. They observed the intensified device voltage, with rising temperature and/or pressure, as well as a sensitivity of ~0.048 V/°C and /or 0.004 k/Pa, respectively. They proposed a potential application era of such devices in intelligent compliant structure, smart sensors, and human–machine interactions, as well.

Ji et al. [46] reported a multi-effect coupled NG by integrating pyro–piezo–tribo–photo effects simultaneously in multifunctional BTO ceramic with only one electrode pair, piezo–tribo TPiENG, which harvests mechanical energy by strain-induced deformations in BTO and between FEP and nylon film, as well as PDMS as a protective layer for sustained confrontment of ITO and Ag electrodes (see Figure 12a). The individual pyroelectric nanogenerator (PyENG) produced current and voltage peaks of 15.2 nA and 1.5 V, at a heating rate of 0.34 K/s for the temperature change of 302 K to 307 K, cooling the device back to 302 K. Similar current and voltage peaks (but in the opposite direction) and maximum output power of 7.8 nW at 150 MΩ were observed. Individual PVC outputs were examined under light illumination of LED with a wavelength 405 nm, stable current of 10.1 nA, voltage of 0.6 V, and maximum power of 7.6 nW, obtained at 25 MΩ load resistance. To quantify individual TPiENG, an airflow of 15 m/s was blown from an air blower at 12 cm from a multi-coupled NG. An AC output current of 3.5 μA was obtained, followed by the oscillation of nylon film. A baseline of voltage was swinging, so it was not easy to detect voltage. The multi-effect coupled NG under the conditions of the heating rate 0.98 K/s, a 405 nm light illumination by LED, and an airflow speed ~15 m/s, which proved to be a compatible power source with a current, voltage, and platform voltage of 1.5 μA, 7 V, and 6 V, respectively. The charging capability of the 0.33 μF capacitor to 1.1 V in 10 s, only as compared to individual units of PyENG, PVC, and TiPENG, are shown in Figure 12b. Zhang et al. [51] prepared a multifunctional-based PZT multi-effect coupled nanogenerator by integrating PyENG, PVC, and TPiENG units. The one device structure had three parts, with PZT being a piezo–pyro–photoelectric active material and flexible nylon vibrating film, which, with FEP, introduce triboelectrification and apply strains to PZT, Ag, and ITO/AgNws/PDMS, as bottom and top electrodes, see Figure 12c. The investigated performance of coupled nanogenerator, with different combinations of individual units with PyENG+PVC, PVC+TPiENG, PyENG+TPiENG, and PyENG + PVC + TPiENG, are shown in Figure 12d. The output of PyENG+PVC acquired the peak values of current, voltage, and platform to 1 μA, 91 V, and 200 nA; all other coupled nanogenerators, comprised of TPiENG, showed similar AC output current peaks, greater than 5 μA. However, the individual TPiENG, under wind flow, showed a higher output current than PyENG and PVC but output voltages are smaller. The coupled nanogenerator gathered the fringe benefits of piezo–pyro–tribo–photoelectric modules and aroused high electric peak current, voltage, and platform voltage of 5 μA, 80 V, and 50 V. The coupled nanogenerator showed outstanding charging response by charging 10 μF capacitors only in 90 s up to 5.1 V. Singh et al. [244] has reported piezo–tribo based HNG with combined piezoelectric and triboelectric effect in ZnO-PVDF film coupled with PTFE (polytetrafluoroethylene) as shown in Figure 12e, the determined that incorporating ZnO nanorod in ferroelectric PVDF polymer uplifted the triboelectric and piezoelectric response of PVDF and given the maximum output instantaneous power of 24.5 μW/cm^2, while combined current and voltage of 0.46 μA and 78 V, without subjecting any surface treatment and poling.

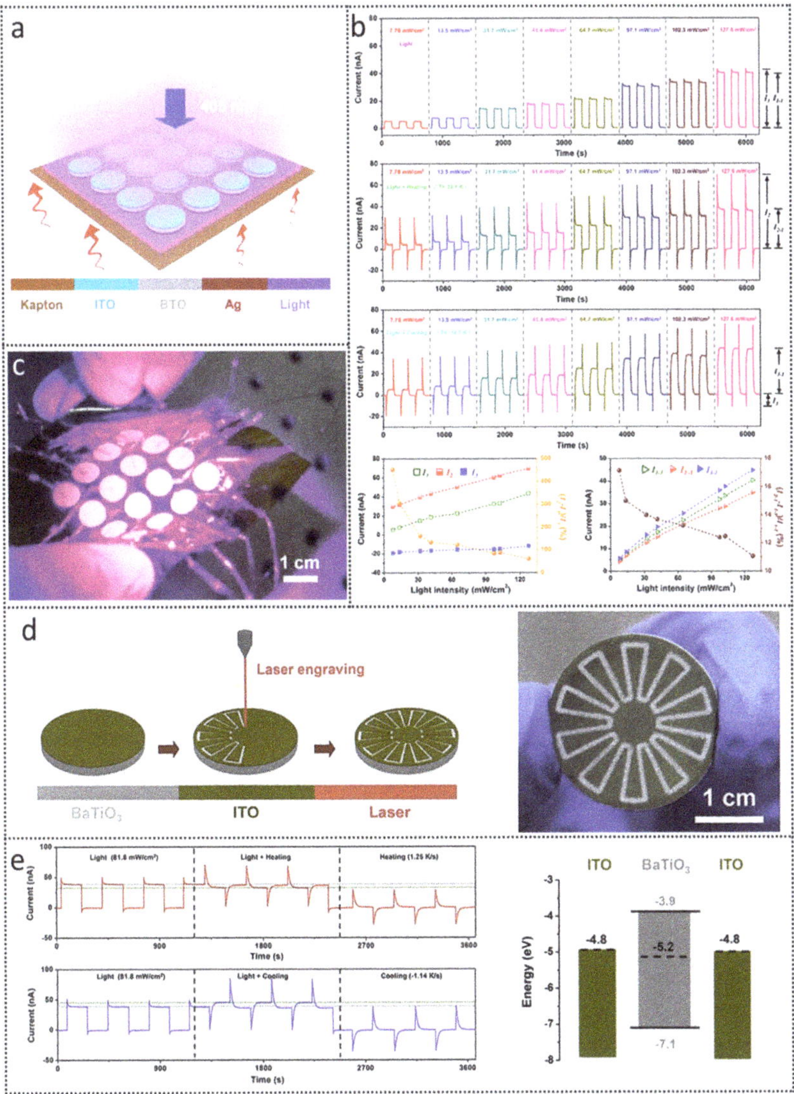

Figure 11. Schematic of ferro–pyro–phototronic effect coupled nanogenerator. (**a**) Schematic illustration of Kapton/ITO/BTO/Ag/light fabricated device. (**b**) Detector current performance I$_1$, I$_2$, and I$_3$, under different light intensities and conditions of "light", "light + heating", "light + cooling", current peak and current peak enhanced ratios, and current plateau and current plateau enhanced ratios. (**c**) The flexible photodetector 4 × 4 array, consisting of 16 BTO units. (**a**–**c**) Reprinted with permission from reference [239], Copyrights 2020, Elsevier. (**d**) Laser engraved ITO electrode at the surface of BTO coupled ferro–pyro–phototronic NG device. (**e**) Current measurements of ITO/BTO/ITO device, under illumination, heating, cooling, heating + illumination, and cooling + illumination. (**d**,**e**) Reprinted with permission from reference [240], Copyrights 2018, Elsevier.

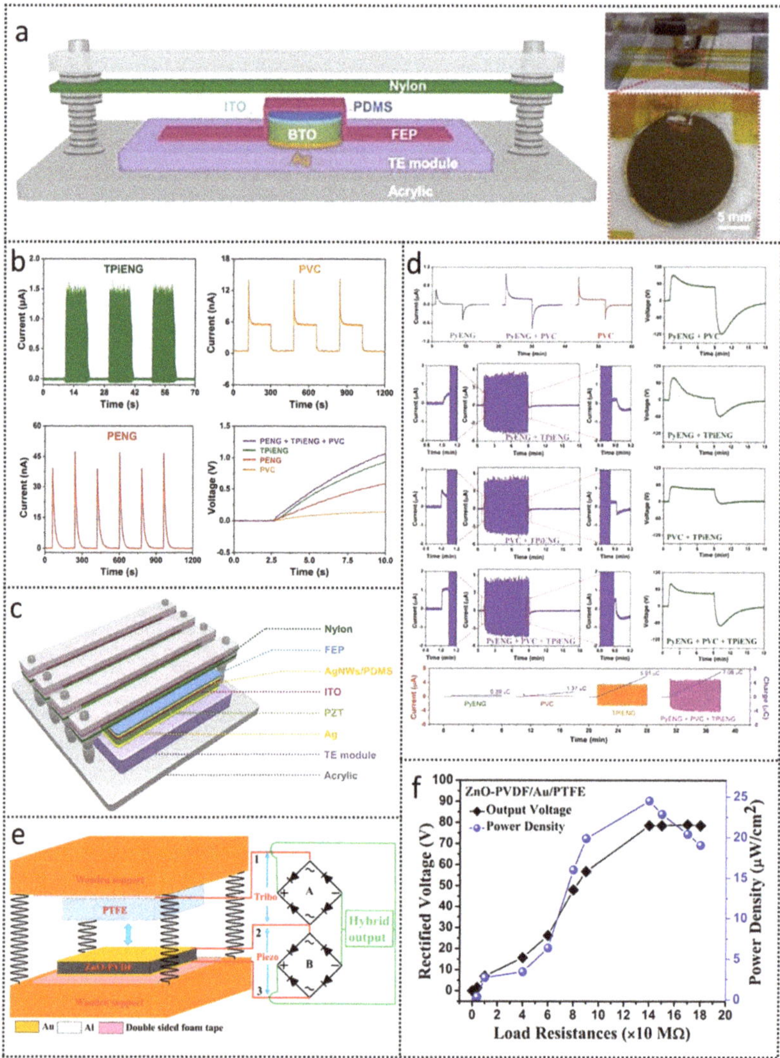

Figure 12. Multi-effects coupled nanogenerators. (**a**) Schematic of coupled nanogenerator with piezo–tribo–pyro–photoelectric effect, along with photo image of a coupled device, with BTO as a functional layer. (**b**) Rectified output current performances of discrete TPiENG, PyENG, PVC, and voltage measurement of capacitor charging for individual and coupled nanogenerator, comprising of TPiENG + PyENG+PVC. (**a**,**b**) Reprinted with permission from reference [46], Copyrights 2018, John Wiley and Sons. (**c**) Schematic diagram of a one-structure, multi-effects coupled nanogenerator, with PZT as a functional layer. (**d**) Output characteristics current and voltages of coupled nanogenerator in different combinations. (**c**,**d**) Reprinted with permission from reference [51], Copyrights, 2017, John Wiley and Sons. (**e**) Schematic of hybrid piezo–tribo nanogenerator based on ZnO-PVDF semiconductor–ferroelectric material. (**f**) Output electric power and voltage, as a function of the resistance of the ZnO-PVDF hybrid nanogenerator. (**e**,**f**) Reprinted with permission from reference [244], Copyrights 2018, Elsevier.

Ji et al. [245] demonstrated the electricity induction by employing the photovoltaic–pyroelectric–piezoelectric coupled effect for BTO-based sensors, utilized to spotlight intensities, temperature differential, and vibrational frequency, individually and all-together, with high susceptivity. The nanogenerator device for couple sensing is shown in Figure 13a. The BTO ceramic slice is interlocked between the ITO top and Ag bottom electrode. One end of TO slice was fixed on a commercial TE module for providing temperature variations and masses on another end. The right-hand photograph is the optical image of the fabricated device. The individual examinations for sensing light, temperature, and vibration marked out the sensitivity of ~0.17 nA/(mW/cm^2), 4.80 nA/K, and 1.75 nA/K for cooling and heating condition, 21.02 nA/Hz, and -14.07 nA/Hz for low ($\leq$17 Hz) and high (>17 Hz) vibrations, respectively, without any external power supply. The device sensitivity for monitoring light-temperature, temperature-vibrations, and light-temperature-vibrations coupled effects were found to be dependent on variations in the spontaneous polarization of BTO. The light + temperature + vibration current ($I_{photo} + I_{pyro} + I_{piezo}$), as a real-time function, observed at a constant light intensity of ~211 mW/cm^2 and constant temperature change of 12.9 K, the total current was an AC-type piezoelectric current, as in Figure 13b(i). The total current, as a function of different vibration conditions under various light intensities and at constant temperature change of 12.9 K, can be seen in Figure 13b(ii). The total current increased under low-frequency vibrations ($\leq$17 Hz) but started to decrease as the frequency vibrations enhanced above 17 Hz, as a response to increasing light intensities from 20–211 mW/cm^2. Additionally, the corresponding sensitivities K$'$ and K$''$ of low and high vibration fall to a minimum of 15.33 and -8.57 nAcm^2/mW, respectively, at 211 mW/cm^2. The trend is attributed to the dwindling of spontaneous polarization, convinced by thermionic and photo-thermionic effects. Their study revealed motivation into commercial and industrial applications of multifunctional sensors. Zhao et al. [246] designed a multi-effects coupled nanogenerator and implicated it in self-powered, multifunctional coupled sensor applications. The schematic of the device consisted of BTO ceramic wafers, with transparent ITO and Ag electrodes deposited on the upper and lower surfaces, as well as a total of the BTO array, encapsulated in a PDMS matrix to achieve flexibility, see Figure 13c. Fabricated device characteristics were conducted for the light of 405 nm, with intensities 8.39–83.2 mW/cm^2, temperature fluctuations of 9.1 K to 35.1 K, and pressure variations of 7.6 kPa to 33.7 kPa. The coupled NG gave an improved current of ~387.3% under simultaneous light gleaming (83.2 mW/cm^2) and pressure (7.1 kPa), as well as for a temperature difference of -19.5 K, the current enhancement of ~375% was observed, rather than only light. They employed the fabricated coupled NG, as the realization of the multi-functional coupled sensor, at the benefit of external power, see Figure 13d. The coupled nanogenerator showed powerful coupling enhancement with the detection sensitivity of 0.42 nA/mWcm^{-2}, 1.43 nA/kPa for light and pressure detection at a constant temperature change of -19.5 K, and 8.85 nA/K for temperature sensing at light irradiance of 83.2 mW/cm^2. Their study opened a way to sense electronic skin evolutions. Shi et al. [247] demonstrated a flexible one structure hybrid nanogenerator, based on coupled piezoelectricity and triboelectricity, using cellulose/BaTiO$_3$ aerogel paper PDMS nanocomposites. The PENG unit, under mechanical impact force, separately showed maximum voltage and power of 15.5 V and 11.8 µW. The PENG unit was coupled to a single electrode TENG. By adding the electrode in the PENG system, the polarization direction of PENG positively couples with triboelectricity and enhances the output voltage and power to ~48 V and 85 µW, at the mechanical impact of 80 kPa. The study revealed an easy coupling approach for the enhancement of device performance and practical applications. Ma et al. [248] also demonstrated sandwich layer structure NG, based on coupled pyroelectric and photovoltaic effects. The BTO functional material was embedded between ITO and Ag electrodes. BTO acts as a pyroelectric and photovoltaic material, deposited on Ag film, with a thickness of ~0.3 mm. The Schottky contact was formed at the interface of ITO/BTO, which created the built-in potential. The photo-pyroelectric coupled effect in ITO/BTO/Ag film was induced by 405 nm light illumination, with the intensity of 111.1 mW/cm^2 by more than 260%. The coupled current of 79 nA was generated with a maximum power output of 7.1 nW for 20 MΩ. The incident light irradiation cause BTO to absorb photons, promoting electrons to hop from valence to conduction band, creating free excitons, which further divide by the built-in and depolarized fields generating a photocurrent. The photons absorbed interact with the lattice phonons of BTO, producing heat that reduces the polarization of BTO, resulting in a pyroelectric current signal. The pyroelectric current signals are mediated through light irradiance, so the NG performance is promoted via the pyro-photoelectric coupled effect. They postulated that coupled photovoltaic and pyroelectric effects have potential sensing/detecting the near UV-light; for this purpose, they fabricated a photodetector system composed of 3×3 matrix arrays of an ITO/BTO/Ag device and utilized it to detect dispersion of 405 nm light intensity at zero cost of external power. The light information was determined by scrutinizing the electric signals, which were self-powered and showed a high photoconductive gain, responsivity, and specific detectivity of the photodetector. Their toil efforts gave a novel plan to attain fast light spotting by self-powered sensors, based on coupled photovoltaic–pyroelectric effects in ferroelectric BTO. The studies above let us make the statement that coupling the various effects in a single device structure increased the effectiveness of energy harvesting from ambient temperature by utilizing ferroelectric materials. Qi et al. [249] devised a pyroelectric–photovoltaic coupled nanogenerator, employing $BiFeO_3$ as a functional ferroelectric material layer and ITO and Ag electrodes, owing to a narrow bandgap of 2.67 eV BFO, which can actively absorb light of even wavelengths less than 465 nm. Additionally, owing to the non-centrosymmetric nature, the light-induced temperature in BFO lattice can instigate pyroelectricity, and light-induced electric signals match the pyroelectric signal and coupled nanogenerator exhibits high-output performance; additionally, the fabricated nanogenerator can be used as a self-powered photodetector. Under the illumination of 450 nm light, the ITO/BFO/Ag coupled NG produced an output current and voltage of 8.8 nA and 0.13 V. The photoconductive gain, responsivity, and detectivity of coupled NG was 9.7 times higher at $0.86 \ mW/cm^2$, compared to the utilization of photovoltaic effect. Until now, many other coupled nanogenerators, based on hybrid and one structure, have been fabricated and more are in progress. Coupled effect devices are highly innovative, as they are rich in power generation capability.

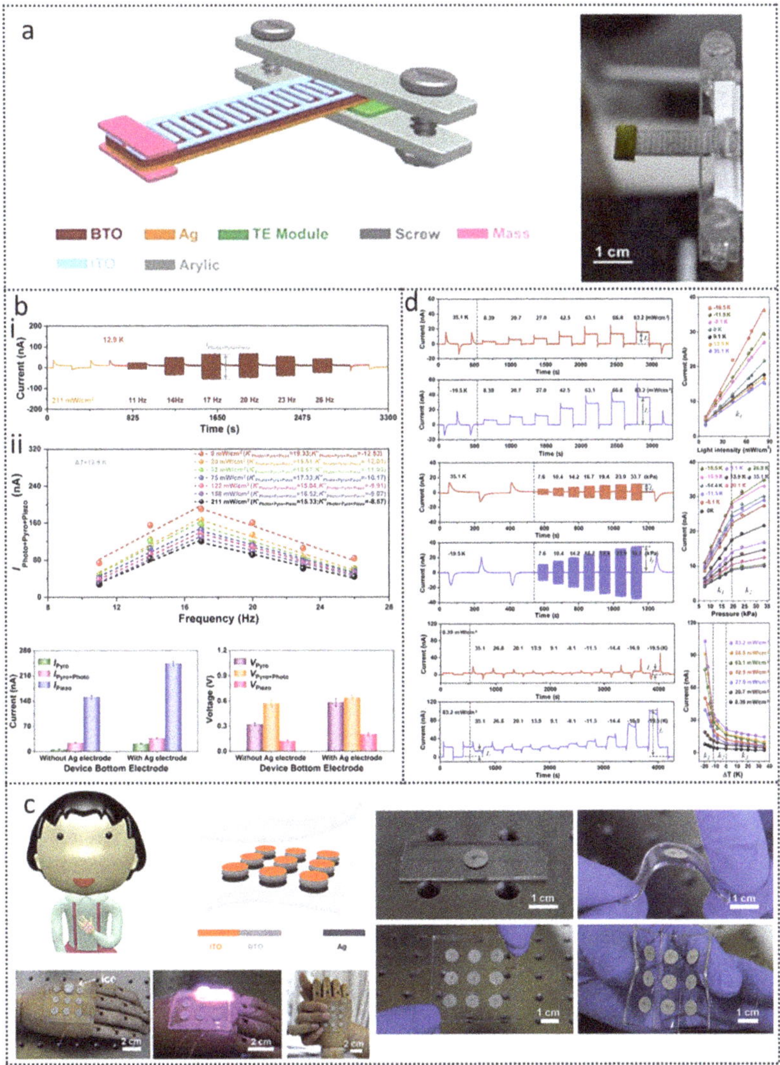

Figure 13. Schematic of coupled nanogenerators for self-powered multifunctional sensing. (**a**) Design structure optical image of the fabricated, BTO-based sensor for coupled sensing. (**b**) Sensing performance under light + heating + vibration condition, a real-time output current, at a constant light intensity of 211 mW/cm^2 and temperature change of 12.9 K (**i**); current, as a function of vibration frequency under different light intensities, at a constant temperature change of 12.9 K (**ii**). (**a**,**b**) Reprinted with permission from reference [245], Copyrights 2019, John Wiley and Sons. (**c**) Schematic of coupled nanogenerator for human body sensing. (**d**) Performance characteristics of multi-effect coupled nanogenerator. (**c**,**d**) Reprinted with permission from reference [246], Copyrights 2020, Elsevier.

5. Applications

The technology of nanogenerators has promoted plenty of advanced modern devices in self-powered micro/nanosystems, comprising of sensors like pressure sensors, tactile sensors, infrared sensors, actuators, flexible and wearable devices for energy harvesting, medical implantations (such as biosensors), health care monitoring through human-computer interfaces, electrochemical systems (such as supercapacitors and batteries), as a power source for many tiny energy harvesting devices, and many others. The nano-generators can drive small power/self-powered electronic devices, such as a lead-free BTO/PDMS-based PENG, which can charge 1.0, 2.2, and 4.7 μF capacitors to 1.26, 1.68, and 2.58 V in only 55, 170, and 527 s [168]. All-fiber wearable nanogenerators, based on nanofiber, nonwoven fabric, employing the $NaNbO_3$ nanoparticles embedded with PVDF and covered with PDMS, successfully harvested the low-frequency mechanical energy of human activity and motion into electricity [250]. Wang et al. [251] prepared a piezoelectric pressure sensor, based on cellular fluorocarbon with a high sensitivity of 7380 pC/N, in the pressure regime of <1 kPa and with a very fast response time of 5 ms, to check physiological motions of the body by attaching it to human skin, e.g., facial muscle contraction, wrist motion, eye blinking, and breathing; their current signals are shown in Figure 14a. The breathing frequency was recorded at ~21 times/minute, which is the normal range of a healthy 36-year-old man. Lee et al. employed a stretched PyENG to charge two capacitors of ~3.3 μF, along with a rectification circuit, and can light up their red, yellow, and green commercial LEDs simultaneously, as well as the LCD, see Figure 14b [186]. Zhang et al. used a PLZT–TENG self-powered wearable sensor as a harvester of biomechanical energies, in order to sense elbow joint motions for different angles of ~130°, 110°, and 70° at a humidity rate of ~20% and action time of Δt = 300 ms. Under dynamics of ultrafast response times of 7 ms, the PLZT–TENG sensor is found extremely suitable for AMP test, as estimated for volunteer AMP was 345, lighting up 119 LEDs by gentle hand taping and recording the footsteps of a running man (Figure 14c) [202]. A multilayer triboelectric wearable biomedical and acoustic sensor (TESs) for biomedical interaction interface for diagnosing pulse pressure and human breathing rates, the acoustic sensors were capable of detecting the high-frequency of Fourier transform signal (STFT) sound sources (shown in Figure 14d), which were proposed by Park et al. [196]. Sahu et al. proposed a hybrid, multi-stacked (MS-HG) BCZT-BHO TENG self-powered IR communication module (see Figure 15a(i,ii)) to analyze the calorie burned from yoga poses of the body, consisting of a receiver and transmitter, voltage regulator circuit, two capacitors, and VGA camera. In order to see IR light, the remote of the LED transmitter was powered via the hybrid system, and IR LED with red LED was lightened by the solar cell, so pressing the remote button glowing light indicated the established communication among the IR LED receiver and transmitter. The low powered appliances, used in this way, were lightened by capacitor charging through hybrid TENG [64]. Sahu et al. established a piezo–tribo hybrid structure (P-DMHS) that showed potential biomechanical energy harvesting by finger tapping and hand and leg motion. The generated currents were proportional to the weight of the body part (see Figure 15b), which can power a 50 μF capacitor, light commercial LEDs, and power wristwatch (Figure 15c). P-DMHS also was demonstrated as an impact sensor and recorded impact by ball dropping through certain heights from 10 cm to 60 cm, with a device sensitivity of 2.101 V/m and maximum impact of 370 mJ, shown in Figure 15d inset. For the impact sensitivity, they further launched the device on a bicycle helmet and found a maximum impact of 810.7 mJ, stating the sensitivity of the impact sensor is as much that they can respond to an impact of <1 J. This study evokes interest in battery-free sensors to sense impacts/shocks in helmets, cars, and other protective equipment [252]. You et al. [253] developed a hybrid piezoelectric–pyroelectric nanogenerator, using non-woven nanofiber membranes of PVDF, with versatile applications in self-powered electronic textiles. By implanting the NG insole of the shoe, they observed the maximum output current of 2.83 nA and 78.55 nA, while walking and running. Coupled nanogenerators have great potential to be used in photodetectors, e.g., a BTO based pyro–photoelectric nanogenerator can efficiently detect light of 405 nm, with a responsivity of 10–7 A/W [45]. A pyro–photoelectric nanogenerator that is based on BFO can detect 450 nm light of intensity 65 mW/cm^2 [215]. Coupled nanogenerators have shown marvelous applications as multifunctional sensors, too, e.g., an Ag/BTO/Ag piezo–pyroelectric coupled nanogenerator has marked to sense pressure and temperature simultaneously [51]. A multifunctional sensor to detect light intensity, pressure, and the temperature has also been constructed, with light and pressure detection sensitivities of 0.42 nA/mWcm^2 and 1.43 nA/kPa, respectively, via an ITO/BTO/Ag structured coupled photo–piezo–pyroelectric nanogenerator [246]. The multi-effect coupled nanogenerators have great potential in the Internet of Things, as well as many other industrial and commercial applications.

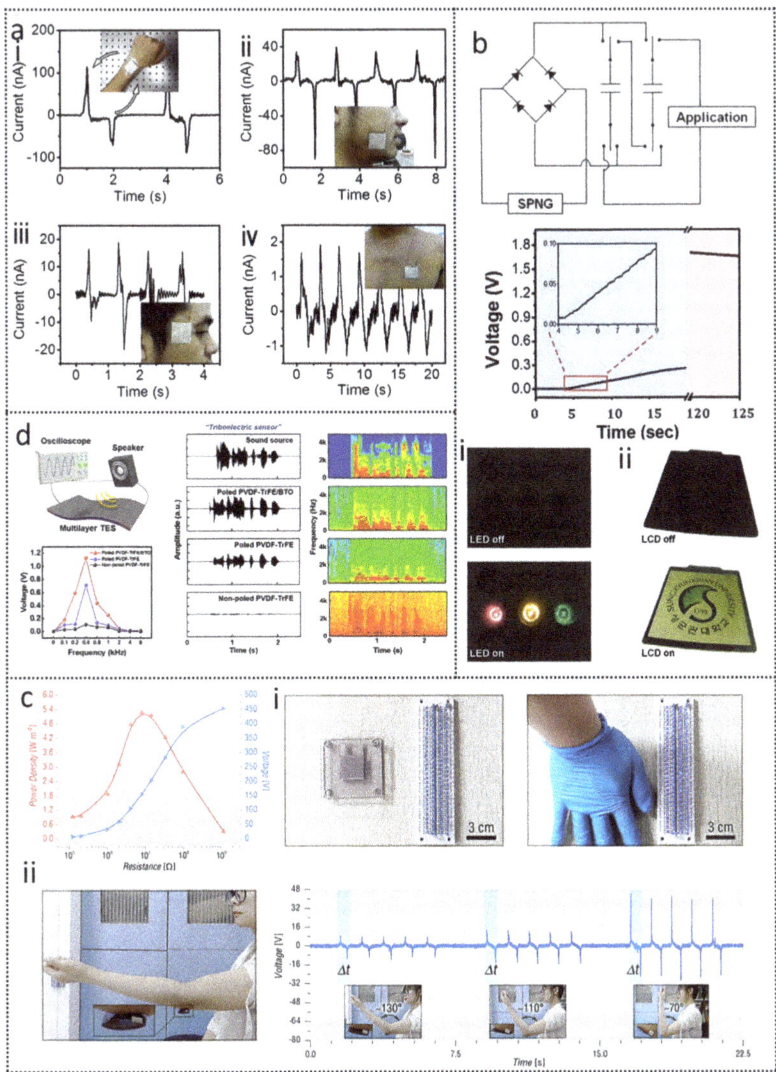

Figure 14. Application of NGs. (**a**) Short-circuit current and image positions of piezoelectric pressure sensor, attached on the skin to check wrist motion (**i**), facial muscle contraction (**ii**), eye blink (**iii**), and breathing (**iv**). Reprinted with permission from reference [251], Copyright 2017, Elsevier. (**b**) Capacitor charging circuit and curve through pyroelectric NG; a photo image of turning red, yellow, and green LEDs off and on, as well as a photo image of turning an LCD off and on. Reprinted with permission from reference [186], Copyright 2015, John Wiley and Sons. (**c**) Photograph of a PLZT–TENG biomechanical energy harvester to light 119 LEDs (**i**) and as a self-powered wearable sensor to detect elbow joint motions at different angles (**ii**). Reprinted with permission from reference [202], Copyright 2021, Elsevier. (**d**) Acoustic sound wave sensor with three configurations of poled, unpoled P(VDF-TrFE), and poled P(VDF-TrFE)/BTO, as a function of sound frequency, alongside the time-dependent sound waveforms and short-time STFT signal. Reprinted with permission from reference [196], Copyright 2020, ACS Publications.

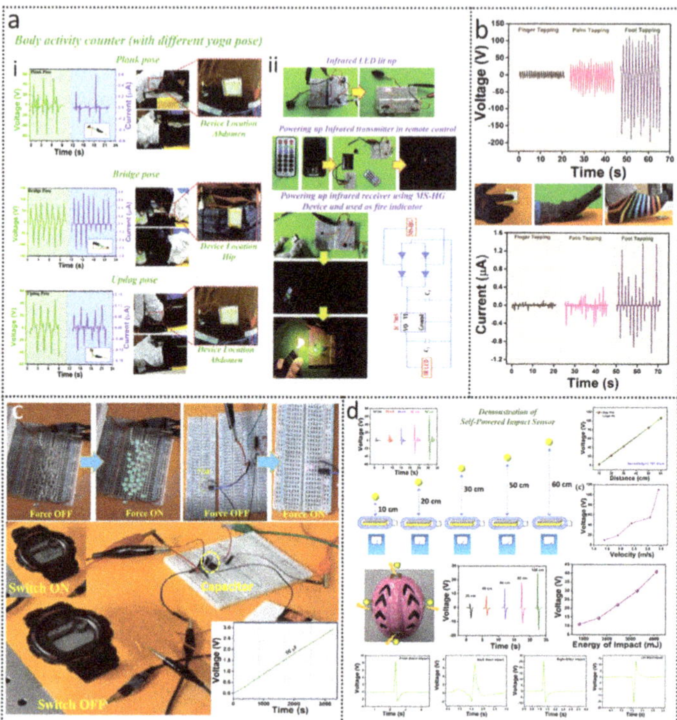

Figure 15. (**a**) Multi-stacked TENG, as a self-powered sensor, body activity counter for various yoga configurations, from count calorie burnt (**i**) to wireless IR communication module (**ii**). Reprinted with permission from reference [64], Copyrights 2020, Royal Society of Chemistry. (**b**) Biomechanical energy harvesting from hybrid piezo–tribo P-DHMG; voltage and current response of human body motion finger tapping, as well as hand and leg motion. (**c**) As a response to compressive force lightening commercial LEDs, powering of the UV LED wristwatch for sports purposes, with inset profile of capacitor charging. (**d**) Self-powered impact sensor, voltage response, as a function of time for various ball impacts, sensitivity as a function of distance, digital photograph of the helmet, with installed impact sensor, as well as voltage amplitudes and electrical properties. (**b–d**) Reprinted with permission from reference [252], Copyrights 2018, Elsevier.

6. Conclusions and Future Prospect

In the present era, power consumption has reached its peak, while energy generation and storage technology is at high risk to meet the power consumption demand. So, there is a need to improve the harvesting of units, while the most tedious part is selecting the appropriate energy materials. Ferroelectric materials have emerged and proved that their properties are promising for harvesting ambient energies and auxiliary power systems in upcoming small technology. In this review, we pictorially visualized the energy harvesting mechanisms. Their output functionalities are mainly the current progressions by various ferroelectric-based individuals and coupled nanogenerators. The detailed analysis helped us to realize the different enhanced nanogenerator features, as a function of material nature organic polymers or inorganic perovskite oxide ceramics, compositions, and morphology control, by doping and device geometry tailored to nanowires, thin films, multilayers, etc. Detailed review let us embrace that all ferroelectric NGs are hooked with spontaneous polarization connected to the configuration of domain walls, as determined by the crystal microstructure and electric poling for energy conversions. High piezoelectric potential and surface potential are produced by strong ferroelectric polarization. The outputs can be optimized by strengthening the switching polarization property of ferroelectric materials and hybridizing and coupling multi-effects in one device, resulting in superimposing of electrical signals. The synergy of various effects imitates the control over production, separation, movement, and recombination of carriers and, hence, strongly influences the energy harvesting and versatility of coupled nanogenerators, in order to make their integration to diverse electric arrays. Emerging wearable and self-powered devices need to be flexible, so the active materials are polymers, such as PVDF or copolymerized PVDF materials with TrFE, nylon, etc., and polymer-ceramic-embedded materials. The nanogenerators are intelligent, flexible, simple, and low cost, with smart applications in self-powered devices, multifunctional sensors and detectors, and wearable and portable devices, with the potential for surveilling the physiological activities, motion detection, tactile sensors for human-environment interactions, examination of respiratory tracks, and photovoltaic response to photodetectors. Multifunctional sensors have also been processed by utilizing multi-effect coupling. Even with a lot of advantages of ferroelectric-based nanogenerators, there is still a long route to go for practical and commercial applications, as a comprehensive revealing of the interactions of various coupled effects is needed for charge modulations to better output performances and promote multi-energies harvesting.

Author Contributions: Y.Y. planned and supervised the whole process. J.Z.M. has surveyed the literature, prepared the outline, and wrote the manuscript. Y.Y. and M.A.M.H. revised the outline. All authors revised the manuscript. All authors have read and agreed to the published version of the manuscript.

Funding: This work was sponsored by the ANSO Scholarship for International Students, National Key R&D Project from Minister of Science and Technology in China (No. 2016YFA0202701), the University of Chinese Academy of Sciences (Grant No. Y8540XX2D2), and the National Natural Science Foundation of China (No. 52072041).

Conflicts of Interest: The authors declare no conflict of interest.

References

1. Kim, T.Y.; Kim, S.K.; Kim, S.-W. Application of ferroelectric materials for improving output power of energy harvesters. *Nano Converg.* **2018**, *5*, 30. [CrossRef]
2. Kumar, S.; Tiwari, P.; Zymbler, M. Internet of Things is a revolutionary approach for future technology enhancement: A review. *J. Big Data* **2019**, *6*, 111. [CrossRef]
3. Zhang, X.-S.; Han, M.-D.; Wang, R.-X.; Zhu, F.-Y.; Li, Z.-H.; Wang, W.; Zhang, H. Frequency-Multiplication High-Output Triboelectric Nanogenerator for Sustainably Powering Biomedical Microsystems. *Nano Lett.* **2013**, *13*, 1168–1172. [CrossRef] [PubMed]
4. Chang, C.; Tran, V.H.; Wang, J.; Fuh, Y.-K.; Lin, L. Direct-Write Piezoelectric Polymeric Nanogenerator with High Energy Conversion Efficiency. *Nano Lett.* **2010**, *10*, 726–731. [CrossRef] [PubMed]
5. Wang, Z.L.; Song, J. Piezoelectric Nanogenerators Based on Zinc Oxide Nanowire Arrays. *Science* **2006**, *312*, 242–246. [CrossRef]
6. Yang, Y. *Hybridized and Coupled Nanogenerators: Design, Performance, and Applications*; Wiley-VCH GmbH: Weinheim, Germany, 2020.
7. Kasap, S.; Capper, P. *Springer Handbook of Electronic and Photonic Materials*, 2nd ed.; Springer: New York, NY, USA, 2007.
8. Jin, W.; Wang, Z.; Huang, H.; Hu, X.; He, Y.; Li, M.; Li, L.; Gao, Y.; Hu, Y.; Gu, H. High-performance piezoelectric energy harvesting of vertically aligned Pb(Zr,Ti)O$_3$ nanorod arrays. *RSC Adv.* **2018**, *8*, 7422–7427. [CrossRef]
9. Mikolajick, T.; Schroeder, U.; Slesazeck, S. The Past, the Present, and the Future of Ferroelectric Memories. *IEEE Trans. Electron Devices* **2020**, *67*, 1434–1443. [CrossRef]
10. Scott, J.F. *Ferroelectric Memories*, 1st ed.; Springer: Berlin/Heidelberg, Germany, 2000. [CrossRef]
11. Martienssen, W.; Warlimont, H. *Springer Handbook of Condensed Matter and Materials Data*, 1st ed.; Springer: Berlin/Heidelberg, Germany, 2005.
12. Milano, G.; Porro, S.; Valov, I.; Ricciardi, C. Nanowire Memristors: Recent Developments and Perspectives for Memristive Devices Based on Metal Oxide Nanowires. *Adv. Electron. Mater.* **2019**, *5*, 1970044. [CrossRef]
13. Sun, L.; Shi, Z.; Wang, H.; Zhang, K.; Dastan, D.; Sun, K.; Fan, R. Ultrahigh discharge efficiency and improved energy density in rationally designed bilayer polyetherimide–BaTiO$_3$/P(VDF-HFP) composites. *J. Mater. Chem. A* **2020**, *8*, 5750–5757. [CrossRef]
14. Yang, J.; Zhu, X.; Wang, H.; Wang, X.; Hao, C.; Fan, R.; Dastan, D.; Shi, Z. Achieving excellent dielectric performance in polymer composites with ultralow filler loadings via constructing hollow-structured filler frameworks. *Compos. Part A Appl. Sci. Manuf.* **2020**, *131*, 105814. [CrossRef]

15. Patterson, E.A.; Staruch, M.; Matis, B.R.; Young, S.; Lofland, S.E.; Antonelli, L.; Blackmon, F.; Damjanovic, D.; Cain, M.G.; Thompson, P.B.J.; et al. Dynamic piezoelectric response of relaxor single crystal under electrically driven inter-ferroelectric phase transformations. *Appl. Phys. Lett.* **2020**, *116*, 222903. [CrossRef]
16. Butler, K.T.; Frost, J.M.; Walsh, A. Ferroelectric materials for solar energy conversion: Photoferroics revisited. *Energy Environ. Sci.* **2014**, *8*, 838–848. [CrossRef]
17. Almadhoun, M.N.; Bhansali, U.S.; Alshareef, H.N. Nanocomposites of ferroelectric polymers with surface-hydroxylated BaTiO$_3$ nanoparticles for energy storage applications. *J. Mater. Chem.* **2012**, *22*, 11196–11200. [CrossRef]
18. Kumar, A.; Mukherjee, S.; Roy, A. Solar energy harvesting with ferroelectric materials. In *Ferroelectric Materials for Energy Harvesting and Storage*; Woodhead Publishing: Sawston, UK, 2021; pp. 43–84. [CrossRef]
19. Wang, S.; Wang, X.; Wang, Z.L.; Yang, Y. Efficient Scavenging of Solar and Wind Energies in a Smart City. *ACS Nano* **2016**, *10*, 5696–5700. [CrossRef] [PubMed]
20. Li, Q.; Li, S.; Pisignano, D.; Persano, L.; Yang, Y.; Su, Y. On the evaluation of output voltages for quantifying the performance of pyroelectric energy harvesters. *Nano Energy* **2021**, *86*, 106045. [CrossRef]
21. Lee, S.; Hinchet, R.; Lee, Y.; Yang, Y.; Lin, Z.-H.; Ardila, G.; Montès, L.; Mouis, M.; Wang, Z.L. Ultrathin Nanogenerators as Self-Powered/Active Skin Sensors for Tracking Eye Ball Motion. *Adv. Funct. Mater.* **2013**, *24*, 1163–1168. [CrossRef]
22. Zhu, G.; Lin, Z.-H.; Jing, Q.; Bai, P.; Pan, C.; Yang, Y.; Zhou, Y.; Wang, Z.L. Toward Large-Scale Energy Harvesting by a Nanoparticle-Enhanced Triboelectric Nanogenerator. *Nano Lett.* **2013**, *13*, 847–853. [CrossRef]
23. Lin, Z.-H.; Yang, Y.; Wu, J.M.; Liu, Y.; Zhang, F.; Wang, Z.L. BaTiO$_3$ Nanotubes-Based Flexible and Transparent Nanogenerators. *J. Phys. Chem. Lett.* **2012**, *3*, 3599–3604. [CrossRef]
24. Kim, P.; Doss, N.M.; Tillotson, J.P.; Hotchkiss, P.J.; Pan, M.-J.; Marder, S.R.; Li, J.; Calame, J.P.; Perry, J.W. High Energy Density Nanocomposites Based on Surface-Modified BaTiO3 and a Ferroelectric Polymer. *ACS Nano* **2009**, *3*, 2581–2592. [CrossRef]
25. Mikolajick, T. Ferroelectric Nonvolatile Memories. In *Encyclopedia of Materials: Science and Technology*; Elsevier: Amsterdam, The Netherlands, 2002; Volume 2, pp. 1–5. [CrossRef]
26. Lee, H.; Kim, H.; Kim, D.Y.; Seo, Y. Pure Piezoelectricity Generation by a Flexible Nanogenerator Based on Lead Zirconate Titanate Nanofibers. *ACS Omega* **2019**, *4*, 2610–2617. [CrossRef]
27. Xue, H.; Yang, Q.; Wang, D.; Luo, W.; Wang, W.; Lin, M.; Liang, D.; Luo, Q. A wearable pyroelectric nanogenerator and self-powered breathing sensor. *Nano Energy* **2017**, *38*, 147–154. [CrossRef]
28. Seung, W.; Yoon, H.-J.; Kim, T.Y.; Ryu, H.; Kim, J.; Lee, J.-H.; Lee, J.H.; Kim, S.; Park, Y.K.; Park, Y.J.; et al. Boosting Power-Generating Performance of Triboelectric Nanogenerators via Artificial Control of Ferroelectric Polarization and Dielectric Properties. *Adv. Energy Mater.* **2016**, *7*, 1600988. [CrossRef]
29. Mistewicz, K.; Nowak, M.; Stróż, D. A Ferroelectric-Photovoltaic Effect in SbSI Nanowires. *Nanomaterials* **2019**, *9*, 580. [CrossRef]
30. Yang, Y.; Zhang, H.; Liu, R.; Wen, X.; Hou, T.-C.; Wang, Z.L. Fully Enclosed Triboelectric Nanogenerators for Applications in Water and Harsh Environments. *Adv. Energy Mater.* **2013**, *3*, 1563–1568. [CrossRef]
31. Jiang, Q.; Chen, B.; Yang, Y. Wind-Driven Triboelectric Nanogenerators for Scavenging Biomechanical Energy. *ACS Appl. Energy Mater.* **2018**, *1*, 4269–4276. [CrossRef]
32. Yang, Y.; Zhang, H.; Lin, Z.-H.; Zhou, Y.; Jing, Q.; Su, Y.; Yang, J.; Chen, J.; Hu, C.; Wang, Z.L. Human Skin Based Triboelectric Nanogenerators for Harvesting Biomechanical Energy and as Self-Powered Active Tactile Sensor System. *ACS Nano* **2013**, *7*, 9213–9222. [CrossRef] [PubMed]
33. Hanani, Z.; Izanzar, I.; Amjoud, M.; Mezzane, D.; Lahcini, M.; Uršič, H.; Prah, U.; Saadoune, I.; El Marssi, M.; Luk'Yanchuk, I.A.; et al. Lead-free nanocomposite piezoelectric nanogenerator film for biomechanical energy harvesting. *Nano Energy* **2020**, *81*, 105661. [CrossRef]
34. Bairagi, S.; Ali, S.W. Flexible lead-free PVDF/SM-KNN electrospun nanocomposite based piezoelectric materials: Significant enhancement of energy harvesting efficiency of the nanogenerator. *Energy* **2020**, *198*, 117385. [CrossRef]
35. Yuan, H.; Lei, T.; Qin, Y.; Yang, R. Flexible electronic skins based on piezoelectric nanogenerators and piezotronics. *Nano Energy* **2019**, *59*, 84–90. [CrossRef]
36. Yang, Y.; Zhou, Y.; Wu, J.M.; Wang, Z.L. Single Micro/Nanowire Pyroelectric Nanogenerators as Self-Powered Temperature Sensors. *ACS Nano* **2012**, *6*, 8456–8461. [CrossRef] [PubMed]
37. Yang, Y.; Wang, S.; Zhang, Y.; Wang, Z.L. Pyroelectric Nanogenerators for Driving Wireless Sensors. *Nano Lett.* **2012**, *12*, 6408–6413. [CrossRef] [PubMed]
38. Yang, Y.; Guo, W.; Pradel, K.C.; Zhu, G.; Zhou, Y.; Zhang, Y.; Hu, Y.; Lin, L.; Wang, Z.L. Pyroelectric Nanogenerators for Harvesting Thermoelectric Energy. *Nano Lett.* **2012**, *12*, 2833–2838. [CrossRef] [PubMed]
39. Zhang, K.; Wang, Y.; Wang, Z.L.; Yang, Y. Standard and figure-of-merit for quantifying the performance of pyroelectric nanogenerators. *Nano Energy* **2018**, *55*, 534–540. [CrossRef]
40. Liu, X.; Zhao, K.; Yang, Y. Effective polarization of ferroelectric materials by using a triboelectric nanogenerator to scavenge wind energy. *Nano Energy* **2018**, *53*, 622–629. [CrossRef]
41. Chen, B.; Yang, Y.; Wang, Z.L. Scavenging Wind Energy by Triboelectric Nanogenerators. *Adv. Energy Mater.* **2018**, *8*, 1702649. [CrossRef]
42. Nechache, R.; Harnagea, C.; Licoccia, S.; Traversa, E.; Ruediger, A.; Pignolet, A.; Rosei, F. Photovoltaic properties of Bi2FeCrO6 epitaxial thin films. *Appl. Phys. Lett.* **2011**, *98*, 202902. [CrossRef]

43. Ma, N.; Yang, Y. Boosted photocurrent via cooling ferroelectric BaTiO$_3$ materials for self-powered 405 nm light detection. *Nano Energy* **2019**, *60*, 95–102. [CrossRef]
44. Zhao, R.; Ma, N.; Qi, J.; Mishra, Y.; Adelung, R.; Yang, Y. Optically Controlled Abnormal Photovoltaic Current Modulation with Temperature in BiFeO$_3$. *Adv. Electron. Mater.* **2019**, *5*, 1800791. [CrossRef]
45. Ma, N.; Yang, Y. Boosted photocurrent in ferroelectric BaTiO$_3$ materials via two dimensional planar-structured contact configurations. *Nano Energy* **2018**, *50*, 417–424. [CrossRef]
46. Zhu, Q.; Dong, L.; Zhang, J.; Xu, K.; Zhang, Y.; Shi, H.; Lu, H.; Wu, Y.; Zheng, H.; Wang, Z. All-in-one hybrid tribo/piezoelectric nanogenerator with the point contact and its adjustable charge transfer by ferroelectric polarization. *Ceram. Int.* **2020**, *46*, 28277–28284. [CrossRef]
47. Ji, Y.; Zhang, K.; Yang, Y. A One-Structure-Based Multieffects Coupled Nanogenerator for Simultaneously Scavenging Thermal, Solar, and Mechanical Energies. *Adv. Sci.* **2017**, *5*, 1700622. [CrossRef] [PubMed]
48. Zhong, X.; Yang, Y.; Wang, X.; Wang, Z.L. Rotating-disk-based hybridized electromagnetic-triboelectric nanogenerator for scavenging biomechanical energy as a mobile power source. *Nano Energy* **2015**, *13*, 771–780. [CrossRef]
49. Quan, T.; Wang, X.; Wang, Z.L.; Yang, Y. Hybridized Electromagnetic–Triboelectric Nanogenerator for a Self-Powered Electronic Watch. *ACS Nano* **2015**, *9*, 12301–12310. [CrossRef] [PubMed]
50. Zhang, K.; Yang, Y. Linear-grating hybridized electromagnetic-triboelectric nanogenerator for sustainably powering portable electronics. *Nano Res.* **2016**, *9*, 974–984. [CrossRef]
51. Zhang, K.; Wang, S.; Yang, Y. A One-Structure-Based Piezo-Tribo-Pyro-Photoelectric Effects Coupled Nanogenerator for Simultaneously Scavenging Mechanical, Thermal, and Solar Energies. *Adv. Energy Mater.* **2016**, *7*, 1601852. [CrossRef]
52. Liu, X.; Zhao, K.; Wang, Z.L.; Yang, Y. Unity Convoluted Design of Solid Li-Ion Battery and Triboelectric Nanogenerator for Self-Powered Wearable Electronics. *Adv. Energy Mater.* **2017**, *7*, 1701629. [CrossRef]
53. Sahu, M.; Hajra, S.; Lee, K.; Deepti, P.; Mistewicz, K.; Kim, H.J. Piezoelectric Nanogenerator Based on Lead-Free Flexible PVDF-Barium Titanate Composite Films for Driving Low Power Electronics. *Crystals* **2021**, *11*, 85. [CrossRef]
54. Shin, S.-H.; Kim, Y.-H.; Lee, M.H.; Jung, J.-Y.; Nah, J. Hemispherically Aggregated BaTiO$_3$ Nanoparticle Composite Thin Film for High-Performance Flexible Piezoelectric Nanogenerator. *ACS Nano* **2014**, *8*, 2766–2773. [CrossRef]
55. Zhang, J.-H.; Hao, X. Enhancing output performances and output retention rates of triboelectric nanogenerators via a design of composite inner-layers with coupling effect and self-assembled outer-layers with superhydrophobicity. *Nano Energy* **2020**, *76*, 105074. [CrossRef]
56. Bairagi, S.; Ali, S.W. A unique piezoelectric nanogenerator composed of melt-spun PVDF/KNN nanorod-based nanocomposite fibre. *Eur. Polym. J.* **2019**, *116*, 554–561. [CrossRef]
57. Liu, J.; Yu, D.; Zheng, Z.; Huangfu, G.; Guo, Y. Lead-free BiFeO$_3$ film on glass fiber fabric: Wearable hybrid piezoelectric-triboelectric nanogenerator. *Ceram. Int.* **2021**, *47*, 3573–3579. [CrossRef]
58. Gu, L.; Liu, J.; Cui, N.; Xu, Q.; Du, T.; Zhang, L.; Wang, Z.; Long, C.; Qin, Y. Enhancing the current density of a piezoelectric nanogenerator using a three-dimensional intercalation electrode. *Nat. Commun.* **2020**, *11*, 1030. [CrossRef]
59. Shaikh, M.O.; Huang, Y.-B.; Wang, C.-C.; Chuang, C.-H. Wearable Woven Triboelectric Nanogenerator Utilizing Electrospun PVDF Nanofibers for Mechanical Energy Harvesting. *Micromachines* **2019**, *10*, 438. [CrossRef] [PubMed]
60. Chen, X.; Ma, X.; Ren, W.; Gao, L.; Lu, S.; Tong, D.; Wang, F.; Chen, Y.; Huang, Y.; He, H.; et al. A Triboelectric Nanogenerator Exploiting the Bernoulli Effect for Scavenging Wind Energy. *Cell Rep. Phys. Sci.* **2020**, *1*, 100207. [CrossRef]
61. Gao, L.; Hu, D.; Qi, M.; Gong, J.; Zhou, H.; Chen, X.; Chen, J.; Cai, J.; Wu, L.; Hu, N.; et al. A double-helix-structured triboelectric nanogenerator enhanced with positive charge traps for self-powered temperature sensing and smart-home control systems. *Nanoscale* **2018**, *10*, 19781–19790. [CrossRef] [PubMed]
62. Hu, X.; Ding, Z.; Fei, L.; Xiang, Y.; Lin, Y. Wearable piezoelectric nanogenerators based on reduced graphene oxide and in situ polarization-enhanced PVDF-TrFE films. *J. Mater. Sci.* **2019**, *54*, 6401–6409. [CrossRef]
63. Horiuchi, S.; Tokura, Y. Organic ferroelectrics. *Nat. Mater.* **2008**, *7*, 357–366. [CrossRef]
64. Sahu, M.; Vivekananthan, V.; Hajra, S.; Abisegapriyan, K.S.; Maria Joseph Raj, N.P.; Kim, S.J. Synergetic enhancement of energy harvesting performance in triboelectric nanogenerator using ferroelectric polarization for self-powered IR signaling and body activity monitoring. *J. Mater. Chem. A* **2020**, *8*, 22257–22268. [CrossRef]
65. Ma, S.W.; Fan, Y.J.; Li, H.Y.; Su, L.; Wang, Z.L.; Zhu, G. Flexible Porous Polydimethylsiloxane/Lead Zirconate Titanate-Based Nanogenerator Enabled by the Dual Effect of Ferroelectricity and Piezoelectricity. *ACS Appl. Mater. Interfaces* **2018**, *10*, 33105–33111. [CrossRef]
66. Promsawat, N.; Promsawat, M.; Janphuang, P.; Luo, Z.; Beeby, S.; Rojviriya, C.; Pakawanit, P.; Pojprapai, S. CNTs-added PMNT/PDMS flexible piezoelectric nanocomposite for energy harvesting application. *Integr. Ferroelectr.* **2018**, *187*, 70–79. [CrossRef]
67. Acosta, M.; Novak, N.; Rojas, V.; Patel, S.; Vaish, R.; Koruza, J.; Jrossetti, G.A.R.; Rödel, J. BaTiO$_3$-based piezoelectrics: Fundamentals, current status, and perspectives. *Appl. Phys. Rev.* **2017**, *4*, 041305. [CrossRef]
68. Martin, L.W.; Rappe, A.M. Thin-film ferroelectric materials and their applications. *Nat. Rev. Mater.* **2016**, *2*, 16087. [CrossRef]
69. Feuersanger, A.E.; Hagenlocher, A.K.; Solomon, A.L. Preparation and Properties of Thin Barium Titanate Films. *J. Electrochem. Soc.* **1964**, *111*, 1387. [CrossRef]

70. Rose, T.L.; Kelliher, E.M.; Scoville, A.N.; Stone, S.E. Characterization of rf-sputtered $BaTiO_3$ thin films using a liquid electrolyte for the top contact. *J. Appl. Phys.* **1998**, *55*, 3706. [CrossRef]

71. Shintani, Y.; Tada, O. Preparation of Thin $BaTiO_3$ Films by dc Diode Sputtering. *J. Appl. Phys.* **2003**, *41*, 2376. [CrossRef]

72. Ishiwara, H.; Okuyama, M.; Arimoto, Y. *Ferroelectric Random Access Memories: Fundamentals and Applications*, 1st ed.; Springer: Berlin/Heidelberg, Germany, 2004.

73. Slater, J.C. Theory of the Transition in KH_2PO_4. *Ferroelectrics* **1987**, *71*, 25–42. [CrossRef]

74. Bogdanov, A.; Mysovsky, A.; Pickard, C.; Kimmel, A.V. Multiphase modelling of $Pb(Zr_{1-x}Ti_x)O_3$ structure. *Ferroelectrics* **2017**, *520*, 1–9. [CrossRef]

75. Shen, D.; Park, J.-H.; Ajitsaria, J.; Choe, S.-Y.; Wikle, H.; Kim, D.-J. The design, fabrication and evaluation of a MEMS PZT cantilever with an integrated Si proof mass for vibration energy harvesting. *J. Micromech. Microeng.* **2008**, *18*, 055017. [CrossRef]

76. Cui, X.; Ni, X.; Zhang, Y. Theoretical study of output of piezoelectric nanogenerator based on composite of PZT nanowires and polymers. *J. Alloys Compd.* **2016**, *675*, 306–310. [CrossRef]

77. Liu, H.; Lin, X.; Zhang, S.; Huan, Y.; Huang, S.; Cheng, X. Enhanced performance of piezoelectric composite nanogenerator based on gradient porous PZT ceramic structure for energy harvesting. *J. Mater. Chem. A* **2020**, *8*, 19631–19640. [CrossRef]

78. Liu, Y.; Ji, Y.; Yang, Y. Growth, Properties and Applications of $Bi_{0.5}Na_{0.5}TiO_3$ Ferroelectric Nanomaterials. *Nanomaterials* **2021**, *11*, 1724. [CrossRef] [PubMed]

79. Wang, N.; Luo, X.; Han, L.; Zhang, Z.; Zhang, R.; Olin, H.; Yang, Y. Structure, Performance, and Application of $BiFeO_3$ Nanomaterials. *Nano Micro Lett.* **2020**, *12*, 81. [CrossRef]

80. Kirillov, V.V.; Isupov, V.A. Relaxation polarization of $PbMg_{1/3}Nb_{2/3}O_3$(PMN)-A ferroelectric with a diffused phase transition. *Ferroelectrics* **1973**, *5*, 3–9. [CrossRef]

81. You, Y.F.; Xu, C.; Wang, J.Z.; Wang, J.P. Preparation of PST Ferroelectric Thin Films by Sol-Gel Process. *Adv. Mater. Res.* **2013**, *734–737*, 2328–2331. [CrossRef]

82. Inbar, I.; Cohen, R.E. Origin of ferroelectricity in $LiNbO_3$ and $LiTaO_3$. *Ferroelectrics* **1997**, *194*, 83–95. [CrossRef]

83. Kiraci, A.; Yurtseven, H. Order–disorder transition in the ferroelectric $LiTaO_3$. *Ferroelectrics* **2019**, *551*, 235–244. [CrossRef]

84. Samuelsen, E.J.; Grande, A.P. The ferroelectric phase transition in $LiTaO_3$ studied by neutron scattering. *Eur. Phys. J. B* **1976**, *24*, 207–210. [CrossRef]

85. Ji, Y.; Liu, Y.; Yang, Y. Multieffect Coupled Nanogenerators. *Research* **2020**, *2020*, 1–24. [CrossRef]

86. Wang, J.L.; Meng, X.J.; Chu, J.H. New Properties and Applications of Polyvinylidene-Based Ferroelectric Polymer. In *Ferroelectric Materials—Synthesis and Characterization*; InTech Open: Rijeka, Croatia, 2015. [CrossRef]

87. Chung, T.C.M.; Petchsuk, A. Polymers, Ferroelectric. In *Encyclopedia of Physical Science and Technology*; Academic Press: Cambridge, MA, USA, 2003; pp. 659–674. [CrossRef]

88. Levanyuk, A.P.; Sannikov, D.G. Improper ferroelectrics. *Sov. Phys. Uspekhi* **1974**, *17*, 199–214. [CrossRef]

89. Li, W.; Tang, G.; Zhang, G.; Jafri, H.M.; Zhou, J.; Liu, D.; Liu, Y.; Wang, J.; Jin, K.; Hu, Y.; et al. Improper molecular ferroelectrics with simultaneous ultrahigh pyroelectricity and figures of merit. *Sci. Adv.* **2021**, *7*, eabe3068. [CrossRef] [PubMed]

90. Hao, X.; Zhai, J.; Yao, X. Improved Energy Storage Performance and Fatigue Endurance of Sr-Doped $PbZrO_3$ Antiferroelectric Thin Films. *J. Am. Ceram. Soc.* **2009**, *92*, 1133–1135. [CrossRef]

91. Park, M.H.; Kim, H.J.; Kim, Y.J.; Moon, T.; Kim, K.D.; Hwang, C.S. Thin $Hf_xZr_{1-x}O_2$ Films: A New Lead-Free System for Electrostatic Supercapacitors with Large Energy Storage Density and Robust Thermal Stability. *Adv. Energy Mater.* **2014**, *4*, 1400610. [CrossRef]

92. Lukasiewicz, T.; Swirkowicz, M.; Dec, J.; Hofman, W.; Szyrski, W. Strontium–barium niobate single crystals, growth and ferroelectric properties. *J. Cryst. Growth* **2007**, *310*, 1464–1469. [CrossRef]

93. Jing-Ya, S.; Yang, Y.; Ke, Z.; Yu-Long, L.; Siu, G.G.; Xu, Z.K. Polarized Micro-Raman Study of the Temperature-Induced Phase Transition In $0.67Pb(Mg_{1/3}Nb_{2/3})O_3$-$0.33PbTiO_3$. *Chin. Phys. Lett.* **2008**, *25*, 290–393. [CrossRef]

94. Hu, X.; Yi, K.; Liu, J.; Chu, B. High Energy Density Dielectrics Based on PVDF-Based Polymers. *Energy Technol.* **2018**, *6*, 849–864. [CrossRef]

95. Zhu, H.; Pruvost, S.; Cottinet, P.J.; Guyomar, D. Energy harvesting by nonlinear capacitance variation for a relaxor ferroelectric poly(vinylidene fluoride-trifluoroethylene-chlorofluoroethylene) terpolymer. *Appl. Phys. Lett.* **2011**, *98*, 222901. [CrossRef]

96. Ali, A.; Hassen, A.; Khang, N.C.; Kim, Y.S. Ferroelectric, and piezoelectric properties of $BaTi_{1-x}Al_xO_3$, $0 \leq x \leq 0.015$. *AIP Adv.* **2015**, *5*, 097125. [CrossRef]

97. Saradhi, B.B.; Srinivas, K.; Prasad, G.; Suryanarayana, S.; Bhimasankaram, T. Impedance spectroscopic studies in ferroelectric $(Na_{1/2}Bi_{1/2})TiO_3$. *Mater. Sci. Eng. B* **2003**, *98*, 10–16. [CrossRef]

98. Ji, S.H.; Cho, J.H.; Jeong, Y.H.; Paik, J.-H.; Yun, J.D.; Yun, J.S. Flexible lead-free piezoelectric nanofiber composites based on BNT-ST and PVDF for frequency sensor applications. *Sens. Actuators A Phys.* **2016**, *247*, 316–322. [CrossRef]

99. Achuthan, A.; Sun, C.T. Domain switching in ferroelectric ceramic materials under combined loads. *J. Appl. Phys.* **2005**, *97*, 114103. [CrossRef]

100. Kao, K.C. Ferroelectrics, Piezoelectrics, and Pyroelectrics. In *Dielectric Phenomena in Solids*; Elsevier: San Diego, CA, USA, 2004; pp. 213–282. [CrossRef]

101. Huibregtse, E.J.; Young, D.R. Triple Hysteresis Loops and the Free-Energy Function in the Vicinity of the 5 °C Transition in $BaTiO_3$. *Phys. Rev.* **1956**, *103*, 1705–1711. [CrossRef]

102. Merz, W.J. Double Hysteresis Loop of BaTiO$_3$ at the Curie Point. *Phys. Rev.* **1953**, *91*, 513–517. [CrossRef]
103. Borderon, C.; Brunier, A.E.; Nadaud, K.; Renoud, R.; Alexe, M.; Gundel, H.W. Domain wall motion in Pb(Zr0.20Ti0.80)O$_3$ epitaxial thin films. *Sci. Rep.* **2017**, *7*, 3444. [CrossRef]
104. Xu, F.; Trolier-McKinstry, S.; Ren, W.; Xu, B.; Xie, Z.-L.; Hemker, K.J. Domain wall motion and its contribution to the dielectric and piezoelectric properties of lead zirconate titanate films. *J. Appl. Phys.* **2001**, *89*, 1336–1348. [CrossRef]
105. Bassiri-Gharb, N.; Fujii, I.; Hong, E.; Trolier-McKinstry, S.; Taylor, D.V.; Damjanovic, D. Domain wall contributions to the properties of piezoelectric thin films. *J. Electroceramics* **2007**, *19*, 49–67. [CrossRef]
106. Fancher, C.; Brewer, S.; Chung, C.; Röhrig, S.; Rojac, T.; Esteves, G.; Deluca, M.; Bassiri-Gharb, N.; Jones, J. The contribution of 180° domain wall motion to dielectric properties quantified from in situ X-ray diffraction. *Acta Mater.* **2016**, *126*, 36–43. [CrossRef]
107. Huan, Y.; Wang, X.; Fang, J.; Li, L. Grain size effect on piezoelectric and ferroelectric properties of BaTiO$_3$ ceramics. *J. Eur. Ceram. Soc.* **2014**, *34*, 1445–1448. [CrossRef]
108. Seidel, J.; Martin, L.W.; He, Q.; Zhan, Q.; Chu, Y.-H.; Rother, A.; Hawkridge, M.E.; Maksymovych, P.; Yu, P.; Gajek, M.; et al. Conduction in domain walls in oxide ferroelectrics. *Nat. Mater.* **2009**, *8*, 229–234. [CrossRef]
109. Ghara, S.; Geirhos, K.; Kuerten, L.; Lunkenheimer, P.; Tsurkan, V.; Fiebig, M.; Kézsmárki, I. Giant conductivity of mobile non-oxide domain walls. *Nat. Commun.* **2021**, *12*, 3975. [CrossRef]
110. Werner, C.S.; Herr, S.J.; Buse, K.; Sturman, B.; Soergel, E.; Razzaghi, C.; Breunig, I. Large and accessible conductivity of charged domain walls in lithium niobate. *Sci. Rep.* **2017**, *7*, 9862. [CrossRef]
111. Mason, W.P. Piezoelectricity, its history and applications. *J. Acoust. Soc. Am.* **1981**, *70*, 1561–1566. [CrossRef]
112. Li, H.; Bowen, C.R.; Yang, Y. Scavenging Energy Sources Using Ferroelectric Materials. *Adv. Funct. Mater.* **2021**, *31*, 2100905. [CrossRef]
113. Yang, W.; Li, P.; Li, F.; Liu, X.; Shen, B.; Zhai, J. Enhanced piezoelectric performance and thermal stability of alkali niobate-based ceramics. *Ceram. Int.* **2019**, *45*, 2275–2280. [CrossRef]
114. Pandey, R.; Narayan, B.; Khatua, D.K.; Tyagi, S.; Mostaed, A.; Abebe, M.; Sathe, V.; Reaney, I.M.; Ranjan, R. High electromechanical response in the non morphotropic phase boundary piezoelectric system PbTiO$_3$−Bi(Zr1/2Ni1/2)O$_3$. *Phys. Rev. B* **2018**, *97*, 224109. [CrossRef]
115. Ge, W.; Li, J.; Viehland, D.; Chang, Y.; Messing, G.L. Electric-field-dependent phase volume fractions and enhanced piezoelectricity near the polymorphic phase boundary of (K0.5Na0.5)1−xLixNbO$_3$ textured ceramics. *Phys. Rev. B* **2011**, *83*, 224110. [CrossRef]
116. Pan, M.; Yuan, C.; Liang, X.; Zou, J.; Zhang, Y.; Bowen, C. Triboelectric and Piezoelectric Nanogenerators for Future Soft Robots and Machines. *iScience* **2020**, *23*, 101682. [CrossRef]
117. Guo, Y.; Suzuki, K.; Nishizawa, K.; Miki, T.; Kato, K. Dielectric and piezoelectric properties of highly (100)-oriented BaTiO$_3$ thin film grown on a Pt/TiOx/SiO$_2$/Si substrate using LaNiO$_3$ as a buffer layer. *J. Cryst. Growth* **2005**, *284*, 190–196. [CrossRef]
118. Du, H.; Li, Z.; Tang, F.; Qu, S.; Pei, Z.; Zhou, W. Preparation and piezoelectric properties of (K0.5Na0.5)NbO$_3$ lead-free piezoelectric ceramics with pressure-less sintering. *Mater. Sci. Eng. B* **2006**, *131*, 83–87. [CrossRef]
119. Matsubara, M.; Yamaguchi, T.; Kikuta, K.; Hirano, S.-I. Sinterability and Piezoelectric Properties of (K,Na)NbO$_3$ Ceramics with Novel Sintering Aid. *Jpn. J. Appl. Phys.* **2004**, *43*, 7159–7163. [CrossRef]
120. Chen, Z.-H.; Li, Z.-W.; Ding, J.-N.; Qiu, J.-H.; Yang, Y. Piezoelectric and ferroelectric properties of Ba$_{0.9}$Ca$_{0.1}$Ti$_{0.9}$Sn$_{0.1}$O$_3$ lead-free ceramics with La$_2$O$_3$ addition. *J. Alloys Compd.* **2017**, *704*, 193–196. [CrossRef]
121. Rana, S.M.S.; Rahman, M.T.; Salauddin, M.; Sharma, S.; Maharjan, P.; Bhatta, T.; Cho, H.; Park, C.; Park, J.Y. Electrospun PVDF-TrFE/MXene Nanofiber Mat-Based Triboelectric Nanogenerator for Smart Home Appliances. *ACS Appl. Mater. Interfaces* **2021**, *13*, 4955–4967. [CrossRef]
122. Mistewicz, K.; Jesionek, M.; Nowak, M.; Kozioł, M. SbSeI pyroelectric nanogenerator for a low temperature waste heat recovery. *Nano Energy* **2019**, *64*, 103906. [CrossRef]
123. Sidney, L.B. Pyroelectricity: From Ancient Curiosity to Modern Imaging Tool. *Phys. Today* **2005**, *58*, 31. Available online: http://www.physicstoday.org (accessed on 18 November 2021).
124. Bowen, C.R.; Taylor, J.; LeBoulbar, E.; Zabek, D.; Chauhan, A.; Vaish, R. Pyroelectric materials and devices for energy harvesting applications. *Energy Environ. Sci.* **2014**, *7*, 3836–3856. [CrossRef]
125. Erhart, J. Experiments to demonstrate piezoelectric and pyroelectric effects. *Phys. Educ.* **2013**, *48*, 438–447. [CrossRef]
126. Li, X.; Lu, S.-G.; Chen, X.-Z.; Gu, H.; Qian, X.-S.; Zhang, Q.M. Pyroelectric and electrocaloric materials. *J. Mater. Chem. C* **2012**, *1*, 23–37. [CrossRef]
127. Ploss, B.; Ploss, B.; Shin, F.; Chan, H.; Choy, C. Pyroelectric activity of ferroelectric PT/PVDF-TRFE. *IEEE Trans. Dielectr. Electr. Insul.* **2000**, *7*, 517–522. [CrossRef]
128. Meirzadeh, E.; Azuri, I.; Qi, Y.; Ehre, D.; Rappe, A.M.; Lahav, M.; Kronik, L.; Lubomirsky, I. Origin and structure of polar domains in doped molecular crystals. *Nat. Commun.* **2016**, *7*, 13351. [CrossRef]
129. Leng, Q.; Chen, L.; Guo, H.; Liu, J.; Liu, G.; Hu, C.; Xi, Y. Harvesting heat energy from hot/cold water with a pyroelectric generator. *J. Mater. Chem. A* **2014**, *2*, 11940–11947. [CrossRef]
130. Song, K.; Ma, N.; Mishra, Y.; Adelung, R.; Yang, Y. Achieving Light-Induced Ultrahigh Pyroelectric Charge Density Toward Self-Powered UV Light Detection. *Adv. Electron. Mater.* **2018**, *5*, 1800413. [CrossRef]
131. Fan, F.-R.; Tian, Z.-Q.; Wang, Z.L. Flexible triboelectric generator. *Nano Energy* **2012**, *1*, 328–334. [CrossRef]

132. Wang, Z.L. Triboelectric nanogenerators as new energy technology and self-powered sensors—Principles, problems and perspectives. *Faraday Discuss.* **2014**, *176*, 447–458. [CrossRef]

133. Song, Y.; Wang, N.; Hu, C.; Wang, Z.L.; Yang, Y. Soft triboelectric nanogenerators for mechanical energy scavenging and self-powered sensors. *Nano Energy* **2021**, *84*, 105919. [CrossRef]

134. Wang, Z.L.; Lin, L.; Chen, J.; Niu, S.; Zi, Y. Triboelectric Nanogenerator: Vertical Contact-Separation Mode. In *Triboelectric Nanogenerators*; Springer: Cham, Switzerland, 2016; pp. 23–47. [CrossRef]

135. Zhu, G.; Chen, J.; Liu, Y.; Bai, P.; Zhou, Y.S.; Jing, Q.; Pan, C.; Wang, Z.L. Linear-Grating Triboelectric Generator Based on Sliding Electrification. *Nano Lett.* **2013**, *13*, 2282–2289. [CrossRef] [PubMed]

136. Yang, Y.; Zhang, H.; Chen, J.; Jing, Q.; Zhou, Y.S.; Wen, X.; Wang, Z.L. Single-Electrode-Based Sliding Triboelectric Nanogenerator for Self-Powered Displacement Vector Sensor System. *ACS Nano* **2013**, *7*, 7342–7351. [CrossRef] [PubMed]

137. Wang, S.; Xie, Y.; Niu, S.; Lin, L.; Wang, Z.L. Freestanding Triboelectric-Layer-Based Nanogenerators for Harvesting Energy from a Moving Object or Human Motion in Contact and Non-contact Modes. *Adv. Mater.* **2014**, *26*, 2818–2824. [CrossRef]

138. Wang, J.; Wu, C.; Dai, Y.; Zhao, Z.; Wang, A.; Zhang, T.; Wang, Z.L. Achieving ultrahigh triboelectric charge density for efficient energy harvesting. *Nat. Commun.* **2017**, *8*, 88. [CrossRef] [PubMed]

139. Eom, K.; Shin, Y.-E.; Kim, J.-K.; Joo, S.H.; Kim, K.; Kwak, S.K.; Ko, H.; Jin, J.; Kang, S.J. Tailored Poly(vinylidene fluoride-co-trifluoroethylene) Crystal Orientation for a Triboelectric Nanogenerator through Epitaxial Growth on a Chitin Nanofiber Film. *Nano Lett.* **2020**, *20*, 6651–6659. [CrossRef]

140. Huang, L.; Wei, M.; Gui, C.; Jia, L. Ferroelectric photovoltaic effect and resistive switching behavior modulated by ferroelectric/electrode interface coupling. *J. Mater. Sci. Mater. Electron.* **2020**, *31*, 20667–20687. [CrossRef]

141. Fridkin, V.M. Bulk photovoltaic effect in noncentrosymmetric crystals. *Crystallogr. Rep.* **2001**, *46*, 654–658. [CrossRef]

142. Ogden, T.R.; Gookin, D.M. Bulk photovoltaic effect in polyvinylidene fluoride. *Appl. Phys. Lett.* **1984**, *45*, 995. [CrossRef]

143. Fridkin, V.M. Ferroelectricity and Giant Bulk Photovoltaic Effect in $BaTiO_3$ Films at the Nanoscale. *Ferroelectrics* **2015**, *484*, 1–13. [CrossRef]

144. Lee, J.-H.; Lee, K.Y.; Gupta, M.K.; Kim, T.Y.; Lee, D.-Y.; Oh, J.; Ryu, C.; Yoo, W.J.; Kang, C.-Y.; Yoon, S.-J.; et al. Highly Stretchable Piezoelectric-Pyroelectric Hybrid Nanogenerator. *Adv. Mater.* **2013**, *26*, 765–769. [CrossRef] [PubMed]

145. Ji, Y.; Zhang, K.; Wang, Z.L.; Yang, Y. Piezo–pyro–photoelectric effects induced coupling enhancement of charge quantity in $BaTiO_3$ materials for simultaneously scavenging light and vibration energies. *Energy Environ. Sci.* **2019**, *12*, 1231–1240. [CrossRef]

146. Wankhade, S.H.; Tiwari, S.; Gaur, A.; Maiti, P. PVDF-PZT nanohhybrid based nanogenerator for energy harvesting applications. *Energy Rep.* **2020**, *6*, 358–364. [CrossRef]

147. Zhu, J.; Qian, J.; Hou, X.; He, J.; Niu, X.; Geng, W.; Mu, J.; Zhang, W.; Chou, X. High-performance stretchable PZT particles/Cu@Ag branch nanofibers composite piezoelectric nanogenerator for self-powered body motion monitoring. *Smart Mater. Struct.* **2019**, *28*, 095014. [CrossRef]

148. Vivekananthan, V.; Chandrasekhar, A.; Alluri, N.R.; Purusothaman, Y.; Kim, W.J.; Kang, C.-N.; Kim, S.-J. A flexible piezoelectric composite nanogenerator based on doping enhanced lead-free nanoparticles. *Mater. Lett.* **2019**, *249*, 73–76. [CrossRef]

149. Ganesh, R.S.; Sharma, S.; Abinnas, N.; Durgadevi, E.; Raji, P.; Ponnusamy, S.; Muthamizhchelvan, C.; Hayakawa, Y.; Kim, D.Y. Fabrication of the flexible nanogenerator from BTO nanopowders on graphene coated PMMA substrates by sol-gel method. *Mater. Chem. Phys.* **2017**, *192*, 274–281. [CrossRef]

150. Park, K.-I.; Xu, S.; Liu, Y.; Hwang, G.-T.; Kang, S.-J.L.; Wang, Z.L.; Lee, K.J. Piezoelectric $BaTiO_3$ Thin Film Nanogenerator on Plastic Substrates. *Nano Lett.* **2010**, *10*, 4939–4943. [CrossRef]

151. Wu, F.; Cai, W.; Yeh, Y.-W.; Xu, S.; Yao, N. Energy scavenging based on a single-crystal PMN-PT nanobelt. *Sci. Rep.* **2016**, *6*, 22513. [CrossRef]

152. Chen, Y.; Zhang, Y.; Zhang, L.; Ding, F.; Schmidt, O.G. Scalable single crystalline PMN-PT nanobelts sculpted from bulk for energy harvesting. *Nano Energy* **2016**, *31*, 239–246. [CrossRef]

153. Li, C.; Luo, W.; Liu, X.; Xu, D.; He, K. PMN-PT/PVDF Nanocomposite for High Output Nanogenerator Applications. *Nanomaterials* **2016**, *6*, 67. [CrossRef]

154. Moorthy, B.; Baek, C.; Wang, J.E.; Jeong, C.K.; Moon, S.; Park, K.-I.; Kim, D.K. Piezoelectric energy harvesting from a PMN–PT single nanowire. *RSC Adv.* **2016**, *7*, 260–265. [CrossRef]

155. Jiang, C.; Zhou, X.; Zhou, K.; Wu, M.; Zhang, D. Synthesis of $Na_{0.5}Bi_{0.5}TiO_3$ whiskers and their nanoscale piezoelectricity. *Ceram. Int.* **2017**, *43*, 11274–11280. [CrossRef]

156. Zhang, Y.; Liu, X.; Wang, G.; Li, Y.; Zhang, S.; Wang, D.; Sun, H. Enhanced mechanical energy harvesting capability in sodium bismuth titanate based lead-free piezoelectric. *J. Alloys Compd.* **2020**, *825*, 154020. [CrossRef]

157. Zhou, X.; Wu, Z.; Jiang, C.; Luo, H.; Yan, Z.; Zhang, D. Molten salt synthesis and characterization of lead-free $(1-x)Na_{0.5}Bi_{0.5}TiO_3$-$xSrTiO_3$ (x = 0, 0.10, 0.26) whiskers. *Ceram. Int.* **2018**, *44*, 9174–9180. [CrossRef]

158. Zhou, D.; Zhou, Y.; Tian, Y.; Tu, Y.; Zheng, G.; Gu, H. Structure and Piezoelectric Properties of Lead-Free $Na_{0.5}Bi_{0.5}TiO_3$ Nanofibers Synthesized by Electrospinning. *J. Mater. Sci. Technol.* **2015**, *31*, 1181–1185. [CrossRef]

159. Liu, B.; Lu, B.; Chen, X.; Wu, X.; Shi, S.; Xu, L.; Liu, Y.; Wang, F.; Zhao, X.; Shi, W. A high-performance flexible piezoelectric energy harvester based on lead-free $(Na_{0.5}Bi_{0.5})TiO_3$–$BaTiO_3$ piezoelectric nanofibers. *J. Mater. Chem. A* **2017**, *5*, 23634–23640. [CrossRef]

160. Xia, M.; Luo, C.; Su, X.; Li, Y.; Li, P.; Hu, J.; Li, G.; Jiang, H.; Zhang, W. KNN/PDMS/C-based lead-free piezoelectric composite film for flexible nanogenerator. *J. Mater. Sci. Mater. Electron.* **2019**, *30*, 7558–7566. [CrossRef]

161. Bairagi, S.; Ali, S.W. Poly (vinylidine fluoride) (PVDF)/Potassium Sodium Niobate (KNN) nanorods based flexible nanocomposite film: Influence of KNN concentration in the performance of nanogenerator. *Org. Electron.* **2020**, *78*, 105547. [CrossRef]
162. Azimi, S.; Golabchi, A.; Nekookar, A.; Rabbani, S.; Amiri, M.H.; Asadi, K.; Abolhasani, M.M. Self-powered cardiac pacemaker by piezoelectric polymer nanogenerator implant. *Nano Energy* **2021**, *83*, 105781. [CrossRef]
163. Hu, P.; Yan, L.; Zhao, C.; Zhang, Y.; Niu, J. Double-layer structured PVDF nanocomposite film designed for flexible nanogenerator exhibiting enhanced piezoelectric output and mechanical property. *Compos. Sci. Technol.* **2018**, *168*, 327–335. [CrossRef]
164. Yan, J.; Liu, M.; Jeong, Y.G.; Kang, W.; Li, L.; Zhao, Y.; Deng, N.; Cheng, B.; Yang, G. Performance enhancements in poly(vinylidene fluoride)-based piezoelectric nanogenerators for efficient energy harvesting. *Nano Energy* **2018**, *56*, 662–692. [CrossRef]
165. Li, R.; Zhou, J.; Liu, H.; Pei, J. Effect of Polymer Matrix on the Structure and Electric Properties of Piezoelectric Lead Zirconate titanate/Polymer Composites. *Materials* **2017**, *10*, 945. [CrossRef]
166. Hao, Y.N.; Wang, X.H.; O'Brien, S.; Lombardi, J.; Li, L.T. Flexible BaTiO$_3$/PVDF gradated multilayer nanocomposite film with enhanced dielectric strength and high energy density. *J. Mater. Chem. C* **2015**, *3*, 9740–9747. [CrossRef]
167. Yaqoob, U.; Uddin, A.I.; Chung, G.-S. A novel tri-layer flexible piezoelectric nanogenerator based on surface- modified graphene and PVDF-BaTiO$_3$ nanocomposites. *Appl. Surf. Sci.* **2017**, *405*, 420–426. [CrossRef]
168. Ren, X.; Fan, H.; Zhao, Y.; Liu, Z. Flexible Lead-Free BiFeO$_3$/PDMS-Based Nanogenerator as Piezoelectric Energy Harvester. *ACS Appl. Mater. Interfaces* **2016**, *8*, 26190–26197. [CrossRef] [PubMed]
169. Lee, Y.B.; Han, J.K.; Noothongkaew, S.; Kim, S.K.; Song, W.; Myung, S.; Lee, S.S.; Lim, J.; Bu, S.D.; An, K.-S. Toward Arbitrary-Direction Energy Harvesting through Flexible Piezoelectric Nanogenerators Using Perovskite PbTiO$_3$ Nanotube Arrays. *Adv. Mater.* **2016**, *29*, 1604500. [CrossRef] [PubMed]
170. Yan, J.; Jeong, Y.G. High Performance Flexible Piezoelectric Nanogenerators based on BaTiO$_3$ Nanofibers in Different Alignment Modes. *ACS Appl. Mater. Interfaces* **2016**, *8*, 15700–15709. [CrossRef] [PubMed]
171. Jung, J.H.; Lee, M.; Hong, J.-I.; Ding, Y.; Chen, C.-Y.; Chou, L.-J.; Wang, Z.L. Lead-Free NaNbO$_3$ Nanowires for a High Output Piezoelectric Nanogenerator. *ACS Nano* **2011**, *5*, 10041–10046. [CrossRef]
172. Isakov, D.; Gomes, E.D.M.; Almeida, B.; Kholkin, A.; Zelenovskiy, P.; Neradovskiy, M.; Shur, V. Energy harvesting from nanofibers of hybrid organic ferroelectric dabcoHReO$_4$. *Appl. Phys. Lett.* **2014**, *104*, 32907. [CrossRef]
173. Zelenovskii, P.S.; Romanyuk, K.; Liberato, M.S.; Brandão, P.; Ferreira, F.F.; Kopyl, S.; Mafra, L.M.; Alves, W.A.; Kholkin, A.L. 2D Layered Dipeptide Crystals for Piezoelectric Applications. *Adv. Funct. Mater.* **2021**, *31*, 2102524. [CrossRef]
174. Nguyen, V.; Zhu, R.; Jenkins, K.; Yang, R. Self-assembly of diphenylalanine peptide with controlled polarization for power generation. *Nat. Commun.* **2016**, *7*, 13566. [CrossRef] [PubMed]
175. Lee, J.-H.; Heo, K.; Schulz-Schönhagen, K.; Desai, M.S.; Jin, H.-E.; Lee, S.-W. Diphenylalanine Peptide Nanotube Energy Harvesters. *ACS Nano* **2018**, *12*, 8138–8144. [CrossRef]
176. Huan, Y.; Zhang, X.; Song, J.; Zhao, Y.; Wei, T.; Zhang, G.; Wang, X. High-performance piezoelectric composite nanogenerator based on Ag/(K,Na)NbO$_3$ heterostructure. *Nano Energy* **2018**, *50*, 62–69. [CrossRef]
177. Ding, R.; Zhang, X.; Chen, G.; Wang, H.; Kishor, R.; Xiao, J.; Gao, F.; Zeng, K.; Chen, X.; Sun, X.W.; et al. High-performance piezoelectric nanogenerators composed of formamidinium lead halide perovskite nanoparticles and poly(vinylidene fluoride). *Nano Energy* **2017**, *37*, 126–135. [CrossRef]
178. Zhai, W.; Zhu, L.; Berbille, A.; Wang, Z.L. Flexible and wearable piezoelectric nanogenerators based on P(VDF-TrFE)/SnS nanocomposite micropillar array. *J. Appl. Phys.* **2021**, *129*, 095501. [CrossRef]
179. Yang, J.; Zhang, Y.; Li, Y.; Wang, Z.; Wang, W.; An, Q.; Tong, W. Piezoelectric Nanogenerators based on Graphene Oxide/PVDF Electrospun Nanofiber with Enhanced Performances by In-Situ Reduction. *Mater. Today Commun.* **2020**, *26*, 101629. [CrossRef]
180. Guan, X.; Xu, B.; Gong, J. Hierarchically architected polydopamine modified BaTiO$_3$@P(VDF-TrFE) nanocomposite fiber mats for flexible piezoelectric nanogenerators and self-powered sensors. *Nano Energy* **2020**, *70*, 104516. [CrossRef]
181. Zhai, W.; Lai, Q.; Chen, L.; Zhu, L.; Wang, Z.L. Flexible Piezoelectric Nanogenerators Based on P(VDF–TrFE)/GeSe Nanocomposite Films. *ACS Appl. Electron. Mater.* **2020**, *2*, 2369–2374. [CrossRef]
182. Liu, J.; Yang, B.; Lu, L.; Wang, X.; Li, X.; Chen, X.; Liu, J. Flexible and lead-free piezoelectric nanogenerator as self-powered sensor based on electrospinning BZT-BCT/P(VDF-TrFE) nanofibers. *Sens. Actuators A Phys.* **2020**, *303*, 111796. [CrossRef]
183. Petrovic, M.V.; Cordero, F.; Mercadelli, E.; Brunengo, E.; Ilic, N.; Galassi, C.; Despotovic, Z.; Bobic, J.; Dzunuzovic, A.; Stagnaro, P.; et al. Flexible lead-free NBT-BT/PVDF composite films by hot pressing for low-energy harvesting and storage. *J. Alloy. Compd.* **2021**, *884*, 161071. [CrossRef]
184. Yu, D.; Zheng, Z.; Liu, J.; Xiao, H.; Huangfu, G.; Guo, Y. Superflexible and Lead-Free Piezoelectric Nanogenerator as a Highly Sensitive Self-Powered Sensor for Human Motion Monitoring. *Nano Micro Lett.* **2021**, *13*, 117. [CrossRef]
185. Kim, J.; Lee, J.H.; Ryu, H.; Lee, J.-H.; Khan, U.; Kim, H.; Kwak, S.S.; Kim, S.-W. High-Performance Piezoelectric, Pyroelectric, and Triboelectric Nanogenerators Based on P(VDF-TrFE) with Controlled Crystallinity and Dipole Alignment. *Adv. Funct. Mater.* **2017**, *27*, 1700702. [CrossRef]
186. Lee, J.-H.; Ryu, H.; Kim, T.-Y.; Kwak, S.S.; Yoon, H.-J.; Seung, W.; Kim, S.-W. Thermally Induced Strain-Coupled Highly Stretchable and Sensitive Pyroelectric Nanogenerators. *Adv. Energy Mater.* **2015**, *5*, 1500704. [CrossRef]
187. Ghosh, S.K.; Xie, M.; Bowen, C.R.; Davies, P.R.; Morgan, D.J.; Mandal, D. A hybrid strain and thermal energy harvester based on an infra-red sensitive Er^{3+} modified poly(vinylidene fluoride) ferroelectret structure. *Sci. Rep.* **2017**, *7*, 16703. [CrossRef]

188. Yang, Y.; Jung, J.H.; Kil Yun, B.; Zhang, F.; Pradel, K.C.; Guo, W.; Wang, Z.L. Flexible Pyroelectric Nanogenerators using a Composite Structure of Lead-Free KNbO$_3$ Nanowires. *Adv. Mater.* **2012**, *24*, 5357–5362. [CrossRef]

189. Gao, F.; Li, W.; Wang, X.; Fang, X.; Ma, M. A self-sustaining pyroelectric nanogenerator driven by water vapor. *Nano Energy* **2016**, *22*, 19–26. [CrossRef]

190. Ko, Y.J.; Park, Y.K.; Kil Yun, B.; Lee, M.; Jung, J.H. High pyroelectric power generation of 0.7Pb(Mg$_{1/3}$Nb$_{2/3}$)O$_3$–0.3PbTiO$_3$ single crystal. *Curr. Appl. Phys.* **2014**, *14*, 1486–1491. [CrossRef]

191. Shen, M.; Hu, L.; Li, L.; Zhang, C.; Xiao, W.; Zhang, Y.; Zhang, Q.; Zhang, G.; Jiang, S.; Chen, Y. High pyroelectric response over a broad temperature range in NBT-BZT: SiO$_2$ composites for energy harvesting. *J. Eur. Ceram. Soc.* **2021**, *41*, 3379–3386. [CrossRef]

192. Shen, M.; Qin, Y.; Zhang, Y.; Marwat, M.A.; Zhang, C.; Wang, W.; Li, M.; Zhang, H.; Zhang, G.; Jiang, S. Enhanced pyroelectric properties of lead-free BNT-BA-KNN ceramics for thermal energy harvesting. *J. Am. Ceram. Soc.* **2018**, *102*, 3990–3999. [CrossRef]

193. Roy, K.; Ghosh, S.K.; Sultana, A.; Garain, S.; Xie, M.; Bowen, C.R.; Henkel, K.; Schmeißer, D.; Mandal, D. A Self-Powered Wearable Pressure Sensor and Pyroelectric Breathing Sensor Based on GO Interfaced PVDF Nanofibers. *ACS Appl. Nano Mater.* **2019**, *2*, 2013–2025. [CrossRef]

194. Wang, Q.; Zhang, X.; Bowen, C.; Li, M.-Y.; Ma, J.; Qiu, S.; Liu, H.; Jiang, S. Effect of Zr/Ti ratio on microstructure and electrical properties of pyroelectric ceramics for energy harvesting applications. *J. Alloy. Compd.* **2017**, *710*, 869–874. [CrossRef]

195. Pandya, S.; Wilbur, J.; Kim, J.; Gao, R.; Dasgupta, A.; Dames, C.; Martin, L.W. Pyroelectric energy conversion with large energy and power density in relaxor ferroelectric thin films. *Nat. Mater.* **2018**, *17*, 432–438. [CrossRef] [PubMed]

196. Park, Y.; Shin, Y.-E.; Park, J.; Lee, Y.; Kim, M.; Kim, Y.-R.; Na, S.; Ghosh, S.K.; Ko, H. Ferroelectric Multilayer Nanocomposites with Polarization and Stress Concentration Structures for Enhanced Triboelectric Performances. *ACS Nano* **2020**, *14*, 7101–7110. [CrossRef] [PubMed]

197. Fang, H.; Li, Q.; He, W.; Li, J.; Xue, Q.; Xu, C.; Zhang, L.; Ren, T.; Dong, G.; Chan, H.L.W.; et al. A high performance triboelectric nanogenerator for self-powered non-volatile ferroelectric transistor memory. *Nanoscale* **2015**, *7*, 17306–17311. [CrossRef] [PubMed]

198. Kim, K.L.; Lee, W.; Hwang, S.K.; Joo, S.H.; Cho, S.M.; Song, G.; Cho, S.H.; Jeong, B.; Hwang, I.; Ahn, J.-H.; et al. Epitaxial Growth of Thin Ferroelectric Polymer Films on Graphene Layer for Fully Transparent and Flexible Nonvolatile Memory. *Nano Lett.* **2015**, *16*, 334–340. [CrossRef]

199. Park, Y.J.; Kang, S.J.; Park, C.; Lotz, B.; Thierry, A.; Kim, K.J.; Huh, J. Molecular and Crystalline Microstructure of Ferroelectric Poly(vinylidene fluoride-co-trifluoroethylene) Ultrathin Films on Bare and Self-Assembled Monolayer-Modified Au Substrates. *Macromolecules* **2007**, *41*, 109–119. [CrossRef]

200. Kang, S.J.; Bae, I.; Shin, Y.J.; Park, Y.J.; Huh, J.; Park, S.-M.; Kim, H.-C.; Park, C. Nonvolatile Polymer Memory with Nanoconfinement of Ferroelectric Crystals. *Nano Lett.* **2010**, *11*, 138–144. [CrossRef]

201. Park, Y.J.; Kang, S.J.; Lotz, B.; Brinkmann, M.; Thierry, A.; Kim, K.J.; Park, C. Ordered Ferroelectric PVDF−TrFE Thin Films by High Throughput Epitaxy for Nonvolatile Polymer Memory. *Macromolecules* **2008**, *41*, 8648–8654. [CrossRef]

202. Zhang, J.-H.; Zhang, Y.; Sun, N.; Li, Y.; Du, J.; Zhu, L.; Hao, X. Enhancing output performance of triboelectric nanogenerator via large polarization difference effect. *Nano Energy* **2021**, *84*, 105892. [CrossRef]

203. Singh, H.H.; Khare, N. Improved performance of ferroelectric nanocomposite flexible film based triboelectric nanogenerator by controlling surface morphology, polarizability, and hydrophobicity. *Energy* **2019**, *178*, 765–771. [CrossRef]

204. Kim, M.; Park, D.; Alam, M.; Lee, S.; Park, P.; Nah, J. Remarkable Output Power Density Enhancement of Triboelectric Nanogenerators via Polarized Ferroelectric Polymers and Bulk MoS$_2$ Composites. *ACS Nano* **2019**, *13*, 4640–4646. [CrossRef]

205. Lee, J.H.; Hinchet, R.; Kim, S.K.; Kim, S.; Kim, S.-W. Shape memory polymer-based self-healing triboelectric nanogenerator. *Energy Environ. Sci.* **2015**, *8*, 3605–3613. [CrossRef]

206. Yu, Y.; Li, Z.; Wang, Y.; Gong, S.; Wang, X. Sequential Infiltration Synthesis of Doped Polymer Films with Tunable Electrical Properties for Efficient Triboelectric Nanogenerator Development. *Adv. Mater.* **2015**, *27*, 4938–4944. [CrossRef] [PubMed]

207. Huang, L.-B.; Bai, G.; Wong, M.-C.; Yang, Z.; Xu, W.; Hao, J. Magnetic-Assisted Noncontact Triboelectric Nanogenerator Converting Mechanical Energy into Electricity and Light Emissions. *Adv. Mater.* **2016**, *28*, 2744–2751. [CrossRef]

208. Kim, H.S.; Kim, D.Y.; Kim, J.; Kim, J.H.; Kong, D.S.; Murillo, G.; Lee, G.; Park, J.Y.; Jung, J.H. Ferroelectric-Polymer-Enabled Contactless Electric Power Generation in Triboelectric Nanogenerators. *Adv. Funct. Mater.* **2019**, *29*, 1905816. [CrossRef]

209. Park, I.W.; Choi, J.; Kim, K.Y.; Jeong, J.; Gwak, D.; Lee, Y.; Ahn, Y.H.; Choi, Y.J.; Hong, Y.J.; Chung, W.-J.; et al. Vertically aligned cyclo-phenylalanine peptide nanowire-based high-performance triboelectric energy generator. *Nano Energy* **2019**, *57*, 737–745. [CrossRef]

210. Singh, H.H.; Kumar, D.; Khare, N. Tuning the performance of ferroelectric polymer-based triboelectric nanogenerator. *Appl. Phys. Lett.* **2021**, *119*, 53901. [CrossRef]

211. Zheng, Z.; Yu, D.; Guo, Y. Dielectric Modulated Glass Fiber Fabric-Based Single Electrode Triboelectric Nanogenerator for Efficient Biomechanical Energy Harvesting. *Adv. Funct. Mater.* **2021**, *31*, 2102431. [CrossRef]

212. Liu, L.; Yang, X.; Zhao, L.; Xu, W.; Wang, J.; Yang, Q.; Tang, Q. Nanowrinkle-patterned flexible woven triboelectric nanogenerator toward self-powered wearable electronics. *Nano Energy* **2020**, *73*, 104797. [CrossRef]

213. Kim, Y.; Wu, X.; Lee, C.; Oh, J.H. Characterization of PI/PVDF-TrFE Composite Nanofiber-Based Triboelectric Nanogenerators Depending on the Type of the Electrospinning System. *ACS Appl. Mater. Interfaces* **2021**, *13*, 36967–36975. [CrossRef]

214. Sharma, S.; Kumar, M. Band gap tuning and optical properties of BiFeO$_3$ nanoparticles. *Mater. Today Proc.* **2020**, *28*, 168–171. [CrossRef]

215. Qi, J.; Ma, N.; Ma, X.; Adelung, R.; Yang, Y. Enhanced Photocurrent in BiFeO$_3$ Materials by Coupling Temperature and Thermo-Phototronic Effects for Self-Powered Ultraviolet Photodetector System. *ACS Appl. Mater. Interfaces* **2018**, *10*, 13712–13719. [CrossRef]

216. Dhar, A.; Mansingh, A. Optical properties of reduced lithium niobate single crystals. *J. Appl. Phys.* **1990**, *68*, 5804. [CrossRef]

217. Wiesendanger, E. Dielectric, mechanical and optical properties of orthorhombic KNbO$_3$. *Ferroelectrics* **1973**, *6*, 263–281. [CrossRef]

218. Grinberg, I.; West, D.V.; Torres, M.F.; Gou, G.; Stein, D.M.; Wu, L.; Chen, G.; Gallo, E.M.; Akbashev, A.R.; Davies, P.K.; et al. Perovskite oxides for visible-light-absorbing ferroelectric and photovoltaic materials. *Nature* **2013**, *503*, 509–512. [CrossRef] [PubMed]

219. Fridkin, V.M. Photoferroelectrics. *Acta Cryst.* **1980**, *36*, 1091–1092. [CrossRef]

220. Fridkin, V.M. Review of recent work on the bulk photovoltaic effect in ferro and piezoelectrics. *Ferroelectrics* **1984**, *53*, 169–187. [CrossRef]

221. Fridkin, V.M. Boltzmann principle violation and bulk photovoltaic effect in a crystal without symmetry center. *Ferroelectrics* **2016**, *503*, 15–18. [CrossRef]

222. Spanier, J.; Fridkin, J.E.S.V.M.; Rappe, A.M.; Akbashev, A.R.; Polemi, A.; Qi, Y.; Gu, Z.; Young, S.M.; Hawley, C.J.; Imbrenda, J.E.S.Z.G.D.; et al. Power conversion efficiency exceeding the Shockley–Queisser limit in a ferroelectric insulator. *Nat. Photonics* **2016**, *10*, 611–616. [CrossRef]

223. Bai, Y.; Vats, G.; Seidel, J.; Jantunen, H.; Juuti, J. Boosting Photovoltaic Output of Ferroelectric Ceramics by Optoelectric Control of Domains. *Adv. Mater.* **2018**, *30*, 1803821. [CrossRef] [PubMed]

224. He, J.; Franchini, C.; Rondinelli, J.M. Ferroelectric Oxides with Strong Visible-Light Absorption from Charge Ordering. *Chem. Mater.* **2016**, *29*, 2445–2451. [CrossRef]

225. Han, F.; Zhang, Y.; Yuan, C.; Liu, X.; Zhu, B.; Liu, F.; Xu, J.; Zhou, C.; Wang, J.; Rao, G. Photocurrent and dielectric/ferroelectric properties of KNbO$_3$–BaFeO$_3$-δ ferroelectric semiconductors. *Ceram. Int.* **2020**, *46*, 14567–14572. [CrossRef]

226. Alkathy, M.S.; Lente, M.H.; Eiras, J. Bandgap narrowing of Ba$_{0.92}$Na$_{0.04}$Bi$_{0.04}$TiO$_3$ ferroelectric ceramics by transition metals doping for photovoltaic applications. *Mater. Chem. Phys.* **2020**, *257*, 123791. [CrossRef]

227. Jiménez, R.; Ricote, J.; Bretos, I.; Riobóo, R.J.J.; Mompean, F.; Ruiz, A.; Xie, H.; Lira-Cantú, M.; Calzada, M.L. Stress-mediated solution deposition method to stabilize ferroelectric BiFe1-xCrxO$_3$ perovskite thin films with narrow bandgaps. *J. Eur. Ceram. Soc.* **2021**, *41*, 3404–3415. [CrossRef]

228. Chakrabartty, J.; Harnagea, C.; Celikin, M.; Rosei, F.; Nechache, R. Improved photovoltaic performance from inorganic perovskite oxide thin films with mixed crystal phases. *Nat. Photonics* **2018**, *12*, 271–276. [CrossRef]

229. Jiang, C.-S.; Noufi, R.; Abushama, J.A.; Ramanathan, K.; Moutinho, H.R.; Pankow, J.; Al-Jassim, M.M. Local built-in potential on grain boundary of Cu(In,Ga)Se$_2$ thin films. *Appl. Phys. Lett.* **2004**, *84*, 3477–3479. [CrossRef]

230. Tian, M.; Li, Y.; Wang, G.; Hao, X. Large photocurrent density in polycrystalline hexagonal YMnO$_3$ thin film induced by ferroelectric polarization and the positive driving effect of grain boundary. *Sol. Energy Mater. Sol. Cells* **2021**, *224*, 111009. [CrossRef]

231. Tu, C.-S.; Chen, P.-Y.; Jou, Y.-S.; Chen, C.-S.; Chien, R.; Schmidt, V.H.; Haw, S.-C. Polarization-modulated photovoltaic conversion in polycrystalline bismuth ferrite. *Acta Mater.* **2019**, *176*, 1–10. [CrossRef]

232. Mai, H.; Lu, T.; Li, Q.; Sun, Q.; Vu, K.; Chen, H.; Wang, G.; Humphrey, M.G.; Kremer, F.; Li, L.; et al. Photovoltaic Effect of a Ferroelectric-Luminescent Heterostructure under Infrared Light Illumination. *ACS Appl. Mater. Interfaces* **2018**, *10*, 29786–29794. [CrossRef]

233. Pei, W.; Chen, J.; You, D.; Zhang, Q.; Li, M.; Lu, Y.; Fu, Z.; He, Y. Enhanced photovoltaic effect in Ca and Mn co-doped BiFeO$_3$ epitaxial thin films. *Appl. Surf. Sci.* **2020**, *530*, 147194. [CrossRef]

234. Hassan, A.K.; Ali, G.M. Thin film NiO/BaTiO$_3$/ZnO heterojunction diode-based UVC photodetectors. *Superlattices Microstruct.* **2020**, *147*, 106690. [CrossRef]

235. Yan, J.-M.; Wang, K.; Xu, Z.-X.; Ying, J.-S.; Chen, T.-W.; Yuan, G.-L.; Zhang, T.; Zheng, H.-W.; Chai, Y.; Zheng, R.-K. Large ferroelectric-polarization-modulated photovoltaic effects in bismuth layered multiferroic/semiconductor heterostructure devices. *J. Mater. Chem. C* **2021**, *9*, 3287–3294. [CrossRef]

236. Guo, K.; Zhang, R.; Fu, Z.; Zhang, L.; Wang, X.; Deng, C. Regulation of Photovoltaic Response in ZSO-Based Multiferroic BFCO/BFCNT Heterojunction Photoelectrodes via Magnetization and Polarization. *ACS Appl. Mater. Interfaces* **2021**, *13*, 35657–35663. [CrossRef]

237. Renuka, H.; Joshna, P.; Venkataraman, B.H.; Ramaswamy, K.; Kundu, S. Understanding the efficacy of electron and hole transport layers in realizing efficient chromium doped BiFeO$_3$ ferroelectric photovoltaic devices. *Sol. Energy* **2020**, *207*, 767–776. [CrossRef]

238. Zhang, J.; Xue, W.; Chen, X.-Y.; Hou, Z.-L. Sm doped BiFeO$_3$ nanofibers for improved photovoltaic devices. *Chin. J. Phys.* **2020**, *66*, 301–306. [CrossRef]

239. Zhao, K.; Ouyang, B.; Bowen, C.R.; Yang, Y. Enhanced photocurrent via ferro-pyro-phototronic effect in ferroelectric BaTiO$_3$ materials for a self-powered flexible photodetector system. *Nano Energy* **2020**, *77*, 105152. [CrossRef]

240. Zhao, K.; Ouyang, B.; Yang, Y. Enhancing Photocurrent of Radially Polarized Ferroelectric BaTiO$_3$ Materials by Ferro-Pyro-Phototronic Effect. *iScience* **2018**, *3*, 208–216. [CrossRef]

241. Zhang, Y.; Su, H.; Li, H.; Xie, Z.; Zhang, Y.; Zhou, Y.; Yang, L.; Lu, H.; Yuan, G.; Zheng, H. Enhanced photovoltaic-pyroelectric coupled effect of BiFeO$_3$/Au/ZnO heterostructures. *Nano Energy* **2021**, *85*, 105968. [CrossRef]

the efficiency of TENGs that are modified through chemical and physical methods, with material selection and structural design.

Within the review, various engineering strategies that improve output performance of TENG will be introduced accordingly, as shown in Figure 1. The first and second parts are a brief introduction and performance of TENG. The third part includes surface engineering methods that play a big role in the TENGs output, which are divided into two parts: surface morphology and surface modification. The strategies for surface morphology and surface modification to enhance the TENG output will be introduced. Finally, perspective and outline for future application in designing high performance of TENGs will be introduced in part four.

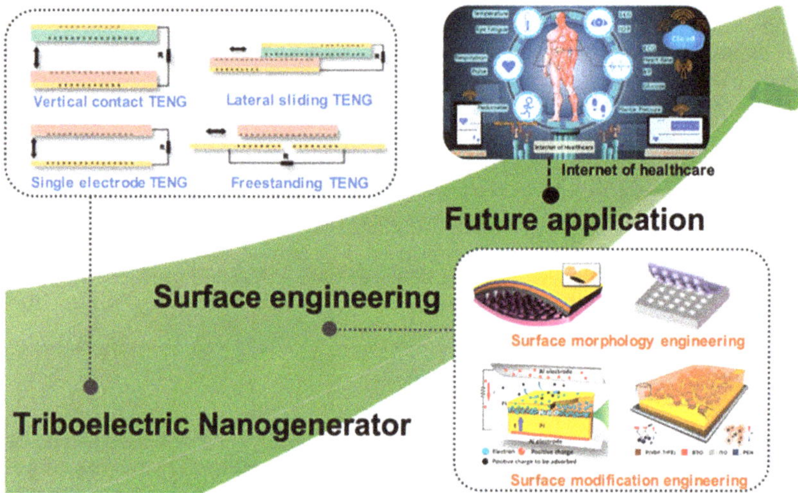

Figure 1. Schematic diagram showing the Triboelectric Nanogenerator (TENG), surface engineering, and application [65–69]. Copyright, 2012 American Chemical Society. Copyright, 2012 American Chemical Society. Copyright, 2016 WILEY-VCH Verlag GmbH & Co. KGaA, Weinheim, Germany. Copyright, 2017 Elsevier Ltd., Amsterdam, The Netherlands.

2. Triboelectric Nanogenerator

2.1. Origin of Triboelectric Nanogenerator

The scientific term for TE is contact electrification (CE) phenomenon, in which it results from physical contact between two materials without rubbing against each other and the transfer of electrons from one material to another. With this process of charge transfer during contact and separation, the surface charge density, polarization, and strength of the charges are strongly dependent on the materials, which helps in understanding the TE mechanism. Since the invention of TENG, Wang et al. have devoted many efforts in the analysis of the origin of triboelectric nanogenerator [18]. Research indicates that electron transfer is the dominant mechanism for CE between solid-solid pairs, and the electron transfer model can be extended to other pairs.

2.2. Choice of Materials

All known materials such as wood, polymers, metals, and silk show the effect of triboelectrification. All materials can be applied for TENG fabrication. The material selection for TENG is large because the original reason is gaining/losing electrons that are dependent on the polarity difference of material pairs. John Carl Wilcke published a paper in 1757 on static charges about the first triboelectric series [70]. The material at the bottom of the chain, when touching a material close to the upper of the chain, will get a negative charge. As such, materials that accept electrons tend to obtain a negative charge, while those that lose electrons become positively charged [71]. When choosing triboelectric material, it is important to consider the surface charge matching of two tribo-layers. This means that we need to choose two materials with quite different triboelectric series, and if we choose the materials with similar triboelectric series, the output performance will be influenced to a large extent.

3. Surface Engineering Methods

The surface charge density of triboelectric materials is the key point that leads to the high output performance of TENG [72]. The surface charge density is determined in the triboelectric layer via the electron-donation or accepting ability. This section reviews many functionalization techniques for improving the TENGs output performance.

3.1. Surface Morphology

Triboelectric material morphology engineering refers to the modification of the material surface. To increase the contact area between friction layers in micro/nano surface patterns, morphological engineering is used to improve the output performances in triboelectric energy harvesters, therefore increase the effective charge density. The physical surface modification is done through the introduction of the micro/nano surface on the electro-frictional layers by using different mechanical methods, such as soft lithography [73–75], photolithography [76–78], and ultrafast laser patterning [79–81].

3.1.1. Soft Lithography

Researchers have developed micro/nano structures for growing the roughness, which has been introduced to improve the TENG output.

Wang et al. [67] proposed the rational design of an arch-shaped structure based on the contact electrification between a polymer and a metal film. In Figure 2a, the structural process of the arch-shaped triboelectric nanogenerator relies on the electrical contact between patterned polydimethylsiloxane (PDMS) and patterned Al foil on the top and the bottom plate, respectively. The patterned surfaces of PDMS film and Al foil are fabricated to enhance the triboelectric charging, and they are characterized using scanning electron microscopy (SEM). Both arrays are uniform and regular across a very large area in Figure 2b. The working mechanism of the TENG was studied by finite element simulation. According to the TENG output voltage and current at a range of frequency (2–10 Hz) in Figure 2c, noted that with the triggering frequency of 6 Hz and controlled amplitude, the accumulation of the triboelectric charges increases and reaches equilibrium in a certain time after multiple cycles. Then, the output will gradually go up in the first stage, where generated 230 V, 15.5 μAcm^{-2}, and energy volume density reached 128 $mWcm^{-3}$, while the energy efficiency is produced 10−39%.

To further improve the output performance, Zhang et al. [82] proposed dual scale structures on TENG output. As schematically presented in Figure 2d, a sandwich-shaped triboelectric nanogenerator is composed of a 20 μm thick aluminum film with 450 μm thick PDMS film fixed with elastic tape between two surface micro/nanostructured to form a sandwich-shaped structure. On the top, the PDMS film fabricated 125 μm thick PET/ITO thin film. The well-designed micro/nano dual-scale structure (i.e., pyramids and V-shape grooves) is fabricated by using photolithography and KOH wet etching at the top of the PDMS surface. As shown in Figure 2e, the output peak voltage, current density, and power density achieved 465 V, 13.4 μAcm^{-2}, and 53.4 $mWcm^{-3}$, respectively. Noted that the pure micro-structures or nanostructures can decrease the roughness faster than the dual-scale structure, hence strengthening the performance of TENG. Moreover, the voltage and the current micro/nano dual-scale with pyramid, compared to that of flat PDMS film, enhanced by 100% and 157%. Likewise, Kim et al. [61] suggested a large area nano patterning technique, where block copolymer (BCP) by lithography technique was introduced on a flexible gold substrate. As illustrated in Figure 2f, Au (100 nm) is deposited on cleaned Kapton film by thermal evaporation of the metal layer. After lifting the BCP template, an Au nanodot was formed on the surface of the Au. As shown in Figure 2g, the generated short-circuit currents and open-circuit voltages from the flat Au are 22 µA and 55 V, respectively. Comparatively, the generated short-circuit currents and open-circuit voltages from nanopatterned Au are 82 µA and 225 V, respectively. The Nanopatterned Au contact surface has a larger effective contact area than the flat Au surface; more electric charge is induced by contact-electrification, dramatically enhancing the output performance with an output power density of 93.2 Wm^{-2}. After BCP nanopatterning, TENG output currents improved up to 16 times due to the increased contact area.

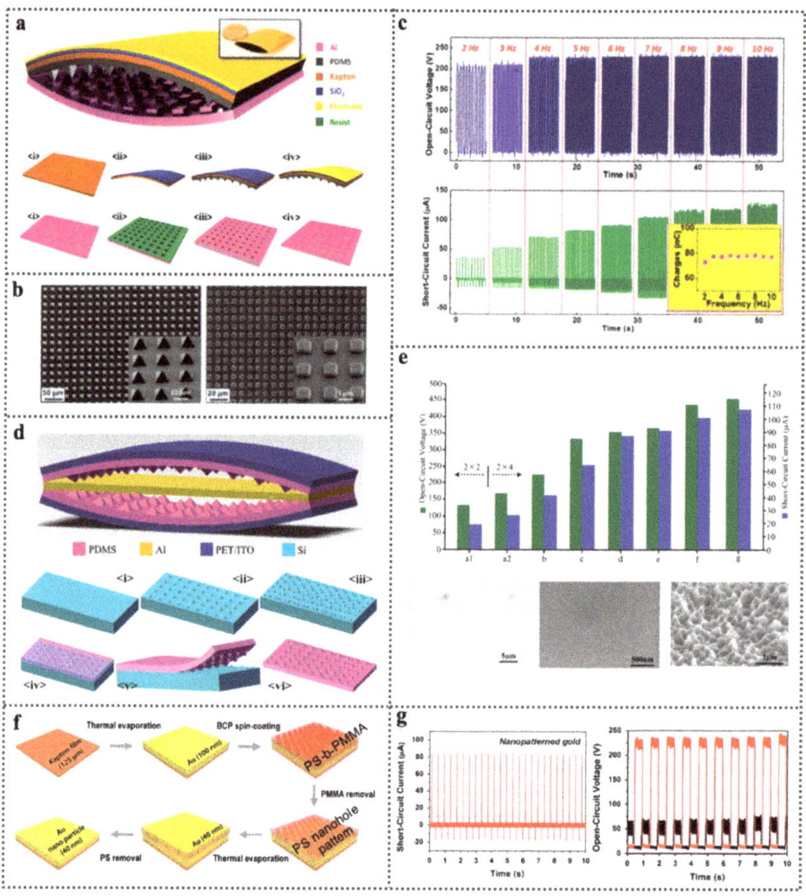

Figure 2. (**a**) Schematic illustration of arch-shape based on TENG. (**b**) SEM images of pyramid and cubic patterns of polydimethylsiloxane (PDMS) with Al surface. (**c**) The electrical voltage and the current from each frequency of the TENG, which gives the total charges transferred in a half cycle [67] Copyright, 2012 American Chemical Society. (**d**) Structural design of the sandwich-shaped triboelectric nanogenerator. (**e**) The output performance of the sandwich-shape based on TENG with flat PDMS film, surface-nanostructured PDMS film, pyramids and V-shape grooves [82] Copyright, 2013 American Chemical Society. (**f**) Au nanopatterning process and fabrication process of TENGs by using block copolymer (BCP) lithography. (**g**) Open-circuit current and voltage comparison between Au nanopatterned and flat Au [61] Copyright, 2015 Elsevier Ltd.

A new design can develop the output of the nanogenerator efficiency by increasing the triboelectric effect and the capacitance change, where Fen ta al [65] proposed the first thin film micro patterned PDMS with different features such as patterned lines, cubes, and pyramids as triboelectric contact layers to improve the output of TENG. As illustrated in Figure 3a–c, during the fabrication process, Si wafer molds were fabricated by the traditional photolithography method to make patterned PDMS as a friction layer with various features including lines, cubes, and pyramids. The PDMS film was fixed on the surface of a clean ITO-coated polyester (PET) substrate and wrapped with the other ITO-coated PET film to form a structured sandwich device. Comparison of the open-circuit voltage and current is as illustrated in Figure 3d. The results reveal that the maximum output voltage is up to 18 V and current is 0.7 µA for pyramid-surface structure, which is four times compared to TENG using flat films. Similar to this work, various physical surface modification has been used in micro/nano structures [27,83–86]. For example, Sun et al. made micro-nano structures on the PDMS by leaves mold with rich surface textures [85]. Besides, the micro-structured PDMS film can also be made through low surface energy sandpaper template without the use of surfactant coating or high vacuum [86].

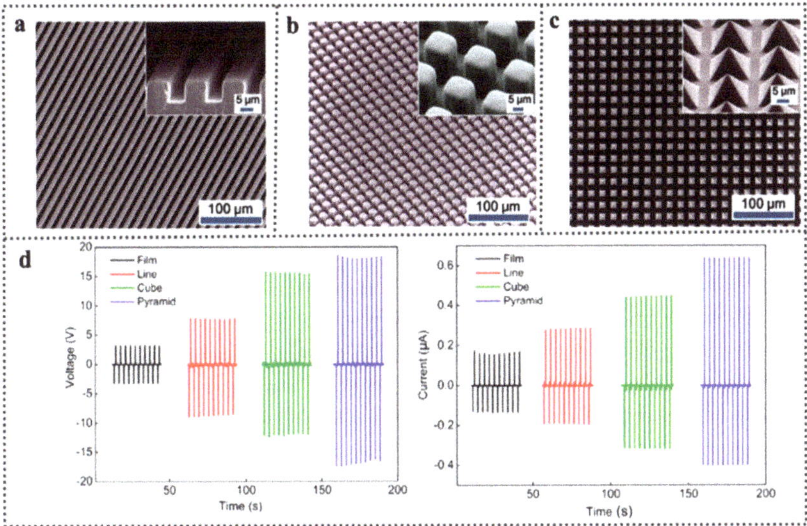

Figure 3. (**a–c**) Scanning electron microscope images of photolithography traditional patterned PDMS as a friction layer with (**a**) lines, (**b**) cubes, and (**c**) pyramids. (**d**) Triboelectric voltage and current with a flat surface and various patterned features device [65] Copyright, 2012 American Chemical Society.

3.1.2. Ultrafast Laser Patterning

Ultrafast laser irradiation is used to control in the PDMS surface morphology by using it for micro/nano hierarchical structures fabrication. Femtosecond laser offers a flexible method in open air for mask-free fabrication of micro/nano structures, due to its short irradiation period and superior intensity [87–89].

The femtosecond laser was first used in producing micro/nano structures on polydimethylsiloxane (PDMS) film by Kim et al. [90]. After femtosecond laser direct writing, the effective contact area between the friction layer is improved, thus improving the output performance of TENG. As illustrated in Figure 4a, a schematic illustration of the ultrafast laser irradiation method was used on the PDMS surface patterning. The TENG is composed of PDMS and aluminum. Aluminum is used as the top electrode. The counter-electrode in the PDMS is another aluminum, which is connected from the back side of PDMS. In this method, the LI-TENG using a laser power of 29 mW in the PDMS patterned to produce a maximum voltage and current with a power density of 42.5 V and 10.1 µA, 107.3 µWcm^{-2}, respectively, as shown in Figure 4b. The largest enhancement in output power, based on the micro structured PDMS, was raised more than two times from the controlled TENG (bare PDMS). Such a fast, direct-writing approach has the potential to make controlled hierarchical micro/nanostructures on different materials.

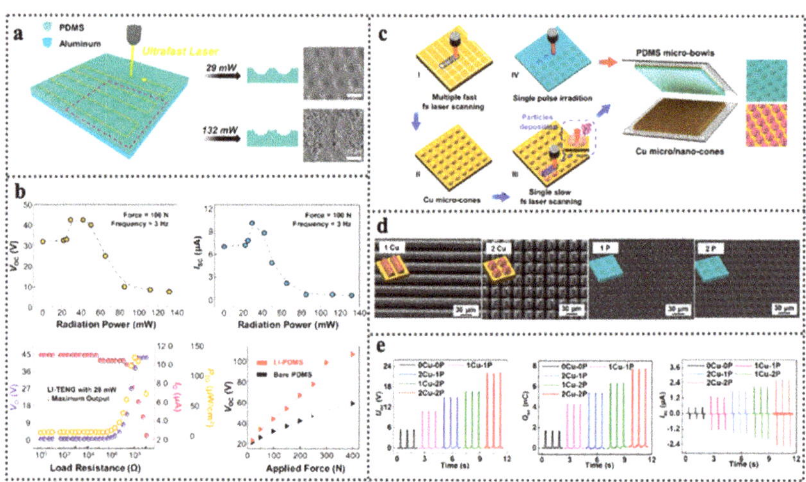

Figure 4. (**a**) Fabrication process of PDMS by using a laser power of 29 mW. (**b**) Electrical output performance of LI-TENGs using laser power from 0 to 132 mW [90] Copyright, 2017 Elsevier Ltd. (**c**) Schematic illustration of TENGs by Femtosecond laser direct writing processes. (**d**) SEM images of Cu and PDMS films with micro/nano-stripes, cones and small/large size micro-bowls, structures. (**e**) Transferred charge density, voltage, and current density with and without various micro/nanostructures [91] Copyright, 2019 Elsevier Ltd.

Except for flexible PDMS, writing directly on Cu can also enhance the output performance. Huang et al. [91] proposed micro/nano structures optimized for TENGs using the method of writing directly on both triboelectric layers by a femtosecond laser such as Cu and PDMS. As shown in Figure 4c, to make micro-bowl structures single-pulse irradiation in different sizes was used on PDMS surfaces. The micro/nano dual-scale structures on Cu film surfaces are fabricated using laser scanning technology in cones and stripes, as illustrated in Figure 4d. TENGs are fabricated with the contact area and contact distance of 8×8 mm^2 and 1 cm, respectively. The fabricated TENG is tested under the frequency of 1.5 Hz and contact forces at 2 N. Figure 4e shows the voltage (V_{OC}), charge (Q_{SC}), and current (I_{SC}) of the TENGs. The results showed that all TENG output performances with different micro/nano structures significantly increased, compared with TENG (0Cu-0PDMS) without micro/nano structures that follows the sequence: TENG$_{2Cu–2P}$ > TENG$_{1Cu–2P}$ > TENG$_{2Cu–1P}$ > TENG$_{1Cu–1P}$ > TENG$_{0Cu–0P}$. The TENG$_{2Cu–2P}$ with cones and bowls micro/nano structures on Cu surface in a larger size on PDMS surface has the optimal performances. The TENG$_{2Cu–2P}$ produced a voltage of 22.04 V, which is improved by 4.13 times of the TENG$_{0Cu–0P}$, indicating the effective surface morphology engineering.

Except for lithography and laser patterning, there are other physical methods that can fabricate microstructures such as reactive ion etching (RIE) and electrospinning. These methods can also improve the output performance. For example, Zhai et al. [92] enhanced the output performance of freestanding mode TENG by treating the triboelectric PTFE film with RIE, but these methods have the drawback that it is difficult to control the shape of the microstructure. So, it is just regarded as a pre-processing step in most research, but it is undeniable that this can indeed improve the output performance of TENG.

3.2. Surface Modification

3.2.1. Chemical Functional Groups

In the literature, researchers have used various techniques of surface functionalization to improve the TENG output. To study the effect of chemical functional groups on the output performance of TENG, several studies have focused on introducing functional groups that improve the triboelectric charge transfer, such as self-assembled monolayer (SAM) techniques. As shown in Figure 5a, Wang et al. [62] analyzed chemical surface functionalization by using scanning Kelvin probe microscopy (SKPM), which demonstrated across the major group of SAMs by varying the surface potential. There were four head functional groups (OH, COOCH$_3$, NH$_2$, and Cl) through thiol- based SAM treatment for 12 h, forming in solutions containing different types of thiols on the Au surface. As a result, shown in Figure 5b, the largest enhancement in electrical output with amine groups was raised more than four times from the contact electrification with FEP. The increased output power, after functionalization with these groups, follows the sequence order: hydroxyl, ester, and amine from 4-amino thiophenol. The chlorinate functional group lead to a decrease of the TENG's output compared to the pristine material, due to the lower charge density. Besides, as shown in Figure 5c, SAM surface functionalization of SiO$_2$ acts as the triboelectric layer in TENGs. After being deposited by plasma chemical vapor deposition on a glass substrate, one kind of silane molecule(3-aminopropyl) triethoxysilane (APTES) was used to form a SAM via three Si-O bonds for each molecule. After functionalization in 10% APTES concentration, transferred charge density, the open-circuit voltage, and short-circuit current density reached 51 μCm^{-2}, 240 V, and 1.75 mAm^{-2}, respectively, as illustrated in Figure 5d. Compared with that produced by 1% APTES concentration, SiO$_2$ has higher chemical potential and can generate more positive charge during contact. So, a higher concentration from APTES is better to improve the output performance.

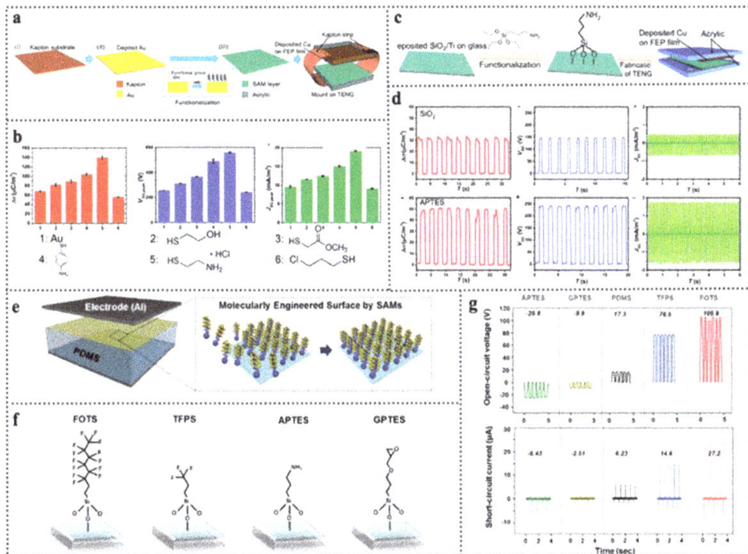

Figure 5. (**a**) Schematic illustration of TENGs built using SAM layer based on Au films. (**b**) The output comparisons of TENG based on each thiol SAM functionalized Au films. (**c**) Fabrication process of TENG using silane -SAM modified SiO$_2$. (**d**) the output of TENG with and without functionalized silica [62] Copyright, 2016 Royal Society of Chemistry. (**e**) Schematic illustration of TENG (inset figure shows the molecularly engineered surface by self-assembled monolayer (SAMs)). (**f**) Chemical structures of self-assembled monolayers (METS). (**g**) The voltage and current with and without METS [66] Copyright, 2015 American Chemical Society.

After demonstrating the effect of SMA on output performance, researchers began to perform more tests on the effect of different kinds of SAMs. Song et al. [66] modified the triboelectric surfaces molecularly engineered by different self-assembled monolayers, such as 3-aminopropyltriethoxysilane (APTES), 3-glycidoxypropyl- triethoxysilane (GPTES), 1H,1H,2H,2H-perfluorooctyltrichloro-silane (FOTS), and trichloro (3,3,3-trifluoropropyl) silane (TFPS) to study the effect of SAMs on the triboelectric properties, shown in Figure 5e. After contact with aluminum foil at the same frequency and force, open circuit voltage (V_{OC}) and the short circuit current (I_{SC}) of APTES, GPTES, FOTS, and TFPS were measured, which is illustrated in Figure 5f. For the PDMS surface treated with either TFPS or FOTS, the open circuit voltage is (105 V) and open circuit current is (27.2 μA) based on FOTS-METS device are six and four times higher than that of pure PDMS/Al device. However, after comparing the V_{OC} and I_{SC} of FOTS, both highest values of the TFPS-METS device decreased. This is due to the fluorine atoms that have lower surface density. On the other hand, when the PDMS surface was treated with either APTES and GPTES, V_{OC} and I_{SC} have negative values for APTES and GPTES confirmed positive charge on the surface of PDMS with these SAMs. As a result, the output performance of TENG was dependent upon end-functional groups of SAMs that formed on PDMS substrate.

Likewise, Byun et al. [93] also proposed the triboelectric series modified by SAM-materials through controlling the surface dipoles and surface electronic states with different electron-donating and withdrawing functional groups. As shown in Figure 6a, the substrate was engineered with four SAM functional groups such as -NH$_2$, -SH, -CH$_3$, and -CF$_3$. After chemical surface modification, the surface potential increased when using the highly electronegative atom F, and decreased when using -NH$_2$, -SH, and -CH$_3$ groups because they electron donated to the metal. KPFM mode in atomic force microscopy (AFM) measurements were used to verify the triboelectricity affected by the surface modification [94]. The charge numbers on the various SAM-modified substrates were explained by contact potential difference (CPD). The surface potential of the substrate, $\Phi_{substrate}$, is defined as

$$\Phi_{substrate} = \Phi_{probe} - e \cdot V_{CPD} \tag{1}$$

where Φ_{probe} is the work function of the probe, V_{CPD} is the measured CPD, and e is the electronic charge. Then, CPD images of various SAM-modified substrates are contracted by an Rh-coated AFM probe. As shown in Figure 6b, the surface potential of NH$_2$-SiO$_2$ is four times higher than that of CH$_3$-SiO$_2$, due to withdrawing group, and the CF$_3$-SiO$_2$ had the opposite polarity of that of CH$_3$-SiO$_2$. Figure 6c shows that when the surface potential increased, also the triboelectric potential increased before contact situation increased. Therefore, the amount and polarity of the electric friction charge were affected by the surface potential. Moreover, the surface modification affected the charge diffusion and thus improved the output of the TENG. In summary, surface modification by these SAM-based contact materials is an effective way of materials application.

In another example, Shin et al. [95] reported the effect of surface modification on PET films after being modified with oxygen plasma and forming reactive –OH groups. Then, the surface was functionalized with nonpolar (–CF$_3$) groups from trichloro (1H,1H,2H,2H-perfluorooctyl) silane (FOTS) vapor which introduced negatively charged and (-NH^{3+}) groups coming from poly-l-lysine solution by using two different techniques (Figure 6d). V_{OC} and J_{SC} of the TENG with PLL treated PET (P-PET) and FOTs treated PET (F-PET) contact pair shown in Figure 6e, showing high V_{OC} and J_{SC} reaching to 330 V and 270 mAm^{-2}, respectively, compared to that between PET and the PET contact pair. These improvements can be explained as follows: -CF$_3$ in F-PET improved negative triboelectric charges and was able to gain electrons while -NH^{3+} in P-PET lost electrons and increased positive triboelectric charges. X-ray photoelectron spectroscopy (XPS), atomic force microscopy (AFM), and scanning Kelvin probe force microscopy (KPFM) methods were used to analyze the surface modification on PET surfaces for a deeper understanding and perfect the performance of TENGs. The analysis of surface potential is useful to characterize the surface materials modified through chemical treatment [96–99].

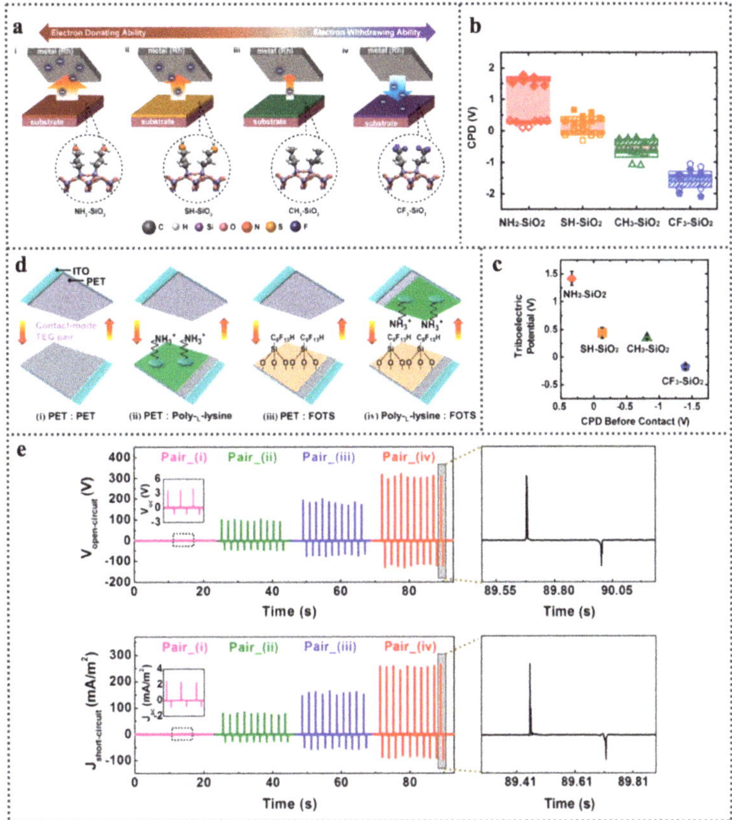

Figure 6. (**a**) Schematic illustration of the SAM-modified substrates from donor electron to acceptor electron layers. (**b**) Contact potential difference (CPD) analysis of each substrate. (**c**) Triboelectric potentials and CPD before contact of the different surfaces [93] Copyright, 2016 American Chemical Society. (**d**) The effect of surface modification on PET films with various contact pairs of the TENGs. (**e**) Voltage output and current densities of the contact pairs of the TENG [95] Copyright, 2015 American Chemical Society.

3.2.2. Ion Injection

Ion injection is an effective way for chemical surface modification via injecting negative ions onto a fluorinated ethylene propylene (FEP) surface to improve output performance of TENG and systematically study the maximum surface charge density (MSCD) of FEP polymer. The electric friction charge density is limited to the MSCD level, which is determined by the electric field of air breakdown in the nearby region, due to the fact that both theoretical analysis and experimental study of the MSCD are highly desirable.

Wang et al. [60] reported a technique using an air ionization gun to inject positive and negative charges into the friction layer and hence enhance TENG performance. As shown in Figure 7a, ionized-air gun produced ions with both poles (negative ions), such as CO_3^-, NO_3^-, NO_2^-, O_3^-, and O_2^- into the top surface of fluorinated ethylene propylene (FEP). Coulomb scale was used in the measurement of the charge flow on the surface of the injected polymer, and it can be monitored to study the MSCD of FEP film. As shown in Figure 7b, the density of charges on the FEP surface reached ~40 μCm^{-2} during the ion injection period, and the process of charge transfer indicated the same amount of negative charge on the FEP surface accumulation. When the ion injection was repeated multiple times, the negative charges injected onto the FEP surface accumulated step by step. After 17 times of injection, the charge density reached to ~630 μCm^{-2}. As shown in Figure 7c, ion injection steps in a TENG to enhance the surface charge density has also been studied. With no injection, the initial short-circuit charge density was about 50 μCm^{-2} and charge density increased to ~100 μCm^{-2} after one ion injection process. After that, the short circuit charge density reached ~240 μCm^{-2} after five injections. Further increase of the times of the ion injection only resulted in a very small amount of the short-circuit charge density enhancement, but after nine injections, the charge suddenly transfers differently. When Al was separated from the FEP film, the short circuit charge density dropped to ~230 μCm^{-2} because the of the air breakdown that was caused by a voltage between FEP polymer and Al plate. As shown in Figure 7d, the FEP layer was in contact with the Al sheet before injection, and the open-circuit voltage of the device only produced 200 V. After ion injection, the surface charge density arrived at the maximum; the open circuit voltage increased to ~1000 V. For comparison, it is clear that the enhancement of electrical output V_{OC} and J_{SC} before and after ion injection increased around four to five times.

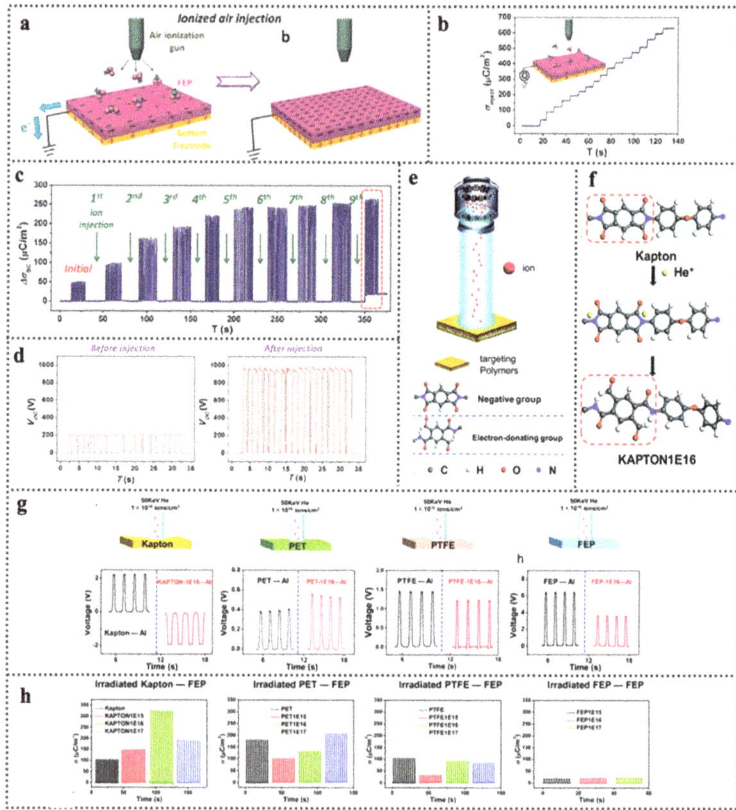

Figure 7. (**a**) Basic processes on FEP surface with ionized air injection. (**b**) The charge flow measurement in the FEP film during the ion injection. (**c**) The short-circuit charge density ($\Delta\sigma_{SC}$) generated during the step-by-step ion injection process. (**d**) Open-circuit voltages of the TENG before and after the process of ion injection [60] Copyright, 2014 WILEY-VCH Verlag GmbH & Co. KGaA, Weinheim. (**e**,**f**) Schematic presentation of the ion irradiation process and corresponding to the molecular formula of Kapton structure change. (**g**,**h**) The output voltages and transferred charge densities of the pre and post -irradiation polymers with various materials [97] Copyright, 2016 Royal Society of Chemistry.

In a more recent study, Li et al. [97] proposed a surface modification method induced by the ion irradiation., and the functional groups of triboelectric materials at the molecular level won't arise change to the surface roughness of polymers. As shown in Figure 7e,f, the ion irradiation process was carried out by using a 50 keV ion beam to irradiate the four polymers which are Kapton, PET, PTFE, and FEP. The Helium ion was chosen as the implantation ion. The electrification performances of the various irradiated polymers were studied by using an irradiation dose applied to four polymers. The triboelectric materials were irradiated polymers and contacted with Al foil, which acts as the second triboelectric material, with a contact area of 7 mm × 7 mm. As shown in Figure 7g, the output voltage of Kapton–Al changes from positive to negative after the treatment with He-irradiation due to the donor electrons from the KAPTON1E16 polymer. There is a slight difference in voltages of the other devices (PET-Al, PTFE-Al, and FEP-Al). As illustrated in Figure 7h, the transferred charge became maximum with KAPTON1E16 (1 × 1016 ions per cm^2). There was no enhancement in the charge density of the other irradiated polymers. So, the irradiated Kapton was the best polymer to enhance the performance of TENG due to the electron-donating capability of Kapton after He-ion irradiation. This approach is the highest compared to reported results using other methods such as ICP etching [98], charge pump [99], or a prior charge injection method [100].

3.2.3. Fluorinated Polymers

Another way of surface modification is using fluorinated polymers on the surface of materials because it has many advantages, such as low surface energy and good electrical properties [101]. Engineering the dielectric properties of fluorinated materials could enhance the triboelectric measurements [102,103]. Polyvinylidene fluoride (PVDF), or polytetrafluoroethylene (PTFE) derivatives, are examples of fluorinated polymers and are used as one of tribo-negative materials for TENG, and many studies also used these materials to enhance the TENG output [54,104–106]. Kim et al. [103] demonstrated that the molecular structure engineering of fluorinated polymers with a controlled fluorine unit from zero to three fluorine units and molecular weight (M_w) such as for [poly (ethyl methacrylate) to poly(2,2,2-trifluoroethyl methacrylate)], and fluorine unit is over three such as [poly (2,2,3,3,3-pentafluoropropyl methacrylate and poly (2,2,3,3,4,4,4-heptafluorobutyl methacrylate)]. As shown in Figure 8a, the effect of the dielectric constants of materials with polarity is investigated. The expression for the dielectric constant (relative permittivity) of the PET-ITO substrates is given by

$$\varepsilon_r = Cd/\varepsilon_0 \cdot A \tag{2}$$

where C is the capacitance, d is thickness of the dielectric layer, A is measured area, and ε_0 is vacuum permittivity. According to this equation, as shown in Figure 8b, when the fluorine units are zero to three in the polymer, the dielectric constant of the fluorinated polymers increase and when the fluorine units are more than three, the dielectric constant of the fluorinated polymers slightly decreases. The slight decrease of the relative dielectric constant can be attributed to the additional polar group in the polymer chain and the efficient polymer chain packing within the polymer films. Figure 8c illustrates the highest triboelectric output performances of PTF containing three fluorine units, which is much bigger than the PEMA polymer without fluorine units. These results are attributed to the effective polymer chain packing structure. As shown in Figure 8d, the annealed PTF polymer possessed a higher triboelectric performance than the PTF polymer without annealing. Therefore, the annealing temperature is critical in improving the triboelectric output performance resulting from polymer chain packing. A similar output enhancement was also demonstrated by silane-based SAMs fluorinated molecules that improve the triboelectric charge.

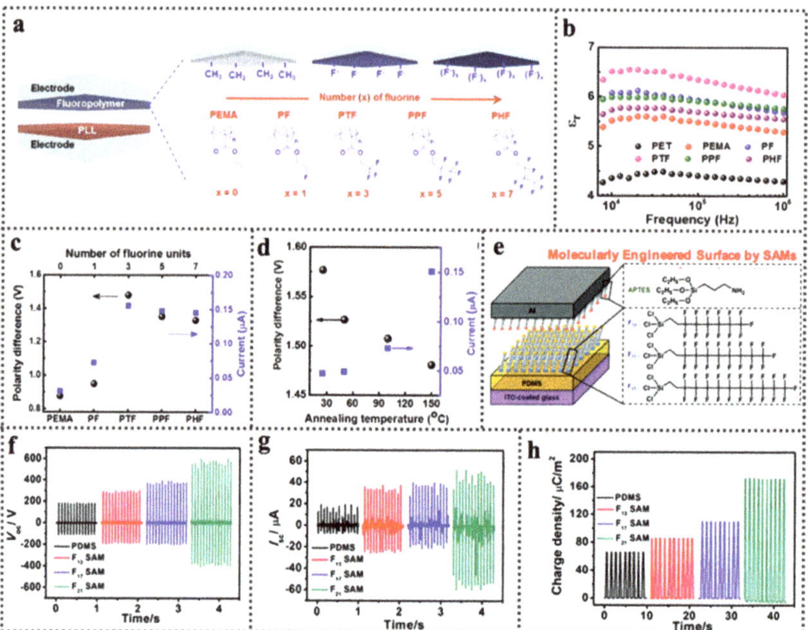

Figure 8. (**a**) Schematic illustration of TENGs built using fluorinated polymers with different numbers of fluorine. (**b**) Dielectric constants (ε_r) of different fluorinated polymers. (**c**) The output performance of TENG based on the effect of PEMA, PF, PTF, PPF, PHF, and number of fluorine units. (**d**) Effect of annealing temperature on electrical measurements [103] Copyright, 2019 Elsevier Ltd. (**e**) Schematic illustration of the TENG based on silane-SAM formation on a hydrated PDMS. (**f**) Output voltage, (**g**) current, and (**h**) charge density of the TENG with (F21, F17, and F13) compared to pristine PDMS [107] Copyright, 2020 Royal Society of Chemistry.

Recently, Wang et al. [107] proposed surface functionalization of the polydimethylsiloxane (PDMS) layer, which was formed by Silane-based SAM having fluorinated molecules, for example: [1H,1H,2H,2H-perfluorooctyl trichlorosilane (F13), 1H,1H,2H,2H-perfluorodecyl (F17), 1H,1H,2H,2H-perfluorododecyl (F21) and 3-aminopropyl triethoxysilane and the aluminum (Al) electrode, as illustrated in Figure 8e. Then the silane-based SAM treatment processing in solutions containing various types of fluorinated molecule solution and APTES solutions for 2h. The results showed that the output performance showed apparent enhancement after functionalization with fluorinated molecules, and the output performance increased with the increase of the number of fluorine atoms. Figure 8f-h show the peak output voltage, current density, and charge transfer of four TENGs with and without fluorinated-based SAM modification. The results show that 1H,1H,2H,2H-perfluorododecyltrichlorosilane (F21) increased the electrical measurements because there is a difference in the attract and transfer electrons from the Al layer to PDMS. In addition to the comparison of the electrical measurements of the TENG with and without fluorinated-based SAM, it follows the order (F21, F17, and F13) compared to pristine PDMS.

3.2.4. Element Doping

Another type of chemical modification includes doping the element on or inside the contact materials to modify the output performance of TENGs. For example, $Mg_xZn_{1-x}O$ is a new triboelectric material prepared via doping Mg in ZnO thin film made by radiofrequency (RF) magnetron sputtering used as an active layer to enhance the difference in work function [108]. As shown in Figure 9a, the Mg content was changed from 0% to 26.5% in $Mg_xZn_{1-x}O$. The work function first raises with increasing Mg composition, which is measured by Kelvin probe microscopy. TENG was fabricated using different $Mg_xZn_{1-x}O$ and PDMS as contacting layers. X-ray photoelectron spectroscopy (XPS), atomic force microscopy (AFM), and scanning Kelvin probe force microscopy (KPFM) methods were used to determine the content of Mg and $Mg_xZn_{1-x}O$ films work function. The results showed that there is a proportional relationship between the concentration of Mg and the work function, which have a maximum value of 5.14 eV, and then decreased to 4.85 eV. TENGs are fabricated from different $Mg_xZn_{1-x}O$ films and PDMS contact pairs. As shown in Figure 9b–c, the TENG made with ZnO film has a triboelectric current and voltage less than the TENG made of $Mg_xZn_{1-x}O$ with 26.5% Mg, due to the larger work function difference between $Mg_xZn_{1-x}O$ and PDMS. This approach confirmed the ability of MgZnO film work function to boost the TENG output.

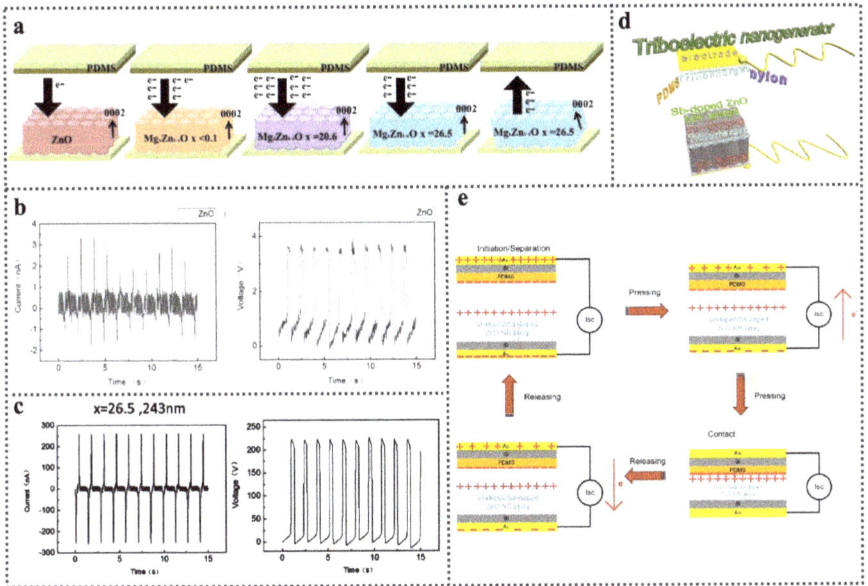

Figure 9. (a) Schematic illustration of TENGs composed of ZnO doping with Mg, triboelectric output current, and voltage of TENG of (b) pure ZnO (c) $Mg_xZn_{1-x}O$ films [108]. (d) Schematic diagram of the TENG based on ZnO doping by Sb. (e) TENG device mechanism of PDMS against ZnO nanorod arrays by Sb doping [109].

Chen et al. [109] showed the modifying of the ZnO NR surface through Sb doped into p-type to improve the TENG output performance, as illustrated in Figure 9d. As a result, for comparison, it is obvious that the output voltage and current of the P-type ZnO NR array reached 24 and 5.5 times that of a TENG with un-doped N-type ZnO, due to delivering electrons to negatively charged PDMS as illustrated in Figure 9e. This approach is very important in designing TENG to select the best combination of materials easily.

3.2.5. Nanomaterial Doping into Triboelectric Materials

Engineering of high dielectric nanomaterials doping into the TENG triboelectric layer is another method to enhance the TENG output performance, by using large charge tapping sites and work function which store triboelectric charges during triboelectrification. Triboelectric charge density increases with increasing the capacitance of the triboelectric material, which increases with the increase in the relative permittivity. Chen et al. [54] used SiO_2, TiO_2, $BaTiO_3$, and $SrTiO_3$ as dielectric materials to fill the PDMS sponge layer to increase the relative permittivity and reduce the contact material thickness. Figure 10a shows the fabrication process of the CS-TENG with different filler. The capacitance (C_{max}) of the device is determined by

$$C_{max} = \varepsilon_0 S \varepsilon_r / d \tag{3}$$

where ε_0 is the vacuum permittivity, ε_r is the relative permittivity of the PDMS, S is device area, and d_{PDMS} is the thickness of PDMS film.

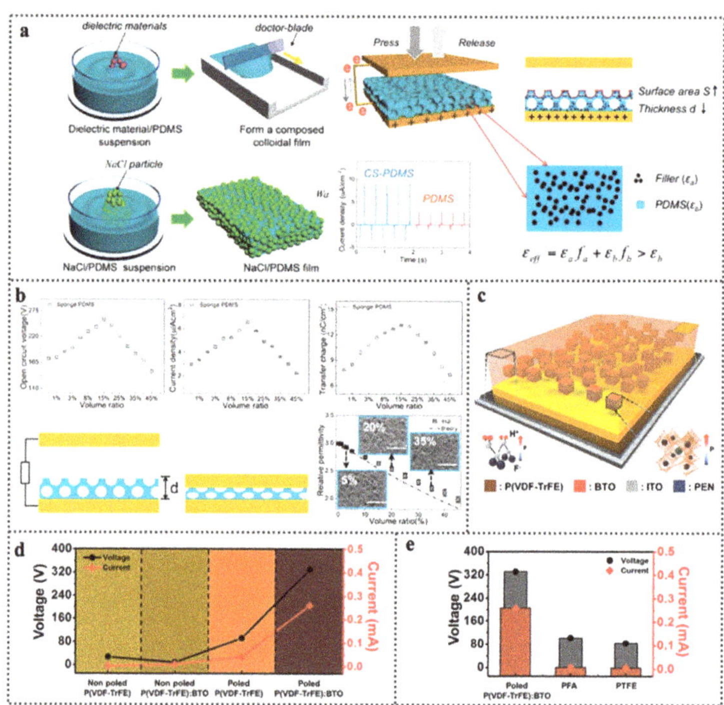

Figure 10. (**a**) Experimental design and diagram of PDMS sponge film-based TENG. (**b**) Electrical measurements of the device [54] Copyright, 2016 American Chemical Society; (**c**) Schematic diagram of (FC-TENG). (**d**) Output voltage and current of poled (PVDF-TrFE): BaTiO$_3$ with PTFE and PFA as a triboelectric layer. (**e**) Output comparison with traditional TENGs [68] Copyright, 2016 WILEY-VCH Verlag GmbH & Co. KGaA, Weinheim.

The output voltage is proportional with the number of charges from the triboelectric layer

$$V_{OC} = (\sigma_0 \cdot \chi(t)) / \varepsilon_0 \tag{4}$$

where σ_0, $\chi(t)$, ε_0 and V_{OC} are the charge density on the PDMS, interlayer distance, vacuum permittivity, and open circuit voltage, respectively [107]. According to the equation, the V_{OC} is proportional with the relative permittivity ε_r or is inversely proportional with the thickness of PDMS film or to both. In the meantime, high dielectric SrTiO$_3$ filling into PDMS film increases the relative permittivity. There is a positive relationship between the PDMS relative permittivity and the permittivity of the filling materials. As shown in Figure 10b, comparing different permittivity composite PDMS films are compared by filling dielectric nanoparticles, which are the SiO$_2$, TiO$_2$, BaTiO$_3$, and SrTiO$_3$ that have ε_r = 3, 80, 150, and 300. As a result, no change in both the open-circuit voltage and short current density after the PDMS are filled with SiO$_2$ because PDMS and SiO$_2$ having the same relative permittivity. The highest enhancement in electrical output for the triboelectric nanogenerator by using SrTiO$_3$ PDMS composite film is because permittivity is high and the reduction in the contact materials thickness, which is induced by pores NaCl salt. While adding 15 vol% pores and 10 vol% SrTiO$_3$ NPs in the PDMS film, the voltage and current reached over five-fold power enhancement with the pure PDMS. Similar work was done by Kang et al. [110]; they proposed that the BaTiO$_3$ nanoparticles doping into the PVDF matrix with a high dielectric constant can increase the output performance of TENG. As a result, with 11.25 vol%, BaTiO$_3$, the dielectric constant increased to 25, while the thickness of composite films decreased, and the transferred charge increased to 114 mCm^{-2}, a 200% improvement compared to the TENG with 45 mm thick composite electrification layer.

Chun et al. [111] presented TENG based on the Au nanoparticles-embedded porous film for enhancing the nanogenerator performance. On the other hand, Seung et al. [68] investigated the P(VDF-TrFE) matrix mixing with a negative triboelectric layer BaTiO$_3$ contact layer to study the permittivity and the polarization effects on the output power of TENGs (Figure 10c). Comparison of the electrical output behaviors between aluminum (Al) and each P(VDF-TrFE)-based surface are shown in Figure 10d. The output voltage and current increased due to the high permittivity and charge trap of BaTiO$_3$. Moreover, Figure 10e shows the electrical measurement of poled (PVDF-TrFE): BaTiO$_3$ with PTFE and PFA as a triboelectric layer. As a result, the poled P(VDF-TrFE): BaTiO$_3$ composite film lead to a higher power generating performance, which is about 150 times that of PTFE-based TEGs due to charge-attracting and transport properties that improved the output performance of the TENG.

3.2.6. Composite Materials Trapping into Friction Materials

There are various composite materials for application in triboelectric energy harvesting devices such as 2D layered structure materials and so on. They can be used to provide electron-trapping sites in TENGs and capture electrons readily, which can give different chemical properties compared to the original material surface. Two-dimensional structured materials are crystalline materials containing covalent bonds, providing in-plane stability, and are used as charge trapping layer, resulting from a specific surface area that can attract electrons readily. Since the 2D materials emerged, researchers have achieved great progress in TENG that are made of 2D material [112–114]. Seol et al. investigated the triboelectric series of various 2D layered materials such as MoSe$_2$, MoS$_2$, WSe$_2$, WS$_2$, graphene and graphene oxide [114]. This work provided a new sight to utilize 2D materials in TENG. Besides, Han et al. discussed patchable and implantable devices using TENGs based on 2D materials. It is suitable for implantable devices due to its flexibility, transparency, and mechanical stability [12]. Wu et al. [69] proposed a new methodology by using an electron-acceptor layer, such as reduced graphene oxide sheet with a polyimide layer, to enhance the TENG output performance (Figure 11a). TENG with a PI: rGO films produced the highest output performance because it can capture electrons in the PI: rGO layer and prevent the accumulation between triboelectric electrons and positive charges, which led to increasing the electron density. The PI-rGO sublayer structure has a power density of 6.3 MWm^{-2}, which is 30 times bigger than that of TENG without PI-rGO. The authors fabricated PI: rGO nanocomposites to form a floating-gate metal-insulator-semiconductor (MIS) device, which was placed in the middle of the PI insulator in a charge-storage area with an Al/p-Si/PI/PI: rGO/PI/Al structure. This is illustrated in Figure 11b. The C-V curves investigate the electrical properties and are measured at 5 MHz. A clockwise hysteresis indicated the presence of sites by the carriers; these sites exist due to the rGO sheets that have an electron-trapping effect on the PI layer.

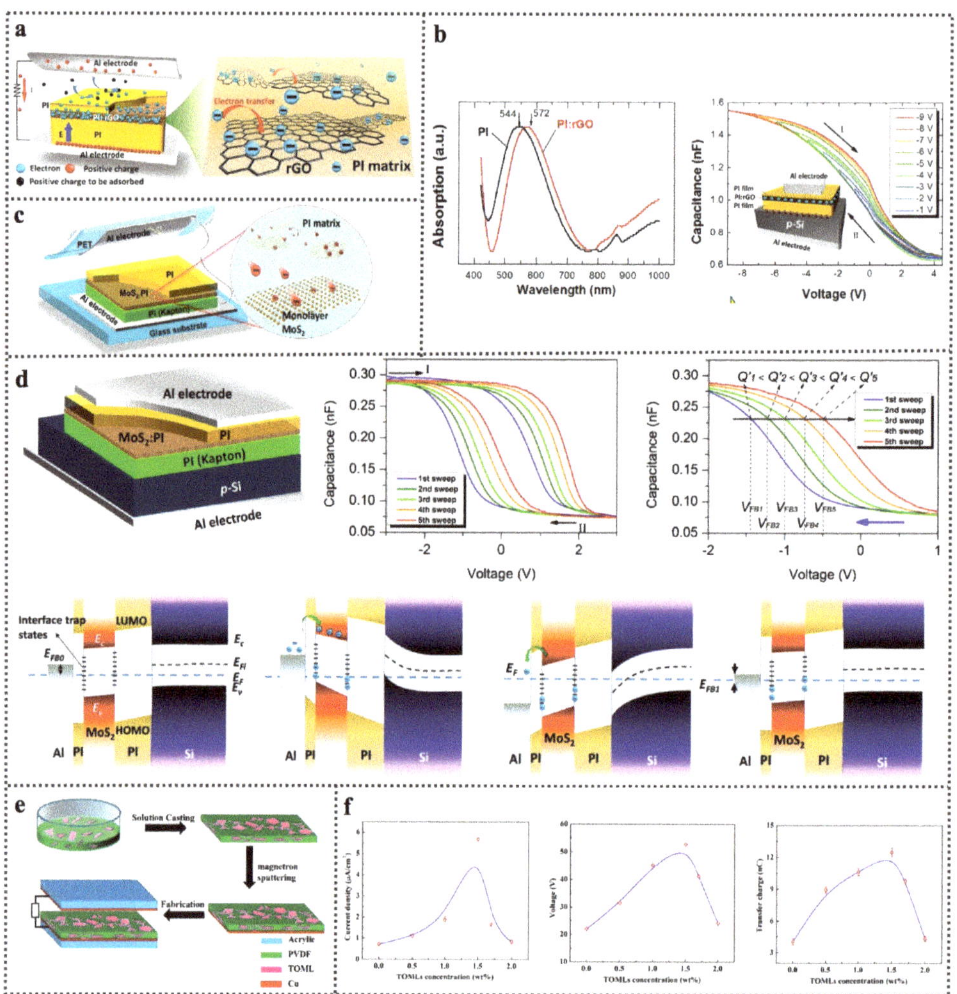

Figure 11. (**a**) TENG made by a composite PI:rGO. (**b**) C-V curves of the Al/p-Si/PI/PI:rGO/PI/Al device [69] Copyright, 2017 Elsevier Ltd. (**c**) TENG with a MoS$_2$-monolayer film as electron trapping layer. (**d**) Schematic of PI: MoS$_2$ device [115] Copyright, 2017 American Chemical Society. (**e**) TENG with the PVDF/TOML nanocomposite films. (**f**) Maximum electrical output TENG device with 1.5% concentration of TOML [116] Copyright, 2018 Elsevier Ltd.

Similar work has been done by Wu et al. [115]. As shown in Figure 11c, the MoS$_2$ monolayer sheet is a functional material as an electron-trapping material inside PI layer to improve the TENGs performance. Owing to the monolayer MoS$_2$, the TENG possesses a maximum peak power density being 120 times bigger than of the pristine device. The authors prepared a floating-gate metal-insulator-semiconductor (MIS) device, in which the PI: MoS$_2$ nanocomposites were used for the C-V measurements, which are shown in Figure 11d. Considering that the MoS$_2$ monolayer bandgap energy is more than 1.8 eV, the electrons occupied with both bottom energy states of the conduction band and MoS$_2$ trap states at the top electrode interface can be released. So as the voltage loads increase, the number of electrons captured at the MoS$_2$ trap interface states increases, resulting in a shift in the C-V curves. Generally, the other 2D materials such as titania, Au, InZnO, MoSe$_2$, and thin layer graphite are used as electron accepting inside dielectric layers to enhance the performance of TENGs.

Wen et al. [116] made transparent TENG of a PVDF and titania monolayer (TOML). The TOMLs have the advantage to enhance the TENGs performance by trapping charge material and a high dielectric constant. The process of the device fabrication is shown in Figure 11e. The TENG current density, voltage, and transfer charge signals with various concentrations of TOML as shown in Figure 11f. As a result, the voltage and current of the TENG, by using various concentrations of TOML in nanocomposite film, can reach the high value of 52.8 V and 5.7 μA when the weight of the TOML is 1.5%, which are nearly 2.4 and 7.8 times than those of pristine PVDF. The electrical output of TOML/PVDF device increased due to the synergy between efficient electron capture and the high dielectric constant of TOML.

4. Summary and Perspectives

In this review, summarized approaches, including surface morphology and surface modification, to the enhancement of the TENG performance are summarized. In the first approach, various surface morphology engineering methods have been developed to improve the performance of TENGs. Soft lithography, Photolithography, and ultrafast laser patterning were applied to regulate the surface morphology of the triboelectric materials. These methods are physical methods that won't change the chemical structure or chemical element of the material. The advantage of this method is the high selectivity of the material, and almost all the materials can be applied in surface morphology engineering. The drawback of this method is the limitation of the surface charge density due to the limitation of the contact area. The other approach is surface modification, which will change the structure or element of the material such as functional groups modification, ion injection, and elemental doping. The surface modification of the triboelectric materials can improve the limitation of the surface charge density, but not all materials are suitable or able to be modified with functional groups. Whether surface morphology engineering or surface modification engineering can enhance the output performance, It is decided by the physical and chemical properties of the material when choosing a suitable surface engineering method. Figure 12 shows the diagram indicating the future direction and challenges of enhanced TENG. The future directions include more effective micro-structures, new functional groups, enhancing mechanism analysis, surface functionalization techniques and surface characterization techniques. The challenges include influence of humidity, stability of the performance, lack of high currents, quick charge decay and suitable packaging techniques. To be more specific, a roadmap for future research is shown as following:

(1) More studies need to be done to discuss the mechanisms of enhancing the TENG performance, and the mechanism study should be taken from the perspective of structure design and material science.

(2) More micro/nano structures that can enhance the output performance need to be detected. Although there have been various micro/nano structures to apply in the triboelectric material, there are still many materials that lack a suitable surface structure for improving the output performance. Besides, more theoretical simulation and structural observation methods need to be developed.

(3) More research on the enhancement of TENG by semiconductors. Most surface modifications focused on polymers as insulation materials where the charges accumulate on the surface of the material, and metals are used as the most common friction materials resulting in the output of the TENG. However, semiconductor is also important but still largely unexplored.

(4) More practical surface engineering methods towards all modes of TENG. The TENG has wide applications in four modes. However, the surface engineering is mainly conducted in the contact-separation mode TENG. Hence, it is important to make the surface engineering practical.

(5) More direct and effective methods to improve the power output of TENGs with the electron-accepting (tend to obtain a negative charge) or donating (tend to obtain a positive charge) ability of functional groups, increasing surface properties of both triboelectric materials. The properties of triboelectric materials should be designed when the difference in the surface potential of triboelectric materials charges is large.

(6) Overcome the influence of humidity through the designed packaging techniques. Besides, it is important to stabilize the enhanced output performance of TENG and avoid quick charge decay. More importantly, the other limitation for TENG is the low current. More efforts still need to be put in for the commercialization of TENG.

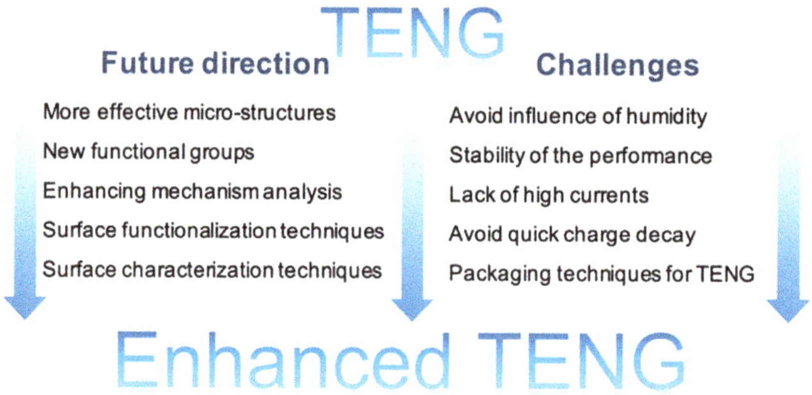

Figure 12. Diagram indicating the future direction and challenges of enhanced TENG.

Author Contributions: Conceptualization, Z.W. and X.S.; writing—original draft preparation, M.I.; writing—review and editing, J.J. All authors have read and agreed to the published version of the manuscript.

Funding: This work was funded by the National Natural Science Foundation of China (NSFC) (No. 61804103), the Suzhou Science and Technology Development Planning Project: Key Industrial Technology Innovation (No. SYG201924) and China Postdoctoral Science Foundation (No. 2021T140494).

Acknowledgments: Mervat Ibrahim would like to acknowledge the fund from the scholarship (CSC type A) under the joint program (Executive program between the Arab Republic of Egypt and China). The authors also acknowledge the support from the Collaborative Innovation Center of Suzhou Nano Science & Technology, the 111 Project and Joint International Research Laboratory of Carbon-Based Functional Materials and Devices.

Conflicts of Interest: The authors declare no conflict of interest.

References

1. Yang, W.; Chen, J.; Jing, Q.; Yang, J.; Wen, X.; Su, Y.; Zhu, G.; Bai, P.; Wang, Z.L. 3D Stack Integrated Triboelectric Nanogenerator for Harvesting Vibration Energy. *Adv. Funct. Mater.* **2014**, *24*, 4090–4096. [CrossRef]
2. Pourrahimi, A.M.; Olsson, R.T.; Hedenqvist, M.S. The Role of Interfaces in Polyethylene/Metal-Oxide Nanocomposites for Ultrahigh-Voltage Insulating Materials. *Adv. Mater.* **2018**, *30*, 1703624. [CrossRef] [PubMed]
3. Zhu, G.; Yang, R.; Wang, S.; Wang, Z.L. Flexible High-Output Nanogenerator Based on Lateral ZnO Nanowire Array. *Nano Lett.* **2010**, *10*, 3151–3155. [CrossRef]
4. Suzuki, Y.; Miki, D.; Edamoto, M.; Honzumi, M. A MEMS electret generator with electrostatic levitation for vibration-driven energy-harvesting applications. *J. Micromech. Microeng.* **2010**, *20*, 104002. [CrossRef]
5. Xu, S.; Hansen, B.J.; Wang, Z.L. Piezoelectric-nanowire-enabled power source for driving wireless microelectronics. *Nat. Commun.* **2010**, *1*, 93. [CrossRef]

6. Kumar, B.; Lee, K.Y.; Park, H.-K.; Chae, S.J.; Lee, Y.H.; Kim, S.-W. Controlled Growth of Semiconducting Nanowire, Nanowall, and Hybrid Nanostructures on Graphene for Piezoelectric Nanogenerators. *ACS Nano* **2011**, *5*, 4197–4204. [CrossRef] [PubMed]

7. Khan, A.; Abbasi, M.A.; Hussain, M.; Ibupoto, Z.H.; Wissting, J.; Nur, O.; Willander, M. Piezoelectric nanogenerator based on zinc oxide nanorods grown on textile cotton fabric. *Appl. Phys. Lett.* **2012**, *101*, 193506. [CrossRef]

8. Yang, Y.; Pradel, K.C.; Jing, Q.; Wu, J.M.; Zhang, F.; Zhou, Y.; Zhang, Y.; Wang, Z.L. Thermoelectric Nanogenerators Based on Single Sb-Doped ZnO Micro/Nanobelts. *ACS Nano* **2012**, *6*, 6984–6989. [CrossRef]

9. Wang, X. Piezoelectric nanogenerators—Harvesting ambient mechanical energy at the nanometer scale. *Nano Energy* **2012**, *1*, 13–24. [CrossRef]

10. Yang, Y.; Guo, W.; Pradel, K.C.; Zhu, G.; Zhou, Y.; Zhang, Y.; Hu, Y.; Lin, L.; Wang, Z.L. Pyroelectric Nanogenerators for Harvesting Thermoelectric Energy. *Nano Lett.* **2012**, *12*, 2833–2838. [CrossRef] [PubMed]

11. Fan, F.-R.; Tian, Z.-Q.; Lin Wang, Z. Flexible triboelectric generator. *Nano Energy* **2012**, *1*, 328–334. [CrossRef]

12. Wang, Z.L. Triboelectric Nanogenerators as New Energy Technology for Self-Powered Systems and as Active Mechanical and Chemical Sensors. *ACS Nano* **2013**, *7*, 9533–9557. [CrossRef] [PubMed]

13. Whiter, R.A.; Narayan, V.; Kar-Narayan, S. A Scalable Nanogenerator Based on Self-Poled Piezoelectric Polymer Nanowires with High Energy Conversion Efficiency. *Adv. Energy Mater.* **2014**, *4*, 1400519. [CrossRef]

14. Crossley, S.; Kar-Narayan, S. Energy harvesting performance of piezoelectric ceramic and polymer nanowires. *Nanotechnology* **2015**, *26*, 344001. [CrossRef] [PubMed]

15. Zhao, J.; Li, Y.; Yang, G.; Jiang, K.; Lin, H.; Ade, H.; Ma, W.; Yan, H. Efficient organic solar cells processed from hydrocarbon solvents. *Nat. Energy* **2016**, *1*, 15027. [CrossRef]

16. Invernizzi, F.; Dulio, S.; Patrini, M.; Guizzetti, G.; Mustarelli, P. Energy harvesting from human motion: Materials and techniques. *Chem. Soc. Rev.* **2016**, *45*, 5455–5473. [CrossRef] [PubMed]

17. Boughey, C.; Davies, T.; Datta, A.; Whiter, R.A.; Sahonta, S.-L.; Kar-Narayan, S. Vertically aligned zinc oxide nanowires electrodeposited within porous polycarbonate templates for vibrational energy harvesting. *Nanotechnology* **2016**, *27*, 28LT02. [CrossRef]

18. Wang, Z.L. On Maxwell's displacement current for energy and sensors: The origin of nanogenerators. *Mater. Today* **2017**, *20*, 74–82. [CrossRef]

19. Ahmed, A.; Hassan, I.; Hedaya, M.; El-Yazid, T.A.; Zu, J.; Wang, Z.L. Farms of triboelectric nanogenerators for harvesting wind energy: A potential approach towards green energy. *Nano Energy* **2017**, *36*, 21–29. [CrossRef]

20. Wang, Z.L. Nanogenerators, self-powered systems, blue energy, piezotronics and piezo-phototronics—A recall on the original thoughts for coining these fields. *Nano Energy* **2018**, *54*, 477–483. [CrossRef]

21. Chen, J.; Wang, Z.L. Reviving Vibration Energy Harvesting and Self-Powered Sensing by a Triboelectric Nanogenerator. *Joule* **2017**, *1*, 480–521. [CrossRef]

22. Wang, Z.L. Triboelectric nanogenerators as new energy technology and self-powered sensors—Principles, problems and perspectives. *Faraday Discuss.* **2014**, *176*, 447–458. [CrossRef] [PubMed]

23. Wen, Z.; Shen, Q.; Sun, X. Nanogenerators for Self-Powered Gas Sensing. *Nano-Micro Lett.* **2017**, *9*, 1–19. [CrossRef] [PubMed]

24. Cheng, P.; Liu, Y.; Wen, Z.; Shao, H.; Wei, A.; Xie, X.; Chen, C.; Yang, Y.; Peng, M.; Zhuo, Q.; et al. Atmospheric pressure difference driven triboelectric nanogenerator for efficiently harvesting ocean wave energy. *Nano Energy* **2018**, *54*, 156–162. [CrossRef]

25. Huang, L.-B.; Xu, W.; Tian, W.; Han, J.-C.; Zhao, C.-H.; Wu, H.-L.; Hao, J. Ultrasonic-assisted ultrafast fabrication of polymer nanowires for high performance triboelectric nanogenerators. *Nano Energy* **2020**, *71*, 104593. [CrossRef]

26. Wang, M.; Pan, J.; Wang, M.; Sun, T.; Ju, J.; Tang, Y.; Wang, J.; Mao, W.; Wang, Y.; Zhu, J. High-Performance Triboelectric Nanogenerators Based on a Mechanoradical Mechanism. *ACS Sustain. Chem. Eng.* **2020**, *8*, 3865–3871. [CrossRef]

27. Tantraviwat, D.; Buarin, P.; Suntalelat, S.; Sripumkhai, W.; Pattamang, P.; Rujijanagul, G.; Inceesungvorn, B. Highly dispersed porous polydimethylsiloxane for boosting power-generating performance of triboelectric nanogenerators. *Nano Energy* **2020**, *67*, 104214. [CrossRef]

28. Kim, J.; Ryu, H.; Lee, J.H.; Khan, U.; Kwak, S.S.; Yoon, H.; Kim, S. High Permittivity $CaCu_3Ti_4O_{12}$ Particle-Induced Internal Polarization Amplification for High Performance Triboelectric Nanogenerators. *Adv. Energy Mater.* **2020**, *10*, 1903524. [CrossRef]

29. Yan, S.; Dong, K.; Lu, J.; Song, W.; Xiao, R. Amphiphobic triboelectric nanogenerators based on silica enhanced thermoplastic polymeric nanofiber membranes. *Nanoscale* **2020**, *12*, 4527–4536. [CrossRef]

30. Rodrigues, C.; Nunes, D.; Clemente, D.; Mathias, N.; Correia, J.M.; Rosa-Santos, P.; Taveira-Pinto, F.; Morais, T.; Pereira, A.; Ventura, J. Emerging triboelectric nanogenerators for ocean wave energy harvesting: State of the art and future perspectives. *Energy Environ. Sci.* **2020**, *13*, 2657–2683. [CrossRef]

31. Gao, Q.; Li, Y.; Xie, Z.; Yang, W.; Wang, Z.; Yin, M.; Lu, X.; Cheng, T.; Wang, Z.L. Robust Triboelectric Nanogenerator with Ratchet-like Wheel-Based Design for Harvesting of Environmental Energy. *Adv. Mater. Technol.* **2019**, *5*, 1900801. [CrossRef]

32. Shi, K.; Zou, H.; Sun, B.; Jiang, P.; He, J.; Huang, X. Dielectric Modulated Cellulose Paper/PDMS-Based Triboelectric Nanogenerators for Wireless Transmission and Electropolymerization Applications. *Adv. Funct. Mater.* **2019**, *30*, 1904536. [CrossRef]

33. Zhang, Y.; Wu, J.; Cui, S.; Wei, W.; Chen, W.; Pang, R.; Wu, Z.; Mi, L. Organosulfonate Counteranions—A Trapped Coordination Polymer as a High-Output Triboelectric Nanogenerator Material for Self-Powered Anticorrosion. *Chem.—A Eur. J.* **2020**, *26*, 584–591. [CrossRef] [PubMed]

34. Khandelwal, G.; Raj, N.P.M.J.; Kim, S. Zeolitic Imidazole Framework: Metal–Organic Framework Subfamily Members for Triboelectric Nanogenerators. *Adv. Funct. Mater.* **2020**, *30*, 1910162. [CrossRef]
35. Khandelwal, G.; Raj, N.P.M.J.; Kim, S.-J. ZIF-62: A mixed linker metal–organic framework for triboelectric nanogenerators. *J. Mater. Chem. A* **2020**, *8*, 17817–17825. [CrossRef]
36. Cui, X.; Zhang, Y.; Hu, G.; Zhang, L.; Zhang, Y. Dynamical charge transfer model for high surface charge density triboelectric nanogenerators. *Nano Energy* **2020**, *70*, 104513. [CrossRef]
37. Li, Y.; Zheng, W.; Zhang, H.; Wang, H.; Cai, H.; Zhang, Y.; Yang, Z. Electron transfer mechanism of graphene/Cu heterostructure for improving the stability of triboelectric nanogenerators. *Nano Energy* **2020**, *70*, 104540. [CrossRef]
38. Guo, Y.; Cao, Y.; Chen, Z.; Li, R.; Gong, W.; Yang, W.; Zhang, Q.; Wang, H. Fluorinated metal-organic framework as bifunctional filler toward highly improving output performance of triboelectric nanogenerators. *Nano Energy* **2020**, *70*, 104517. [CrossRef]
39. Zhou, L.; Liu, D.; Wang, J.; Wang, Z.L. Triboelectric nanogenerators: Fundamental physics and potential applications. *Friction* **2020**, *8*, 481–506. [CrossRef]
40. Wang, Y.; Duan, J.; Yang, X.; Liu, L.; Zhao, L.; Tang, Q. The unique dielectricity of inorganic perovskites toward high-performance triboelectric nanogenerators. *Nano Energy* **2020**, *69*, 104418. [CrossRef]
41. Feng, Y.; Zheng, Y.; Ma, S.; Wang, D.; Zhou, F.; Liu, W. High output polypropylene nanowire array triboelectric nanogenerator through surface structural control and chemical modification. *Nano Energy* **2016**, *19*, 48–57. [CrossRef]
42. Cao, V.A.; Lee, S.; Kim, M.; Alam, M.; Park, P.; Nah, J. Output power density enhancement of triboelectric nanogenerators via ferroelectric polymer composite interfacial layers. *Nano Energy* **2020**, *67*, 104300. [CrossRef]
43. Zhao, P.; Soin, N.; Kumar, A.; Shi, L.; Guan, S.; Tsonos, C.; Yu, Z.; Ray, S.C.; McLaughlin, J.A.; Zhu, Z.; et al. Expanding the portfolio of tribo-positive materials: Aniline formaldehyde condensates for high charge density triboelectric nanogenerators. *Nano Energy* **2020**, *67*, 104291. [CrossRef]
44. Xu, S.; Ding, W.; Guo, H.; Wang, X.; Wang, Z.L. Boost the Performance of Triboelectric Nanogenerators through Circuit Oscillation. *Adv. Energy Mater.* **2019**, *9*, 1900772. [CrossRef]
45. Stanford, M.G.; Li, J.; Chyan, Y.; Wang, Z.; Wang, W.; Tour, J.M. Laser-Induced Graphene Triboelectric Nanogenerators. *ACS Nano* **2019**, *13*, 7166–7174. [CrossRef] [PubMed]
46. Šutka, A.; Mālnieks, K.; Lapčinskis, L.; Kaufelde, P.; Linarts, A.; Bērziņa, A.; Zābels, R.; Jurķāns, V.; Gorņevs, I.; Blūms, J.; et al. The role of intermolecular forces in contact electrification on polymer surfaces and triboelectric nanogenerators. *Energy Environ. Sci.* **2019**, *12*, 2417–2421. [CrossRef]
47. Gao, L.; Chen, X.; Lu, S.; Zhou, H.; Xie, W.; Chen, J.; Qi, M.; Yu, H.; Mu, X.; Wang, Z.L.; et al. Triboelectric Nanogenerators: Enhancing the Output Performance of Triboelectric Nanogenerator via Grating-Electrode-Enabled Surface Plasmon Excitation (Adv. Energy Mater. 44/2019). *Adv. Energy Mater.* **2019**, *9*, 1970177. [CrossRef]
48. Zhao, X.J.; Zhu, G.; Wang, Z.L. Coplanar Induction Enabled by Asymmetric Permittivity of Dielectric Materials for Mechanical Energy Conversion. *ACS Appl. Mater. Interfaces* **2015**, *7*, 6025–6029. [CrossRef] [PubMed]
49. Shin, S.-H.; Bae, Y.E.; Moon, H.K.; Kim, J.; Choi, S.-H.; Kim, Y.; Yoon, H.J.; Lee, M.H.; Nah, J. Formation of Triboelectric Series via Atomic-Level Surface Functionalization for Triboelectric Energy Harvesting. *ACS Nano* **2017**, *11*, 6131–6138. [CrossRef]
50. Liu, C.-Y.; Bard, A.J. Electrons on dielectrics and contact electrification. *Chem. Phys. Lett.* **2009**, *480*, 145–156. [CrossRef]
51. Kim, Y.J.; Lee, J.; Park, S.; Park, C.; Park, C.; Choi, H.J. Effect of the relative permittivity of oxides on the performance of triboelectric nanogenerators. *RSC Adv.* **2017**, *7*, 49368–49373. [CrossRef]
52. Żenkiewicz, M.; Żuk, T.; Markiewicz, E. Triboelectric series and electrostatic separation of some biopolymers. *Polym. Test.* **2015**, *42*, 192–198. [CrossRef]
53. Lee, B.-Y.; Kim, S.-U.; Kang, S.; Lee, S.-D. Transparent and flexible high power triboelectric nanogenerator with metallic nanowire-embedded tribonegative conducting polymer. *Nano Energy* **2018**, *53*, 152–159. [CrossRef]
54. Chen, J.; Guo, H.; He, X.; Liu, G.; Xi, Y.; Shi, H.; Hu, C. Enhancing Performance of Triboelectric Nanogenerator by Filling High Dielectric Nanoparticles into Sponge PDMS Film. *ACS Appl. Mater. Interfaces* **2016**, *8*, 736–744. [CrossRef] [PubMed]
55. Lin, W.-C.; Lee, S.-H.; Karakachian, M.; Yu, B.-Y.; Chen, Y.-Y.; Lin, Y.-C.; Kuo, C.-H.; Shyue, J.-J. Tuning the surface potential of gold substrates arbitrarily with self-assembled monolayers with mixed functional groups. *Phys. Chem. Chem. Phys.* **2009**, *11*, 6199–6204. [CrossRef] [PubMed]
56. Lin, Z.-H.; Xie, Y.; Yang, Y.; Wang, S.; Zhu, G.; Wang, Z.L. Enhanced Triboelectric Nanogenerators and Triboelectric Nanosensor Using Chemically Modified TiO$_2$ Nanomaterials. *ACS Nano* **2013**, *7*, 4554–4560. [CrossRef] [PubMed]
57. Zhu, G.; Lin, Z.-H.; Jing, Q.; Bai, P.; Pan, C.; Yang, Y.; Zhou, Y.; Wang, Z.L. Toward Large-Scale Energy Harvesting by a Nanoparticle-Enhanced Triboelectric Nanogenerator. *Nano Lett.* **2013**, *13*, 847–853. [CrossRef]
58. Niu, S.; Wang, S.; Liu, Y.; Zhou, Y.; Lin, L.; Hu, Y.; Pradel, K.C.; Wang, Z.L. A theoretical study of grating structured triboelectric nanogenerators. *Energy Environ. Sci.* **2014**, *7*, 2339–2349. [CrossRef]
59. Jeong, C.K.; Baek, K.M.; Niu, S.; Nam, T.W.; Hur, Y.H.; Park, D.Y.; Hwang, G.-T.; Byun, M.; Wang, Z.L.; Jung, Y.S.; et al. Topographically-Designed Triboelectric Nanogenerator via Block Copolymer Self-Assembly. *Nano Lett.* **2014**, *14*, 7031–7038. [CrossRef] [PubMed]
60. Wang, S.; Xie, Y.; Niu, S.; Lin, L.; Liu, C.; Zhou, Y.S.; Wang, Z.L. Maximum Surface Charge Density for Triboelectric Nanogenerators Achieved by Ionized-Air Injection: Methodology and Theoretical Understanding. *Adv. Mater.* **2014**, *26*, 6720–6728. [CrossRef]

61. Kim, D.; Jeon, S.-B.; Kim, J.Y.; Seol, M.-L.; Kim, S.O.; Choi, Y.-K. High-performance nanopattern triboelectric generator by block copolymer lithography. *Nano Energy* **2015**, *12*, 331–338. [CrossRef]
62. Wang, S.; Zi, Y.; Zhou, Y.S.; Li, S.; Fan, F.; Lin, L.; Wang, Z.L. Molecular surface functionalization to enhance the power output of triboelectric nanogenerators. *J. Mater. Chem. A* **2016**, *4*, 3728–3734. [CrossRef]
63. Wang, H.S.; Jeong, C.K.; Seo, M.-H.; Joe, D.; Han, J.H.; Yoon, J.-B.; Lee, K.J. Performance-enhanced triboelectric nanogenerator enabled by wafer-scale nanogrates of multistep pattern downscaling. *Nano Energy* **2017**, *35*, 415–423. [CrossRef]
64. Zhu, G.; Pan, C.; Guo, W.; Chen, C.-Y.; Zhou, Y.; Yu, R.; Wang, Z.L. Triboelectric-Generator-Driven Pulse Electrodeposition for Micropatterning. *Nano Lett.* **2012**, *12*, 4960–4965. [CrossRef]
65. Fan, F.-R.; Lin, L.; Zhu, G.; Wu, W.; Zhang, R.; Wang, Z.L. Transparent Triboelectric Nanogenerators and Self-Powered Pressure Sensors Based on Micropatterned Plastic Films. *Nano Lett.* **2012**, *12*, 3109–3114. [CrossRef]
66. Song, G.; Kim, Y.; Yu, S.; Kim, M.-O.; Park, S.-H.; Cho, S.M.; Velusamy, D.B.; Cho, S.H.; Kim, K.L.; Kim, J.; et al. Molecularly Engineered Surface Triboelectric Nanogenerator by Self-Assembled Monolayers (METS). *Chem. Mater.* **2015**, *27*, 4749–4755. [CrossRef]
67. Wang, S.; Lin, L.; Wang, Z.L. Nanoscale Triboelectric-Effect-Enabled Energy Conversion for Sustainably Powering Portable Electronics. *Nano Lett.* **2012**, *12*, 6339–6346. [CrossRef] [PubMed]
68. Seung, W.; Yoon, H.J.; Kim, T.Y.; Ryu, H.; Kim, J.; Lee, J.H.; Lee, J.H.; Kim, S.; Park, Y.K.; Park, Y.J.; et al. Boosting Power-Generating Performance of Triboelectric Nanogenerators via Artificial Control of Ferroelectric Polarization and Dielectric Properties. *Adv. Energy Mater.* **2017**, *7*, 1600988. [CrossRef]
69. Wu, C.; Kim, T.W.; Choi, H.Y. Reduced graphene-oxide acting as electron-trapping sites in the friction layer for giant triboelectric enhancement. *Nano Energy* **2017**, *32*, 542–550. [CrossRef]
70. Wang, Z.L.; Wang, A.C. On the origin of contact-electrification. *Mater. Today* **2019**, *30*, 34–51. [CrossRef]
71. Elahi, H.; Mughal, M.R.; Eugeni, M.; Qayyum, F.; Israr, A.; Ali, A.; Munir, K.; Praks, J.; Gaudenzi, P. Characterization and Implementation of a Piezoelectric Energy Harvester Configuration: Analytical, Numerical and Experimental Approach. *Integr. Ferroelectr.* **2020**, *212*, 39–60. [CrossRef]
72. Wang, Z.L. On the first principle theory of nanogenerators from Maxwell's equations. *Nano Energy* **2020**, *68*, 104272. [CrossRef]
73. Huang, T.; Lu, M.; Yu, H.; Zhang, Q.; Wang, H.; Zhu, M. Enhanced Power Output of a Triboelectric Nanogenerator Composed of Electrospun Nanofiber Mats Doped with Graphene Oxide. *Sci. Rep.* **2015**, *5*, srep13942. [CrossRef]
74. Li, W.; Zhang, Y.; Liu, L.; Li, D.; Liao, L.; Pan, C. A high energy output nanogenerator based on reduced graphene oxide. *Nanoscale* **2015**, *7*, 18147–18151. [CrossRef]
75. Park, S.-J.; Seol, M.-L.; Kim, D.; Jeon, S.-B.; Choi, Y.-K. Triboelectric nanogenerator with nanostructured metal surface using water-assisted oxidation. *Nano Energy* **2016**, *21*, 258–264. [CrossRef]
76. Lee, J.H.; Yu, I.; Hyun, S.; Kim, J.K.; Jeong, U. Remarkable increase in triboelectrification by enhancing the conformable contact and adhesion energy with a film-covered pillar structure. *Nano Energy* **2017**, *34*, 233–241. [CrossRef]
77. Zhao, L.; Zheng, Q.; Ouyang, H.; Li, H.; Yan, L.; Shi, B.; Li, Z. A size-unlimited surface microstructure modification method for achieving high performance triboelectric nanogenerator. *Nano Energy* **2016**, *28*, 172–178. [CrossRef]
78. Zhang, H.; Yang, Y.; Su, Y.; Chen, J.; Adams, K.; Lee, S.; Hu, C.; Wang, Z.L. Triboelectric Nanogenerator for Harvesting Vibration Energy in Full Space and as Self-Powered Acceleration Sensor. *Adv. Funct. Mater.* **2014**, *24*, 1401–1407. [CrossRef]
79. Cui, N.; Gu, L.; Lei, Y.; Liu, J.; Qin, Y.; Ma, X.-H.; Hao, Y.; Wang, Z.L. Dynamic Behavior of the Triboelectric Charges and Structural Optimization of the Friction Layer for a Triboelectric Nanogenerator. *ACS Nano* **2016**, *10*, 6131–6138. [CrossRef]
80. Kaur, N.; Bahadur, J.; Panwar, V.; Singh, P.; Rathi, K.; Pal, K. Effective energy harvesting from a single electrode based triboelectric nanogenerator. *Sci. Rep.* **2016**, *6*, 38835. [CrossRef]
81. Guo, H.; Li, T.; Cao, X.; Xiong, J.; Jie, Y.; Willander, M.; Cao, X.; Wang, N.; Wang, Z.L. Self-Sterilized Flexible Single-Electrode Triboelectric Nanogenerator for Energy Harvesting and Dynamic Force Sensing. *ACS Nano* **2017**, *11*, 856–864. [CrossRef] [PubMed]
82. Zhang, X.-S.; Han, M.-D.; Wang, R.-X.; Zhu, F.-Y.; Li, Z.-H.; Wang, W.; Zhang, H. Frequency-Multiplication High-Output Triboelectric Nanogenerator for Sustainably Powering Biomedical Microsystems. *Nano Lett.* **2013**, *13*, 1168–1172. [CrossRef] [PubMed]
83. Nafari, A.; Sodano, H.A. Surface morphology effects in a vibration based triboelectric energy harvester. *Smart Mater. Struct.* **2018**, *27*, 015029. [CrossRef]
84. Gong, J.; Xu, B.; Tao, X. Breath Figure Micromolding Approach for Regulating the Microstructures of Polymeric Films for Triboelectric Nanogenerators. *ACS Appl. Mater. Interfaces* **2017**, *9*, 4988–4997. [CrossRef] [PubMed]
85. Sun, J.-G.; Yang, T.N.; Kuo, I.-S.; Wu, J.-M.; Wang, C.-Y.; Chen, L.-J. A leaf-molded transparent triboelectric nanogenerator for smart multifunctional applications. *Nano Energy* **2017**, *32*, 180–186. [CrossRef]
86. Rasel, M.S.U.; Park, J.Y. A sandpaper assisted micro-structured polydimethylsiloxane fabrication for human skin based triboelectric energy harvesting application. *Appl. Energy* **2017**, *206*, 150–158. [CrossRef]
87. Jiang, L.; Wang, A.; Li, B.; Cui, T.; Lu, Y.-F. Electrons dynamics control by shaping femtosecond laser pulses in micro/nanofabrication: Modeling, method, measurement and application. *Light. Sci. Appl.* **2018**, *7*, 17134. [CrossRef] [PubMed]

88. Malinauskas, M.; Žukauskas, A.; Hasegawa, S.; Hayasaki, Y.; Mizeikis, V.; Buividas, R.; Juodkazis, S. Ultrafast laser processing of materials: From science to industry. *Light. Sci. Appl.* **2016**, *5*, e16133. [CrossRef]

89. Vorobyev, A.Y.; Guo, C.L. Direct femtosecond laser surface nano/microstructuring and its applications. *Laser Photonics Rev.* **2013**, *7*, 385–407. [CrossRef]

90. Kim, D.; Tcho, I.-W.; Jin, I.K.; Park, S.-J.; Jeon, S.-B.; Kim, W.-G.; Cho, H.-S.; Lee, H.-S.; Jeoung, S.C.; Choi, Y.-K. Direct-laser-patterned friction layer for the output enhancement of a triboelectric nanogenerator. *Nano Energy* **2017**, *35*, 379–386. [CrossRef]

91. Huang, J.; Fu, X.; Liu, G.; Xu, S.; Li, X.; Zhang, C.; Jiang, L. Micro/nano-structures-enhanced triboelectric nanogenerators by femtosecond laser direct writing. *Nano Energy* **2019**, *62*, 638–644. [CrossRef]

92. Zhai, N.; Wen, Z.; Chen, X.; Wei, A.; Sha, M.; Fu, J.; Liu, Y.; Zhong, J.; Sun, X. Blue Energy Collection toward All-Hours Self-Powered Chemical Energy Conversion. *Adv. Energy Mater.* **2020**, *10*, 2001041. [CrossRef]

93. Byun, K.-E.; Cho, Y.; Seol, M.; Kim, S.; Kim, S.-W.; Shin, H.-J.; Park, S.; Hwang, S.W. Control of Triboelectrification by Engineering Surface Dipole and Surface Electronic State. *ACS Appl. Mater. Interfaces* **2016**, *8*, 18519–18525. [CrossRef] [PubMed]

94. Zhou, Y.S.; Liu, Y.; Zhu, G.; Lin, Z.-H.; Pan, C.; Jing, Q.; Wang, Z.L. In Situ Quantitative Study of Nanoscale Triboelectrification and Patterning. *Nano Lett.* **2013**, *13*, 2771–2776. [CrossRef] [PubMed]

95. Shin, S.-H.; Kwon, Y.H.; Kim, Y.-H.; Jung, J.-Y.; Lee, M.H.; Nah, J. Triboelectric Charging Sequence Induced by Surface Functional-ization as a Method to Fabricate High Performance Triboelectric Generators. *ACS Nano* **2015**, *9*, 4621–4627. [CrossRef]

96. Busolo, T.; Ura, D.; Kim, S.K.; Marzec, M.M.; Bernasik, A.; Stachewicz, U.; Kar-Narayan, S. Surface potential tailoring of PMMA fibers by electrospinning for enhanced triboelectric performance. *Nano Energy* **2019**, *57*, 500–506. [CrossRef]

97. Li, S.; Fan, Y.; Chen, H.; Nie, J.; Liang, Y.; Tao, X.; Zhang, J.; Chen, X.; Fu, E.; Wang, Z.L. Manipulating the triboelectric surface charge density of polymers by low-energy helium ion irradiation/implantation. *Energy Environ. Sci.* **2020**, *13*, 896–907. [CrossRef]

98. Liu, L.; Tang, W.; Wang, Z.L. Inductively-coupled-plasma-induced electret enhancement for triboelectric nanogenerators. *Nanotechnology* **2016**, *28*, 035405. [CrossRef]

99. Xu, L.; Bu, T.Z.; Yang, X.D.; Zhang, C.; Wang, Z.L. Ultrahigh charge density realized by charge pumping at ambient conditions for triboelectric nanogenerators. *Nano Energy* **2018**, *49*, 625–633. [CrossRef]

100. Lin, Z.-H.; Cheng, G.; Lee, S.; Pradel, K.C.; Wang, Z.L. Harvesting Water Drop Energy by a Sequential Contact-Electrification and Electrostatic-Induction Process. *Adv. Mater.* **2014**, *26*, 4690–4696. [CrossRef]

101. Babudri, F.; Farinola, G.M.; Naso, F.; Ragni, R. Fluorinated organic materials for electronic and optoelectronic applications: The role of the fluorine atom. *Chem. Commun.* **2007**, *10*, 1003–1022. [CrossRef] [PubMed]

102. Shao, J.; Tang, W.; Jiang, T.; Chen, X.; Xu, L.; Chen, B.; Zhou, T.; Deng, C.R.; Wang, Z.L. A multi-dielectric-layered triboelectric nanogenerator as energized by corona discharge. *Nanoscale* **2017**, *9*, 9668–9675. [CrossRef]

103. Kim, M.P.; Lee, Y.; Hur, Y.H.; Park, J.; Kim, J.; Lee, Y.; Ahn, C.W.; Song, S.W.; Jung, Y.S.; Ko, H. Molecular structure engineering of dielectric fluorinated polymers for enhanced performances of triboelectric nanogenerators. *Nano Energy* **2018**, *53*, 37–45. [CrossRef]

104. Park, S.; Kim, H.; Vosgueritchian, M.; Cheon, S.; Kim, H.; Koo, J.H.; Kim, T.R.; Lee, S.; Schwartz, G.; Chang, H.; et al. Stretchable Energy-Harvesting Tactile Electronic Skin Capable of Differentiating Multiple Mechanical Stimuli Modes. *Adv. Mater.* **2014**, *26*, 7324–7332. [CrossRef]

105. Lee, K.Y.; Chun, J.; Lee, J.-H.; Kim, K.N.; Kang, N.-R.; Kim, J.-Y.; Kim, M.H.; Shin, K.-S.; Gupta, M.K.; Baik, J.M.; et al. Hydrophobic Sponge Structure-Based Triboelectric Nanogenerator. *Adv. Mater.* **2014**, *26*, 5037–5042. [CrossRef] [PubMed]

106. Thakur, V.K.; Tan, E.J.; Lin, M.-F.; Lee, P.S. Polystyrene grafted polyvinylidenefluoride copolymers with high capacitive perfor-mance. *Polym. Chem.* **2011**, *2*, 2000–2009. [CrossRef]

107. Wang, C.C.; Chang, C.Y. Enhanced output performance and stability of triboelectric nanogenerators by employing silane-based self-assembled monolayers. *J. Mater. Chem. C* **2020**, *8*, 4542–4548. [CrossRef]

108. Guo, Q.Z.; Yang, L.C.; Wang, R.C.; Liu, C.P. Tunable Work Function of $Mg_xZn_{1-x}O$ as a Viable Friction Material for a Triboelectric Nanogenerator. *ACS Appl. Mater. Interfaces* **2019**, *11*, 1420–1425. [CrossRef]

109. Chen, S.-N.; Chen, C.-H.; Lin, Z.-H.; Tsao, Y.-H.; Liu, C.-P. On enhancing capability of tribocharge transfer of ZnO nanorod arrays by Sb doping for anomalous output performance improvement of triboelectric nanogenerators. *Nano Energy* **2018**, *45*, 311–318. [CrossRef]

110. Kang, X.; Pan, C.; Chen, Y.; Pu, X. Boosting performances of triboelectric nanogenerators by optimizing dielectric properties and thickness of electrification layer. *RSC Adv.* **2020**, *10*, 17752–17759. [CrossRef]

111. Chun, J.; Kim, J.W.; Jung, W.-S.; Kang, C.-Y.; Kim, S.-W.; Wang, Z.L.; Baik, J.M. Mesoporous pores impregnated with Au nanoparticles as effective dielectrics for enhancing triboelectric nanogenerator performance in harsh environments. *Energy Environ. Sci.* **2015**, *8*, 3006–3012. [CrossRef]

112. Han, S.A.; Lee, J.H.; Seung, W.; Lee, J.; Kim, S.W.; Kim, J.H. Patchable and Implantable 2D Nanogenerator. *Small* **2021**, *17*, e1903519. [CrossRef]

113. Mariappan, V.K.; Krishnamoorthy, K.; Pazhamalai, P.; Natarajan, S.; Sahoo, S.; Nardekar, S.S.; Kim, S.-J. Antimonene dendritic nanostructures: Dual-functional material for high-performance energy storage and harvesting devices. *Nano Energy* **2020**, *77*, 105248. [CrossRef]

114. Seol, M.; Kim, S.; Cho, Y.; Byun, K.E.; Kim, H.; Kim, J.; Kim, S.K.; Kim, S.W.; Shin, H.J.; Park, S. Triboelectric Series of 2D Layered Materials. *Adv. Mater.* **2018**, *30*, e1801210. [CrossRef]

115. Wu, C.; Kim, T.W.; Park, J.H.; Whan, K.T.; Shao, J.; Chen, X.; Wang, Z.L. Enhanced Triboelectric Nanogenerators Based on MoS2 Monolayer Nanocomposites Acting as Electron-Acceptor Layers. *ACS Nano* **2017**, *11*, 8356–8363. [CrossRef] [PubMed]
116. Wen, R.; Guo, J.; Yu, A.; Zhang, K.; Kou, J.; Zhu, Y.; Zhang, Y.; Li, B.-W.; Zhai, J. Remarkably enhanced triboelectric nanogenerator based on flexible and transparent monolayer titania nanocomposite. *Nano Energy* **2018**, *50*, 140–147. [CrossRef]

 *nanoenergy advances*

Review

Progress in the Triboelectric Human–Machine Interfaces (HMIs)-Moving from Smart Gloves to AI/Haptic Enabled HMI in the 5G/IoT Era

Zhongda Sun [1,2,3], Minglu Zhu [1,2,3] and Chengkuo Lee [1,2,3,4,*]

[1] Department of Electrical & Computer Engineering, National University of Singapore, Singapore 117576, Singapore; e0320823@u.nus.edu (Z.S.); e0267925@u.nus.edu (M.Z.)
[2] Center for Intelligent Sensors and MEMS (CISM), National University of Singapore, Singapore 117608, Singapore
[3] Singapore Institute of Manufacturing Technology and National University of Singapore (SIMTech-NUS) Joint Laboratory on Large-Area Flexible Hybrid Electronics, National University of Singapore, Singapore 117576, Singapore
[4] NUS Graduate School—Integrative Sciences and Engineering Program (ISEP), National University of Singapore, Singapore 119077, Singapore
* Correspondence: elelc@nus.edu.sg

Abstract: Entering the 5G and internet of things (IoT) era, human–machine interfaces (HMIs) capable of providing humans with more intuitive interaction with the digitalized world have experienced a flourishing development in the past few years. Although the advanced sensing techniques based on complementary metal-oxide-semiconductor (CMOS) or microelectromechanical system (MEMS) solutions, e.g., camera, microphone, inertial measurement unit (IMU), etc., and flexible solutions, e.g., stretchable conductor, optical fiber, etc., have been widely utilized as sensing components for wearable/non-wearable HMIs development, the relatively high-power consumption of these sensors remains a concern, especially for wearable/portable scenarios. Recent progress on triboelectric nanogenerator (TENG) self-powered sensors provides a new possibility for realizing low-power/self-sustainable HMIs by directly converting biomechanical energies into valuable sensory information. Leveraging the advantages of wide material choices and diversified structural design, TENGs have been successfully developed into various forms of HMIs, including glove, glasses, touchpad, exoskeleton, electronic skin, etc., for sundry applications, e.g., collaborative operation, personal healthcare, robot perception, smart home, etc. With the evolving artificial intelligence (AI) and haptic feedback technologies, more advanced HMIs could be realized towards intelligent and immersive human–machine interactions. Hence, in this review, we systematically introduce the current TENG HMIs in the aspects of different application scenarios, i.e., wearable, robot-related and smart home, and prospective future development enabled by the AI/haptic-feedback technology. Discussion on implementing self-sustainable/zero-power/passive HMIs in this 5G/IoT era and our perspectives are also provided.

Keywords: human–machine interface (HMI); triboelectric nanogenerator (TENG); artificial intelligence (AI); robot perception; wearable sensor; Internet of things (IoT)

Citation: Sun, Z.; Zhu, M.; Lee, C. Progress in the Triboelectric Human–Machine Interfaces (HMIs)-Moving from Smart Gloves to AI/Haptic Enabled HMI in the 5G/IoT Era. *Nanoenergy Adv.* **2021**, 1, 81–120. https://doi.org/10.3390/nanoenergyadv1010005

Academic Editors: Christian Falconi and Ya Yang

Received: 28 August 2021
Accepted: 15 September 2021
Published: 19 September 2021

Publisher's Note: MDPI stays neutral with regard to jurisdictional claims in published maps and institutional affiliations.

Copyright: © 2021 by the authors. Licensee MDPI, Basel, Switzerland. This article is an open access article distributed under the terms and conditions of the Creative Commons Attribution (CC BY) license (https://creativecommons.org/licenses/by/4.0/).

1. Introduction

With the rapid development of the Internet of things (IoT) and 5G communication technology in recent years, human–machine interfaces (HMIs) have gradually evolved from traditional computer peripherals, e.g., keyboard, mouse, and joystick as illustrated in Figure 1, to a more intuitive interface that directly collects human's original signals [1–3], such as voice and basic body motions, providing users with a more intuitive and easier interaction with computers and intelligent robots in the applications of healthcare, rehabilitation, industrial automation, smart home, virtual reality (VR) game control, etc. [3–5]. Current commercialized advanced HMIs include non-wearable solutions based on voice and vision recognition [6–8], and wearable solutions that use inertial measurement units (IMUs) [9,10], i.e., microelectromechanical system (MEMS) based accelerometers and gyroscopes, for the somatosensory information collection. However, these kinds of HMIs require complicated sensory information and high-performance signal collection/processing units, thus resulting in great power consumption in the system. Additionally, the MEMS-based sensing components are commonly bulky and rigid [11], making them relatively incompatible for wearable scenarios. For these issues, stretchable and flexible HMI solutions with minimalistic sensor design emerge recently, including sensors based on the sensing mechanisms of resistive [12–20], capacitive [21–24], and optical fiber [25–27], etc. Though the signal complexity and processing cost are greatly reduced to save energy and enhance the timeliness of the system, these sensors themselves still require a small amount of driving energy, and the power consumption may be large considering the massive number of sensor nodes in a sensor network. Moreover, repetitive charging is also annoying, especially for wearable or portable HMIs. So to address these issues, self-powered sensors using nanogenerator technologies of piezoelectric [28,29], triboelectric [30,31] have been developed frequently, to build low-power/self-sustainable human–machine interactive systems.

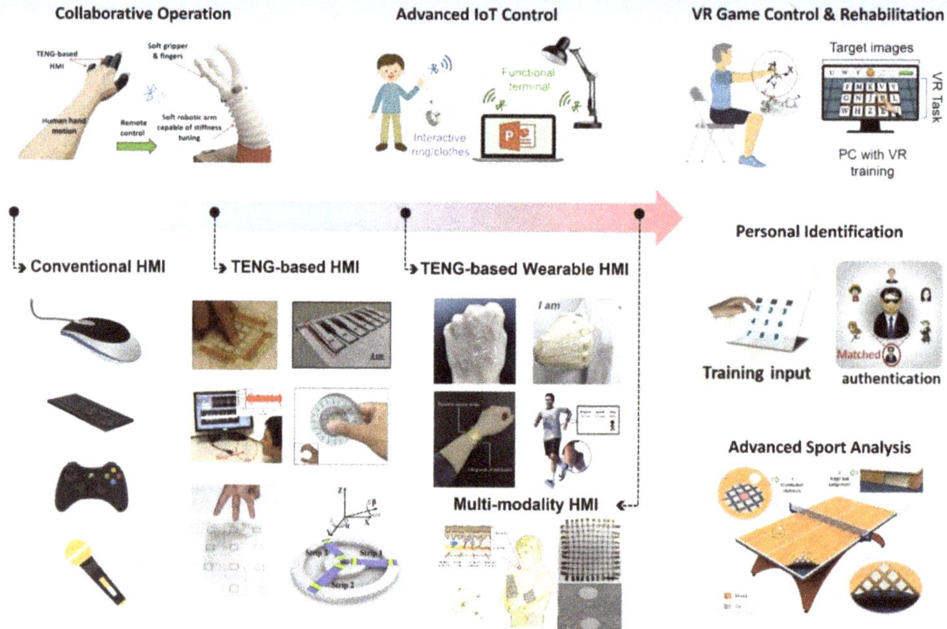

Figure 1. Schematic illustration for the development progress of triboelectric human–machine interfaces and their applications in the 5G/IoT era. Reprinted with permission from Reference [32], Copyright 2021, Wiley. Reprinted with permission from Reference [33], Copyright 2020, Wiley. Reprinted with permission from Reference [34], Copyright 2019, Springer Nature. Reprinted with permission from Reference [35], Copyright 2019, Elsevier. Reprinted with permission from Reference [36], Copyright 2019, Elsevier. Reprinted with permission from Reference [37], Copyright 2013, American Chemical Society. Reprinted with permission from Reference [38], Copyright 2019, Elsevier. Reprinted with permission from Reference [39], Copyright 2015, American Chemical Society. Reprinted with permission from Reference [40], Copyright 2018, Elsevier. Reprinted with permission from Reference [41], Copyright 2019, Wiley. Reprinted with permission from Reference [42], Copyright 2019, Elsevier. Reprinted with permission from Reference [43], Copyright 2017, AAAS. Reprinted with permission from Reference [44], Copyright 2018, American Chemical Society. Reprinted with permission from Reference [45], Copyright 2021, Wiley. Reprinted with permission from Reference [46], Copyright 2018, Elsevier. Reprinted with permission from Reference [47], Copyright 2018, Springer Nature.

Triboelectric nanogenerator (TENG), first reported by Prof. Zhong Lin Wang in 2012 [48], has been widely developed as energy harvesters for mechanical energy scavenging, ranging from natural wind energy [49–53], blue energy [54–59] to the human body's biomechanical energy [37,43,60–64], thanks to its exceptional merits of good output performance, wide material choices, good scalability, simple fabrication and low cost. Considering that the kinetic energies generated from the daily activities of humans, e.g., hand motion, joint rotation, foot motion, etc., contain valuable information of the corresponding motions, so such kinds of motion-induced energies could be collected by the nanogenerators towards a fully self-powered sensing strategy for human–machine interactions as well as health status monitoring purpose [65]. Compared with piezoelectric-based sensors that are commonly difficult for design customization due to the limitation of materials and complexity of the fabrication process [66–69], TENGs show the advantages of wide choices of stretchable and flexible materials, e.g., fabric, silicone rubber, plastic thin film, etc., and versatile operation modes, i.e., contact-separation mode, liner-sliding mode, single electrode mode and freestanding triboelectric-layer mode [70,71]. Therefore, TENGs have been successfully designed into various structures for different interactions (Figure 1), such as touchpad interface [35–38,41,72], auditory-based interface [39,73,74], 3D motion manipulator [33,40,42], etc., and can be further designed as self-powered wearable HMIs, e.g., electronic skin (e-skin) [43,75–77], data glove [32,44], wearable band [45,46], intelligent sock [78,79], breath-driven mask [80], etc., for advanced robotic manipulation, IoT control, VR game control/rehabilitation, personal identification and advanced sport analysis, showing the wide application prospects of triboelectric in HMIs area.

The new era of artificial intelligence (AI) provides a new possibility to enhance the functionalities of HMIs via machine learning (ML) enabled data analytics [81,82], where the subtle features hidden behind the real-time signal spectrum could be extracted automatically towards more advanced human–machine interactions, e.g., gesture recognition [83], voice recognition [84], pose estimation [85], personal identification [86], object classification [87,88], etc. Thus, due to these benefits, combining TENGs with ML-enabled analytics reveals a promising research direction for the development of HMIs with enhanced intelligence and low power consumption, which has attracted great attention in the past few years [73,89–93]. Besides, integrating TENGs with other sensing mechanisms, such as piezoresistive [76,94], pyroelectric [95–99], etc., to implement a multimodal sensing system capable of perceiving different sensory information simultaneously, i.e., tactile, strain, temperature, etc., is also a good strategy to broaden the applications of TENG-based HMIs. In addition, to build a complete human–machine interaction system, the haptic feedback function enabled by mechanical actuators [100,101], i.e., wire actuator, pneumatic actuator, vibration motor, etc., is indispensable to provide users with a more immersive experience for specific applications, e.g., robotic collaborative operation [102,103], VR game manipulation [104–108], etc., and deserves to be further integrated into current TENG-based HMI solutions to boost up the capability of information interpretation. Furthermore, by integrating self-powered TENG sensors with energy harvesters or passive wireless techniques, e.g., near-field communication (NFC) [109], surface acoustic wave (SAW) [110], etc., battery-free systems can also be achieved towards fully self-sustainable/zero-power HMI terminals under the future IoT framework.

Herein, in this review, we systematically introduce the recent progress in the TENG-based HMIs from the following sections: (1) glove-based HMIs for advanced manipulation, gesture recognition and tactile sensing; (2) wearable HMIs for other biomechanical signal collection, e.g., eye motion, facial expressions, voice, posture, etc.; self-powered HMIs for (3) robotic perception and (4) smart home applications; (5) ML-enabled intelligent HMIs, and the possible future research direction enabled by the (6) haptic-feedback technology and towards (7) self-sustainable/zero-power/passive HMI terminals. In the end, current issues and the potential development trends for TENG-based HMIs are also provided for future research in this 5G/IoT era.

2. Glove-Based HMIs

As we all know, nearly all the conventional HMIs are inseparable from the fine operation of our hands: the keyboard needs ten-finger tapping, the mouse needs to be moved and clicked by the hand, and the gamepad needs the finger to press the button and operate the joystick, etc. Compared with the proportion of nerve areas in other parts of the body, the sensory and motor nerve areas hidden in hands and fingers are huge [111,112], making each finger can bend and stretch independently to a certain extent, and can also expand and shrink the lateral intervals. This degree of flexibility enables them to complete quite complex gestures to achieve diversified operations. A glove, as a common wearable item, is quite suitable to be further designed into glove-based HMI by integrating sensors for highly sensitive finger motion tracking. One of the most mature sensing techniques for data gloves is to use IMUs [113,114], consisting of accelerometers and gyroscopes. However, the rigid sensing elements, complex data format and relatively high energy consumption remain concerns. Other flexible solutions based on resistive [13–15,115–117], optical fiber [25,27], etc., also reveal their own drawbacks, e.g., temperature effect, limited sensing range, etc. Thus, the emerging sensing technology based on triboelectric, with the advantages of diversified material selection, extremely simplified design and self-powered characteristics, provides a new research direction for designing the next generation of low-power data gloves [118–124].

He et al. proposed a glove-based HMI using TENG textile as the sensing unit with the minimalist design as shown in Figure 2a [125]. The conducting polymer made of poly (3,4-ethylene dioxythiophene): poly (styrene sulfonate) (PEDOT: PSS), due to its advantages of good physical and chemical stability, was chosen as the coating material to fabricate the highly stretchable electrodes as well as the positive triboelectric parts in this fabric-based contact-separation-mode TENG sensor. Additionally, a layer of silicone rubber thin film was coated on the textile glove serving as the negative triboelectric contacting layer. There are two kinds of sensor configurations in this design, where the arch-shaped sensor is utilized to measure the finger bending motions, while the sensors mounted at the sides of the fingers are to detect the contacts between the index finger and its adjacent fingers. Based on these two configurations, the movement of the index finger in all directions can be recognized to realize the intuitive control of aircraft minigame in cyberspace, and the car/drone movement in the real 3D space. Moreover, the function of a mouse can also be mimicked by this glove-based HMI for web surfing and alphabet writing, providing a simpler, power-compatible interactive method in daily life. However, for the actual applications of such a kind of textile glove, the humidity in the environment or the sweat generated from the human body may negatively affect the triboelectric output [126,127]. So to solve this issue, a more advanced design with superhydrophobicity of the triboelectric textile for performance improvement enabled by a facile carbon nanotubes/thermoplastic elastomer (CNTs/TPE) coating approach was reported by Wen et al. as shown in Figure 2b [128]. The anti-sweat capability enables the TENG sensor to remain at 80% voltage output even after a 1 h exercise. By leveraging machine learning technology, complex gestures' recognition function could be realized with the minimal sweat effect, which was successfully demonstrated to be further utilized for real-time VR/AR control applications, i.e., shooting games, baseball pitching and floral arrangement.

The abovementioned two works are all based on the arch-shaped TENG strain sensors [129,130], meaning that a large air space needs to be reserved between two contact layers and a relatively limited sensing range: obvious sensor response only occurs at the moment of contact and separation of the friction layers. For this problem, Zhou et al. proposed a TENG strain sensor based on a unique yarn structure as illustrated in Figure 2c [131]. The core of the sensing unit is composed of a conductive yarn coiled around a rubber microfiber, with the entire body sheathed by a polydimethylsiloxane (PDMS) sleeve. Varying degrees of deformation will result in a constant and continuous change in the contact area between the PDMS sleeve and the coiled conductive yarn, enabling the sensor with good linearity and sensitivity within a large strain range (20–90%). After integrating a wireless printed circuit board (PCB) for signal collection, processing and transmission, a wearable sign-to-speech translation system could be achieved with the multi-class support vector machine (SVM) machine-learning algorithm, whose overall accuracy could be maintained higher than 98.63% with fast response time (<1 s), showing a cost-effective approach for assisted communication between signers and non-signers, as well as the prospect of TENG-based HMI in the field of healthcare.

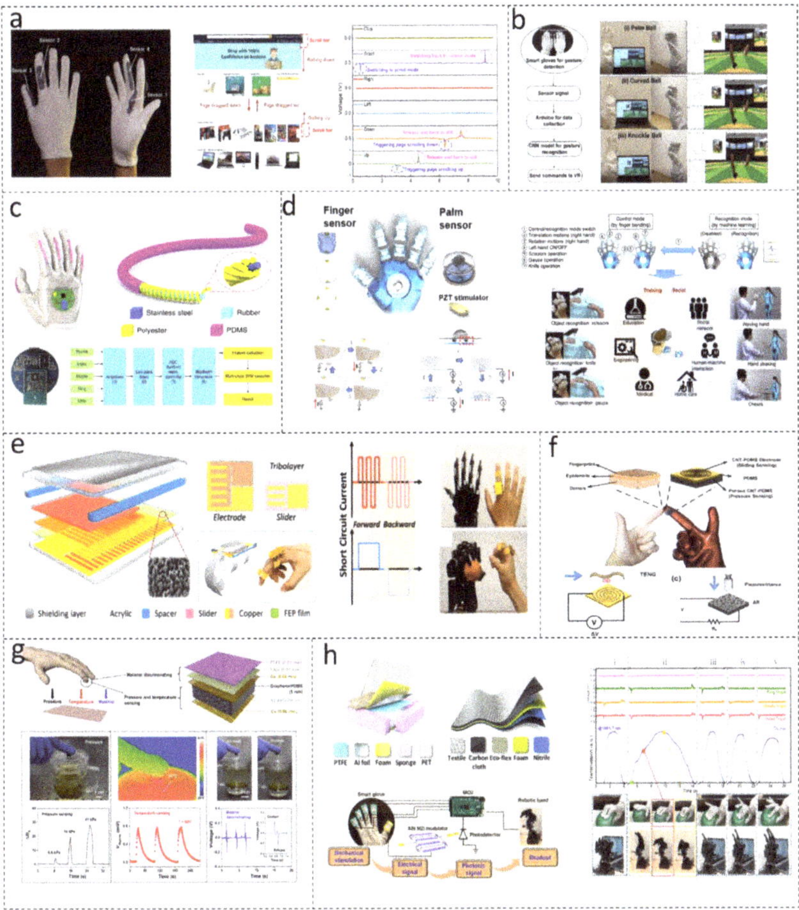

Figure 2. Glove-based HMIs. (**a**) A smart glove using TENG textile sensor for cursor control and web surfing application. Reprinted with permission from Reference [125], Copyright 2019, Elsevier. (**b**) A superhydrophobic textile glove enabled by carbon nanotubes/thermoplastic elastomer (CNTs/TPE) coating for VR/AR applications. Reprinted with permission from Reference [128], Copyright 2020, Wiley. (**c**) A yarn structural TENG strain sensor enabled glove for sign language recognition. Reprinted with permission from Reference [131], Copyright 2020, Springer Nature. (**d**) A multifunctional glove capable of bending sensing, sliding event detecting, as well as haptic stimulation for augmented AR/VR experiences. Reprinted with permission from Reference [132], Copyright 2020, AAAS. (**e**) A joint motion TENG quantization sensor enabled glove for the robotic collaborative operation. Reprinted with permission from Reference [133], Copyright 2018, Elsevier. (**f**) An electronic skin integrating triboelectric and piezoresistive sensing mechanisms for grasping tactile perception. Reprinted with permission from Reference [94], Copyright 2017, Elsevier. (**g**) A multifunctional fingertip tactile sensor capable of pressure sensing, temperature perception and material identification. Reprinted with permission from Reference [134], Copyright 2020, AAAS. (**h**) Nanophotonic modulator enabled readout strategy for TENG-based continuous pressure sensing. Reprinted with permission from Reference [135], Copyright 2020, American Chemical Society.

To further enrich the function of the glove-based HMI, Zhu et al. integrated TENG-based finger bending sensors, palm sliding sensors, and also piezoelectric mechanical stimulators onto one 3D-printed glove which realizes the multidirectional bending sensing, sliding event detecting, as well as haptic stimulation simultaneously for augmented AR/VR experiences (Figure 2d) [132]. By attaching multiple elastomer-based TENG tactile sensors onto different joints of each finger, the perception of motions of each phalanx with multiple degrees of freedoms (DOFs) could be achieved, where these sensors provide more useful subtle features regarding the finger bending compared with other common solutions that installing one sensor node per finger [136–139]. In the meanwhile, the sensory information of the normal and shear force could also be collected by the TENG palm sliding sensor to realize diversified sensing, especially for grasping-related tasks, revealing a new possibility of multifunctional HMI solution based on TENGs for VR entertainment and training applications.

Though these works based on the contact-separation mode TENG strain sensors have shown the great potential of TENG enabled glove-based HMI as effective gesture interfaces, limitations are still there and need to be further improved. One of the main issues is that most of these studies use the generated output peaks' amplitudes to judge the bending degree [120,140], which is unstable and could be easily influenced by the varying environmental factors, e.g., humidity, etc. Due to the fast decline of the stimuli-induced electrical states caused by electrostatic equilibrium, the generated pulse-like signals can only reflect the momentary motion of finger bending. Other valuable information, e.g., bending speed, the degree at a certain moment during the entire bending movement, etc., will be lost in such kind of signal format. One possible solution is to use grating-sliding mode TENG for better quantifying the bending degree/speed by counting the generated output peak number. A joint motion TENG quantization sensor for the robotic collaborative operation application was developed by Pu et al. as illustrated in Figure 2e [133]. When the slider is driven by the finger to slid forward/backward across the well-designed interdigitated electrodes, the alternating output signal in a series of periodic narrow pulses will be generated, where the bending degree and speed could be easily distinguished based on the pulse number and width respectively. The minimum resolution of the TENG joint sensor can reach 3.8° and could be further improved with finer grating segments. With such kind of measurement method, the real-time continuous robot bending control can be realized with high precision, demonstrating a more stable and reliable solution of TENG-based strain/displacement sensor with strong tolerance to environmental interference.

Besides the strain sensors for the finger bending monitoring, the tactile sensory information also plays a key role in glove-based HMI for mimicking the biological perception system of human skin to provide a more complete anthropomorphic feedback in contact force detecting, roughness recognition, as well as temperature sensing applications [141–143]. Due to the high sensitivity and fast response to tiny stimuli, TENGs have been frequently investigated to simulate the fast adapting (FA) sensory cells of the skin that respond to dynamic touch and vibration [144,145] and can be further integrated with other traditional sensors based on resistance or capacitance that mimic the slow adapting (SA) cells due to the signal maintenance capability, to form a multimodal tactile sensory system [76,146]. As depicted in Figure 2f, an electronic skin that simultaneously perceives the lateral and vertical movements of the fingertip during grasping tasks was reported by Chen et al. [94]. In this design, the carbon nanotube-poly-dimethylsiloxane (CNT-PDMS) electrode layer works as a freestanding TENG sensor for roughness differentiation according to the generated output peaks when the device slides across the object surface, where rougher objects commonly contribute more peaks during the sliding motion. The porous CNT-PDMS layer serves as the static pressure sensor based on the mechanism of piezoresistive, making the tactile HMI also capable of real-time status monitoring, e.g., holding or releasing, etc., during the grasping/tapping operation process. A similar but more advanced multifunctional tactile sensor was proposed by Wang et al. as shown in Figure 2g [134]. The main difference between this device and previous tactile designs lies in the functionality of the TENG layer. The electrification layer made of hydrophobic polytetrafluoroethylene (PTFE) film here is utilized for material identification based on the varying electron affinity ability of different materials [147], and 10 different flat materials have been successfully demonstrated to be inferred with a simple lookup table algorithm. Moreover, apart from the pressure sensing ability brought by the piezoresistive property, the sponge-like graphene/polydimethylsiloxane (PDMS) composite also shows thermoelectric effects, which can be used for detecting the temperature of contacted objects with a high resolution of 1 kelvin, reveling the possibility to further enhance the diversity of the functions for TENG-integrated tactile HMIs.

Limited by the pulse-like output signals as mentioned above, the TENG-based tactile sensing units developed in most works were used for dynamic touch/vibration sensing and cannot be used to detect the continuous force variation due to the fast shift of electrical states. In this case, combining with resistive/capacitive sensors becomes necessary, and a fully self-powered sensing system could not be achieved. Actually, the problem of the electrical state shifts could be suppressed by using a high-impedance readout circuit to maintain the triboelectric output [148,149]. However, a complicated sensing system with an amplifying circuit for small current collection is needed. To address this issue, Dong et al. proposed a strategy by using the robust nanophotonic aluminum nitride (AlN) modulators to read the TENG output, as shown in Figure 2h [135]. The TENG sensor can work in the open-circuit condition with negligible charge flows due to the electrically capacitive nature of AlN modulators, and the stimuli-induced triboelectric voltage can be transformed to AlN modulators' optical output based on the electro-optic Pockels effect. Owing to the negligible charge flow and the high-speed optical information carrier, continuous force sensing with good linearity and stability was successfully achieved based on the TENG tactile sensor with nanophotonic readout circuits, demonstrating the potential to replace resistive/capacitive sensors as SA sensing units in anthropomorphic skin, and to realize a fully TENG-based self-powered multifunctional tactile HMI.

3. Other Wearable HMIs

In addition to the glove-based HMIs, due to the advantages of high output, lightweight, high flexibility/stretchability and applicability of various structural designs [150,151], TENGs have also been developed into sundry biomechanical sensors for other biosignal collection, e.g., eye motion, facial expressions, voice, posture, etc., as self-powered wearable HMIs to bring great convenience to people [152–154], especially those with disabilities, in the era of information and IoT.

Among these mechanical motions, eye blinking has been proven as a new, simple and effective triggering method for handicapped people to realize convenient electrical appliance control for smart healthcare/home purposes [155–157]. A TENG-based micromotion sensor for eye blink motion monitoring with high sensitivity was reported by Pu et al. as shown in Figure 3a [158]. This sensor works in single-electrode mode and has a multilayered structure, where a fluorinated ethylene propylene (FEP) thin film and a natural latex thin film serve as the tribo-layers in this device, with an acrylic thin annulus as the spacer to reserve the necessary space for the contact and separation process. This tiny sensor could be easily mounted on the arms of glasses by fixators to capture the mechanical micromotion of the skin around the eyes, and function as an intuitive HMI to control electrical appliances, e.g., table lamp, electric fan and doorbell, via the simple trigger signal generated while eye blinking. By using a wireless module for data transmission, a hands-free typing system based on the input method of adjusting the number of blinks in a period was also successfully demonstrated, which may bring great convenience to our daily life, especially for the disabled, or people whose hands are fully occupied while working. Similarly, a non-attached electrode-dielectric TENG sensor for eye blinking sensing was reported by Anaya et al. as illustrated in Figure 3b [159]. This sensor is fully made of soft materials, i.e., PEDOT:PSS coated conductive textile and silicone rubber, and shows good comfortability when placed on the lateral skin tissue of the eye to detect the orbicularis oculi muscle motion. Due to the near-field transmission of signals based on the non-contact electrostatic induction, the output electrode could be mounted on the temple of the eyeglasses without a wire connected to the main sensor unit, which significantly improves the simplicity of the whole system. By analyzing the trigger signal induced by eye blinking, diversified hands-free human–machine interaction applications, including cursor control, car/drone control, etc., could be realized to assist people with mobility impairment. Additionally, the developed HMI could also be used for eye fatigue monitoring to evaluate the tiredness of a driver and give proper warnings to avoid danger.

Apart from the eye movements, fluctuations induced by the movements of other facial muscles, e.g., masseter muscle, etc., are also good choices to be translated into control command as communication aid HMI for the disabled [160]. Inspired by the frogs' croaking behavior, Zhou et al. reported a bionic TENG-based sensor for masseter muscle motion monitoring as illustrated in Figure 3c [161]. By imitating the oral structure and acoustic capsule, the flexible PDMS elastomer was made into a sensing membrane and a deformable vibrating membrane, with an air layer as the spacer, to amplify the small fluctuations of the masseter muscle into a significant movement of the vibrating membrane, due to the varying deformations of films with different radiuses under the same volume change. TENG technology was then integrated to convert the vibration of the film into an electrical signal, which could be further utilized in a Morse code communication system, a hands-free typing tool, as well as an intelligent authentication system, to achieve the efficient collaboration between the disabled and the digitalized world.

Another biosignal that can be effectively utilized for the human–machine interaction purpose is the human voice, which has attracted great attention recently due to the rapid development of voice recognition based on artificial intelligence technology [162–164]. TENG-based acoustic devices have also been investigated a lot towards self-powered microphones [39,73,74,165–168]. Although these flexible acoustic devices show good performance and functionality, they are still rigid, which hinders their applications in the wearable device platform that can be comfortably integrated with the human skin. For this issue, Kang et al. developed a skin-attachable microphone with high transparency and adhesion by using hybrid freestanding nanomembranes (NMs) combined with a micropatterned PDMS and a holey PDMS film in a sandwich structure as illustrated in Figure 3d [169]. The holey PDMS film effectively enhances the vibration of the freestanding NM membrane compared with a planar PDMS film, result in a larger output voltage and excellent sensitivity. The time-dependent waveform of a speech recorded by the wearable TENG microphone was also demonstrated, which was very much in line with the original sound waveform, proving the good acoustic sensing capability of the developed microphone especially when attached to a person's neck. Then a personal voice security system was successfully established to recognize the users' identity with high accuracy by perceiving the collected voiceprint, showing its potential as HMI for biometric purposes.

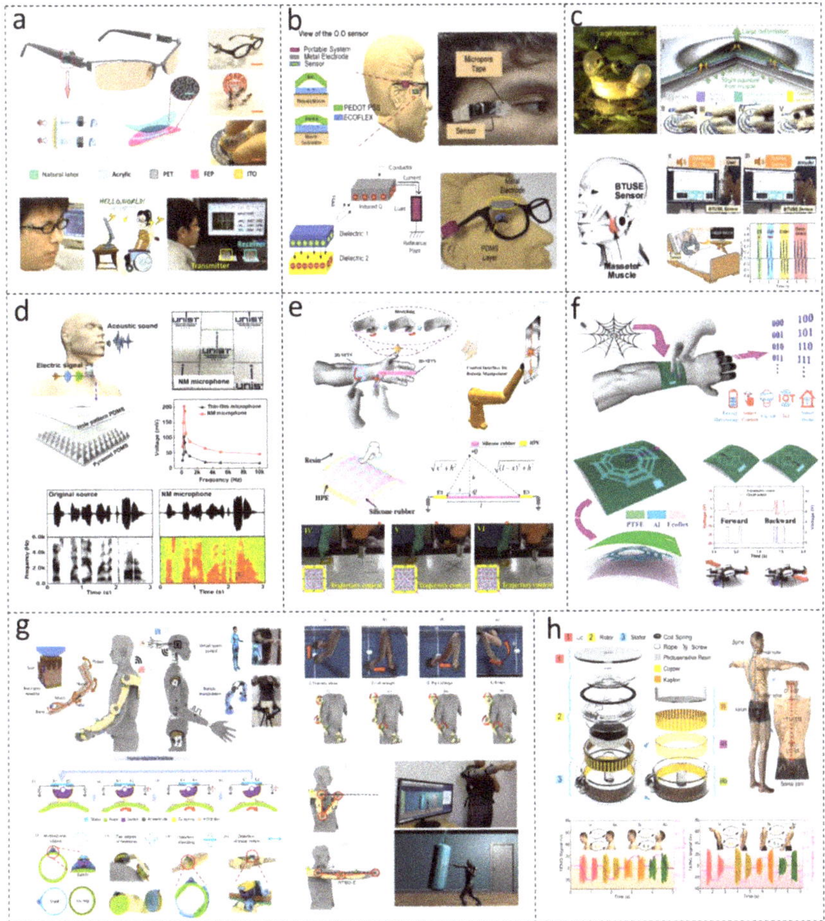

Figure 3. Other wearable HMIs. (**a**) A TENG-based micromotion sensor for eye blink motion monitoring. Reprinted with permission from Reference [158], Copyright 2017, AAAS. (**b**) A non-attached electrode-dielectric TENG sensor for eye blinking sensing. Reprinted with permission from Reference [159], Copyright 2020, Elsevier. (**c**) A bionic TENG-based sensor for masseter muscle motion monitoring. Reprinted with permission from Reference [161], Copyright 2021, Wiley. (**d**) A skin-attachable TENG microphone with high transparency and adhesion. Reprinted with permission from Reference [169], Copyright 2018, AAAS. (**e**) A wearable two-dimensional TENG touchpad for robotic arm manipulation. Reprinted with permission from Reference [170], Copyright 2018, American Chemical Society. (**f**) A bioinspired spider-net-coding (BISNC) TENG patch for multidirectional drone control. Reprinted with permission from Reference [171], Copyright 2019, Wiley. (**g**) A minimalistic exoskeleton enabled by triboelectric bidirectional sensors for upper limbs' joint motion monitoring. Reprinted with permission from Reference [172], Copyright 2021, Springer Nature. (**h**) A badge-reel-like TENG stretch sensor for spinal information collection. Reprinted with permission from Reference [173], Copyright 2021, Springer Nature.

In addition, a tactile sensing patch is also a common form of wearable HMI by collecting finger touching/sliding movements to realize simple manipulation commands [174–179] Currently, TENG tactile HMIs commonly consist of a large number of sensing pixels, where each pixel is connected to an independent output, resulting in complex readout circuits and output signals [180–187]. In order to simplify the outputs to effectively reduce the difficulty and cost of data collection and processing, some novel solutions with minimalistic electrode design have been demonstrated recently [188–193]. A two-dimensional TENG patch with a

5×5 pixel matrix for finger trajectory sensing was developed by Chen et al. as shown in Figure 3e [170]. With only four edge electrodes, the accurate position of the finger sliding or touching along the x and y axis of the patch surface could be achieved with a minimum resolution of ~1.6 mm based on the output ratio of two pairs of opposite electrodes, which greatly reduce the total number of output channels for multi-pixel sensing. By utilizing another one-dimensional TENG patch to detect the position along the z-axis, the real-time three-dimensional manipulation of a robotic arm was demonstrated and could be applied in various complicated operations, e.g., welding, handling, spraying, etc. A more advanced design was proposed by Shi et al., as depicted in Figure 3f [171], where a bio-inspired spider-net-coding TENG interface was developed for multi-directional sliding sensing with only one output electrode. By adjusting the electrode widths and positions along with different directions, the output signal patterns could be differentiated according to the relative amplitude and positions of the generated peaks in the time domain and could be further transferred into binary codes for more straightforward signal processing. Such a coding method enables the minimalistic TENG tactile interface with good reliability and robustness, and makes it applicable for diversified human–machine interaction applications, e.g., drone control, security code, etc.

Though some of the abovementioned HMIs have shown the possibility to realize the robotic control by defining specific commands according to the sensor signals, the number of commands is limited by the sensor design and data format, and as a result, diversified operations for multi-tasks may not be achieved. Additionally, some customized interaction commands also require a certain learning cost for the users. To realize a more intuitive HMI for parallel manipulation with enhanced degrees of freedom in the applications of advanced industrial automation or virtual reality interactions, the sensory information of the human pose is of great significance [194–197]. Zhu et al. proposed a customized exoskeleton enabled by triboelectric bidirectional sensors for upper limbs' joint motion monitoring as depicted in Figure 3g [172]. With the well-designed grating patterns and the bistable switch integrated into the sliding-mode TENG rotational sensor, the rotating degree, direction, and speed can be achieved simultaneously to accurately reflect the real-time status of shoulder rotations, wrist twisting, as well as finger bending with a minimum resolution of $4°$, and can be further utilized to collaboratively control the robotic arms and virtual character in both real and cyber space. Moreover, a ping-pong/boxing game was demonstrated to verify the capability of the control system for complex and coherent movements, revealing its potential for virtual sports training and rehabilitation applications. Apart from the movements of upper limbs, spinal bending is also essential towards a more complete pose monitoring. A badge-reel-like stretch sensor based on a similar grating-structured TENG was reported by Li et al. as shown in Figure 3h [173]. By analyzing the generated peaks' number, high sensitivity of 8 V mm^{-1} and a minimum resolution of 0.6 mm with good robustness and low hysteresis can be achieved for this wearable badge reel, which can be used for patients' spinal shape change monitoring when serving as a rehabilitation brace. In addition, such a kind of spinal information is valuable when combined with other limb motions towards a whole-body movement detection, which can be widely used for human motion capture and reconstruction in the 3D animation/game industry.

4. Robotic-Related HMIs

With the gradual rollout of AI technology around the world, intelligent robots will play a more important role in our society and will gradually replace humans in labor-intensive or dangerous tasks, from the industries of service, manufacturing and medical, to daily life assistant, as well as future scientific-related space exploration [198]. As the medium to perceive the external world and enhance the interactions with humans, sensors based on TENGs have also been investigated a lot to mimic the bionic sensory system for robots' tactile sensing [199–203], gesture/motion monitoring [204,205], gait analysis [206], etc., thanks to their good compatibility brought by TENGs' wide material choices.

As shown in Figure 4a, a typical TENG tactile sensor for robot perception was reported by Yao et al. [207]. With the novel interlocking microstructures enabled by the surface morphology of natural plants, and the polytetrafluoroethylene (PTFE) tinny burrs fabricated on the tribo-layers, the pressure sensing sensitivity was effectively enhanced by 14-fold compared with sensors using flat tribo-surfaces. Owing to the high flexibility, the developed sensor can be easily attached to a robotic hand, to measure the pressure distribution, as well as finger bending, during handshaking with humans. Additionally, the capability of the sensor for surface roughness and object hardness recognition was also demonstrated, showing its potential for more advanced robotic dexterous operation and human–machine interactions.

In addition to the perception of external tactile stimulation, auditory information is also a straightforward communication strategy for robots to receive feedback information from the outside world and interact with humans [208,209]. A TENG-based auditory system for social robotics was reported by Guo et al. as illustrated in Figure 4b [210]. This developed tiny auditory sensor shows ultrahigh sensitivity of 110 mV/dB based on the triboelectricity generated by contact-separation motions of the Kapton and FEP layers during the air flow-induced vibrations. With the special design of the annular or sectional inner boundary architecture, a broadband response from 100 to 5000 Hz could be achieved, which almost covers the human voice's frequency range. After being integrated onto a smart robot, the auditory sensor was successfully utilized to capture the music sound waveform with high quality, which can be further used for high-accuracy voice recognition in human–robot interaction applications.

During the parallel control process of robots, the sensing of the robot's pose and motions is also critical, which can be used as feedback information to achieve more precise control and status monitoring [211,212]. As depicted in Figure 4c, Wang et al. reported a self-powered angle sensor that can be mounted on robotic arms for high-resolution angular monitoring [213]. Two rotary sliding TENGs are integrated with the difference in the overlaps of electrodes, to form a detectable phase difference between the two output electrodes for rotating direction detecting, i.e., clockwise or anti-clockwise. By counting the generated pulse number and calculating the corresponding consumption time, the overall rotation angles and the angular velocity can be obtained respectively, with a high resolution (2.03 nano-radian), high sensitivity (5.16 V/0.01°), and good signal-to-noise ratio (98.68 dB). A robotic arm equipped with this angle sensor was successfully controlled to reproduce the traditional Chinese calligraphy, proving the effectiveness of the collected signals for accurately reflecting the movement trajectory of the robot.

Thanks to the merits of good flexibility and multi-degree deformation [214–216], soft robots [217–219] made of soft materials, e.g., silicone rubber, thermoplastic polyurethane (TPU), etc., can perform conformal contact with external objects and environments, which make them applicable to various scenarios instead of making specific designs for different product lines like the widely used rigid robotic manipulators, thereby greatly reducing the costs. Considering that many commonly used TENG and soft robot materials have similar Young's modulus, flexible TENG-based sensors have been frequently developed recently to realize the tactile/deformation perception for soft robots with good compatibility [220–223].

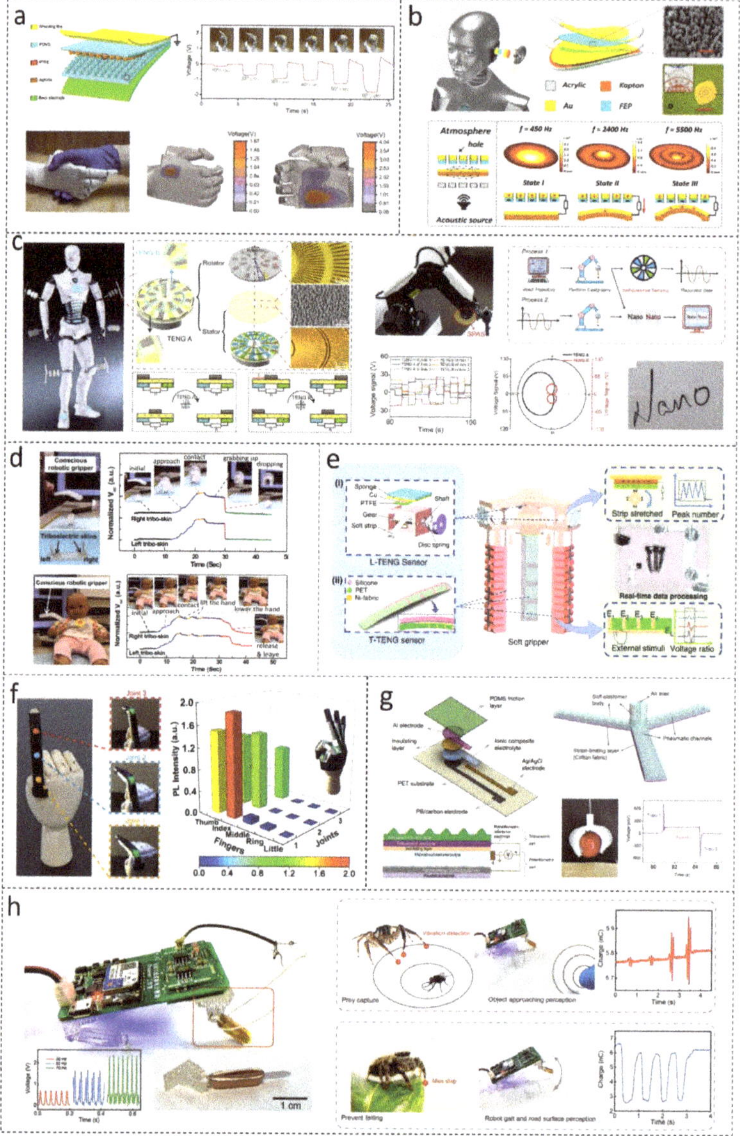

Figure 4. Robotic-related HMIs. (**a**) A TENG tactile sensor with interlocking microstructures for touch pressure perception. Reprinted with permission from Reference [207], Copyright 2019, Wiley. (**b**) A TENG-based auditory system for social robotics. Reprinted with permission from Reference [210], Copyright 2018, AAAS. (**c**) A TENG angle sensor for high-resolution angular monitoring of the robotic arm. Reprinted with permission from Reference [213], Copyright 2020, Wiley. (**d**) A flexible/stretchable TENG skin for soft robot tactile perception. Reprinted with permission from Reference [224], Copyright 2018, Wiley. (**e**) An intelligent soft gripper enabled by TENG-based tactile and bending sensor for grasped object recognition. Reprinted with permission from Reference [225], Copyright 2020, Springer Nature. (**f**) A triboelectric-photonic hybridized smart skin for robot tactile and gesture sensing. Reprinted with permission from Reference [226], Copyright 2018, Wiley. (**g**) A self-powered potentiometric–triboelectric hybridized mechanoreceptor for soft robot tactile sensing. Reprinted with permission from Reference [227], Copyright 2020, Wiley. (**h**) A quadruped robot equipped with TENG-enabled biomimetic whisker mechanoreceptors for exploration applications. Reprinted with permission from Reference [228], Copyright 2021, Wiley.

A TENG robotic skin for soft robot tactile perception was proposed by Lai et al. as shown in Figure 4d [224]. The sensing skin is made of silicone rubber with triangular micro prisms patterned on the surface to enhance the pressure sensitivity (9.54 V kPa^{-1}). With the high stretchability of 100%, the sensor could be easily integrated into a soft gripper to sense the different motions during grasping tasks, including approaching, grabbing, lifting, lowering and dropping, according to the variation of the open-circuit voltage in different stages. The self-muscle motion perception of a robotic crawler could also be achieved by integrating multiple tribo-skins, showing its scalability for large-area soft robot tactile sensing. Besides the tactile sensing skin, Jin et al. also integrated a gear-structural TENG bending sensor into a soft gripper to form a multifunctional sensory system to detect the tactile and deformation related information simultaneously as indicated in Figure 4e [225]. With the distributed electrodes along with the tactile sensor patch, the accurate position of the external stimuli on the sensor surface could be extracted based on the output ratio of different channels. Moreover, the continuous bending monitoring of the pneumatic finger could be realized by counting the generated peak numbers of the TENG bending sensor, with a minimum resolution of about 12°. By fusing the data from these two types of sensors, more valuable information, including grasping position, contact area, object shape and size, could be achieved for grasping tasks, providing the possibility for the soft manipulator to implement high-accuracy object recognition with the help of machine learning analysis.

However, as mentioned in the tactile sensor part, the pulse-like output signals make the TENGs more sensitive to the external dynamic stimuli, and the real-time static status monitoring, i.e., deformation or pressure, of the soft robots can be achieved by fusing with other sensing mechanisms, e.g., capacitive, resistive, photonic, etc., towards a multifunctional sensory system [229,230]. As shown in Figure 4f, Bu et al. proposed a triboelectric-photonic smart skin for robots' tactile and gesture sensing [226]. By doping the aggregation-induced emission (AIE) powder into the flexible silicone rubber substrate with the grating-structured metal film for exposure area adjustment, the photoluminescence and photocurrent are tunable for continuous tensile measurement under the lateral stretching range of 0–160%. Due to the high electron affinity ability of the silicone rubber, the triboelectric output can also be generated when external stimuli occur, which can be used for vertical static pressure detection with a maximum sensitivity of 34 mV Pa^{-1}. After integrating onto a robotic hand, the precise joint bending and touch pressure monitoring can be achieved simultaneously, demonstrating the applicability of the triboelectric-photonic fused sensory system for soft robot-related HMI applications. Another hybridizing strategy based on the triboelectric and potentiometric for soft robot perception was reported by Wu et al. as illustrated in Figure 4g [227]. Inspired by the slow adapting (SA) and fast adapting (FA) capabilities of human skin, the potentiometric sensing mechanism that is sensitive to the static or slowly varying stimuli is utilized to detect the compressive strain induced by the external force, while the triboelectric mechanism suitable for dynamic stimuli sensing can provide the instantaneous signal information at the beginning or ending moments of the stimulation. With this multifunctional sensing capability brought by the complementary effects of these two mechanisms, more valuable information, including the pressure and duration during the process of approaching, touching, holding and releasing, can be obtained towards a more detailed object manipulation monitoring for soft robotic grippers.

Expert for these humanoid robots, animal-like exploration robots have also become a hot topic recently. By sending these robots to natural or harsh environments where people cannot go, valuable sensory data or physical tasks could be achieved for environmental monitoring or space exploration purposes [231–234]. To endow the robots with the capability to respond to complex environmental situations, e.g., avoiding obstacles, etc., varying advanced sensing technologies based on the mechanisms of piezoresistive, piezoelectric, optical and magnetic have been investigated [235–238]. However, thanks to the advantages of lightweight and low power consumption, TENGs have become a new strategy for developing self-sustainable exploration robots [230]. As shown in Figure 4h, inspired by the hair-based sensory system that animals used to explore the environment, An et al. developed a TENG-enabled self-powered biomimetic whisker mechanoreceptor for robotic tactile sensing [228]. The sensor consists of a fluorinated ethylene propylene (FEP) layer serving as the animal whisker, covered by a biomimetic hair follicle made of two metal electrodes. When the whisker swings between the two electrodes, the potential distribution will change due to the triboelectricity and electrostatic induction, which can be used to reflect the deflection direction and amplitude based on the signs and magnitude of the transferred charges, respectively. When integrated into a quadruped robot, the biomimetic whisker can help robots detect the movements of surrounding objects by analyzing the amplitude and frequency of the vibrational signal. The real-time pressure applied on its feet can also be achieved to reflect the gait and ground environment information, which can effectively help it pass complex roads in harsh environments for exploration applications.

5. HMIs for Smart Home Applications

The rapid development of numerous smart electronics enables a large variety of applications in the ambient environment to realize the smart home, with the intelligent monitoring and response systems for healthcare monitoring, elderly/children care, fall detection, body motion monitoring, automation, and security [239,240]. The current mainstream technologies for smart home applications include camera-based image recognition and commercial humidity/temperature sensors [241]. Although these technologies are continuously developing, drawbacks still exist such as video-based privacy concerns, large power consumption, bulky volume, and rigidity. Considering the fast development of the energy harvesting/self-powered sensing technology (triboelectric, etc.), lowpower HMIs can be realized with high convenience, low cost, and self-powered sensing ability to reduce the overall power consumption and extend the lifetime of the system [5,242,243]. Additionally, because of the energy harvesting capabilities, these HMIs can also scavenge the ambient energy and convert it into electricity for potential wireless communication.

As a good substitute for mechanical switches, flexible touchpads with multiple arrays/units have been investigated frequently by using TENG technology to achieve self-powered HMIs to interact with household appliances [244–247]. For these multi-array touchpads, there are two common ways for wire connection. One is connecting each unit to an output channel, which will lead to a messy wiring layout with complex processing circuits. Another method is to build cross nodes of X-axial channels and Y-axial channels, by which the total output channels will be reduced greatly, but the issues of crosstalk between the intersected channels could not be avoided. To solve these problems, Pu et al. developed a subdivision-structural 3D touchpad as illustrated in Figure 5a [248]. With the novel subdivision structure, the overlapping area between the intersected electrodes is significantly reduced, which can effectively suppress the crosstalk problems and improve the position sensing resolution. Additionally, the pressure sensing function was also realized by integrating another three-layer TENG unit for the touch or press differentiation purpose. This sensing array was successfully further designed into an anti-peek built-in code lock, where the information of the position and pressure can be integrated to form a more complex access password compared with the methods that are only based on the number location, effectively improving the safety factor of the authorization system in the smart home.

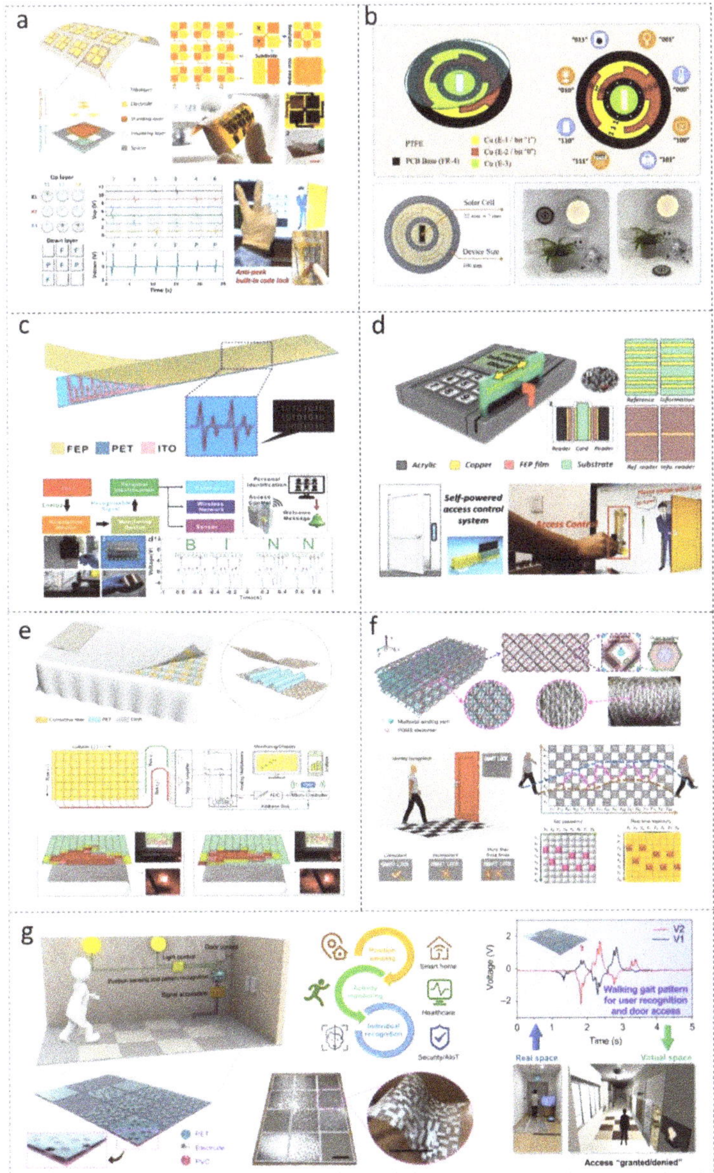

Figure 5. HMIs for smart home applications. (**a**) A subdivision-structural 3D touchpad enabled authorization system. Reprinted with permission from Reference [248], Copyright 2020, Elsevier. (**b**) A sliding-mode TENG-based control disk interface. Reprinted with permission from Reference [249], Copyright 2020, Elsevier. (**c**) A triboelectric-based transparent secret code. Reprinted with permission from Reference [250], Copyright 2018, Wiley. (**d**) A double-sided information card with a reference barcode component for reliability improvement. Reprinted with permission from Reference [251], Copyright 2017, Elsevier. (**e**) A large-scale and washable TENG textile enabled bedsheet for sleep behavior monitoring. Reprinted with permission from Reference [252], Copyright 2017, Wiley. (**f**) A self-powered identity recognition carpet system using TENG-based e-textile for safeguarding entrance. Reprinted with permission from Reference [253], Copyright 2020, Springer Nature. (**g**) TENG enabled smart mats as a scalable floor monitoring system. Reprinted with permission from Reference [254], Copyright 2020, Springer Nature.

Another smart method to simplify the output channels of touchpads is using binary codes to define more functions with as few electrodes as possible. As shown in Figure 5b, Qiu et al. reported a sliding-mode TENG-enabled control disk interface based on the binary coding mechanism [249]. Two sensing electrodes with specific patterns are attached to the panel, where one serves as the binary bit "1", and the other serves as the binary bit "0". Due to the variation in the width and location of these two electrodes along with different sliding directions, eight sensing transitions can be achieved based on the 3-bit binary code, e.g., "001","010", etc., for smart appliances control. Additionally, the sliding speed can also be utilized to adjust the light brightness or fan speed. After integrating a soler cell to form a hybrid energy harvester, the self-sustainable capability could be achieved by scavenging both the light energy and the mechanical energy from hand tapping. Moreover, this control disk can also be designed into a password input interface with more than 262,000 potential combinations for smart home authentication applications.

Apart from the password input panel, the barcode is also a widely-used strategy as the information carrier for personal identification/authentication applicable in smart home applications [255,256], and can be developed with the TENG technology towards a self-powered identification system. As depicted in Figure 5c, a triboelectric-based transparent secret code was proposed by Yuan et al. [250]. The information is hidden in the patterned indium tin oxide (ITO) electrodes, with a fluorinated ethylene propylene (FEP) film covered on top serving as the tribo-layer. The elongated design of the ITO stripes can effectively enlarge the contacting area within a short sliding time, and the different lengths will contribute to the variation of the output amplitudes, which can be further translated into the binary information of "1" or "0" with a reasonable threshold value by a sliding-check or roll-to-roll reader for security defense purposes. However, the peak height enabled sensing method is sliding velocity-dependent, and the non-uniform sliding speed may result in the error in the peak amplitudes, as well as the corresponding identified information. To solve this issue, Chen et al. proposed to use a reference barcode component to improve the reliability of the identification system [251]. As illustrated in Figure 5d, the information card is double-sided, where one side is patterned with the standard barcode electrodes with equal intervals as the reference component, and the other side is patterned with the information barcode that is aligned, swiped and measured simultaneously with the reference one. By comparing the number and positions of the generated positive and negative peaks from two sides, the coded information could be easily converted into the digital information of "1" and "0" with high accuracy even under non-uniform sliding speed by human hands, demonstrating a more reliable coding method for self-powered access system in practical usage.

In this fast-paced era, people are under increasing pressure from work and life, leading to more and more people being annoyed by the sleeping disorder, which may further increase the risk of other health problems, e.g., obesity, heart disease, and diabetes [257]. To realize the long-term sleep behavior monitoring for sleep quality assessment, many works have been done based on real-time pressure sensing [258,259]. However, most of these technologies are still restricted by the issues of low sensitivity, high fabrication cost, and non-washability, limiting their practical applications for large-area sleeping monitoring in daily usage. Recently, the fast development of TENG-based textile technology shows the great potential of TENGs to be further designed into smart clothes for wearable or household scenarios with high sensitivity and low cost [260–262]. By utilizing this technology, a large-scale and washable smart textile based on TENG arrays was developed by Lin et al. as shown in Figure 5e [252]. The proposed bedsheet has three layers, where a layer of wave-shaped PET film is sandwiched by two layers of Ag-coating conductive fabrics. When external stimuli occur, the applied pressure will enlarge the contact area between the PET layer and the two conductive layers, thus resulting in the electrical potential change and generating the outputs. With this optimal design, the smart textile shows a good sensitivity of 0.77 V Pa^{-1} with a fast response time (<80 ms), as well as high durability and stability even after being washed in tap water. By connecting multiple TENG units to form a sensing array, the information of the body's posture, position and pressure distribution over an entire night could be collected, providing a reliable dataset for analyzing the sleep quality. Additionally, the smart bedsheet can also serve as a warning system to prevent the elderly from falling off the bed, demonstrating a new possibility to realize the real-time remote healthcare service in the smart home.

To avoid the privacy concerns introduced by cameras, floor-embedded sensors can be implemented to extract the abundant sensory information associated with human activities in the smart home, such as the indoor position and gait-based individual identity [263–265]. Thanks to the advantages of low cost, easy fabrication and wide choices for triboelectric materials, TENG-based sensors commonly show good scalability and are compatible with large-scale manufacturing, making them suitable to be further designed into the floor-based HMIs [266–268]. A self-powered identity recognition carpet system enabled by TENG-based e-textile for safeguarding entrance was reported by Dong et al. as illustrated in Figure 5f [253]. The carpet consists of 128 black and white squares of the same size, where the black blocks are attached with the TENG fabrics serving as the sensing region, and the white blocks are used to reduce the interference between the adjacent sensing regions. According to the generated output peaks in these black sensing blocks, the walking trajectories of a visitor can be mapped with high accuracy and good stability, which can be further compared with the correct password path to validate the authentication. However, in this design, each sensing block is connected to an independent output channel, which means a complex wire connection and high signal processing cost. To simplify the output channels of the multi-array sensing system, a more advanced design of the smart mat was reported by Shi et al. as shown in Figure 5g [254]. Six distinct electrode patterns with varying coverage rates, i.e., 0% to 100% with 20% intervals, were designed and fabricated by screen printing, acting as sensing arrays for this floor mat. Due to that different electrode areas will contribute to different amounts of induced charge, the arrays are self-distinguishable and can be connected in an interval parallel manner to form a 3×4 mat array with minimal two-electrode outputs. So that the indoor positioning and activity monitoring can be achieved with minimal output terminals and minimized system complexity, which also benefits the backend signal processing and data analysis. Furthermore, with deep learning enabled data analytics, the identity information associated with gait patterns can be extracted from the output signals, and high recognition accuracy of 96% could be achieved for 10 persons based on their specific walking gaits, enabling diversified applications in the smart home such as position sensing, activity/healthcare monitoring, and security.

6. ML-Enabled Advanced HMIs

In terms of complex data analysis, the new trend of AI-enabled machine learning has shown a new direction for enhancing the functionalities of sensors [1,269,270]. Due to the powerful feature extraction capabilities of machine learning, more comprehensive/detailed sensory information can be utilized to realize diversified applications, e.g., gesture/pose estimation, voice recognition, object recognition, etc. [81,87,164], for advanced human–machine interactions as illustrated in Figure 6.

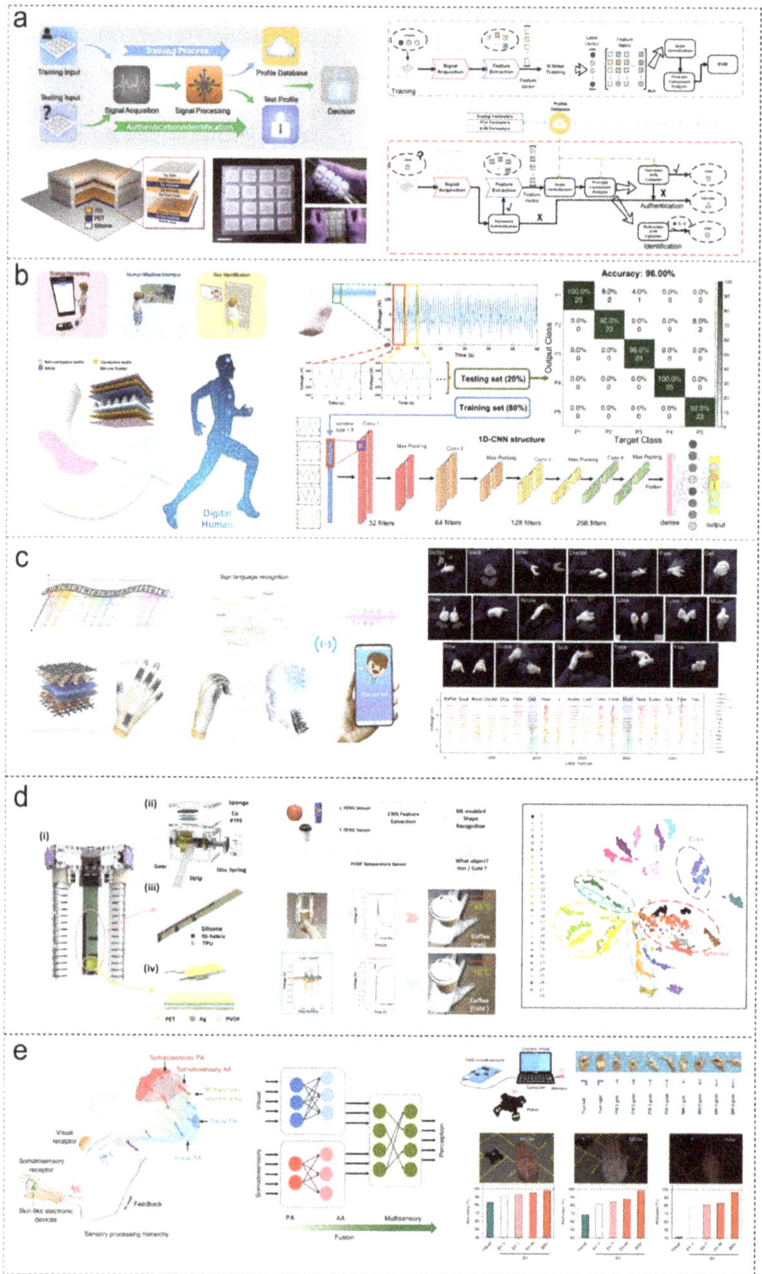

Figure 6. ML-enabled advanced HMIs. (**a**) A TENG-based smart keyboard for keystroke dynamics monitoring. Reprinted with permission from Reference [271], Copyright 2018, Elsevier. (**b**) Deep-learning-enabled TENG socks for gait analysis. Reprinted with permission from Reference [272], Copyright 2020, Springer Nature. (**c**) A deep-learning-enabled TENG glove for sign language translating. Reprinted with permission from Reference [273], Copyright 2021, Springer Nature. (**d**) A TENG-enhanced smart soft robotic manipulator for AIoT virtual shop applications. Reprinted with permission from Reference [274], Copyright 2021, Wiley. (**e**) A bioinspired deep-learning-based data fusion architecture integrating the vision data and somatosensory data for high-accuracy gesture recognition. Reprinted with permission from Reference [275], Copyright 2020, Springer Nature.

Recently, several AI-enabled TENG-based HMIs have been successfully developed [89,92,276–278] based on the algorithms of support vector machine (SVM), neural network (NN), etc., where the subtle features hidden in the triboelectric waveform, including contact sequence, impact vibration, etc., have been proven to effectively enhance the recognition capability of the intelligent sensory system. Compared with state-of-art works [279–281] that using a large number of resistive/capacitive sensor nodes for high-accuracy ML analysis, the minimalistic approach of TENGs shows comparable performance with significantly lower power consumption. Here, in this section, some typical examples of TENG-based intelligent HMIs for diversified applications are reviewed.

For cybersecurity applications, keystroke dynamics enabled authentication system has been proven as an effective approach to enhance the security level based on people's typing attributes with non-invasive monitoring characteristics [282–284]. A TENG-based smart keyboard for keystroke dynamics monitoring was developed by Wu et al. as illustrated in Figure 6a [271]. There are a total of 16 silicone-based keys with high flexibility and stretchability forming a multichannel keypad array, where each key consists of a contact-separation-mode TENG to convert the typing behavior into electrical signals, and a shield electrode to minimize the environmental interference. By using an analog-to-digital converter (ADC) to collect the open-circuit-voltage signals, keystroke-related features, i.e., typing latencies, hold time and signal magnitudes, can be acquired simultaneously with specific signal processing, e.g., denoising, baseline elimination, etc. Following the principal component analysis (PCA) for feature dimensionality reduction, a multi-class SVM classier was utilized to recognize the identities of 5 users based on the established dataset (150 sets of data for each user), and high accuracy of 98.7% could be achieved, showing the feasibility of the keystroke dynamics enabled authentication system for practical usage.

In addition to the keystroke dynamics, identification based on gait analysis is also a promising technology for biometric authentication applications [46,285]. With the help of artificial intelligence, complex personal information regarding the identity, health status as well as real-time activity of the users could be delivered at the same time by analyzing the gait patterns acquired from the floor- or sock-based sensory system. Zhang et al. proposed deep learning-enabled TENG socks for gait analysis as depicted in Figure 6b [272]. A textile-based TENG pressure sensor, consisting of a silicone rubber film with patterned frustum structures as the negative tribo-layer, a nitrile thin film as the positive tribo-layer and two conductive textiles as output electrodes, was fabricated and integrated onto a smart sock for gait monitoring with high sensitivity (0.4 V kPa^{-1}) and large sensing range (>200 kPa). With an optimized four-layer one-dimensional (1D) convolutional neural network (CNN) model for automatic feature extraction from the original walking spectrums, a high recognition accuracy of 96% could be achieved for 5 participants with varying weights, and could still be maintained higher than 93.5% when the number of people increased to 13, making it applicable for most indoor scenarios, e.g., home or office. By combining this intelligent sock with an IoT module for wireless communication, an artificial intelligence of thing (AIoT)-enabled two-stage recognition platform was established at the cloud server to realize the functions of family member identification and real-time indoor activities, i.e., run, walk and jump, monitoring simultaneously, revealing its potential for future smart home applications.

With AI for comprehensive sensory information extraction and autonomous learning, sophisticated hand gestures could be discriminated for glove-based HMIs towards advanced control or sign language interpretation applications [286,287]. Though several works have demonstrated the feasibility of developing TENG-based sign language perception platforms with minimalistic design and high recognition accuracy [131,139,288], most of them are limited to the identification of only several discrete and simple words or letters and are not suitable for real-time sentence recognition. To deal with these issues, Wen et al. developed a more advanced TENG-enabled sign language recognition system with improved glove design and training strategies as shown in Figure 6c [273]. Apart from the strain sensors that are mounted on each finger for finger bending monitoring, some sensors are placed on the wrists, fingertips and palms of the signers to capture subtle features towards more comprehensive sign language gestures. By utilizing a 5-layer CNN for feature extraction, high accuracy of 91.3% and 93.5% could be achieved for 50 words and 20 complete sentences respectively based on a non-segmentation method. To further extend its potential for new/never-seen sentences recognition, a segmentation strategy was used, where the signal spectrum of a whole sentence was split into word units first and then based on the correlation between the word units and the sentence, the original sentence's information could be reconstructed by the AI framework with an accuracy of 85.58%. With this novel learning approach, new/never-seen sentences could also be identified with an average accuracy of 86.67%, showing a new methodology to effectively expand the dataset and improve the practicality of the sign language recognition system.

As mentioned in the section on robotics-related HMIs, TENG-based sensors have been frequently investigated for developing low-power-consumption robot perception systems due to the advantages of high flexibility and self-powered property. However, most of the abovementioned works mainly focus on the simple manual analysis of the collected sensory information [221,223], e.g., deformation, tactile, etc. For realizing more advanced interactive functions, e.g., pose estimation, surface roughness perception and object recognition, AI-enabled analytics are needed, to capture more valuable features and endow the self-discrimination ability [225]. A TENG-enhanced smart soft robotic manipulator for AIoT enabled virtual shop applications was reported by Sun et al. as shown in Figure 6d [274]. A TENG bending sensor consisting of a rotating gear was utilized to monitor the real-time deformation of the pneumatic finger according to the generated peak numbers during the stretching process, and a TENG tactile sensor with distributed electrodes design was used for contact position and area detection. With the aid of 1D CNN ML algorithm for data processing, an intelligent robotic gripper that fuses these two kinds of sensory information could be easily achieved to realize a high recognition accuracy of 97.143% for 28 grasped objects with different shapes and sizes. Moreover, temperature distribution information could also be implemented by a poly(vinylidene fluoride) (PVDF) pyroelectric temperature sensor, towards a comprehensive and fully self-powered perception system. This AI-enhanced smart gripper with multifunctional sensing capability was further applied in a digital-twin-based virtual shop to provide users with real-time feedback information of the goods, as well as a more immersive experience of online shopping, showing great potential to realize advanced human–machine interactions in unmanned working space.

Due to the existence of environmental interferences, the performance of sensors could be affected in different environmental conditions, e.g., humidity effect for triboelectric sensors, etc. [127], which may result in poor stability of the established recognition system. For those TENG HMIs based on the sensing mechanism of grating sling or output ratio as mentioned above, though the output amplitudes will be influenced under varying environmental conditions, the generated peak numbers or the output ratio of distributed electrodes will not change, resulting in strong resistance to environmental interference. So, the ML result achieved from such kinds of output data will show better robustness without the effect of environmental elements. While for the ML results obtained from the real-time output spectrum, the variation in voltage amplitude under different environmental conditions will inevitably affect the results of machine learning. However, if we collect the data under different environmental conditions and combine them into a more generalized data set, i.e., each category contains the data that captured under different environmental conditions, the influence of environmental elements on accuracy will be avoided to a certain extent due to the more generalized trained model. Similarly, the influence from the different human behavior habits could also be avoided by enhancing the generalization ability of the dataset, where the data from different users are collected, thus taking into account the individual differences.

Another possible solution for this issue is to fuse sensors based on different mechanisms to build a more robust sensory system by utilizing the complementary effects of different sensors [289–291]. In Figure 6e, Wang et al. proposed a bioinspired deep-learning-based data fusion architecture that integrates the vision data and somatosensory data for high-accuracy gesture recognition in a harsh/dark environment [275], where the somatosensory information could contain complementary features for maintaining recognition accuracy especially when images are noisy and under- or over-exposed. To ensure high compatibility with visual information, a highly transparent stretchable strain sensor using single-walled carbon nanotubes (SWCNTs) was developed, which can collect high-quality somatosensory data without affecting the visual information. With the visual data captured by commercial cameras, a bioinspired somatosensory-visual (BSV) associated architecture was then reported for further multimodal data fusion, which can mimic the biological visual and somatosensory interactions in the area of the multisensory neuron of the human brain. By using this approach for human gesture recognition, high accuracy of 100% can be maintained even with low-quality images, proving the complementary effect of the somatosensory information on vision-based gesture recognition. Additionally, this architecture has also been demonstrated for more accurate robot manipulation, where 98.3% accuracy could be achieved with enough illumination and 96.7% accuracy could still be maintained in the dark environment, illustrating its applications for human–machine interactions in harsh environmental scenarios. Such a data fusion strategy provides a good development direction for future TENG-based HMIs with better stability and applicability.

7. Haptic-Feedback Enabled HMIs

The completeness of an HMI system not only rely on the dexterous sensing units which can monitor various human physiological signals and motions, but also requires a specialized feedback module that can provide necessary stimulations to assist the cognition of the manipulation status, so that a control-feedback loop is able to be established [292]. Noticeably, the consistency between the artificially stimulated sensation and the real sensation to humans is the key to improve the user experience and the value of the feedback information. Therefore, a majority of the feedback research is also focusing on the biomimetic stimulators, include kinesthetic feedback which can reflect the spatial movement of different body parts, and cutaneous feedback which is in charge of performing the tactile and thermal stimulations to various skin's receptors [293–296].

With the concern of conformability to the human body, wire and pneumatic actuators, as commonly used actuation techniques, are also adopted in feedback systems for achieving kinesthetic actuation [297,298]. Kang et al. reported a polymer-based soft wearable exo-glove using tendon-driven feedback, as shown in Figure 7a [299]. The main functional part consists of a body with stretchable designed finger modules, and the thimbles which are connected to the corresponding finger modules. The actuation wires are then covered by the sheaths embedded in the main body and linked to the motor box. This device can help disabled people to recover the capability of hand grasping. Motor-based vibrators are frequently used in mobile phones and joysticks as feedback units. Yu et al. developed a skin-integrated wireless haptic interface for VR and remote interactions by making the electromagnetic vibrator array (Figure 7b) [300]. This elastomer packaged device can be directly attached to the human skin, and it is powered wirelessly via an NFC coil. The design of serpentine Cu connectors ensures the stable performance under strain.

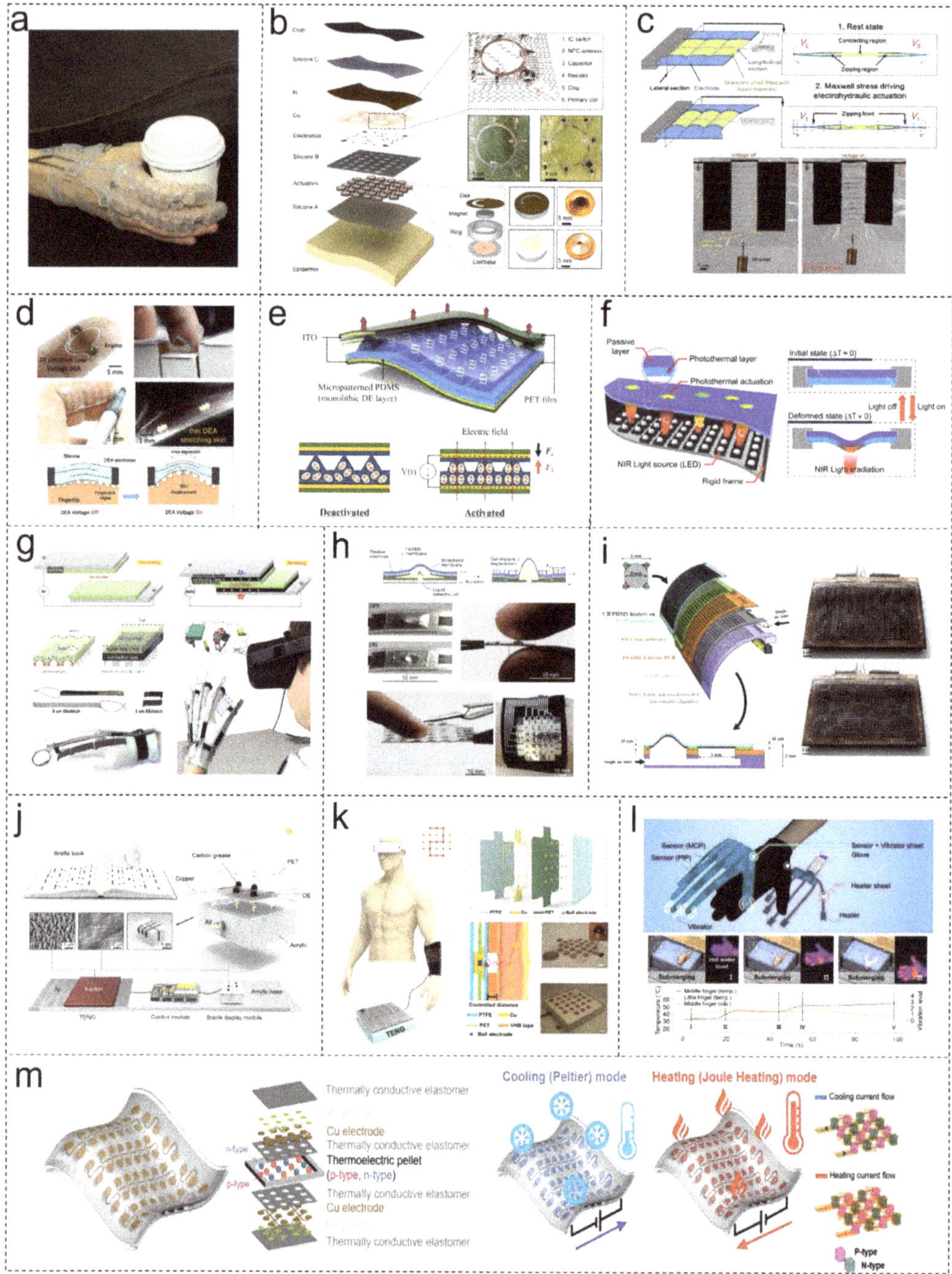

Figure 7. Haptic-feedback enhanced HMIs. (**a**) Soft wearable exo-glove using a tendon driven feedback for assisting people with

disabilities. Reprinted with permission from Reference [299], Copyright 2019, Mary Ann Liebert. (**b**) skin-integrated electromagnetic vibrator based wireless haptic interface for VR and remote interactions. Reprinted with permission from Reference [300], Copyright 2019, Springer Nature. (**c**) Mimicking the muscle motions via hydraulically amplified self-healing electrostatic actuator. Reprinted with permission from Reference [301], Copyright 2019, Wiley. (**d**) Low-voltage dielectric elastomer actuator (DEA) based fingertip haptic feedback can provide feel-through stimulation. Reprinted with permission from Reference [302], Copyright 2020, Wiley. (**e**) Pyramid microstructured DEA for vibrational stimulus under AC voltage. Reprinted with permission from Reference [303], Copyright 2018, AIP Publishing. (**f**) Near-infrared (NIR) light induced thermoelastic deformation for programmable vibrotactile feedback system. Reprinted with permission from Reference [304], Copyright 2021, American Chemical Society. (**g**) A glove with high force density electrostatic clutch for VR feedback. Reprinted with permission from Reference [305], Copyright 2019, Wiley. (**h**) A multimode electrostatic actuator with hydraulically amplified haptic feedback for creating tactile stimulus. Reprinted with permission from Reference [306], Copyright 2020, Wiley. (**i**) A large reconfigurable pneumatic haptic array made by shape memory polymer activated by heating. Reprinted with permission from Reference [307], Copyright 2017, Wiley. (**j**) Refreshable braille display system based on pneumatic actuation and TENG based DEA. Reprinted with permission from Reference [308], Copyright 2020, Wiley. (**k**) Electrical discharge based feedback system using TENG array with ball electrode. Reprinted with permission from Reference [309], Copyright 2020, AAAS. (**l**) A multi-modal sensing and feedback glove with liquid metal based resistive strain sensor, vibrator, and thermal feedback units. Reprinted with permission from Reference [310], Copyright 2020, Wiley. (**m**) Skin-like thermo-haptic device with thermoelectric units using Peltier effect. Reprinted with permission from Reference [311], Copyright 2020, Wiley.

Dielectric elastomer actuator (DEA) is becoming another popular research direction in recent years. The electric field induced deformation of DEA can be reshaped into vibration, stretching, bending and other modes, by applying the different designs [312]. In Figure 7c, Wang et al. presented a hydraulically amplified self-healing electrostatic actuator that can mimic the muscle contraction motions under the applied electric field [301]. The actuator is made of rectangular polymer shells filled with a liquid dielectric, and the electrodes are coated at both ends. Hence, the applied voltage can induce a Maxwell stress cause the electrode to zip together and the shell is then contracted due to the hydraulic pressure. This actuator achieved a maximum linear contraction of about 24% at a loading of 0.2 N. On the other hand, as shown in Figure 7d, a feel-through low-voltage DEA made by multi-layer PDMS based DEA sandwiched by single-walled carbon nanotube (SWCNT) electrodes for fingertip haptic feedback is demonstrated by Ji et al. [302]. The combination of three active layers is only 18 μm thick, which can ensure the mechanoreceptors of the skin remain sensitive to external stimuli. By applying the voltage, the surface area of the elastomer increases or decreases to stretch or compress the skin for feedback. Similarly, Pyo et al. presented a pyramid microstructured DEA layer. The applied AC voltage can also deliver the vibrational stimulus (Figure 7e) [303]. Interestingly, Hwang et al. developed a light-driven and low-power vibrotactile actuator using thin poly(3,4-ethylenedioxythiophene) doped with p-toluenesulfonate (PEDOT-Tos) and PET film, as illustrated in Figure 7f [304]. The irradiation of near-infrared (NIR) light from LED can initiate a thermoelastic bending deformation due to the mismatch of thermal expansion coefficient. Therefore, with the assistance of LED array, a programmable vibrotactile feedback system is demonstrated.

There are also researches focus on electrostatic force-based feedback. The applied voltage can cause the attraction of two electrodes under the difference of electrical potential. In Figure 7g, Hinchet et al. designed a high force density electrostatic clutch for making the VR feedback glove [305]. The clutch is simply made by the conductive textile with poly vinylidene fluoride, trifluoroethylene, 1,1-chlorotrifluoro- ethylene (P(VDF-TrFE-CTFE) based high friction insulation layer. The proposed device can generate frictional shear stress up to 21 N cm^{-2} at 300 V, which can be used for blocking the finger motion during VR events. In the meantime, Leroy et al. presented a multimode electrostatic actuator with hydraulically amplified haptic feedback, as illustrated in Figure 7h [306]. The liquid dielectric is encapsulated within a chamber formed by the flexible membrane and central stretchable membrane. The electrode pairs are then located at the surroundings of the center. The voltage-induced electrostatic force will force the electrode to attract and squeeze the outer region of the chamber, and hence, the central membrane will be expanded under hydraulic pressure. By controlling the electrode designs and activation sequences, the device can realize the multimode haptic feedback.

To enhance the feedback functionality of the pneumatic actuator with minimal air inlet, various techniques are also integrated to tune the activation region of the device. Besse et al. reported a large reconfigurable pneumatic haptic array with the aid of flexible shape memory polymer (SMP) membrane, as depicted in Figure 7i [307]. The heater can change Young's modulus of the SMP membrane and allow it the deform under positive or negative air pressure. As a result, the array design of the chambers and membranes enables the dense reconfigurable tactile system. On the other hand, Qu et al. developed a refreshable braille display system based on pneumatic actuation and TENG-based DEA (Figure 7j) [308]. A similar air chamber with a membrane design is adopted. As mentioned earlier, the DEA can be stretched under applied TENG voltage. Together with the air pressure, the DEA membrane is then raised up to form a braille dot.

Moving forward, TENG is also directly used as a feedback unit via the generated voltage. As shown in Figure 7k, Shi et al. designed a TENG array with ball electrodes to make a feedback system for VR [309]. When the slider is sliding across the TENG array, the electrical output can be generated via the triboelectrification effect. With the direct contact of the ball electrodes to human skin, the TENG electrostatic discharge can be delivered as electrical virtual tactile stimulation. Hence, the sliding trajectory can be sensed on the skin as a feedback function.

Thermal sensation, as another important function of human skin, can bring a new dimension of feedback to enrich the reconstruction of the environment in cyber or remote space. The collection of temperature information is also critical to prevent potential damage during remote control. Oh et al. developed a multimodal sensing and feedback glove with a resistive strain sensor, vibrator, and thermal feedback units (Figure 7l) [310]. By applying the direct ink writing technique, eGaIn liquid metal is printed into meandered shape to make the units with both strain sensing and thermal feedback functions via the power supply. Moving forward, the capabilities of both heating and cooling is the next challenge for those wearable and flexible feedback systems. As illustrated in Figure 7m, a skin-like thermo-haptic device with thermoelectric units is developed by Lee et al. [311]. This device consists of Cu serpentine electrode, thermal conductive elastomer, and n-type/p-type thermoelectric-based pellets. Based on Peltier effect, this flexible thermo haptic skin can alternate the heating and cooling modes with the temperature difference of 15 °C.

In general, haptic feedback technologies can drastically boost up the capability of information interpretation, instead of receiving digitized data. Equipment with wearable feedback systems not only paves the way for immersive interaction, but also improves the efficiency of manipulation via rapid cognition of the target environment [313,314]. The research of more feedback parameters, such as smell and airflow, will eventually establish a fully biomimetic feedback system.

8. Towards Self-Sustainable/Zero-Power/Passive HMI Terminals in the 5G/IoT Era

In this information age, the expansion of the Internet of things not only comes from mobile phones, tablets, and computers, but also thanks to other millions of smart devices [315] that are connected to the Internet through wireless communication technologies, e.g., WI-FI, Bluetooth, etc. Though the sensors based on self-powered mechanisms [154,316], i.e., triboelectric and piezoelectric, in the IoT devices may do not need a power supply, such signal transmission modules still mean high power consumption, especially when there are massive sensor nodes under the IoT framework. In addition, as the traditional power supply for portable devices, batteries experience drawbacks of high contamination and limited lifespan, as well as the annoying replacement or recharging process. To effectively extend the lifespan of portable or remote devices, energy harvesters, which can convert biomechanical energies or wasted energy in the ambient into electricity based on the mechanisms of electromagnetic, piezoelectric, and triboelectric, have been frequently investigated recently to replace batteries as the power supply units towards self-sustainable IoT systems/HMIs [317–319].

As shown in Figure 8a, an intelligent walking stick, consisting of a top press TENG for contact point sensing, a rotational TENG for gait abnormality detection and a rotational electromagnetic generator (EMG) for energy harvesting, was reported by Guo et al. [320]. With the deep-learning-enabled data analytic, complicated sensing functions including disability evaluation, identity recognition and activity status identification could be implemented through the TENG sensor outputs. Additionally, the linear-to-rotary structure enables the integrated EMG to harvest the ultra-low-frequency biomechanical energy from human motions and provide an average output power of 27.5 mW under 1 Hz stimuli. By stacking two EMGs for higher power output and using customized circuits for power management, a self-sustainable system capable of locomotion tracing and temperature/humidity monitoring was successfully achieved, where a temperature/humidity sensing module could be driven continuously with 1 Hz working, and a GPS and wireless module could work 6 s after every 90 s charging. Moreover, by transmitting the TENG sensor signal to an AI cloud/server via the 5G network for further analysis, real-time monitoring of the user's location and well-being status could be achieved in outdoor environments, showing the feasibility of sustainable HMIs enabled by energy harvesters for future IoT applications.

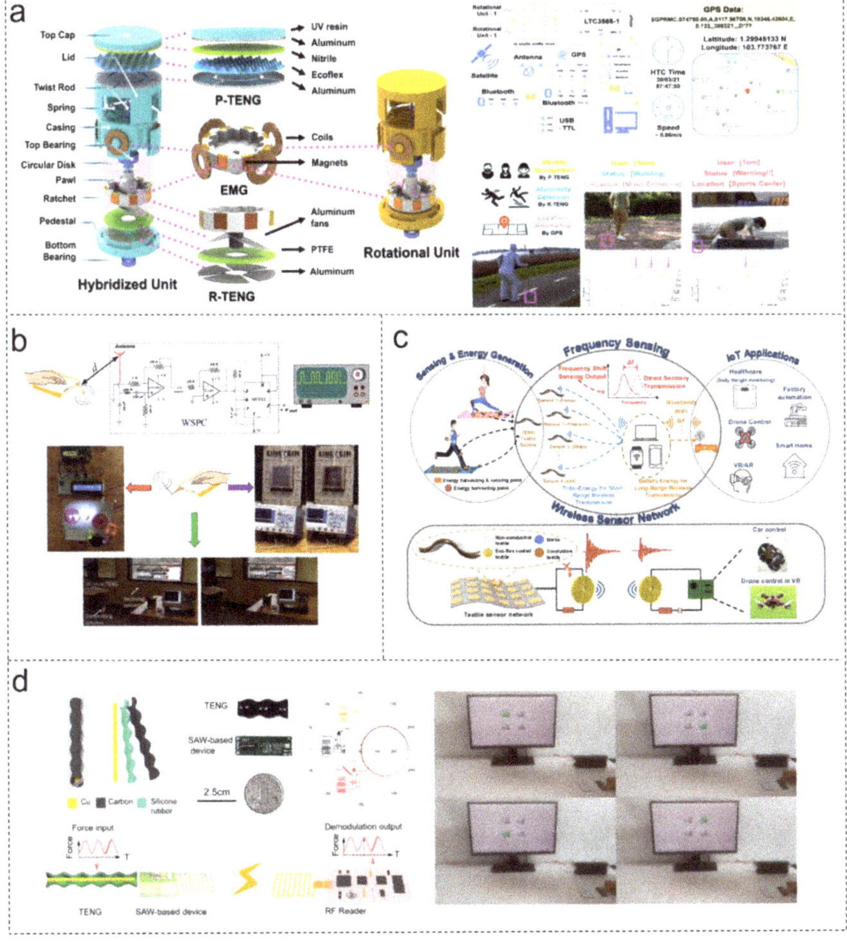

Figure 8. Self-sustainable/zero-power/passive HMIs. (**a**) A sustainable intelligent walking stick for real-time monitoring of

the user's location and well-being status in outdoor environments. Reprinted with permission from Reference [320], Copyright 2021, American Chemical Society. (**b**) A wireless tactile patch enabled by TENG's direct coupling. Reprinted with permission from Reference [321], Copyright 2017, Wiley. (**c**) A zero-power TENG wireless network via the mechanical switch enabled frequency boosting up strategy. Reprinted with permission from Reference [322], Copyright 2019, Elsevier. (**d**) A passive wireless TENG sensor using a surface acoustic wave resonator (SAWR). Reprinted with permission from Reference [323], Copyright 2020, Elsevier.

However, current sustainable HMIs enabled by the integrated energy harvesting units are not suitable for continuous wireless monitoring or fast information exchange due to the inevitable charging phase [61,324–326]. The additional energy harvesters or power management circuits also increase the size and cost of the whole device, reducing the applicability for wearable scenarios. To these issues, some works chose to utilize the high-voltage transient of the TENG to realize wireless coupling, and achieve complete zero power consumption on the sensor terminals with the minimalistic design [327,328]. As shown in Figure 8b [321], Mallineni et al. reported a wireless TENG tactile patch consisting of graphene polylactic acid (gPLA) nanocomposite and Teflon serving as the tribo-layers, which can generate a high electric field in the surrounding air enabling ~3 m wireless sensing, thanks to the extremely high output voltage (>2 kV) through simple mechanical stimuli, e.g., hand tapping. By connecting a customized signal processing circuit to the receiver, the hand tapping signal could be real-time collected wirelessly without a power supply or even signal transmitter at the sensor side, realizing a truly zero-power sensing terminal. Such a self-powered HMI could function as a controller for diversified smart home applications, e.g., activating lights, displays, photo frames, or even security systems via Morse code based passwords, showing great advantages in terms of size and cost when compared with conventional HMIs that need bulky power supply components or wireless circuit modules. However, the wireless sensing solution using signal amplitude as the sensing parameter is easily affected by the ambient environment, so a more advanced self-powered wireless network that detects the frequency of the TENG output via the mechanical switch enabled frequency boosting up strategy was proposed by Wen et al. as shown in Figure 8c [322]. With the enhanced frequency brought by the switch-induced instantaneous discharging, the TENG output signal could be easily transmitted wirelessly through a couple of coils, serving as a reliable reference for force calibration with high sensitivity (434.7 Hz N^{-1}) and a large sensing range thanks to the stable resonant frequency. By changing the wire connection of TENG sensor layers, i.e., in series or parallel, or adjusting the capacitors connected to different pixels in a sensor array, the tunable frequency could be achieved for complicated manipulation, e.g., the multiple-freedom-degree 2D car control or 3D VR drone control, revealing the feasibility of zero-power HMI based on the TENG-enabled frequency shift wireless sensing for IoT applications.

Another strategy to realize battery-free sensor nodes with continuous sensing capability under the IoT framework is to use passive wireless technologies, including near-field communication (NFC) [329,330], radio frequency identification (RFID) [331,332], and surface acoustic wave (SAW) sensors [333,334]. Compared with the abovementioned wireless transmission methods based on the TENG output enabled electric/magnetic coupling, these passive wireless approaches commonly show better stability in varying environmental circumstances, as well as smaller wireless sensing and receiving units. A passive wireless TENG sensor using a surface acoustic wave resonator (SAWR) was reported by Tan et al. as illustrated in Figure 8d [323]. By connecting a TENG force sensor to a SAWR, the frequency and amplitude of the SAWR's response signal can be modulated via the TENG output, making the radio frequency (RF) signal contain the tactile related sensory information which can be further extracted by the RF reader through the demodulating process. All the energy in this wireless transmission system is provided by the remote RF reader, making the sensor terminal a completely passive sensing node capable of continuous monitoring with a transmission distance larger than 2 m and high sensitivity of 23.75 kHz V^{-1}. This system could also act as a passive controller for trigger signal detection and Morse code based wireless manipulation, or as a wireless matrix keyboard with resonators working under different center frequencies, showing the promising prospects of distributed sensing or human–machine interaction in this 5G/IoT era.

9. Conclusions and Prospects

With the advantages of the wide material choices and diversified structural designs, TENG-based sensing technology has been frequently investigated towards low-power HMIs in this 5G/IoT era. In this review, we systematically summarize the key technologies and progress of TENG-based HMIs in terms of different application scenarios, including wearable HMIs, robotic-related HMIs and HMIs for smart home applications. By using the ML analytic for automatic feature extraction, advanced interactions, such as biometric authentication, gait analysis, gesture recognition and object recognition, can be implemented to enhance the functionalities of TENG-based HMIs towards intelligent interactive systems in the age of AI. The state-of-art haptic feedback technologies have also been reviewed and discussed, which yields a promising research direction for developing self-driven immersive HMIs when bringing in the TENG technology. Furthermore, the integration of self-powered TENG sensors and energy harvesters or passive wireless transmission technology also provides the possibility to realize battery-free/self-sustainable HMI terminals.

Despite the viable progress in developing TENG-based HMIs for diversified applications, there are still challenges that remain to be solved for the current solutions. First, the peak-like output signals make the TENG-based sensors only suitable for dynamic stimuli sensing and cannot be used to detect the continuous variation due to the fast shift of electrical states. Though methods based on grating-sliding mode TENG have been frequently reported [133,172], the resolution is limited by the size and spacing of the grating, and is difficult to reach a very high level without advanced fabrication processes [335], e.g., MEMS process, screen printing, micro-machine, etc., which means high fabrication cost. Other methods based on high-impedance readout circuits [149] or nanophotonic modulators [135,336] need complicated measurement systems, which are unlikely to be made into portable sensing devices for daily usage. Therefore, a reliable and convenient way to achieve continuous sensing capability for TENG-based HMIs that is compatible with IoT mobile platforms is needed. Second, though the energy harvesting or passive wireless transmission technology shows the feasibility to realize battery-free/self-sustainable HMI terminals with TENG sensors, limitations still exist, e.g., intermittent operation for energy harvester integrated system due to the charging phase, short transmission distance for electric/magnetic coupling or passive wireless transmission, etc. Further investigation on battery-free/self-sustainable systems with continuous operation capability and long wireless transmission distance is still desired. Third, the durability of TENG-based HMIs for long-term usage, and the robustness of the sensor performance in varying environmental conditions are also major concerns, considering the inevitable friction layer loss and the triboelectric charge loss under the high humidity condition, which may cause instability of the output and bring great challenges to the sensor collaboration. Moreover, TENG sensors can be fused with other sensing mechanisms towards a multimodal sensor fusion with AI analysis [337] for multifunctional purposes or more robust performance under varying environmental conditions. In this regard, the seamless implementation of the low-power/self-sustainable TENG-based HMIs will definitely shed light on the harmonic coexistence of humans and machines in the future of the IoT era, along with the immersive and efficient interactions in numerous scenarios.

Author Contributions: Conceptualization, Z.S., M.Z. and C.L.; investigation, Z.S. and M.Z.; resources, Z.S. and M.Z., writing—review and editing, Z.S., M.Z., and C.L.; supervision, C.L.; funding acquisition, C.L. All authors have read and agreed to the published version of the manuscript.

Funding: This work was supported by the Advanced Research and Technology Innovation Centre (ARTIC), the National University of Singapore under Grant (project number: R-261-518-009-720); "Intelligent monitoring system based on smart wearable sensors and artificial technology for the treatment of adolescent idiopathic scoliosis", the "Smart sensors and artificial intelligence (AI) for health" seed grant (R-263-501-017133) at NUS Institute for Health Innovation & Technology (NUS iHealthtech); the Collaborative Research Project under the SIMTech-NUS Joint Laboratory, "SIMTechNUS Joint Lab on Large-area Flexible Hybrid Electronics"; and National Research Funding—Competitive Research Program (NRF-CRP) (R-719-000-001-281).

Conflicts of Interest: The authors declare no conflict of interest.

References

1. Zhu, M.; He, T.; Lee, C. Technologies toward next Generation Human Machine Interfaces: From Machine Learning Enhanced Tactile Sensing to Neuromorphic Sensory Systems. *Appl. Phys. Rev.* **2020**, *7*, 031305. [CrossRef]
2. Yin, R.; Wang, D.; Zhao, S.; Lou, Z.; Shen, G. Wearable Sensors-Enabled Human–Machine Interaction Systems: From Design to Application. *Adv. Funct. Mater.* **2021**, *31*, 2008936. [CrossRef]
3. Wang, H.; Ma, X.; Hao, Y. Electronic Devices for Human-Machine Interfaces. *Adv. Mater. Interfaces* **2017**, *4*, 1600709. [CrossRef]
4. Arab Hassani, F.; Shi, Q.; Wen, F.; He, T.; Haroun, A.; Yang, Y.; Feng, Y.; Lee, C. Smart Materials for Smart Healthcare– Moving from Sensors and Actuators to Self-Sustained Nanoenergy Nanosystems. *Smart Mater. Med.* **2020**, *1*, 92–124. [CrossRef]
5. Dong, B.; Shi, Q.; Yang, Y.; Wen, F.; Zhang, Z.; Lee, C. Technology Evolution from Self-Powered Sensors to AIoT Enabled Smart Homes. *Nano Energy* **2021**, *79*, 105414. [CrossRef]
6. Principi, E.; Squartini, S.; Bonfigli, R.; Ferroni, G.; Piazza, F. An Integrated System for Voice Command Recognition and Emergency Detection Based on Audio Signals. *Expert Syst. Appl.* **2015**, *42*, 5668–5683. [CrossRef]
7. Rautaray, S.S.; Agrawal, A. Vision Based Hand Gesture Recognition for Human Computer Interaction: A Survey. *Artif. Intell. Rev.* **2015**, *43*, 1–54. [CrossRef]
8. Ionescu, D.; Suse, V.; Gadea, C.; Solomon, B.; Ionescu, B.; Islam, S. A New Infrared 3D Camera for Gesture Control. In Proceedings of the 2013 IEEE International Instrumentation and Measurement Technology Conference (I2MTC), Minneapolis, MN, USA, 6–9 May 2013; IEEE: Piscataway, NJ, USA, 2013; pp. 629–634.
9. Ivanov, A.V.; Zhilenkov, A.A. The Use of IMU MEMS-Sensors for Designing of Motion Capture System for Control of Robotic Objects. In Proceedings of the 2018 IEEE Conference of Russian Young Researchers in Electrical and Electronic Engineering (EIConRus), Moscow, Russia, 29 January–1 February 2018; IEEE: Piscataway, NJ, USA, 2018; pp. 890–893.
10. Kim, M.; Cho, J.; Lee, S.; Jung, Y. IMU Sensor-Based Hand Gesture Recognition for Human-Machine Interfaces. *Sensors* **2019**, *19*, 3827. [CrossRef] [PubMed]
11. Zhu, J.; Liu, X.; Shi, Q.; He, T.; Sun, Z.; Guo, X.; Liu, W.; Sulaiman, O.B.; Dong, B.; Lee, C. Development Trends and Perspectives of Future Sensors and MEMS/NEMS. *Micromachines* **2020**, *11*, 7. [CrossRef] [PubMed]
12. Zhu, H.; Wang, X.; Liang, J.; Lv, H.; Tong, H.; Ma, L.; Hu, Y.; Zhu, G.; Zhang, T.; Tie, Z.; et al. Versatile Electronic Skins for Motion Detection of Joints Enabled by Aligned Few-Walled Carbon Nanotubes in Flexible Polymer Composites. *Adv. Funct. Mater.* **2017**, *27*, 1606604. [CrossRef]
13. Amjadi, M.; Pichitpajongkit, A.; Lee, S.; Ryu, S.; Park, I. Highly Stretchable and Sensitive Strain Sensor Based on Silver Nanowire–Elastomer Nanocomposite. *ACS Nano* **2014**, *8*, 5154–5163. [CrossRef]
14. Dejace, L.; Laubeuf, N.; Furfaro, I.; Lacour, S.P. Gallium-Based Thin Films for Wearable Human Motion Sensors. *Adv. Intell. Syst.* **2019**, *1*, 1900079. [CrossRef]
15. Gao, Y.; Ota, H.; Schaler, E.W.; Chen, K.; Zhao, A.; Gao, W.; Fahad, H.M.; Leng, Y.; Zheng, A.; Xiong, F.; et al. Wearable Microfluidic Diaphragm Pressure Sensor for Health and Tactile Touch Monitoring. *Adv. Mater.* **2017**, *29*, 1701985. [CrossRef] [PubMed]
16. Kang, T.-H.; Chang, H.; Choi, D.; Kim, S.; Moon, J.; Lim, J.A.; Lee, K.-Y.; Yi, H. Hydrogel-Templated Transfer-Printing of Conductive Nanonetworks for Wearable Sensors on Topographic Flexible Substrates. *Nano Lett.* **2019**, *19*, 3684–3691. [CrossRef] [PubMed]
17. Cao, Y.; Li, T.; Gu, Y.; Luo, H.; Wang, S.; Zhang, T. Fingerprint-Inspired Flexible Tactile Sensor for Accurately Discerning Surface Texture. *Small* **2018**, *14*, 1703902. [CrossRef] [PubMed]
18. Kenry, J.C.Y.; Yu, J.; Shang, M.; Loh, K.P.; Lim, C.T. Highly Flexible Graphene Oxide Nanosuspension Liquid-Based Microfluidic Tactile Sensor. *Small* **2016**, *12*, 1593–1604. [CrossRef]
19. Wang, S.-S.; Liu, H.-B.; Kan, X.-N.; Wang, L.; Chen, Y.-H.; Su, B.; Li, Y.-L.; Jiang, L. Superlyophilicity-Facilitated Synthesis Reaction at the Microscale: Ordered Graphdiyne Stripe Arrays. *Small* **2017**, *13*, 1602265. [CrossRef]
20. Wang, Y.; Wang, Y.; Yang, Y. Graphene-Polymer Nanocomposite-Based Redox-Induced Electricity for Flexible Self-Powered Strain Sensors. *Adv. Energy Mater.* **2018**, *8*, 1800961. [CrossRef]
21. Tutika, R.; Kmiec, S.; Haque, A.B.M.T.; Martin, S.W.; Bartlett, M.D. Liquid Metal–Elastomer Soft Composites with Independently Controllable and Highly Tunable Droplet Size and Volume Loading. *ACS Appl. Mater. Interfaces* **2019**, *11*, 17873–17883. [CrossRef]
22. Navaraj, W.; Dahiya, R. Fingerprint-Enhanced Capacitive-Piezoelectric Flexible Sensing Skin to Discriminate Static and Dynamic Tactile Stimuli. *Adv. Intell. Syst.* **2019**, *1*, 1900051. [CrossRef]

23. Lee, J.; Kwon, H.; Seo, J.; Shin, S.; Koo, J.H.; Pang, C.; Son, S.; Kim, J.H.; Jang, Y.H.; Kim, D.E.; et al. Conductive Fiber-Based Ultrasensitive Textile Pressure Sensor for Wearable Electronics. *Adv. Mater.* **2015**, *27*, 2433–2439. [CrossRef]

24. Wu, R.; Ma, L.; Hou, C.; Meng, Z.; Guo, W.; Yu, W.; Yu, R.; Hu, F.; Liu, X.Y. Silk Composite Electronic Textile Sensor for High Space Precision 2D Combo Temperature–Pressure Sensing. *Small* **2019**, *15*, 1901558. [CrossRef] [PubMed]

25. Leber, A.; Cholst, B.; Sandt, J.; Vogel, N.; Kolle, M. Stretchable Thermoplastic Elastomer Optical Fibers for Sensing of Extreme Deformations. *Adv. Funct. Mater.* **2019**, *29*, 1802629. [CrossRef]

26. Guo, J.; Liu, X.; Jiang, N.; Yetisen, A.K.; Yuk, H.; Yang, C.; Khademhosseini, A.; Zhao, X.; Yun, S.-H. Highly Stretchable, Strain Sensing Hydrogel Optical Fibers. *Adv. Mater.* **2016**, *28*, 10244–10249. [CrossRef]

27. Bai, H.; Li, S.; Barreiros, J.; Tu, Y.; Pollock, C.R.; Shepherd, R.F. Stretchable Distributed Fiber-Optic Sensors. *Science.* **2020**, *370*, 848–852. [CrossRef]

28. Zhou, H.; Zhang, Y.; Qiu, Y.; Wu, H.; Qin, W.; Liao, Y.; Yu, Q.; Cheng, H. Stretchable Piezoelectric Energy Harvesters and Self-Powered Sensors for Wearable and Implantable Devices. *Biosens. Bioelectron.* **2020**, *168*, 112569. [CrossRef]

29. Park, K.-I.; Jeong, C.K.; Kim, N.K.; Lee, K.J. Stretchable Piezoelectric Nanocomposite Generator. *Nano Converg.* **2016**, *3*, 12. [CrossRef]

30. Ding, W.; Wang, A.C.; Wu, C.; Guo, H.; Wang, Z.L. Human-Machine Interfacing Enabled by Triboelectric Nanogenerators and Tribotronics. *Adv. Mater. Technol.* **2019**, *4*, 1800487. [CrossRef]

31. Pu, X.; An, S.; Tang, Q.; Guo, H.; Hu, C. Wearable Triboelectric Sensors for Biomedical Monitoring and Human-Machine Interface. *iScience* **2021**, *24*, 102027. [CrossRef] [PubMed]

32. Chen, S.; Pang, Y.; Cao, Y.; Tan, X.; Cao, C. Soft Robotic Manipulation System Capable of Stiffness Variation and Dexterous Operation for Safe Human–Machine Interactions. *Adv. Mater. Technol.* **2021**, *6*, 2100084. [CrossRef]

33. Hou, C.; Geng, J.; Yang, Z.; Tang, T.; Sun, Y.; Wang, F.; Liu, H.; Chen, T.; Sun, L. A Delta-Parallel-Inspired Human Machine Interface by Using Self-Powered Triboelectric Nanogenerator Toward 3D and VR/AR Manipulations. *Adv. Mater. Technol.* **2021**, *6*, 2000912. [CrossRef]

34. Luo, J.; Wang, Z.; Xu, L.; Wang, A.C.; Han, K.; Jiang, T.; Lai, Q.; Bai, Y.; Tang, W.; Fan, F.R.; et al. Flexible and Durable Wood-Based Triboelectric Nanogenerators for Self-Powered Sensing in Athletic Big Data Analytics. *Nat. Commun.* **2019**, *10*, 5147. [CrossRef]

35. Huang, J.; Yang, X.; Yu, J.; Han, J.; Jia, C.; Ding, M.; Sun, J.; Cao, X.; Sun, Q.; Wang, Z.L. A Universal and Arbitrary Tactile Interactive System Based on Self-Powered Optical Communication. *Nano Energy* **2020**, *69*, 104419. [CrossRef]

36. He, Q.; Wu, Y.; Feng, Z.; Sun, C.; Fan, W.; Zhou, Z.; Meng, K.; Fan, E.; Yang, J. Triboelectric Vibration Sensor for a Human-Machine Interface Built on Ubiquitous Surfaces. *Nano Energy* **2019**, *59*, 689–696. [CrossRef]

37. Yang, Y.; Zhang, H.; Lin, Z.-H.; Zhou, Y.S.; Jing, Q.; Su, Y.; Yang, J.; Chen, J.; Hu, C.; Wang, Z.L. Human Skin Based Triboelectric Nanogenerators for Harvesting Biomechanical Energy and as Self-Powered Active Tactile Sensor System. *ACS Nano* **2013**, *7*, 9213–9222. [CrossRef]

38. Chen, S.; Jiang, J.; Xu, F.; Gong, S. Crepe Cellulose Paper and Nitrocellulose Membrane-Based Triboelectric Nanogenerators for Energy Harvesting and Self-Powered Human-Machine Interaction. *Nano Energy* **2019**, *61*, 69–77. [CrossRef]

39. Fan, X.; Chen, J.; Yang, J.; Bai, P.; Li, Z.; Wang, Z.L. Ultrathin, Rollable, Paper-Based Triboelectric Nanogenerator for Acoustic Energy Harvesting and Self-Powered Sound Recording. *ACS Nano* **2015**, *9*, 4236–4243. [CrossRef]

40. Chen, T.; Zhao, M.; Shi, Q.; Yang, Z.; Liu, H.; Sun, L.; Ouyang, J.; Lee, C. Novel Augmented Reality Interface Using a Self-Powered Triboelectric Based Virtual Reality 3D-Control Sensor. *Nano Energy* **2018**, *51*, 162–172. [CrossRef]

41. Tang, Y.; Zhou, H.; Sun, X.; Diao, N.; Wang, J.; Zhang, B.; Qin, C.; Liang, E.; Mao, Y. Triboelectric Touch-Free Screen Sensor for Noncontact Gesture Recognizing. *Adv. Funct. Mater.* **2020**, *30*, 1907893. [CrossRef]

42. Chen, T.; Shi, Q.; Zhu, M.; He, T.; Yang, Z.; Liu, H.; Sun, L.; Yang, L.; Lee, C. Intuitive-Augmented Human-Machine Multidimensional Nano-Manipulation Terminal Using Triboelectric Stretchable Strip Sensors Based on Minimalist Design. *Nano Energy* **2019**, *60*, 440–448. [CrossRef]

43. Pu, X.; Liu, M.; Chen, X.; Sun, J.; Du, C.; Zhang, Y.; Zhai, J.; Hu, W.; Wang, Z.L. Ultrastretchable, Transparent Triboelectric Nanogenerator as Electronic Skin for Biomechanical Energy Harvesting and Tactile Sensing. *Sci. Adv.* **2017**, *3*, e1700015. [CrossRef]

44. Ha, M.; Lim, S.; Cho, S.; Lee, Y.; Na, S.; Baig, C.; Ko, H. Skin-Inspired Hierarchical Polymer Architectures with Gradient Stiffness for Spacer-Free, Ultrathin, and Highly Sensitive Triboelectric Sensors. *ACS Nano* **2018**, *12*, 3964–3974. [CrossRef] [PubMed]

45. Yoon, H.; Lee, D.; Kim, Y.; Jeon, S.; Jung, J.; Kwak, S.S.; Kim, J.; Kim, S.; Kim, Y.; Kim, S. Mechanoreceptor-Inspired Dynamic Mechanical Stimuli Perception Based on Switchable Ionic Polarization. *Adv. Funct. Mater.* **2021**, *31*, 2100649. [CrossRef]

46. Han, Y.; Yi, F.; Jiang, C.; Dai, K.; Xu, Y.; Wang, X.; You, Z. Self-Powered Gait Pattern-Based Identity Recognition by a Soft and Stretchable Triboelectric Band. *Nano Energy* **2019**, *56*, 516–523. [CrossRef]

47. Hua, Q.; Sun, J.; Liu, H.; Bao, R.; Yu, R.; Zhai, J.; Pan, C.; Wang, Z.L. Skin-Inspired Highly Stretchable and Conformable Matrix Networks for Multifunctional Sensing. *Nat. Commun.* **2018**, *9*, 244. [CrossRef]

48. Fan, F.-R.; Tian, Z.-Q.; Lin Wang, Z. Flexible Triboelectric Generator. *Nano Energy* **2012**, *1*, 328–334. [CrossRef]

49. Bae, J.; Lee, J.; Kim, S.; Ha, J.; Lee, B.-S.; Park, Y.; Choong, C.; Kim, J.-B.; Wang, Z.L.; Kim, H.-Y.; et al. Flutter-Driven Triboelectrification for Harvesting Wind Energy. *Nat. Commun.* **2014**, *5*, 4929. [CrossRef] [PubMed]

50. Yang, Y.; Zhu, G.; Zhang, H.; Chen, J.; Zhong, X.; Lin, Z.-H.; Su, Y.; Bai, P.; Wen, X.; Wang, Z.L. Triboelectric Nanogenerator for Harvesting Wind Energy and as Self-Powered Wind Vector Sensor System. *ACS Nano* **2013**, *7*, 9461–9468. [CrossRef]

51. Xie, Y.; Wang, S.; Lin, L.; Jing, Q.; Lin, Z.-H.; Niu, S.; Wu, Z.; Wang, Z.L. Rotary Triboelectric Nanogenerator Based on a Hybridized Mechanism for Harvesting Wind Energy. *ACS Nano* **2013**, *7*, 7119–7125. [CrossRef]

52. Chen, B.; Yang, Y.; Wang, Z.L. Scavenging Wind Energy by Triboelectric Nanogenerators. *Adv. Energy Mater.* **2018**, *8*, 1702649. [CrossRef]

53. Jiang, Q.; Chen, B.; Zhang, K.; Yang, Y. Ag Nanoparticle-Based Triboelectric Nanogenerator To Scavenge Wind Energy for a Self-Charging Power Unit. *ACS Appl. Mater. Interfaces* **2017**, *9*, 43716–43723. [CrossRef] [PubMed]

54. Wang, Z.L.; Jiang, T.; Xu, L. Toward the Blue Energy Dream by Triboelectric Nanogenerator Networks. *Nano Energy* **2017**, *39*, 9–23. [CrossRef]

55. Chen, J.; Yang, J.; Li, Z.; Fan, X.; Zi, Y.; Jing, Q.; Guo, H.; Wen, Z.; Pradel, K.C.; Niu, S.; et al. Networks of Triboelectric Nanogenerators for Harvesting Water Wave Energy: A Potential Approach toward Blue Energy. *ACS Nano* **2015**, *9*, 3324–3331. [CrossRef]

56. Liu, L.; Shi, Q.; Lee, C. A Novel Hybridized Blue Energy Harvester Aiming at All-Weather IoT Applications. *Nano Energy* **2020**, *76*, 105052. [CrossRef]

57. Liu, L.; Shi, Q.; Ho, J.S.; Lee, C. Study of Thin Film Blue Energy Harvester Based on Triboelectric Nanogenerator and Seashore IoT Applications. *Nano Energy* **2019**, *66*, 104167. [CrossRef]

58. Hou, C.; Chen, T.; Li, Y.; Huang, M.; Shi, Q.; Liu, H.; Sun, L.; Lee, C. A Rotational Pendulum Based Electromagnetic/Triboelectric Hybrid-Generator for Ultra-Low-Frequency Vibrations Aiming at Human Motion and Blue Energy Applications. *Nano Energy* **2019**, *63*, 103871. [CrossRef]

59. Chen, X.; Gao, L.; Chen, J.; Lu, S.; Zhou, H.; Wang, T.; Wang, A.; Zhang, Z.; Guo, S.; Mu, X.; et al. A Chaotic Pendulum Triboelectric-Electromagnetic Hybridized Nanogenerator for Wave Energy Scavenging and Self-Powered Wireless Sensing System. *Nano Energy* **2020**, *69*, 104440. [CrossRef]

60. Liu, L.; Shi, Q.; Lee, C. A Hybridized Electromagnetic-Triboelectric Nanogenerator Designed for Scavenging Biomechanical Energy in Human Balance Control. *Nano Res.* **2021**, *12*, 1–9.

61. Niu, S.; Wang, X.; Yi, F.; Zhou, Y.S.; Wang, Z.L. A Universal Self-Charging System Driven by Random Biomechanical Energy for Sustainable Operation of Mobile Electronics. *Nat. Commun.* **2015**, *6*, 8975. [CrossRef]

62. Wang, J.; Li, S.; Yi, F.; Zi, Y.; Lin, J.; Wang, X.; Xu, Y.; Wang, Z.L. Sustainably Powering Wearable Electronics Solely by Biomechanical Energy. *Nat. Commun.* **2016**, *7*, 12744. [CrossRef]

63. He, T.; Guo, X.; Lee, C. Flourishing Energy Harvesters for Future Body Sensor Network: From Single to Multiple Energy Sources. *iScience* **2021**, *24*, 101934. [CrossRef]

64. Zhang, K.; Wang, X.; Yang, Y.; Wang, Z.L. Hybridized Electromagnetic–Triboelectric Nanogenerator for Scavenging Biomechanical Energy for Sustainably Powering Wearable Electronics. *ACS Nano* **2015**, *9*, 3521–3529. [CrossRef]

65. Khalifa, S.; Lan, G.; Hassan, M.; Seneviratne, A.; Das, S.K. HARKE: Human Activity Recognition from Kinetic Energy Harvesting Data in Wearable Devices. *IEEE Trans. Mob. Comput.* **2018**, *17*, 1353–1368. [CrossRef]

66. Zhang, Q.; Jiang, T.; Ho, D.; Qin, S.; Yang, X.; Cho, J.H.; Sun, Q.; Wang, Z.L. Transparent and Self-Powered Multistage Sensation Matrix for Mechanosensation Application. *ACS Nano* **2018**, *12*, 254–262. [CrossRef]

67. Chen, Z.; Wang, Z.; Li, X.; Lin, Y.; Luo, N.; Long, M.; Zhao, N.; Xu, J.-B. Flexible Piezoelectric-Induced Pressure Sensors for Static Measurements Based on Nanowires/Graphene Heterostructures. *ACS Nano* **2017**, *11*, 4507–4513. [CrossRef] [PubMed]

68. Chun, K.-Y.; Son, Y.J.; Jeon, E.-S.; Lee, S.; Han, C.-S. A Self-Powered Sensor Mimicking Slow- and Fast-Adapting Cutaneous Mechanoreceptors. *Adv. Mater.* **2018**, *30*, 1706299. [CrossRef]

69. Ha, M.; Lim, S.; Park, J.; Um, D.-S.; Lee, Y.; Ko, H. Bioinspired Interlocked and Hierarchical Design of ZnO Nanowire Arrays for Static and Dynamic Pressure-Sensitive Electronic Skins. *Adv. Funct. Mater.* **2015**, *25*, 2841–2849. [CrossRef]

70. Shi, Q.; He, T.; Lee, C. More than Energy Harvesting—Combining Triboelectric Nanogenerator and Flexible Electronics Technology for Enabling Novel Micro-/Nano-Systems. *Nano Energy* **2019**, *57*, 851–871. [CrossRef]

71. Zhu, J.; Zhu, M.; Shi, Q.; Wen, F.; Liu, L.; Dong, B.; Haroun, A.; Yang, Y.; Vachon, P.; Guo, X.; et al. Progress in TENG Technology—A Journey from Energy Harvesting to Nanoenergy and Nanosystem. *EcoMat* **2020**, *2*, 1–45. [CrossRef]

72. Yang, Y.; Zhou, Y.S.; Zhang, H.; Liu, Y.; Lee, S.; Wang, Z.L. A Single-Electrode Based Triboelectric Nanogenerator as Self-Powered Tracking System. *Adv. Mater.* **2013**, *25*, 6594–6601. [CrossRef]

73. Ji, X.; Zhao, T.; Zhao, X.; Lu, X.; Li, T. Triboelectric Nanogenerator Based Smart Electronics via Machine Learning. *Adv. Mater. Technol.* **2020**, *5*, 1900921. [CrossRef]

74. Yang, J.; Chen, J.; Liu, Y.; Yang, W.; Su, Y.; Wang, Z.L. Triboelectrification-Based Organic Film Nanogenerator for Acoustic Energy Harvesting and Self-Powered Active Acoustic Sensing. *ACS Nano* **2014**, *8*, 2649–2657. [CrossRef] [PubMed]

75. Peng, X.; Dong, K.; Ye, C.; Jiang, Y.; Zhai, S.; Cheng, R.; Liu, D.; Gao, X.; Wang, J.; Wang, Z.L. A Breathable, Biodegradable, Antibacterial, and Self-Powered Electronic Skin Based on All-Nanofiber Triboelectric Nanogenerators. *Sci. Adv.* **2020**, *6*, eaba9624. [CrossRef]

76. Chun, S.; Son, W.; Kim, H.; Lim, S.K.; Pang, C.; Choi, C. Self-Powered Pressure- and Vibration-Sensitive Tactile Sensors for Learning Technique-Based Neural Finger Skin. *Nano Lett.* **2019**, *19*, 3305–3312. [CrossRef] [PubMed]

77. Xiang, S.; Liu, D.; Jiang, C.; Zhou, W.; Ling, D.; Zheng, W.; Sun, X.; Li, X.; Mao, Y.; Shan, C. Liquid-Metal-Based Dynamic Thermoregulating and Self-Powered Electronic Skin. *Adv. Funct. Mater.* **2021**, *31*, 2100940. [CrossRef]

78. Zhu, M.; Shi, Q.; He, T.; Yi, Z.; Ma, Y.; Yang, B.; Chen, T.; Lee, C. Self-Powered and Self-Functional Cotton Sock Using Piezoelectric and Triboelectric Hybrid Mechanism for Healthcare and Sports Monitoring. *ACS Nano* **2019**, *13*, 1940–1952. [CrossRef] [PubMed]

79. Zhang, P.; Zhang, Z.; Cai, J. A Foot Pressure Sensor Based on Triboelectric Nanogenerator for Human Motion Monitoring. *Microsyst. Technol.* **2021**, *27*, 3507–3512. [CrossRef]

80. Zhang, B.; Tang, Y.; Dai, R.; Wang, H.; Sun, X.; Qin, C.; Pan, Z.; Liang, E.; Mao, Y. Breath-Based Human–Machine Interaction System Using Triboelectric Nanogenerator. *Nano Energy* **2019**, *64*, 103953. [CrossRef]

81. Nweke, H.F.; Teh, Y.W.; Al-garadi, M.A.; Alo, U.R. Deep Learning Algorithms for Human Activity Recognition Using Mobile and Wearable Sensor Networks: State of the Art and Research Challenges. *Expert Syst. Appl.* **2018**, *105*, 233–261. [CrossRef]

82. Zhou, Y.; Shen, M.; Cui, X.; Shao, Y.; Li, L.; Zhang, Y. Triboelectric Nanogenerator Based Self-Powered Sensor for Artificial Intelligence. *Nano Energy* **2021**, *84*, 105887. [CrossRef]

83. Berman, S.; Stern, H. Sensors for Gesture Recognition Systems. *IEEE Trans. Syst. Man, Cybern. Part C (Appl. Rev.)* **2012**, *42*, 277–290. [CrossRef]

84. Alhamada, A.I.; Khalifa, O.O.; Abdalla, A.H. Deep Learning for Environmentally Robust Speech Recognition. *AIP Conf. Proc.* **2020**, *2306*, 020025.

85. Moeslund, T.B.; Hilton, A.; Krüger, V. A Survey of Advances in Vision-Based Human Motion Capture and Analysis. *Comput. Vis. Image Underst.* **2006**, *104*, 90–126. [CrossRef]

86. Sundararajan, K.; Woodard, D.L. Deep Learning for Biometrics. *ACM Comput. Surv.* **2018**, *51*, 1–34. [CrossRef]

87. Hughes, J.; Spielberg, A.; Chounlakone, M.; Chang, G.; Matusik, W.; Rus, D. A Simple, Inexpensive, Wearable Glove with Hybrid Resistive-Pressure Sensors for Computational Sensing, Proprioception, and Task Identification. *Adv. Intell. Syst.* **2020**, *2*, 2000002. [CrossRef]

88. Li, G.; Liu, S.; Wang, L.; Zhu, R. Skin-Inspired Quadruple Tactile Sensors Integrated on a Robot Hand Enable Object Recognition. *Sci. Robot.* **2020**, *5*, eabc8134. [CrossRef]

89. Zhang, H.; Cheng, Q.; Lu, X.; Wang, W.; Wang, Z.L.; Sun, C. Detection of Driving Actions on Steering Wheel Using Triboelectric Nanogenerator via Machine Learning. *Nano Energy* **2021**, *79*, 105455. [CrossRef]

90. Zhang, W.; Deng, L.; Yang, L.; Yang, P.; Diao, D.; Wang, P.; Wang, Z.L. Multilanguage-Handwriting Self-Powered Recognition Based on Triboelectric Nanogenerator Enabled Machine Learning. *Nano Energy* **2020**, *77*, 105174. [CrossRef]

91. Zhang, W.; Wang, P.; Sun, K.; Wang, C.; Diao, D. Intelligently Detecting and Identifying Liquids Leakage Combining Triboelectric Nanogenerator Based Self-Powered Sensor with Machine Learning. *Nano Energy* **2019**, *56*, 277–285. [CrossRef]

92. Syu, M.H.; Guan, Y.J.; Lo, W.C.; Fuh, Y.K. Biomimetic and Porous Nanofiber-Based Hybrid Sensor for Multifunctional Pressure Sensing and Human Gesture Identification via Deep Learning Method. *Nano Energy* **2020**, *76*, 105029. [CrossRef]

93. Ji, X.; Fang, P.; Xu, B.; Xie, K.; Yue, H.; Luo, X.; Wang, Z.; Zhao, X.; Shi, P. Biohybrid Triboelectric Nanogenerator for Label-Free Pharmacological Fingerprinting in Cardiomyocytes. *Nano Lett.* **2020**, *20*, 4043–4050. [CrossRef]

94. Chen, H.; Miao, L.; Su, Z.; Song, Y.; Han, M.; Chen, X.; Cheng, X.; Chen, D.; Zhang, H. Fingertip-Inspired Electronic Skin Based on Triboelectric Sliding Sensing and Porous Piezoresistive Pressure Detection. *Nano Energy* **2017**, *40*, 65–72. [CrossRef]

95. Song, K.; Zhao, R.; Wang, Z.L.; Yang, Y. Conjuncted Pyro-Piezoelectric Effect for Self-Powered Simultaneous Temperature and Pressure Sensing. *Adv. Mater.* **2019**, *31*, 1902831. [CrossRef]

96. Wang, S.; Wang, Z.L.; Yang, Y. A One-Structure-Based Hybridized Nanogenerator for Scavenging Mechanical and Thermal Energies by Triboelectric-Piezoelectric-Pyroelectric Effects. *Adv. Mater.* **2016**, *28*, 2881–2887. [CrossRef] [PubMed]

97. Yang, Y.; Guo, W.; Pradel, K.C.; Zhu, G.; Zhou, Y.; Zhang, Y.; Hu, Y.; Lin, L.; Wang, Z.L. Pyroelectric Nanogenerators for Harvesting Thermoelectric Energy. *Nano Lett.* **2012**, *12*, 2833–2838. [CrossRef] [PubMed]

98. Yang, Y.; Jung, J.H.; Yun, B.K.; Zhang, F.; Pradel, K.C.; Guo, W.; Wang, Z.L. Flexible Pyroelectric Nanogenerators Using a Composite Structure of Lead-Free KNbO 3 Nanowires. *Adv. Mater.* **2012**, *24*, 5357–5362. [CrossRef] [PubMed]

99. Zhang, K.; Wang, Y.; Wang, Z.L.; Yang, Y. Standard and Figure-of-Merit for Quantifying the Performance of Pyroelectric Nanogenerators. *Nano Energy* **2019**, *55*, 534–540. [CrossRef]

100. Demain, S.; Metcalf, C.D.; Merrett, G.V.; Zheng, D.; Cunningham, S. A Narrative Review on Haptic Devices: Relating the Physiology and Psychophysical Properties of the Hand to Devices for Rehabilitation in Central Nervous System Disorders. *Disabil. Rehabil. Assist. Technol.* **2013**, *8*, 181–189. [CrossRef]

101. Shahid, T.; Gouwanda, D.; Nurzaman, S.G.; Gopalai, A.A. Moving toward Soft Robotics: A Decade Review of the Design of Hand Exoskeletons. *Biomimetics* **2018**, *3*, 17. [CrossRef]

102. Zubrycki, I.; Granosik, G. Novel Haptic Device Using Jamming Principle for Providing Kinaesthetic Feedback in Glove-Based Control Interface. *J. Intell. Robot. Syst.* **2017**, *85*, 413–429. [CrossRef]

103. Low, J.H.; Lee, W.W.; Khin, P.M.; Thakor, N.V.; Kukreja, S.L.; Ren, H.L.; Yeow, C.H. Hybrid Tele-Manipulation System Using a Sensorized 3-D-Printed Soft Robotic Gripper and a Soft Fabric-Based Haptic Glove. *IEEE Robot. Autom. Lett.* **2017**, *2*, 880–887. [CrossRef]

104. Martinez, J.; Garcia, A.; Oliver, M.; Molina, J.P.; Gonzalez, P. Identifying Virtual 3D Geometric Shapes with a Vibrotactile Glove. *IEEE Comput. Graph. Appl.* **2016**, *36*, 42–51. [CrossRef]

105. Baldi, T.L.; Scheggi, S.; Meli, L.; Mohammadi, M.; Prattichizzo, D. GESTO: A Glove for Enhanced Sensing and Touching Based on Inertial and Magnetic Sensors for Hand Tracking and Cutaneous Feedback. *IEEE Trans. Human-Machine Syst.* **2017**, *47*, 1066–1076. [CrossRef]

106. Liu, H.; Zhang, Z.; Xie, X.; Zhu, Y.; Liu, Y.; Wang, Y.; Zhu, S.-C. High-Fidelity Grasping in Virtual Reality Using a Glove-Based System. In Proceedings of the 2019 International Conference on Robotics and Automation (ICRA), Montreal, QC, Canada, 20–24 May 2019; IEEE: Piscataway, NJ, USA, 2019; pp. 5180–5186.

107. Hinchet, R.; Vechev, V.; Shea, H.; Hilliges, O. DextrES: Wearable Haptic Feedback for Grasping in VR via a Thin Form-Factor Electrostatic Brake. In Proceedings of the 31st Annual ACM Symposium on User Interface Software and Technology, Berlin, Germany, 14–18 October 2018; ACM: New York, NY, USA, 2018; pp. 901–912.

108. Yem, V.; Kajimoto, H. A Fingertip Glove with Motor Rotational Acceleration Enables Stiffness Perception When Grasping a Virtual Object. In Proceedings of the 20th International Conference on Human Interface and the Management of Information, Las Vegas, NV, USA, 15–20 July 2018; Springer: Cham, Switzerland, 2018; pp. 463–473.

109. Pickering, K.L.; Efendy, M.G.A.; Le, T.M. A Review of Recent Developments in Natural Fibre Composites and Their Mechanical Performance. *Compos. Part A Appl. Sci. Manuf.* **2016**, *83*, 98–112. [CrossRef]

110. Voiculescu, I.; Nordin, A.N. Acoustic Wave Based MEMS Devices for Biosensing Applications. *Biosens. Bioelectron.* **2012**, *33*, 1–9. [CrossRef]

111. Cazala, F.; Vienney, N.; Stoléru, S. The Cortical Sensory Representation of Genitalia in Women and Men: A Systematic Review. *Socioaffective Neurosci. Psychol.* **2015**, *5*, 26428. [CrossRef]

112. Schott, G.D. Penfield's Homunculus: A Note on Cerebral Cartography. *J. Neurol. Neurosurg. Psychiatry* **1993**, *56*, 329–333. [CrossRef]

113. Kim, J.-H.; Thang, N.D.; Kim, T.-S. 3-D Hand Motion Tracking and Gesture Recognition Using a Data Glove. In Proceedings of the 2009 IEEE International Symposium on Industrial Electronics, Seoul, Korea, 5–8 July 2009; IEEE: Piscataway, NJ, USA, 2009; pp. 1013–1018.

114. Lei, L.; Dashun, Q. Design of Data-Glove and Chinese Sign Language Recognition System Based on ARM9. In Proceedings of the 2015 12th IEEE International Conference on Electronic Measurement & Instruments (ICEMI), Qingdao, China, 16–18 July 2015; IEEE: Piscataway, NJ, USA, 2015; Volume 3, pp. 1130–1134.

115. Suzuki, K.; Yataka, K.; Okumiya, Y.; Sakakibara, S.; Sako, K.; Mimura, H.; Inoue, Y. Rapid-Response, Widely Stretchable Sensor of Aligned MWCNT/Elastomer Composites for Human Motion Detection. *ACS Sensors* **2016**, *1*, 817–825. [CrossRef]

116. Wei, P.; Yang, X.; Cao, Z.; Guo, X.; Jiang, H.; Chen, Y.; Morikado, M.; Qiu, X.; Yu, D. Flexible and Stretchable Electronic Skin with High Durability and Shock Resistance via Embedded 3D Printing Technology for Human Activity Monitoring and Personal Healthcare. *Adv. Mater. Technol.* **2019**, *4*, 1900315. [CrossRef]

117. Kim, S.; Oh, J.; Jeong, D.; Bae, J. Direct Wiring of Eutectic Gallium–Indium to a Metal Electrode for Soft Sensor Systems. *ACS Appl. Mater. Interfaces* **2019**, *11*, 20557–20565. [CrossRef]

118. Shi, Q.; Dong, B.; He, T.; Sun, Z.; Zhu, J.; Zhang, Z.; Lee, C. Progress in Wearable Electronics/Photonics—Moving toward the Era of Artificial Intelligence and Internet of Things. *InfoMat* **2020**, *2*, 1131–1162. [CrossRef]

119. Lu, C.; Chen, J.; Jiang, T.; Gu, G.; Tang, W.; Wang, Z.L. A Stretchable, Flexible Triboelectric Nanogenerator for Self-Powered Real-Time Motion Monitoring. *Adv. Mater. Technol.* **2018**, *3*, 1800021. [CrossRef]

120. Dong, K.; Deng, J.; Ding, W.; Wang, A.C.; Wang, P.; Cheng, C.; Wang, Y.-C.; Jin, L.; Gu, B.; Sun, B.; et al. Versatile Core-Sheath Yarn for Sustainable Biomechanical Energy Harvesting and Real-Time Human-Interactive Sensing. *Adv. Energy Mater.* **2018**, *8*, 1801114. [CrossRef]

121. Shi, Q.; Wang, H.; Wang, T.; Lee, C. Self-Powered Liquid Triboelectric Microfluidic Sensor for Pressure Sensing and Finger Motion Monitoring Applications. *Nano Energy* **2016**, *30*, 450–459. [CrossRef]

122. Qin, K.; Chen, C.; Pu, X.; Tang, Q.; He, W.; Liu, Y.; Zeng, Q.; Liu, G.; Guo, H.; Hu, C. Magnetic Array Assisted Triboelectric Nanogenerator Sensor for Real-Time Gesture Interaction. *Nano Micro Lett.* **2021**, *13*, 51. [CrossRef]

123. Xie, L.; Chen, X.; Wen, Z.; Yang, Y.; Shi, J.; Chen, C.; Peng, M.; Liu, Y.; Sun, X. Spiral Steel Wire Based Fiber-Shaped Stretchable and Tailorable Triboelectric Nanogenerator for Wearable Power Source and Active Gesture Sensor. *Nano Micro Lett.* **2019**, *11*, 39. [CrossRef]

124. Zhang, M.; Gao, T.; Wang, J.; Liao, J.; Qiu, Y.; Xue, H.; Shi, Z.; Xiong, Z.; Chen, L. Single BaTiO3 Nanowires-Polymer Fiber Based Nanogenerator. *Nano Energy* **2015**, *11*, 510–517. [CrossRef]

125. He, T.; Sun, Z.; Shi, Q.; Zhu, M.; Anaya, D.V.; Xu, M.; Chen, T.; Yuce, M.R.; Thean, A.V.-Y.; Lee, C. Self-Powered Glove-Based Intuitive Interface for Diversified Control Applications in Real/Cyber Space. *Nano Energy* **2019**, *58*, 641–651. [CrossRef]

126. Nguyen, V.; Yang, R. Effect of Humidity and Pressure on the Triboelectric Nanogenerator. *Nano Energy* **2013**, *2*, 604–608. [CrossRef]

127. Nguyen, V.; Zhu, R.; Yang, R. Environmental Effects on Nanogenerators. *Nano Energy* **2015**, *14*, 49–61. [CrossRef]

128. Wen, F.; Sun, Z.; He, T.; Shi, Q.; Zhu, M.; Zhang, Z.; Li, L.; Zhang, T.; Lee, C. Machine Learning Glove Using Self-Powered Conductive Superhydrophobic Triboelectric Textile for Gesture Recognition in VR/AR Applications. *Adv. Sci.* **2020**, *7*, 2000261. [CrossRef]

129. He, T.; Shi, Q.; Wang, H.; Wen, F.; Chen, T.; Ouyang, J.; Lee, C. Beyond Energy Harvesting—Multi-Functional Triboelectric Nanosensors on a Textile. *Nano Energy* **2019**, *57*, 338–352. [CrossRef]

130. Dhakar, L.; Pitchappa, P.; Tay, F.E.H.; Lee, C. An Intelligent Skin Based Self-Powered Finger Motion Sensor Integrated with Triboelectric Nanogenerator. *Nano Energy* **2016**, *19*, 532–540. [CrossRef]

131. Zhou, Z.; Chen, K.; Li, X.; Zhang, S.; Wu, Y.; Zhou, Y.; Meng, K.; Sun, C.; He, Q.; Fan, W.; et al. Sign-to-Speech Translation Using Machine-Learning-Assisted Stretchable Sensor Arrays. *Nat. Electron.* **2020**, *3*, 571–578. [CrossRef]

132. Zhu, M.; Sun, Z.; Zhang, Z.; Shi, Q.; He, T.; Liu, H.; Chen, T.; Lee, C. Haptic-Feedback Smart Glove as a Creative Human-Machine Interface (HMI) for Virtual/Augmented Reality Applications. *Sci. Adv.* **2020**, *6*, eaaz8693. [CrossRef] [PubMed]
133. Pu, X.; Guo, H.; Tang, Q.; Chen, J.; Feng, L.; Liu, G.; Wang, X.; Xi, Y.; Hu, C.; Wang, Z.L. Rotation Sensing and Gesture Control of a Robot Joint via Triboelectric Quantization Sensor. *Nano Energy* **2018**, *54*, 453–460. [CrossRef]
134. Wang, Y.; Wu, H.; Xu, L.; Zhang, H.; Yang, Y.; Wang, Z.L. Hierarchically Patterned Self-Powered Sensors for Multifunctional Tactile Sensing. *Sci. Adv.* **2020**, *6*, eabb9083. [CrossRef]
135. Dong, B.; Yang, Y.; Shi, Q.; Xu, S.; Sun, Z.; Zhu, S.; Zhang, Z.; Kwong, D.-L.; Zhou, G.; Ang, K.-W.; et al. Wearable Triboelectric–Human–Machine Interface (THMI) Using Robust Nanophotonic Readout. *ACS Nano* **2020**, *14*, 8915–8930. [CrossRef] [PubMed]
136. He, Q.; Wu, Y.; Feng, Z.; Fan, W.; Lin, Z.; Sun, C.; Zhou, Z.; Meng, K.; Wu, W.; Yang, J. An All-Textile Triboelectric Sensor for Wearable Teleoperated Human–Machine Interaction. *J. Mater. Chem. A* **2019**, *7*, 26804–26811. [CrossRef]
137. Liao, J.; Zou, Y.; Jiang, D.; Liu, Z.; Qu, X.; Li, Z.; Liu, R.; Fan, Y.; Shi, B.; Li, Z.; et al. Nestable Arched Triboelectric Nanogenerator for Large Deflection Biomechanical Sensing and Energy Harvesting. *Nano Energy* **2020**, *69*, 104417. [CrossRef]
138. Lai, Y.; Lu, H.; Wu, H.; Zhang, D.; Yang, J.; Ma, J.; Shamsi, M.; Vallem, V.; Dickey, M.D. Elastic Multifunctional Liquid–Metal Fibers for Harvesting Mechanical and Electromagnetic Energy and as Self-Powered Sensors. *Adv. Energy Mater.* **2021**, *11*, 2100411. [CrossRef]
139. Maharjan, P.; Bhatta, T.; Salauddin, M.; Rasel, M.S.; Rahman, M.T.; Rana, S.M.S.; Park, J.Y. A Human Skin-Inspired Self-Powered Flex Sensor with Thermally Embossed Microstructured Triboelectric Layers for Sign Language Interpretation. *Nano Energy* **2020**, *76*, 105071. [CrossRef]
140. Yi, F.; Wang, X.; Niu, S.; Li, S.; Yin, Y.; Dai, K.; Zhang, G.; Lin, L.; Wen, Z.; Guo, H.; et al. A Highly Shape-Adaptive, Stretchable Design Based on Conductive Liquid for Energy Harvesting and Self-Powered Biomechanical Monitoring. *Sci. Adv.* **2016**, *2*, e1501624. [CrossRef]
141. Wang, X.; Dong, L.; Zhang, H.; Yu, R.; Pan, C.; Wang, Z.L. Recent Progress in Electronic Skin. *Adv. Sci.* **2015**, *2*, 1500169. [CrossRef]
142. Yang, J.C.; Mun, J.; Kwon, S.Y.; Park, S.; Bao, Z.; Park, S. Electronic Skin: Recent Progress and Future Prospects for Skin-Attachable Devices for Health Monitoring, Robotics, and Prosthetics. *Adv. Mater.* **2019**, *31*, 1904765. [CrossRef] [PubMed]
143. Lee, Y.; Park, J.; Choe, A.; Cho, S.; Kim, J.; Ko, H. Mimicking Human and Biological Skins for Multifunctional Skin Electronics. *Adv. Funct. Mater.* **2020**, *30*, 1904523. [CrossRef]
144. Tao, J.; Bao, R.; Wang, X.; Peng, Y.; Li, J.; Fu, S.; Pan, C.; Wang, Z.L. Self-Powered Tactile Sensor Array Systems Based on the Triboelectric Effect. *Adv. Funct. Mater.* **2019**, *29*, 1806379. [CrossRef]
145. Chen, H.; Song, Y.; Cheng, X.; Zhang, H. Self-Powered Electronic Skin Based on the Triboelectric Generator. *Nano Energy* **2019**, *56*, 252–268. [CrossRef]
146. Chen, H.; Song, Y.; Guo, H.; Miao, L.; Chen, X.; Su, Z.; Zhang, H. Hybrid Porous Micro Structured Finger Skin Inspired Self-Powered Electronic Skin System for Pressure Sensing and Sliding Detection. *Nano Energy* **2018**, *51*, 496–503. [CrossRef]
147. Zou, H.; Zhang, Y.; Guo, L.; Wang, P.; He, X.; Dai, G.; Zheng, H.; Chen, C.; Wang, A.C.; Xu, C.; et al. Quantifying the Triboelectric Series. *Nat. Commun.* **2019**, *10*, 1427. [CrossRef] [PubMed]
148. Wang, H.; Wu, H.; Hasan, D.; He, T.; Shi, Q.; Lee, C. Self-Powered Dual-Mode Amenity Sensor Based on the Water–Air Triboelectric Nanogenerator. *ACS Nano* **2017**, *11*, 10337–10346. [CrossRef]
149. Jin, L.; Tao, J.; Bao, R.; Sun, L.; Pan, C. Self-Powered Real-Time Movement Monitoring Sensor Using Triboelectric Nanogenerator Technology. *Sci. Rep.* **2017**, *7*, 10521. [CrossRef]
150. Liu, L.; Guo, X.; Lee, C. Promoting Smart Cities into the 5G Era with Multi-Field Internet of Things (IoT) Applications Powered with Advanced Mechanical Energy Harvesters. *Nano Energy* **2021**, *88*, 106304. [CrossRef]
151. Zhu, M.; Yi, Z.; Yang, B.; Lee, C. Making Use of Nanoenergy from Human—Nanogenerator and Self-Powered Sensor Enabled Sustainable Wireless IoT Sensory Systems. *Nano Today* **2021**, *36*, 101016. [CrossRef]
152. Zhang, X.-S.; Han, M.; Kim, B.; Bao, J.-F.; Brugger, J.; Zhang, H. All-in-One Self-Powered Flexible Microsystems Based on Triboelectric Nanogenerators. *Nano Energy* **2018**, *47*, 410–426. [CrossRef]
153. Liu, Z.; Li, H.; Shi, B.; Fan, Y.; Wang, Z.L.; Li, Z. Wearable and Implantable Triboelectric Nanogenerators. *Adv. Funct. Mater.* **2019**, *29*, 1808820. [CrossRef]
154. Li, Z.; Zheng, Q.; Wang, Z.L.; Li, Z. Nanogenerator-Based Self-Powered Sensors for Wearable and Implantable Electronics. *Research* **2020**, *2020*, 1–25. [CrossRef] [PubMed]
155. Ameri, S.K.; Kim, M.; Kuang, I.A.; Perera, W.K.; Alshiekh, M.; Jeong, H.; Topcu, U.; Akinwande, D.; Lu, N. Imperceptible Electrooculography Graphene Sensor System for Human–Robot Interface. *npj 2D Mater. Appl.* **2018**, *2*, 19. [CrossRef]
156. Wei, L.; Hu, H.; Zhang, Y. Fusing Emg and Visual Data for Hands-Free Control of An Intelligent Wheelchair. *Int. J. Hum. Robot.* **2011**, *08*, 707–724. [CrossRef]
157. Choudhari, A.M.; Porwal, P.; Jonnalagedda, V.; Mériaudeau, F. An Electrooculography Based Human Machine Interface for Wheelchair Control. *Biocybern. Biomed. Eng.* **2019**, *39*, 673–685. [CrossRef]
158. Pu, X.; Guo, H.; Chen, J.; Wang, X.; Xi, Y.; Hu, C.; Wang, Z.L. Eye Motion Triggered Self-Powered Mechnosensational Communication System Using Triboelectric Nanogenerator. *Sci. Adv.* **2017**, *3*, e1700694. [CrossRef]
159. Vera Anaya, D.; He, T.; Lee, C.; Yuce, M.R. Self-Powered Eye Motion Sensor Based on Triboelectric Interaction and near-Field Electrostatic Induction for Wearable Assistive Technologies. *Nano Energy* **2020**, *72*, 104675. [CrossRef]

160. Fall, C.L.; Gagnon-Turcotte, G.; Dube, J.-F.; Gagne, J.S.; Delisle, Y.; Campeau-Lecours, A.; Gosselin, C.; Gosselin, B. Wireless SEMG-Based Body–Machine Interface for Assistive Technology Devices. *IEEE J. Biomed. Heal. Inform.* **2017**, *21*, 967–977. [CrossRef]

161. Zhou, H.; Li, D.; He, X.; Hui, X.; Guo, H.; Hu, C.; Mu, X.; Wang, Z.L. Bionic Ultra-Sensitive Self-Powered Electromechanical Sensor for Muscle-Triggered Communication Application. *Adv. Sci.* **2021**, *8*, 2101020. [CrossRef]

162. Li, W.; Torres, D.; Díaz, R.; Wang, Z.; Wu, C.; Wang, C.; Lin Wang, Z.; Sepúlveda, N. Nanogenerator-Based Dual-Functional and Self-Powered Thin Patch Loudspeaker or Microphone for Flexible Electronics. *Nat. Commun.* **2017**, *8*, 15310. [CrossRef]

163. Han, J.H.; Bae, K.M.; Hong, S.K.; Park, H.; Kwak, J.-H.; Wang, H.S.; Joe, D.J.; Park, J.H.; Jung, Y.H.; Hur, S.; et al. Machine Learning-Based Self-Powered Acoustic Sensor for Speaker Recognition. *Nano Energy* **2018**, *53*, 658–665. [CrossRef]

164. Liu, H.; Dong, W.; Li, Y.; Li, F.; Geng, J.; Zhu, M.; Chen, T.; Zhang, H.; Sun, L.; Lee, C. An Epidermal SEMG Tattoo-like Patch as a New Human–Machine Interface for Patients with Loss of Voice. *Microsyst. Nanoeng.* **2020**, *6*, 16. [CrossRef]

165. Chen, C.; Wen, Z.; Shi, J.; Jian, X.; Li, P.; Yeow, J.T.W.; Sun, X. Micro Triboelectric Ultrasonic Device for Acoustic Energy Transfer and Signal Communication. *Nat. Commun.* **2020**, *11*, 4143. [CrossRef]

166. Arora, N.; Zhang, S.L.; Shahmiri, F.; Osorio, D.; Wang, Y.-C.; Gupta, M.; Wang, Z.; Starner, T.; Wang, Z.L.; Abowd, G.D. SATURN: A Thin and Flexible Self-Powered Microphone Leveraging Triboelectric Nanogenerator. *Proc. ACM Interact. Mob. Wearable Ubiquitous Technol.* **2018**, *2*, 1–28. [CrossRef]

167. Yuan, M.; Li, C.; Liu, H.; Xu, Q.; Xie, Y. A 3D-Printed Acoustic Triboelectric Nanogenerator for Quarter-Wavelength Acoustic Energy Harvesting and Self-Powered Edge Sensing. *Nano Energy* **2021**, *85*, 105962. [CrossRef]

168. Chen, F.; Wu, Y.; Ding, Z.; Xia, X.; Li, S.; Zheng, H.; Diao, C.; Yue, G.; Zi, Y. A Novel Triboelectric Nanogenerator Based on Electrospun Polyvinylidene Fluoride Nanofibers for Effective Acoustic Energy Harvesting and Self-Powered Multifunctional Sensing. *Nano Energy* **2019**, *56*, 241–251. [CrossRef]

169. Kang, S.; Cho, S.; Shanker, R.; Lee, H.; Park, J.; Um, D.-S.; Lee, Y.; Ko, H. Transparent and Conductive Nanomembranes with Orthogonal Silver Nanowire Arrays for Skin-Attachable Loudspeakers and Microphones. *Sci. Adv.* **2018**, *4*, eaas8772. [CrossRef]

170. Chen, T.; Shi, Q.; Zhu, M.; He, T.; Sun, L.; Yang, L.; Lee, C. Triboelectric Self-Powered Wearable Flexible Patch as 3D Motion Control Interface for Robotic Manipulator. *ACS Nano* **2018**, *12*, 11561–11571. [CrossRef]

171. Shi, Q.; Lee, C. Self-Powered Bio-Inspired Spider-Net-Coding Interface Using Single-Electrode Triboelectric Nanogenerator. *Adv. Sci.* **2019**, *6*, 1900617. [CrossRef]

172. Zhu, M.; Sun, Z.; Chen, T.; Lee, C. Low Cost Exoskeleton Manipulator Using Bidirectional Triboelectric Sensors Enhanced Multiple Degree of Freedom Sensory System. *Nat. Commun.* **2021**, *12*, 2692. [CrossRef]

173. Li, C.; Liu, D.; Xu, C.; Wang, Z.; Shu, S.; Sun, Z.; Tang, W.; Wang, Z.L. Sensing of Joint and Spinal Bending or Stretching via a Retractable and Wearable Badge Reel. *Nat. Commun.* **2021**, *12*, 2950. [CrossRef]

174. Park, J.; Lee, Y.; Hong, J.; Lee, Y.; Ha, M.; Jung, Y.; Lim, H.; Kim, S.Y.; Ko, H. Tactile-Direction-Sensitive and Stretchable Electronic Skins Based on Human-Skin-Inspired Interlocked Microstructures. *ACS Nano* **2014**, *8*, 12020–12029. [CrossRef]

175. Harada, S.; Kanao, K.; Yamamoto, Y.; Arie, T.; Akita, S.; Takei, K. Fully Printed Flexible Fingerprint-like Three-Axis Tactile and Slip Force and Temperature Sensors for Artificial Skin. *ACS Nano* **2014**, *8*, 12851–12857. [CrossRef]

176. Sarwar, M.S.; Dobashi, Y.; Preston, C.; Wyss, J.K.M.; Mirabbasi, S.; Madden, J.D.W. Bend, Stretch, and Touch: Locating a Finger on an Actively Deformed Transparent Sensor Array. *Sci. Adv.* **2017**, *3*, e1602200. [CrossRef]

177. Lou, Z.; Li, L.; Wang, L.; Shen, G. Recent Progress of Self-Powered Sensing Systems for Wearable Electronics. *Small* **2017**, *13*, 1701791. [CrossRef]

178. Wen, Z.; Yang, Y.; Sun, N.; Li, G.; Liu, Y.; Chen, C.; Shi, J.; Xie, L.; Jiang, H.; Bao, D.; et al. A Wrinkled PEDOT:PSS Film Based Stretchable and Transparent Triboelectric Nanogenerator for Wearable Energy Harvesters and Active Motion Sensors. *Adv. Funct. Mater.* **2018**, *28*, 1803684. [CrossRef]

179. Bai, Z.; Xu, Y.; Lee, C.; Guo, J. Autonomously Adhesive, Stretchable, and Transparent Solid-State Polyionic Triboelectric Patch for Wearable Power Source and Tactile Sensor. *Adv. Funct. Mater.* **2021**, *31*, 2104365. [CrossRef]

180. Yuan, F.; Liu, S.; Zhou, J.; Wang, S.; Wang, Y.; Xuan, S.; Gong, X. Smart Touchless Triboelectric Nanogenerator towards Safeguard and 3D Morphological Awareness. *Nano Energy* **2021**, *86*, 106071. [CrossRef]

181. Song, Y.; Wang, N.; Hu, C.; Wang, Z.L.; Yang, Y. Soft Triboelectric Nanogenerators for Mechanical Energy Scavenging and Self-Powered Sensors. *Nano Energy* **2021**, *84*, 105919. [CrossRef]

182. Park, S.; Park, J.; Kim, Y.; Bae, S.; Kim, T.-W.; Park, K.-I.; Hong, B.H.; Jeong, C.K.; Lee, S.-K. Laser-Directed Synthesis of Strain-Induced Crumpled MoS2 Structure for Enhanced Triboelectrification toward Haptic Sensors. *Nano Energy* **2020**, *78*, 105266. [CrossRef]

183. He, J.; Xie, Z.; Yao, K.; Li, D.; Liu, Y.; Gao, Z.; Lu, W.; Chang, L.; Yu, X. Trampoline Inspired Stretchable Triboelectric Nanogenerators as Tactile Sensors for Epidermal Electronics. *Nano Energy* **2021**, *81*, 105590. [CrossRef]

184. Wang, X.; Zhang, Y.; Zhang, X.; Huo, Z.; Li, X.; Que, M.; Peng, Z.; Wang, H.; Pan, C. A Highly Stretchable Transparent Self-Powered Triboelectric Tactile Sensor with Metallized Nanofibers for Wearable Electronics. *Adv. Mater.* **2018**, *30*, 1706738. [CrossRef] [PubMed]

185. An, T.; Anaya, D.V.; Gong, S.; Yap, L.W.; Lin, F.; Wang, R.; Yuce, M.R.; Cheng, W. Self-Powered Gold Nanowire Tattoo Triboelectric Sensors for Soft Wearable Human-Machine Interface. *Nano Energy* **2020**, *77*, 105295. [CrossRef]

186. Lee, Y.; Kim, J.; Jang, B.; Kim, S.; Sharma, B.K.; Kim, J.-H.; Ahn, J.-H. Graphene-Based Stretchable/Wearable Self-Powered Touch Sensor. *Nano Energy* **2019**, *62*, 259–267. [CrossRef]

187. Chen, X.; Wu, Y.; Shao, J.; Jiang, T.; Yu, A.; Xu, L.; Wang, Z.L. On-Skin Triboelectric Nanogenerator and Self-Powered Sensor with Ultrathin Thickness and High Stretchability. *Small* **2017**, *13*, 1702929. [CrossRef] [PubMed]
188. Qiu, C.; Wu, F.; Shi, Q.; Lee, C.; Yuce, M.R. Sensors and Control Interface Methods Based on Triboelectric Nanogenerator in IoT Applications. *IEEE Access* **2019**, *7*, 92745–92757. [CrossRef]
189. Shi, Q.; Zhang, Z.; Chen, T.; Lee, C. Minimalist and Multi-Functional Human Machine Interface (HMI) Using a Flexible Wearable Triboelectric Patch. *Nano Energy* **2019**, *62*, 355–366. [CrossRef]
190. Shi, Q.; Qiu, C.; He, T.; Wu, F.; Zhu, M.; Dziuban, J.A.; Walczak, R.; Yuce, M.R.; Lee, C. Triboelectric Single-Electrode-Output Control Interface Using Patterned Grid Electrode. *Nano Energy* **2019**, *60*, 545–556. [CrossRef]
191. Tang, G.; Shi, Q.; Zhang, Z.; He, T.; Sun, Z.; Lee, C. Hybridized Wearable Patch as a Multi-Parameter and Multi-Functional Human-Machine Interface. *Nano Energy* **2021**, *81*, 105582. [CrossRef]
192. Cao, R.; Pu, X.; Du, X.; Yang, W.; Wang, J.; Guo, H.; Zhao, S.; Yuan, Z.; Zhang, C.; Li, C.; et al. Screen-Printed Washable Electronic Textiles as Self-Powered Touch/Gesture Tribo-Sensors for Intelligent Human–Machine Interaction. *ACS Nano* **2018**, *12*, 5190–5196. [CrossRef] [PubMed]
193. Lin, X.; Mao, Y.; Li, P.; Bai, Y.; Chen, T.; Wu, K.; Chen, D.; Yang, H.; Yang, L. Ultra-Conformable Ionic Skin with Multi-Modal Sensing, Broad-Spectrum Antimicrobial and Regenerative Capabilities for Smart and Expedited Wound Care. *Adv. Sci.* **2021**, *8*, 2004627. [CrossRef]
194. Bouteraa, Y.; Ben Abdallah, I. A Gesture-Based Telemanipulation Control for a Robotic Arm with Biofeedback-Based Grasp. *Ind. Robot An Int. J.* **2017**, *44*, 575–587. [CrossRef]
195. Fall, C.L.; Turgeon, P.; Campeau-Lecours, A.; Maheu, V.; Boukadoum, M.; Roy, S.; Massicotte, D.; Gosselin, C.; Gosselin, B. Intuitive Wireless Control of a Robotic Arm for People Living with an Upper Body Disability. In Proceedings of the 2015 37th Annual International Conference of the IEEE Engineering in Medicine and Biology Society (EMBC), Milan, Italy, 25–29 August 2015; IEEE: Piscataway, NJ, USA, 2015; pp. 4399–4402.
196. Miehlbradt, J.; Cherpillod, A.; Mintchev, S.; Coscia, M.; Artoni, F.; Floreano, D.; Micera, S. Correction for Miehlbradt et Al., Data-Driven Body–Machine Interface for the Accurate Control of Drones. *Proc. Natl. Acad. Sci. USA* **2019**, *116*, 19209.
197. Gao, S.; He, T.; Zhang, Z.; Ao, H.; Jiang, H.; Lee, C. A Motion Capturing and Energy Harvesting Hybridized Lower-Limb System for Rehabilitation and Sports Applications. *Adv. Sci.* **2021**. [CrossRef]
198. Yang, G.-Z.; Bellingham, J.; Dupont, P.E.; Fischer, P.; Floridi, L.; Full, R.; Jacobstein, N.; Kumar, V.; McNutt, M.; Merrifield, R.; et al. The Grand Challenges of Science Robotics. *Sci. Robot.* **2018**, *3*, eaar7650. [CrossRef]
199. Gu, W.; Cao, J.; Dai, S.; Hu, H.; Zhong, Y.; Cheng, G.; Zhang, Z.; Ding, J. Self-Powered Slide Tactile Sensor with Wheel-Belt Structures Based on Triboelectric Effect and Electrostatic Induction. *Sensors Actuators A Phys.* **2021**, *331*, 113022. [CrossRef]
200. Rong, X.; Zhao, J.; Guo, H.; Zhen, G.; Yu, J.; Zhang, C.; Dong, G. Material Recognition Sensor Array by Electrostatic Induction and Triboelectric Effects. *Adv. Mater. Technol.* **2020**, *5*, 2000641. [CrossRef]
201. Liu, H.; Ji, Z.; Xu, H.; Sun, M.; Chen, T.; Sun, L.; Chen, G.; Wang, Z. Large-Scale and Flexible Self-Powered Triboelectric Tactile Sensing Array for Sensitive Robot Skin. *Polymers* **2017**, *9*, 586. [CrossRef] [PubMed]
202. Zhang, C.; Liu, S.; Huang, X.; Guo, W.; Li, Y.; Wu, H. A Stretchable Dual-Mode Sensor Array for Multifunctional Robotic Electronic Skin. *Nano Energy* **2019**, *62*, 164–170. [CrossRef]
203. Shi, M.; Zhang, J.; Chen, H.; Han, M.; Shankaregowda, S.A.; Su, Z.; Meng, B.; Cheng, X.; Zhang, H. Self-Powered Analogue Smart Skin. *ACS Nano* **2016**, *10*, 4083–4091. [CrossRef] [PubMed]
204. Wang, Z.; Zhang, F.; Yao, T.; Li, N.; Li, X.; Shang, J. Self-Powered Non-Contact Triboelectric Rotation Sensor with Interdigitated Film. *Sensors* **2020**, *20*, 4947. [CrossRef]
205. Zhao, X.; Kang, Z.; Liao, Q.; Zhang, Z.; Ma, M.; Zhang, Q.; Zhang, Y. Ultralight, Self-Powered and Self-Adaptive Motion Sensor Based on Triboelectric Nanogenerator for Perceptual Layer Application in Internet of Things. *Nano Energy* **2018**, *48*, 312–319. [CrossRef]
206. Xu, Y.; Wang, Z.; Hao, W.; Zhao, W.; Lin, W.; Jin, B.; Ding, N. A Flexible Multimodal Sole Sensor for Legged Robot Sensing Complex Ground Information during Locomotion. *Sensors* **2021**, *21*, 5359. [CrossRef]
207. Yao, G.; Xu, L.; Cheng, X.; Li, Y.; Huang, X.; Guo, W.; Liu, S.; Wang, Z.L.; Wu, H. Bioinspired Triboelectric Nanogenerators as Self-Powered Electronic Skin for Robotic Tactile Sensing. *Adv. Funct. Mater.* **2020**, *30*, 1907312. [CrossRef]
208. Kanda, T.; Shiomi, M.; Miyashita, Z.; Ishiguro, H.; Hagita, N. An Affective Guide Robot in a Shopping Mall. In Proceedings of the Proceedings of the 4th ACM/IEEE international conference on Human robot interaction—HRI '09, La Jolla, CA, USA, 9–13 March 2009; ACM Press: New York, NY, USA, 2009; p. 173.
209. Norberto Pires, J. Robot-by-voice: Experiments on Commanding an Industrial Robot Using the Human Voice. *Ind. Robot Int. J.* **2005**, *32*, 505–511. [CrossRef]
210. Guo, H.; Pu, X.; Chen, J.; Meng, Y.; Yeh, M.-H.; Liu, G.; Tang, Q.; Chen, B.; Liu, D.; Qi, S.; et al. A Highly Sensitive, Self-Powered Triboelectric Auditory Sensor for Social Robotics and Hearing Aids. *Sci. Robot.* **2018**, *3*, eaat2516. [CrossRef]
211. Lange, F.; Bertleff, W.; Suppa, M. Force and Trajectory Control of Industrial Robots in Stiff Contact. In Proceedings of the 2013 IEEE International Conference on Robotics and Automation, Karlsruhe, Germany, 6–10 May 2013; IEEE: Piscataway, NJ, USA, 2013; pp. 2927–2934.
212. Cheng, P.; Oelmann, B. Joint-Angle Measurement Using Accelerometers and Gyroscopes—A Survey. *IEEE Trans. Instrum. Meas.* **2010**, *59*, 404–414. [CrossRef]

213. Wang, Z.; An, J.; Nie, J.; Luo, J.; Shao, J.; Jiang, T.; Chen, B.; Tang, W.; Wang, Z.L. A Self-Powered Angle Sensor at Nanoradian-Resolution for Robotic Arms and Personalized Medicare. *Adv. Mater.* **2020**, *32*, 2001466. [CrossRef]
214. Shintake, J.; Cacucciolo, V.; Floreano, D.; Shea, H. Soft Robotic Grippers. *Adv. Mater.* **2018**, *30*, 1707035. [CrossRef]
215. Laschi, C.; Mazzolai, B.; Cianchetti, M. Soft Robotics: Technologies and Systems Pushing the Boundaries of Robot Abilities. *Sci. Robot.* **2016**, *1*, eaah3690. [CrossRef]
216. Wang, J.; Gao, D.; Lee, P.S. Recent Progress in Artificial Muscles for Interactive Soft Robotics. *Adv. Mater.* **2021**, *33*, 2003088. [CrossRef]
217. Hines, L.; Petersen, K.; Lum, G.Z.; Sitti, M. Soft Actuators for Small-Scale Robotics. *Adv. Mater.* **2017**, *29*, 1603483. [CrossRef]
218. Ilami, M.; Bagheri, H.; Ahmed, R.; Skowronek, E.O.; Marvi, H. Materials, Actuators, and Sensors for Soft Bioinspired Robots. *Adv. Mater.* **2021**, *33*, 2003139. [CrossRef] [PubMed]
219. Rus, D.; Tolley, M.T. Design, Fabrication and Control of Soft Robots. *Nature* **2015**, *521*, 467–475. [CrossRef] [PubMed]
220. Chen, J.; Chen, B.; Han, K.; Tang, W.; Wang, Z.L. A Triboelectric Nanogenerator as a Self-Powered Sensor for a Soft–Rigid Hybrid Actuator. *Adv. Mater. Technol.* **2019**, *4*, 1900337. [CrossRef]
221. Chen, S.; Pang, Y.; Yuan, H.; Tan, X.; Cao, C. Smart Soft Actuators and Grippers Enabled by Self-Powered Tribo-Skins. *Adv. Mater. Technol.* **2020**, *5*, 1901075. [CrossRef]
222. Zhu, M.; Xie, M.; Lu, X.; Okada, S.; Kawamura, S. A Soft Robotic Finger with Self-Powered Triboelectric Curvature Sensor Based on Multi-Material 3D Printing. *Nano Energy* **2020**, *73*, 104772. [CrossRef]
223. Chen, J.; Han, K.; Luo, J.; Xu, L.; Tang, W.; Wang, Z.L. Soft Robots with Self-Powered Configurational Sensing. *Nano Energy* **2020**, *77*, 105171. [CrossRef]
224. Lai, Y.-C.; Deng, J.; Liu, R.; Hsiao, Y.-C.; Zhang, S.L.; Peng, W.; Wu, H.-M.; Wang, X.; Wang, Z.L. Actively Perceiving and Responsive Soft Robots Enabled by Self-Powered, Highly Extensible, and Highly Sensitive Triboelectric Proximity- and Pressure-Sensing Skins. *Adv. Mater.* **2018**, *30*, 1801114. [CrossRef] [PubMed]
225. Jin, T.; Sun, Z.; Li, L.; Zhang, Q.; Zhu, M.; Zhang, Z.; Yuan, G.; Chen, T.; Tian, Y.; Hou, X.; et al. Triboelectric Nanogenerator Sensors for Soft Robotics Aiming at Digital Twin Applications. *Nat. Commun.* **2020**, *11*, 5381. [CrossRef] [PubMed]
226. Bu, T.; Xiao, T.; Yang, Z.; Liu, G.; Fu, X.; Nie, J.; Guo, T.; Pang, Y.; Zhao, J.; Xi, F.; et al. Stretchable Triboelectric-Photonic Smart Skin for Tactile and Gesture Sensing. *Adv. Mater.* **2018**, *30*, 1800066. [CrossRef]
227. Wu, X.; Zhu, J.; Evans, J.W.; Arias, A.C. A Single-Mode, Self-Adapting, and Self-Powered Mechanoreceptor Based on a Potentiometric–Triboelectric Hybridized Sensing Mechanism for Resolving Complex Stimuli. *Adv. Mater.* **2020**, *32*, 2005970. [CrossRef]
228. An, J.; Chen, P.; Wang, Z.; Berbille, A.; Pang, H.; Jiang, Y.; Jiang, T.; Wang, Z.L. Biomimetic Hairy Whiskers for Robotic Skin Tactility. *Adv. Mater.* **2021**, *33*, 2101891. [CrossRef] [PubMed]
229. Liu, Z.; Wang, Y.; Ren, Y.; Jin, G.; Zhang, C.; Chen, W.; Yan, F. Poly(Ionic Liquid) Hydrogel-Based Anti-Freezing Ionic Skin for a Soft Robotic Gripper. *Mater. Horizons* **2020**, *7*, 919–927. [CrossRef]
230. Jin, G.; Sun, Y.; Geng, J.; Yuan, X.; Chen, T.; Liu, H.; Wang, F.; Sun, L. Bioinspired Soft Caterpillar Robot with Ultra-Stretchable Bionic Sensors Based on Functional Liquid Metal. *Nano Energy* **2021**, *84*, 105896. [CrossRef]
231. Garcia, E.; Jimenez, M.A.; De Santos, P.G.; Armada, M. The Evolution of Robotics Research. *IEEE Robot. Autom. Mag.* **2007**, *14*, 90–103. [CrossRef]
232. Kim, J.-H.; Sharma, G.; Iyengar, S.S. FAMPER: A Fully Autonomous Mobile Robot for Pipeline Exploration. In Proceedings of the 2010 IEEE International Conference on Industrial Technology, Viña del Mar, Chile, 14–17 March 2010; IEEE: Piscataway, NJ, USA, 2010; pp. 517–523.
233. Wang, Y.; Jiang, Y.; Wu, H.; Yang, Y. Floating Robotic Insects to Obtain Electric Energy from Water Surface for Realizing Some Self-Powered Functions. *Nano Energy* **2019**, *63*, 103810. [CrossRef]
234. Wang, Y.; Dai, M.; Wu, H.; Xu, L.; Zhang, T.; Chen, W.; Wang, Z.L.; Yang, Y. Moisture Induced Electricity for Self-Powered Microrobots. *Nano Energy* **2021**, 106499. [CrossRef]
235. Asadnia, M.; Kottapalli, A.G.P.; Miao, J.; Warkiani, M.E.; Triantafyllou, M.S. Artificial Fish Skin of Self-Powered Micro-Electromechanical Systems Hair Cells for Sensing Hydrodynamic Flow Phenomena. *J.R. Soc. Interface* **2015**, *12*, 20150322. [CrossRef] [PubMed]
236. Shih, B.; Shah, D.; Li, J.; Thuruthel, T.G.; Park, Y.-L.; Iida, F.; Bao, Z.; Kramer-Bottiglio, R.; Tolley, M.T. Electronic Skins and Machine Learning for Intelligent Soft Robots. *Sci. Robot.* **2020**, *5*, eaaz9239. [CrossRef]
237. Lu, H.; Hong, Y.; Yang, Y.; Yang, Z.; Shen, Y. Battery-Less Soft Millirobot That Can Move, Sense, and Communicate Remotely by Coupling the Magnetic and Piezoelectric Effects. *Adv. Sci.* **2020**, *7*, 2000069. [CrossRef] [PubMed]
238. Goldoni, R.; Ozkan-Aydin, Y.; Kim, Y.-S.; Kim, J.; Zavanelli, N.; Mahmood, M.; Liu, B.; Hammond, F.L.; Goldman, D.I.; Yeo, W.-H. Stretchable Nanocomposite Sensors, Nanomembrane Interconnectors, and Wireless Electronics toward Feedback–Loop Control of a Soft Earthworm Robot. *ACS Appl. Mater. Interfaces* **2020**, *12*, 43388–43397. [CrossRef] [PubMed]
239. Chan, M.; Campo, E.; Estève, D.; Fourniols, J.-Y. Smart Homes—Current Features and Future Perspectives. *Maturitas* **2009**, *64*, 90–97. [CrossRef]
240. Demiris, G.; Hensel, B.K. Technologies for an Aging Society: A Systematic Review of "Smart Home" Applications. *Yearb. Med. Inform.* **2008**, *17*, 33–40.

241. Ding, D.; Cooper, R.A.; Pasquina, P.F.; Fici-Pasquina, L. Sensor Technology for Smart Homes. *Maturitas* **2011**, *69*, 131–136. [CrossRef] [PubMed]
242. Haroun, A.; Le, X.; Gao, S.; Dong, B.; He, T.; Zhang, Z.; Wen, F.; Xu, S.; Lee, C. Progress in Micro/Nano Sensors and Nanoenergy for Future AIoT-Based Smart Home Applications. *Nano Express* **2021**, *2*, 022005. [CrossRef]
243. Anaya, D.V.; Zhan, K.; Tao, L.; Lee, C.; Yuce, M.R.; Alan, T. Contactless Tracking of Humans Using Non-Contact Triboelectric Sensing Technology: Enabling New Assistive Applications for The Elderly and The Visually Impaired. *Nano Energy* **2021**, 106486. [CrossRef]
244. Yan, Z.; Wang, L.; Xia, Y.; Qiu, R.; Liu, W.; Wu, M.; Zhu, Y.; Zhu, S.; Jia, C.; Zhu, M.; et al. Flexible High-Resolution Triboelectric Sensor Array Based on Patterned Laser-Induced Graphene for Self-Powered Real-Time Tactile Sensing. *Adv. Funct. Mater.* **2021**, *31*, 2100709. [CrossRef]
245. Sala de Medeiros, M.; Chanci, D.; Martinez, R.V. Moisture-Insensitive, Self-Powered Paper-Based Flexible Electronics. *Nano Energy* **2020**, *78*, 105301. [CrossRef]
246. Yuan, Z.; Zhou, T.; Yin, Y.; Cao, R.; Li, C.; Wang, Z.L. Transparent and Flexible Triboelectric Sensing Array for Touch Security Applications. *ACS Nano* **2017**, *11*, 8364–8369. [CrossRef]
247. Liu, Y.; Yang, W.; Yan, Y.; Wu, X.; Wang, X.; Zhou, Y.; Hu, Y.; Chen, H.; Guo, T. Self-Powered High-Sensitivity Sensory Memory Actuated by Triboelectric Sensory Receptor for Real-Time Neuromorphic Computing. *Nano Energy* **2020**, *75*, 104930. [CrossRef]
248. Pu, X.; Tang, Q.; Chen, W.; Huang, Z.; Liu, G.; Zeng, Q.; Chen, J.; Guo, H.; Xin, L.; Hu, C. Flexible Triboelectric 3D Touch Pad with Unit Subdivision Structure for Effective XY Positioning and Pressure Sensing. *Nano Energy* **2020**, *76*, 105047. [CrossRef]
249. Qiu, C.; Wu, F.; Lee, C.; Yuce, M.R. Self-Powered Control Interface Based on Gray Code with Hybrid Triboelectric and Photovoltaics Energy Harvesting for IoT Smart Home and Access Control Applications. *Nano Energy* **2020**, *70*, 104456. [CrossRef]
250. Yuan, Z.; Du, X.; Li, N.; Yin, Y.; Cao, R.; Zhang, X.; Zhao, S.; Niu, H.; Jiang, T.; Xu, W.; et al. Triboelectric-Based Transparent Secret Code. *Adv. Sci.* **2018**, *5*, 1700881. [CrossRef] [PubMed]
251. Chen, J.; Pu, X.; Guo, H.; Tang, Q.; Feng, L.; Wang, X.; Hu, C. A Self-Powered 2D Barcode Recognition System Based on Sliding Mode Triboelectric Nanogenerator for Personal Identification. *Nano Energy* **2018**, *43*, 253–258. [CrossRef]
252. Lin, Z.; Yang, J.; Li, X.; Wu, Y.; Wei, W.; Liu, J.; Chen, J.; Yang, J. Large-Scale and Washable Smart Textiles Based on Triboelectric Nanogenerator Arrays for Self-Powered Sleeping Monitoring. *Adv. Funct. Mater.* **2018**, *28*, 1704112. [CrossRef]
253. Dong, K.; Peng, X.; An, J.; Wang, A.C.; Luo, J.; Sun, B.; Wang, J.; Wang, Z.L. Shape Adaptable and Highly Resilient 3D Braided Triboelectric Nanogenerators as E-Textiles for Power and Sensing. *Nat. Commun.* **2020**, *11*, 2868. [CrossRef]
254. Shi, Q.; Zhang, Z.; He, T.; Sun, Z.; Wang, B.; Feng, Y.; Shan, X.; Salam, B.; Lee, C. Deep Learning Enabled Smart Mats as a Scalable Floor Monitoring System. *Nat. Commun.* **2020**, *11*, 4609. [CrossRef]
255. Kato, H.; Tan, K.T. Pervasive 2D Barcodes for Camera Phone Applications. *IEEE Pervasive Comput.* **2007**, *6*, 76–85. [CrossRef]
256. Houni, K.; Sawaya, W.; Delignon, Y. One-Dimensional Barcode Reading: An Information Theoretic Approach. *Appl. Opt.* **2008**, *47*, 1025. [CrossRef] [PubMed]
257. Sixel-Döring, F.; Zimmermann, J.; Wegener, A.; Mollenhauer, B.; Trenkwalder, C. The Evolution of REM Sleep Behavior Disorder in Early Parkinson Disease. *Sleep* **2016**, *39*, 1737–1742. [CrossRef]
258. Jones, M.H.; Goubran, R.; Knoefel, F. Reliable Respiratory Rate Estimation from a Bed Pressure Array. In Proceedings of the 2006 International Conference of the IEEE Engineering in Medicine and Biology Society, New York, NY, USA, 31 August–3 September 2006; IEEE: Piscataway, NJ, USA, 2006; pp. 6410–6413.
259. Samy, L.; Huang, M.-C.; Liu, J.J.; Xu, W.; Sarrafzadeh, M. Unobtrusive Sleep Stage Identification Using a Pressure-Sensitive Bed Sheet. *IEEE Sens. J.* **2014**, *14*, 2092–2101. [CrossRef]
260. Kwak, S.S.; Yoon, H.-J.; Kim, S.-W. Textile-Based Triboelectric Nanogenerators for Self-Powered Wearable Electronics. *Adv. Funct. Mater.* **2019**, *29*, 1804533. [CrossRef]
261. Dong, K.; Peng, X.; Wang, Z.L. Fiber/Fabric-Based Piezoelectric and Triboelectric Nanogenerators for Flexible/Stretchable and Wearable Electronics and Artificial Intelligence. *Adv. Mater.* **2020**, *32*, 1902549. [CrossRef]
262. Paosangthong, W.; Torah, R.; Beeby, S. Recent Progress on Textile-Based Triboelectric Nanogenerators. *Nano Energy* **2019**, *55*, 401–423. [CrossRef]
263. Kim, K.-B.; Cho, J.Y.; Jabbar, H.; Ahn, J.H.; Hong, S.D.; Woo, S.B.; Sung, T.H. Optimized Composite Piezoelectric Energy Harvesting Floor Tile for Smart Home Energy Management. *Energy Convers. Manag.* **2018**, *171*, 31–37. [CrossRef]
264. Middleton, L.; Buss, A.A.; Bazin, A.; Nixon, M.S. A Floor Sensor System for Gait Recognition. In Proceedings of the Fourth IEEE Workshop on Automatic Identification Advanced Technologies (AutoID'05), Buffalo, NY, USA, 17–18 October 2005; IEEE: Piscataway, NJ, USA, 2005; pp. 171–176.
265. Li, Y.; Gao, Z.; He, Z.; Zhang, P.; Chen, R.; El-Sheimy, N. Multi-Sensor Multi-Floor 3D Localization With Robust Floor Detection. *IEEE Access* **2018**, *6*, 76689–76699. [CrossRef]
266. Jeon, S.-B.; Nho, Y.-H.; Park, S.-J.; Kim, W.-G.; Tcho, I.-W.; Kim, D.; Kwon, D.-S.; Choi, Y.-K. Self-Powered Fall Detection System Using Pressure Sensing Triboelectric Nanogenerators. *Nano Energy* **2017**, *41*, 139–147. [CrossRef]
267. Ma, J.; Jie, Y.; Bian, J.; Li, T.; Cao, X.; Wang, N. From Triboelectric Nanogenerator to Self-Powered Smart Floor: A Minimalist Design. *Nano Energy* **2017**, *39*, 192–199. [CrossRef]
268. Cheng, X.; Song, Y.; Han, M.; Meng, B.; Su, Z.; Miao, L.; Zhang, H. A Flexible Large-Area Triboelectric Generator by Low-Cost Roll-to-Roll Process for Location-Based Monitoring. *Sensors Actuators A Phys.* **2016**, *247*, 206–214. [CrossRef]

269. Ha, N.; Xu, K.; Ren, G.; Mitchell, A.; Ou, J.Z. Machine Learning-Enabled Smart Sensor Systems. *Adv. Intell. Syst.* **2020**, *2*, 2000063. [CrossRef]
270. Krittanawong, C.; Rogers, A.J.; Johnson, K.W.; Wang, Z.; Turakhia, M.P.; Halperin, J.L.; Narayan, S.M. Integration of Novel Monitoring Devices with Machine Learning Technology for Scalable Cardiovascular Management. *Nat. Rev. Cardiol.* **2021**, *18*, 75–91. [CrossRef]
271. Wu, C.; Ding, W.; Liu, R.; Wang, J.; Wang, A.C.; Wang, J.; Li, S.; Zi, Y.; Wang, Z.L. Keystroke Dynamics Enabled Authentication and Identification Using Triboelectric Nanogenerator Array. *Mater. Today* **2018**, *21*, 216–222. [CrossRef]
272. Zhang, Z.; He, T.; Zhu, M.; Sun, Z.; Shi, Q.; Zhu, J.; Dong, B.; Yuce, M.R.; Lee, C. Deep Learning-Enabled Triboelectric Smart Socks for IoT-Based Gait Analysis and VR Applications. *npj Flex. Electron.* **2020**, *4*, 29. [CrossRef]
273. Wen, F.; Zhang, Z.; He, T.; Lee, C. AI Enabled Sign Language Recognition and VR Space Bidirectional Communication Using Triboelectric Smart Glove. *Nat. Commun.* **2021**, *12*, 5378. [CrossRef]
274. Sun, Z.; Zhu, M.; Zhang, Z.; Chen, Z.; Shi, Q.; Shan, X.; Yeow, R.C.H.; Lee, C. Artificial Intelligence of Things (AIoT) Enabled Virtual Shop Applications Using Self-Powered Sensor Enhanced Soft Robotic Manipulator. *Adv. Sci.* **2021**, *8*, 2100230. [CrossRef] [PubMed]
275. Wang, M.; Yan, Z.; Wang, T.; Cai, P.; Gao, S.; Zeng, Y.; Wan, C.; Wang, H.; Pan, L.; Yu, J.; et al. Gesture Recognition Using a Bioinspired Learning Architecture That Integrates Visual Data with Somatosensory Data from Stretchable Sensors. *Nat. Electron.* **2020**, *3*, 563–570. [CrossRef]
276. Yun, J.; Jayababu, N.; Kim, D. Self-Powered Transparent and Flexible Touchpad Based on Triboelectricity towards Artificial Intelligence. *Nano Energy* **2020**, *78*, 105325. [CrossRef]
277. Maharjan, P.; Shrestha, K.; Bhatta, T.; Cho, H.; Park, C.; Salauddin, M.; Rahman, M.T.; Rana, S.S.; Lee, S.; Park, J.Y. Keystroke Dynamics Based Hybrid Nanogenerators for Biometric Authentication and Identification Using Artificial Intelligence. *Adv. Sci.* **2021**, *8*, 2100711. [CrossRef] [PubMed]
278. Zhao, X.; Zhang, Z.; Xu, L.; Gao, F.; Zhao, B.; Ouyang, T.; Kang, Z.; Liao, Q.; Zhang, Y. Fingerprint-Inspired Electronic Skin Based on Triboelectric Nanogenerator for Fine Texture Recognition. *Nano Energy* **2021**, *85*, 106001. [CrossRef]
279. Moin, A.; Zhou, A.; Rahimi, A.; Menon, A.; Benatti, S.; Alexandrov, G.; Tamakloe, S.; Ting, J.; Yamamoto, N.; Khan, Y.; et al. A Wearable Biosensing System with In-Sensor Adaptive Machine Learning for Hand Gesture Recognition. *Nat. Electron.* **2021**, *4*, 54–63. [CrossRef]
280. Luo, Y.; Li, Y.; Sharma, P.; Shou, W.; Wu, K.; Foshey, M.; Li, B.; Palacios, T.; Torralba, A.; Matusik, W. Learning Human–Environment Interactions Using Conformal Tactile Textiles. *Nat. Electron.* **2021**, *4*, 193–201. [CrossRef]
281. Sundaram, S.; Kellnhofer, P.; Li, Y.; Zhu, J.-Y.; Torralba, A.; Matusik, W. Learning the Signatures of the Human Grasp Using a Scalable Tactile Glove. *Nature* **2019**, *569*, 698–702. [CrossRef]
282. Monrose, F.; Rubin, A.D. Keystroke Dynamics as a Biometric for Authentication. *Futur. Gener. Comput. Syst.* **2000**, *16*, 351–359. [CrossRef]
283. Zhao, G.; Yang, J.; Chen, J.; Zhu, G.; Jiang, Z.; Liu, X.; Niu, G.; Wang, Z.L.; Zhang, B. Keystroke Dynamics Identification Based on Triboelectric Nanogenerator for Intelligent Keyboard Using Deep Learning Method. *Adv. Mater. Technol.* **2019**, *4*, 1800167. [CrossRef]
284. Chen, J.; Zhu, G.; Yang, J.; Jing, Q.; Bai, P.; Yang, W.; Qi, X.; Su, Y.; Wang, Z.L. Personalized Keystroke Dynamics for Self-Powered Human–Machine Interfacing. *ACS Nano* **2015**, *9*, 105–116. [CrossRef]
285. Caldas, R.; Mundt, M.; Potthast, W.; Buarque de Lima Neto, F.; Markert, B. A Systematic Review of Gait Analysis Methods Based on Inertial Sensors and Adaptive Algorithms. *Gait Posture* **2017**, *57*, 204–210. [CrossRef] [PubMed]
286. Fang, B.; Sun, F.; Liu, H.; Liu, C. 3D Human Gesture Capturing and Recognition by the IMMU-Based Data Glove. *Neurocomputing* **2018**, *277*, 198–207. [CrossRef]
287. Neto, P.; Pereira, D.; Pires, J.N.; Moreira, A.P. Real-Time and Continuous Hand Gesture Spotting: An Approach Based on Artificial Neural Networks. In Proceedings of the 2013 IEEE International Conference on Robotics and Automation, Karlsruhe, Germany, 6–10 May 2013; IEEE: Piscataway, NJ, USA, 2013; pp. 178–183.
288. Chiu, C.-M.; Chen, S.-W.; Pao, Y.-P.; Huang, M.-Z.; Chan, S.-W.; Lin, Z.-H. A Smart Glove with Integrated Triboelectric Nanogenerator for Self-Powered Gesture Recognition and Language Expression. *Sci. Technol. Adv. Mater.* **2019**, *20*, 964–971. [CrossRef]
289. Debie, E.; Fernandez Rojas, R.; Fidock, J.; Barlow, M.; Kasmarik, K.; Anavatti, S.; Garratt, M.; Abbass, H.A. Multimodal Fusion for Objective Assessment of Cognitive Workload: A Review. *IEEE Trans. Cybern.* **2021**, *51*, 1542–1555. [CrossRef] [PubMed]
290. Imran, J.; Raman, B. Evaluating Fusion of RGB-D and Inertial Sensors for Multimodal Human Action Recognition. *J. Ambient Intell. Humaniz. Comput.* **2020**, *11*, 189–208. [CrossRef]
291. Xue, T.; Wang, W.; Ma, J.; Liu, W.; Pan, Z.; Han, M. Progress and Prospects of Multimodal Fusion Methods in Physical Human–Robot Interaction: A Review. *IEEE Sens. J.* **2020**, *20*, 10355–10370. [CrossRef]
292. Biswas, S.; Visell, Y. Emerging Material Technologies for Haptics. *Adv. Mater. Technol.* **2019**, *4*, 1900042. [CrossRef]
293. Rothemund, P.; Kellaris, N.; Mitchell, S.K.; Acome, E.; Keplinger, C. HASEL Artificial Muscles for a New Generation of Lifelike Robots—Recent Progress and Future Opportunities. *Adv. Mater.* **2021**, *33*, 2003375. [CrossRef]
294. Yin, J.; Hinchet, R.; Shea, H.; Majidi, C. Wearable Soft Technologies for Haptic Sensing and Feedback. *Adv. Funct. Mater.* **2020**, 2007428. [CrossRef]

295. Xiong, J.; Chen, J.; Lee, P.S. Functional Fibers and Fabrics for Soft Robotics, Wearables, and Human–Robot Interface. *Adv. Mater.* **2021**, *33*, 2002640. [CrossRef]

296. Nakamura, T.; Yamamoto, A. Multi-Finger Surface Visuo-Haptic Rendering Using Electrostatic Stimulation with Force-Direction Sensing Gloves. In Proceedings of the 2014 IEEE Haptics Symposium (HAPTICS), Houston, TX, USA, 23–26 February 2014; IEEE: Piscataway, NJ, USA, 2014; pp. 489–491.

297. Zhao, H.; Hussain, A.M.; Israr, A.; Vogt, D.M.; Duduta, M.; Clarke, D.R.; Wood, R.J. A Wearable Soft Haptic Communicator Based on Dielectric Elastomer Actuators. *Soft Robot.* **2020**, *7*, 451–461. [CrossRef]

298. Yang, Y.; Wu, Y.; Li, C.; Yang, X.; Chen, W. Flexible Actuators for Soft Robotics. *Adv. Intell. Syst.* **2020**, *2*, 1900077. [CrossRef]

299. Kang, B.B.; Choi, H.; Lee, H.; Cho, K.-J. Exo-Glove Poly II: A Polymer-Based Soft Wearable Robot for the Hand with a Tendon-Driven Actuation System. *Soft Robot.* **2019**, *6*, 214–227. [CrossRef]

300. Yu, X.; Xie, Z.; Yu, Y.; Lee, J.; Vazquez-Guardado, A.; Luan, H.; Ruban, J.; Ning, X.; Akhtar, A.; Li, D.; et al. Skin-Integrated Wireless Haptic Interfaces for Virtual and Augmented Reality. *Nature* **2019**, *575*, 473–479. [CrossRef]

301. Wang, X.; Mitchell, S.K.; Rumley, E.H.; Rothemund, P.; Keplinger, C. High-Strain Peano-HASEL Actuators. *Adv. Funct. Mater.* **2020**, *30*, 1908821. [CrossRef]

302. Ji, X.; Liu, X.; Cacucciolo, V.; Civet, Y.; El Haitami, A.; Cantin, S.; Perriard, Y.; Shea, H. Untethered Feel-Through Haptics Using 18-μm Thick Dielectric Elastomer Actuators. *Adv. Funct. Mater.* **2020**, *2006639*. [CrossRef]

303. Pyo, D.; Ryu, S.; Kyung, K.-U.; Yun, S.; Kwon, D.-S. High-Pressure Endurable Flexible Tactile Actuator Based on Microstructured Dielectric Elastomer. *Appl. Phys. Lett.* **2018**, *112*, 061902. [CrossRef]

304. Hwang, I.; Kim, H.J.; Mun, S.; Yun, S.; Kang, T.J. A Light-Driven Vibrotactile Actuator with a Polymer Bimorph Film for Localized Haptic Rendering. *ACS Appl. Mater. Interfaces* **2021**, *13*, 6597–6605. [CrossRef] [PubMed]

305. Hinchet, R.; Shea, H. High Force Density Textile Electrostatic Clutch. *Adv. Mater. Technol.* **2020**, *5*, 1900895. [CrossRef]

306. Leroy, E.; Hinchet, R.; Shea, H. Multimode Hydraulically Amplified Electrostatic Actuators for Wearable Haptics. *Adv. Mater.* **2020**, *32*, 2002564. [CrossRef]

307. Besse, N.; Rosset, S.; Zarate, J.J.; Shea, H. Flexible Active Skin: Large Reconfigurable Arrays of Individually Addressed Shape Memory Polymer Actuators. *Adv. Mater. Technol.* **2017**, *2*, 1700102. [CrossRef]

308. Qu, X.; Ma, X.; Shi, B.; Li, H.; Zheng, L.; Wang, C.; Liu, Z.; Fan, Y.; Chen, X.; Li, Z.; et al. Refreshable Braille Display System Based on Triboelectric Nanogenerator and Dielectric Elastomer. *Adv. Funct. Mater.* **2021**, *31*, 2006612. [CrossRef]

309. Shi, Y.; Wang, F.; Tian, J.; Li, S.; Fu, E.; Nie, J.; Lei, R.; Ding, Y.; Chen, X.; Wang, Z.L. Self-Powered Electro-Tactile System for Virtual Tactile Experiences. *Sci. Adv.* **2021**, *7*, eabe2943. [CrossRef]

310. Oh, J.; Kim, S.; Lee, S.; Jeong, S.; Ko, S.H.; Bae, J. A Liquid Metal Based Multimodal Sensor and Haptic Feedback Device for Thermal and Tactile Sensation Generation in Virtual Reality. *Adv. Funct. Mater.* **2020**, *2007772*. [CrossRef]

311. Lee, J.; Sul, H.; Lee, W.; Pyun, K.R.; Ha, I.; Kim, D.; Park, H.; Eom, H.; Yoon, Y.; Jung, J.; et al. Stretchable Skin-Like Cooling/Heating Device for Reconstruction of Artificial Thermal Sensation in Virtual Reality. *Adv. Funct. Mater.* **2020**, *30*, 1909171. [CrossRef]

312. Kellaris, N.; Gopaluni Venkata, V.; Smith, G.M.; Mitchell, S.K.; Keplinger, C. Peano-HASEL Actuators: Muscle-Mimetic, Electrohydraulic Transducers That Linearly Contract on Activation. *Sci. Robot.* **2018**, *3*, eaar3276. [CrossRef] [PubMed]

313. Kim, S.; Kim, T.; Kim, C.S.; Choi, H.; Kim, Y.J.; Lee, G.S.; Oh, O.; Cho, B.J. Two-Dimensional Thermal Haptic Module Based on a Flexible Thermoelectric Device. *Soft Robot.* **2020**, *7*, 736–742. [CrossRef] [PubMed]

314. Kim, S.-W.; Kim, S.H.; Kim, C.S.; Yi, K.; Kim, J.-S.; Cho, B.J.; Cha, Y. Thermal Display Glove for Interacting with Virtual Reality. *Sci. Rep.* **2020**, *10*, 11403. [CrossRef]

315. Souri, H.; Banerjee, H.; Jusufi, A.; Radacsi, N.; Stokes, A.A.; Park, I.; Sitti, M.; Amjadi, M. Wearable and Stretchable Strain Sensors: Materials, Sensing Mechanisms, and Applications. *Adv. Intell. Syst.* **2020**, *2*, 2000039. [CrossRef]

316. Wang, Z.L. Triboelectric Nanogenerators as New Energy Technology and Self-Powered Sensors—Principles, Problems and Perspectives. *Faraday Discuss.* **2014**, *176*, 447–458. [CrossRef]

317. Shi, Q.; Sun, Z.; Zhang, Z.; Lee, C. Triboelectric Nanogenerators and Hybridized Systems for Enabling Next-Generation IoT Applications. *Research* **2021**, *2021*, 1–30.

318. Zhang, K.; Wang, Y.; Yang, Y. Structure Design and Performance of Hybridized Nanogenerators. *Adv. Funct. Mater.* **2019**, *29*, 1806435. [CrossRef]

319. Guo, X.; Liu, L.; Zhang, Z.; Gao, S.; He, T.; Shi, Q.; Lee, C. Technology Evolution from Micro-Scale Energy Harvesters to Nanogenerators. *J. Micromech. Microeng.* **2021**, *31*, 093002. [CrossRef]

320. Guo, X.; He, T.; Zhang, Z.; Luo, A.; Wang, F.; Ng, E.J.; Zhu, Y.; Liu, H.; Lee, C. Artificial Intelligence-Enabled Caregiving Walking Stick Powered by Ultra-Low-Frequency Human Motion. *ACS Nano* **2021**. [CrossRef] [PubMed]

321. Mallineni, S.S.K.; Dong, Y.; Behlow, H.; Rao, A.M.; Podila, R. A Wireless Triboelectric Nanogenerator. *Adv. Energy Mater.* **2018**, *8*, 1702736. [CrossRef]

322. Wen, F.; Wang, H.; He, T.; Shi, Q.; Sun, Z.; Zhu, M.; Zhang, Z.; Cao, Z.; Dai, Y.; Zhang, T.; et al. Battery-Free Short-Range Self-Powered Wireless Sensor Network (SS-WSN) Using TENG Based Direct Sensory Transmission (TDST) Mechanism. *Nano Energy* **2020**, *67*, 104266. [CrossRef]

323. Tan, X.; Zhou, Z.; Zhang, L.; Wang, X.; Lin, Z.; Yang, R.; Yang, J. A Passive Wireless Triboelectric Sensor via a Surface Acoustic Wave Resonator (SAWR). *Nano Energy* **2020**, *78*, 105307. [CrossRef]

324. Lin, Z.; Chen, J.; Li, X.; Zhou, Z.; Meng, K.; Wei, W.; Yang, J.; Wang, Z.L. Triboelectric Nanogenerator Enabled Body Sensor Network for Self-Powered Human Heart-Rate Monitoring. *ACS Nano* **2017**, *11*, 8830–8837. [CrossRef]

325. He, T.; Wang, H.; Wang, J.; Tian, X.; Wen, F.; Shi, Q.; Ho, J.S.; Lee, C. Self-Sustainable Wearable Textile Nano-Energy Nano-System (NENS) for Next-Generation Healthcare Applications. *Adv. Sci.* **2019**, *6*, 1901437. [CrossRef]

326. Wang, L.; He, T.; Zhang, Z.; Zhao, L.; Lee, C.; Luo, G.; Mao, Q.; Yang, P.; Lin, Q.; Li, X.; et al. Self-Sustained Autonomous Wireless Sensing Based on a Hybridized TENG and PEG Vibration Mechanism. *Nano Energy* **2021**, *80*, 105555. [CrossRef]

327. Zhang, C.; Chen, J.; Xuan, W.; Huang, S.; You, B.; Li, W.; Sun, L.; Jin, H.; Wang, X.; Dong, S.; et al. Conjunction of Triboelectric Nanogenerator with Induction Coils as Wireless Power Sources and Self-Powered Wireless Sensors. *Nat. Commun.* **2020**, *11*, 58. [CrossRef]

328. Chen, J.; Xuan, W.; Zhao, P.; Farooq, U.; Ding, P.; Yin, W.; Jin, H.; Wang, X.; Fu, Y.; Dong, S.; et al. Triboelectric Effect Based Instantaneous Self-Powered Wireless Sensing with Self-Determined Identity. *Nano Energy* **2018**, *51*, 1–9. [CrossRef]

329. Bandodkar, A.J.; Gutruf, P.; Choi, J.; Lee, K.; Sekine, Y.; Reeder, J.T.; Jeang, W.J.; Aranyosi, A.J.; Lee, S.P.; Model, J.B.; et al. Battery-Free, Skin-Interfaced Microfluidic/Electronic Systems for Simultaneous Electrochemical, Colorimetric, and Volumetric Analysis of Sweat. *Sci. Adv.* **2019**, *5*, eaav3294. [CrossRef]

330. Kim, J.; Banks, A.; Cheng, H.; Xie, Z.; Xu, S.; Jang, K.-I.; Lee, J.W.; Liu, Z.; Gutruf, P.; Huang, X.; et al. Epidermal Electronics with Advanced Capabilities in Near-Field Communication. *Small* **2015**, *11*, 906–912. [CrossRef] [PubMed]

331. Fiddes, L.K.; Yan, N. RFID Tags for Wireless Electrochemical Detection of Volatile Chemicals. *Sensors Actuators B Chem.* **2013**, *186*, 817–823. [CrossRef]

332. Zhang, J.; Tian, G.; Marindra, A.; Sunny, A.; Zhao, A. A Review of Passive RFID Tag Antenna-Based Sensors and Systems for Structural Health Monitoring Applications. *Sensors* **2017**, *17*, 265. [CrossRef]

333. Jin, H.; Tao, X.; Dong, S.; Qin, Y.; Yu, L.; Luo, J.; Deen, M.J. Flexible Surface Acoustic Wave Respiration Sensor for Monitoring Obstructive Sleep Apnea Syndrome. *J. Micromech. Microeng.* **2017**, *27*, 115006. [CrossRef]

334. Xu, H.; Dong, S.; Xuan, W.; Farooq, U.; Huang, S.; Li, M.; Wu, T.; Jin, H.; Wang, X.; Luo, J. Flexible Surface Acoustic Wave Strain Sensor Based on Single Crystalline LiNbO 3 Thin Film. *Appl. Phys. Lett.* **2018**, *112*, 093502. [CrossRef]

335. Ren, Z.; Xu, J.; Le, X.; Lee, C. Heterogeneous Wafer Bonding Technology and Thin-Film Transfer Technology-Enabling Platform for the Next Generation Applications beyond 5G. *Micromachines* **2021**, *12*, 946. [CrossRef]

336. Dong, B.; Shi, Q.; He, T.; Zhu, S.; Zhang, Z.; Sun, Z.; Ma, Y.; Kwong, D.; Lee, C. Wearable Triboelectric/Aluminum Nitride Nano-Energy-Nano-System with Self-Sustainable Photonic Modulation and Continuous Force Sensing. *Adv. Sci.* **2020**, *7*, 1903636. [CrossRef] [PubMed]

337. Zhang, Z.; Shi, Q.; He, T.; Guo, X.; Dong, B.; Lee, J.; Lee, C. Artificial Intelligence of Toilet (AI-Toilet) for an Integrated Health Monitoring System (IHMS) using Smart Triboelectric Pressure Sensors and Image Sensor. *Nano Energy* **2021**, *90A*, 106517. [CrossRef]

Article

A Near-Zero Power Triboelectric Wake-Up System for Autonomous Beaufort Scale of Wind Force Monitoring

Tong Tong [1,2], Guoxu Liu [1,2], Yuan Lin [1,3], Shaohang Xu [1,2] and Chi Zhang [1,2,4,5,*]

[1]   CAS Center for Excellence in Nanoscience, Beijing Key Laboratory of Micro-Nano Energy and Sensor, Beijing Institute of Nanoenergy and Nanosystems, Chinese Academy of Sciences, Beijing 101400, China; ttongg126@126.com (T.T.); liuguoxu@binn.cas.cn (G.L.); 1911301021@st.gxu.edu.cn (Y.L.); xushaohangucas@163.com (S.X.)

[2]   School of Nanoscience and Technology, University of Chinese Academy of Sciences, Beijing 100049, China

[3]   School of Mechanical Engineering, Guangxi University, Nanning 530004, China

[4]   Center on Nanoenergy Research, School of Physical Science and Technology, Guangxi University, Nanning 530004, China

[5]   Tribotronics Research Group, Beijing Institute of Nanoenergy and Nanosystems, Chinese Academy of Sciences, Beijing 100029, China

*   Correspondence: czhang@binn.cas.cn

Citation: Tong, T.; Liu, G.; Lin, Y.; Xu, S.; Zhang, C. A Near-Zero Power Triboelectric Wake-Up System for Autonomous Beaufort Scale of Wind Force Monitoring. *Nanoenergy Adv.* **2021**, *1*, 121–130. https://doi.org/10.3390/nanoenergyadv1020006

Academic Editors: Dukhyun Choi and Chris R. Bowen

Received: 12 May 2021
Accepted: 17 September 2021
Published: 1 October 2021

Publisher's Note: MDPI stays neutral with regard to jurisdictional claims in published maps and institutional affiliations.

Copyright: © 2021 by the authors. Licensee MDPI, Basel, Switzerland. This article is an open access article distributed under the terms and conditions of the Creative Commons Attribution (CC BY) license (https://creativecommons.org/licenses/by/4.0/).

Abstract: Beaufort scale of wind force monitoring is the basic content of meteorological monitoring, which is an important means to ensure the safety of production and life by timely warning of natural disasters. As there is a limited battery life for sensors, determining how to reduce power consumption and extend system life is still an urgent problem. In this work, a near-zero power triboelectric wake-up system for autonomous Beaufort scale of wind force monitoring is proposed, in which a rotary TENG is used to convert wind energy into a stored electric energy capacitor. When the capacitor voltage accumulates to the threshold voltage of a transistor, it turns on as an electronic switch and the system wakes up. In active mode, the Beaufort scale of wind force can be judged according to the electric energy and the signal is sent out wirelessly. In standby mode, when there is no wind, the power consumption of the system is only 14 nW. When the wind scale reaches or exceeds light air, the system can switch to active mode within one second and accurately judge the Beaufort scale of wind force within 10 s. This work provided a triboelectric sensor-based wake-up system with ultralow static power consumption, which has great prospects for unattended environmental monitoring, hurricane warning, and big data acquisition.

Keywords: Beaufort scale monitoring; near-zero power; wake-up system; triboelectric nanogenerator; triboelectric sensor

1. Introduction

Beaufort scale of wind force monitoring is an important meteorological predicting technology, which has been widely used in guiding production and preventing natural disasters [1]. For example, in the early warning of marine storms, the Beaufort scale is one of the important parameters in predicting typhoons. However, the conventional Beaufort scale monitoring systems of wind force cannot keep running for a long time because batteries continuously power most sensors, leading to a short lifetime and high maintenance costs. In order to improve the lifespan of the monitoring systems, the project of "Near Zero Power RF and Sensor Operations (N-ZERO)" was proposed by the Defense Advanced Research Projects Agency (DARPA) in 2015 [2], which aims to develop near-zero power systems. To further reduce energy consumption, the auto-wakeup system is proposed based on various active sensors [3–7]. In 2017, Robert W. Reger. et al. proposed a near-zero power wakeup system based on a complementary metal–oxide–semiconductor comparator (CMOS) and a piezoelectric microelectromechanical accelerometer, in which the system can be awakened by threshold acceleration [8]. In the same year, a near-zero power sensing system awakened by sound was proposed, by coupling a piezoelectric sensor with low-power CMOS circuits [9]. In 2019 and 2020, Venkateswaran Vivekananthan and his coauthors designed and fabricated a triboelectrification based energy collector, which realized the function of zero-power consuming seismic detection and a zero-power consuming active pressure sensor, the reported result brought the self-power electronic device into zero-power or near-zero power mode [10,11]. In 2021, Chenxi Zhang et, al proposed a near-zero power system awakened by touch based on the triboelectrification effect, which demonstrated an ultralow quiescent power consumption wake-up technology [12]. With the advantages of a long lifespan and low maintenance costs, the auto-wakeup technology is promising for wind force monitoring.

In 2012, the triboelectric nanogenerator (TENG) was invented based on contact electrification and electrostatic induction [13–15], which can convert mechanical energy into electrical energy effectively [16–18]. At present, TENGs are widely used in self-powered systems [19–21], sensor array [22], and big data acquisition [23]. Previous studies have shown that the TENG can be used as active sensors for wind speed monitoring [24], such as the active wind sensor based on the flutter TENG [25,26], the self-powered wind sensing system based on the rotating TENG [27–29], the intelligent self-powered wireless wind sensing systems based on a single TENG [30], and so on. Nevertheless, the existing wind-driven TENGs-based sensing systems always remain in active mode for signal monitoring and measurement [31], which lead to continuous consumption of electrical energy. Constant power consumption neither conserves energy nor ensures that the system maximizes the use of power when transmitting signals. The mechanical energy collected from the natural environment is not enough to convert into electrical energy to power the sensing system all the time. Therefore, to extend the lifespan with low static power consumption, it is necessary to develop the TENG-based auto-wakeup system for wind force monitoring.

In this work, a near-zero power triboelectric wake-up system (NP-TWS) for autonomous Beaufort scale of wind force monitoring is proposed, in which a rotary TENG is used to convert wind energy into a stored electric energy capacitor. When the capacitor voltage accumulates to the threshold voltage of a transistor, it turns on as an electronic switch and the system wakes up. In active mode, the Beaufort scale of wind force can be judged according to the electric energy and the signal is sent out wirelessly. In standby mode when there is no wind, the power consumption of the system is only 14 nW. When the wind scale reaches or exceeds light air, the system can switch to the active mode within one second and accurately judge the Beaufort scale of wind force within 10 s. This work has provided a triboelectric sensor-based wake-up system with ultralow static power consumption, which has great prospects for unattended environmental monitoring, hurricane warning, and big data acquisition.

2. System Design and Characterization

2.1. Design of the System Framework

The NP-TWS for autonomous Beaufort scale of wind force monitoring consists of a rotating TENG, a wake-up circuit, and a stage circuit (SC), which is shown in Figure 1a. The TENG is used to convert wind energy into electric energy and stores it in a capacitor. Then, the wake-up circuit determines whether to wake up the system according to the capacitor voltage. If the system is not awakened, it remains in standby mode and maintains ultra-low static power consumption. Otherwise, the system will shift to active mode, in which the SC judges the Beaufort scale of wind force according to the comparison voltage of the storage capacitor and the signal is sent out wirelessly.

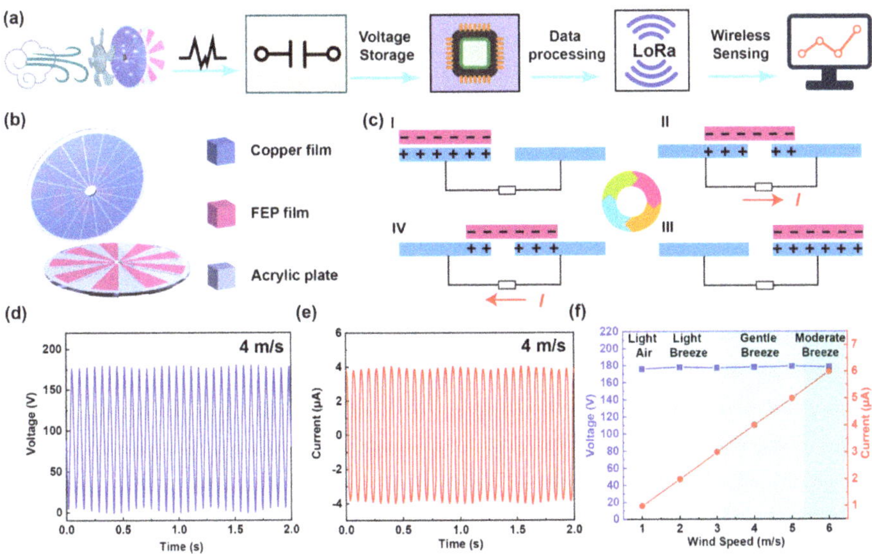

Figure 1. Overview of the wake-up system and the triboelectric nanogenerator (TENG). (**a**) Schematic diagrams showing the near-zero power triboelectric wake-up system (NP-TWS) for autonomous Beaufort scale of wind force monitoring. (**b**) Schematic diagram showing the structure of the TENG. (**c**) Working principle of the TENG. (**d**) The output voltage and (**e**) output current of TENG in moderate breeze. (**f**) The output peak-voltage and peak-current of TENG with the Beaufort scale of wind force at grades 1–4.

2.2. Structure and Characteristics of the Rotating TENG

As depicted in Figure 1b, the TENG in rotating freestanding-mode was designed as a wind energy harvester with nine fixed power generation units, which contain a stator and a rotator. The two 280 mm diameter acrylic disks with a center hole were divided into 18 fan shapes. The 18 fan shapes of the stator disk were covered with copper. A fan-shaped area was left between every two film coated areas, and the 9 fan shapes of the rotator were attached with fluorinated ethylene propylene (FEP). The stator and the rotor were connected through the shaft, and a fan blade was added in front of the rotor. The working principle of the freestanding-mode TENG is shown in Figure 1c. The FEP layer was in contact with the copper electrode in advance, and the same amount of heterogeneous charge was generated. During the rotation, a small gap is left between the two triboelectric layers, which can eliminate friction and extend the service life of the TENG. In the original state, the FEP layer was aligned with the left copper electrode. At this time, since the TENG was in an electrostatic equilibrium state, there was no current in the external circuit, as shown in Figure 1c (I). When the rotor started to rotate, the alignment area between the FEP and the left electrode was reduced, resulting in the generation of a potential difference, so electrons flowed from the right electrode to the left electrode through the external circuit, as shown in Figure 1c (II). This process would continue until the FEP layer is completely aligned with the right copper electrode, as shown in Figure 1c (III). Then, the alignment area between the FEP and the left electrode increased, which created a reverse current, as shown in Figure 1c (IV). The periodic movement of the FEP layer would generate periodic alternating current (AC) signals. When the wind speed was 4 m/s, the output voltage of TENG was 178 V, and the output current was 4 μA, the corresponding waveform is shown in Figure 1d,e, respectively. The output peak–peak open-circuit voltage and peak short-circuit current of TENG with the Beaufort scale of wind force at grades 1–4 is shown in Figure 1f. From light air to moderate breeze, the voltage of TENG hardly changed and stabilized at 178 V, while the current increased from 1 μA to 6 μA, and the current was positively correlated with the wind speed. The output power of TENG is shown in Figure S6. The durability test of TENG is shown in Figure S5. For 2500 s, the output voltage of TENG remained stable.

2.3. Mechanism and Characteristics of the Wake-Up Circuit

The circuit schematic diagram of the wake-up circuit is shown in Figure 2a, including a half-wave rectifier (D1, D2), a storage capacitor C1, a resistor R1, a zener diode D3, a MOSFET, and a relay. The AC signal output by TENG was converted into a DC signal by a half-wave rectifier composed of D1 and D2 and stored in the capacitor C1. When the voltage of C1 accumulated to the threshold voltage of MOSFET-N, it turned on as an electronic switch. Then current flowed through the coil of the relay KA, causing its switch to pull in and the system woke up, as shown in Figure 2b. With the resistor R1, the electrical energy stored in the capacitor C1 could be released to prevent the system from being triggered by mistake. The zener diode D3 was mainly used for overvoltage protection of MOSFET-N. In the same way, the resistor R2 was mainly used for overcurrent protection of relay KA.

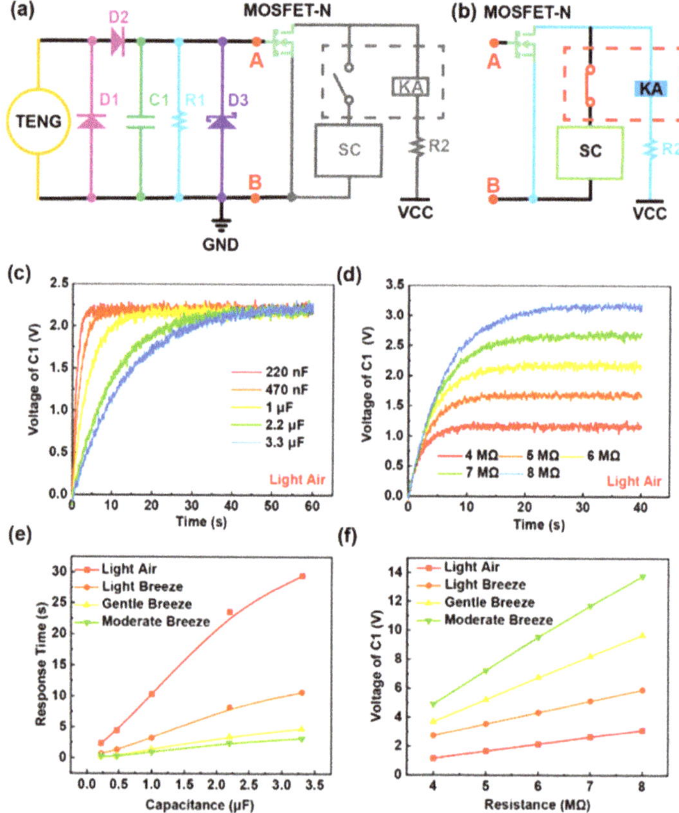

Figure 2. The wake-up circuit using TENG as a sensor. (**a**) The circuit schematic diagram of the wake-up circuit in standby mode. (**b**) The circuit schematic diagram of the wake-up circuit in active mode. (**c**) The voltage of the C1 with different capacitance in light air. (**d**) The voltage of C1 with different resistance in light air. (**e**) The response time of the wake-up circuit with different capacitance on the Beaufort scale of wind force at grades 1–4. (**f**) The voltage of C1 with different resistance on the Beaufort scale of wind force at grades 1–4.

The effects of the capacitance of C1 and the resistance of R1 on the voltage of C1 were studied systematically. The wind speed ranges corresponding to the Beaufort scale of wind force at grades 1–4 are shown in the Figure S1. Figure 2c illustrates the voltage of C1 with different capacitance in light air. The threshold voltage of the MOSFET-N set in the system was 2 V, so the time from 0 to 2 V of capacitor voltage was the response time of the system. The experiments results indicated that the response time of the system decreased along with decreasing the capacitance of C1. The voltage of the C1 with different resistance in light air is shown in Figure 2d, which indicated that the voltage of C1 increased along with increasing the resistance of R1. The voltage of the C1 with different capacitance and resistance on the Beaufort scale of wind force at grades 2–4 had the same result as above, as shown in Figure S2. Figure 2e summarizes the response times of the wake-up circuit with different capacitance on the Beaufort scale of wind force at grades 1–4. The experiment's results indicated that the response time decreased with the decrease in the capacitance and the increase in the Beaufort scale of wind force. In order to avoid a rapid response causing the system to be awakened by mistake, when a burst of strong disturbance suddenly appears, while maintaining a fast response, a 1 µF capacitor was selected as an energy storage unit. In this case, the system can be awakened within one second and the voltage of C1 would tend to smooth within 10 s. The voltage of C1 with different resistance on the Beaufort scale of wind force at grades 1–4 is summarized in Figure 2f, which shows the increase in capacitor voltage not only increased with the resistance but also increased with the Beaufort scale of wind force. The power supply VCC was 10 V. A 6 MΩ resistor could meet the voltage of a capacitor between 2 and 10 V on the Beaufort scale of wind force at grades 1–4.

2.4. Mechanism and Characteristics of the Stage Circuit (SC)

The circuit schematic diagram of the stage circuit (SC) is shown in Figure 3a, including comparators, resistors, the Microprogrammed Control Unit (MCU), and wireless module. In active mode, the power supply VCC connected to the SC. Resistors R1, R2..Rxn + 1 acted as resistor dividers to provide a threshold voltage for comparators, which were compared with the voltage of capacitor C1. The comparators output a signal of 0 or 1 according to the comparison of the two voltages. The MCU collected and processed the results and transmitted signals by wireless mode. In the experiment, a four-stage circuit was designed to monitor the Beaufort scale of wind force at grades 1–4. The comparison voltage on the Beaufort scale of wind force at grades 1–4 with 1 µF capacitor and a 6 MΩ resistor can be seen in Figure 3b, which shows that the comparison voltages tended to smooth at 2 V, 4 V, 6 V, and 8 V respectively. Considering the voltage fluctuation at the boundary of classification and the power supply VCC was 10 V, five resistors of 100 Ω in series could provide threshold voltages of 2 V, 4 V, 6 V, and 8 V for each of the four comparators. The output signal of the comparator is shown in Figure S3, which indicated that the system could distinguish the Beaufort scale of wind force at grades 1–4.

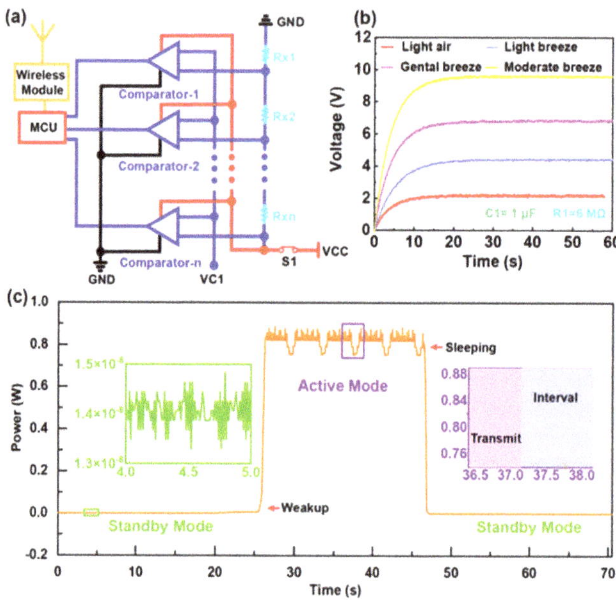

Figure 3. The stage circuit (SC) and the system power consumption. (**a**) The circuit schematic diagram of SC in active mode. (**b**) The comparison voltage with the Beaufort scale of wind force at grades 1–4. (**c**) The power consumption of the wake-up system in standby mode and active mode.

3. The Result and Discussion

3.1. The Power Consumption of the Wake-Up System

Since the purpose of the NP-TWS is to reduce static power consumption and thereby increase the battery life, the characteristic of the power consumption of the system was examined as shown in Figure 3c, and the power consumption was provided by the power supply VCC. In standby mode, there was only the leakage current of the MOSFET in the system, and the power consumption was just 14 nW, which demonstrated that the system had ultra-low static power consumption and could meet the requirements of low power consumption. In active mode, the power consumption was between 0.75 W and 0.88 W, which was caused by the wireless module. The power consumption would be higher when the wireless module sends data. When the system returned to the standby mode from the active mode, it maintained ultra-low static power consumption again.

3.2. Demonstration of Beaufort Scale of Wind Force Monitoring

Verification experiments were demonstrated by utilizing this wake-up system to monitor the Beaufort scale of wind force. With the auto-wakeup technology, the NP-TWS realized the autonomous monitoring with ultra-low static power consumption and indicated potential application prospects for wind force monitoring, as shown in Figure 4a. Driven by wind energy, the NP-TWS can be used as a wireless sensor network node unattended and with a long lifetime. By monitoring the Beaufort scale of wind force, a large amount of data can be remotely sent to the terminal for wind information collection. The photos of the TENG, wake-up circuit, MCU, LoRa transmitter (LTM), and LoRa receiver (LRM) are shown in Figure 4b,c. Moreover, the flow chart of the NP-TWS is shown in Figure S4. As a demonstration, the compared voltage and system power consumption with different Beaufort scale of wind force are summarized in Figure 4d,e. The first part simulated the wind speed from 0 to 4.2 m/s in the external environment. Then the wind speed was changed to 1.1 m/s. Finally, the wind stopped blowing. In this process, the display on the terminal interface was "Calm", "Gentle Breeze", "Light Air" and "Calm".

The compared voltage measured also changed with the Beaufort scale of wind force and corresponded to the system monitoring results one by one. The power consumption of the system increased when the system was awake, and maintained near-zero power when there was no wind. In the second part, the system was kept in windless mode for a period of time, and the measurement results showed that the system could maintain ultralow static power consumption. Subsequently, we simulated the change of wind speed, which was 0, 1.9 m/s, 6.1 m/s, 1.1 m/s, and 0 in this order. The experimental results indicated that the display on the terminal interface, the comparison voltage, and the power consumption of the system were in line with expectations. In addition, the tr1 and tr2 shown in Figure 4e was the response time of the system.

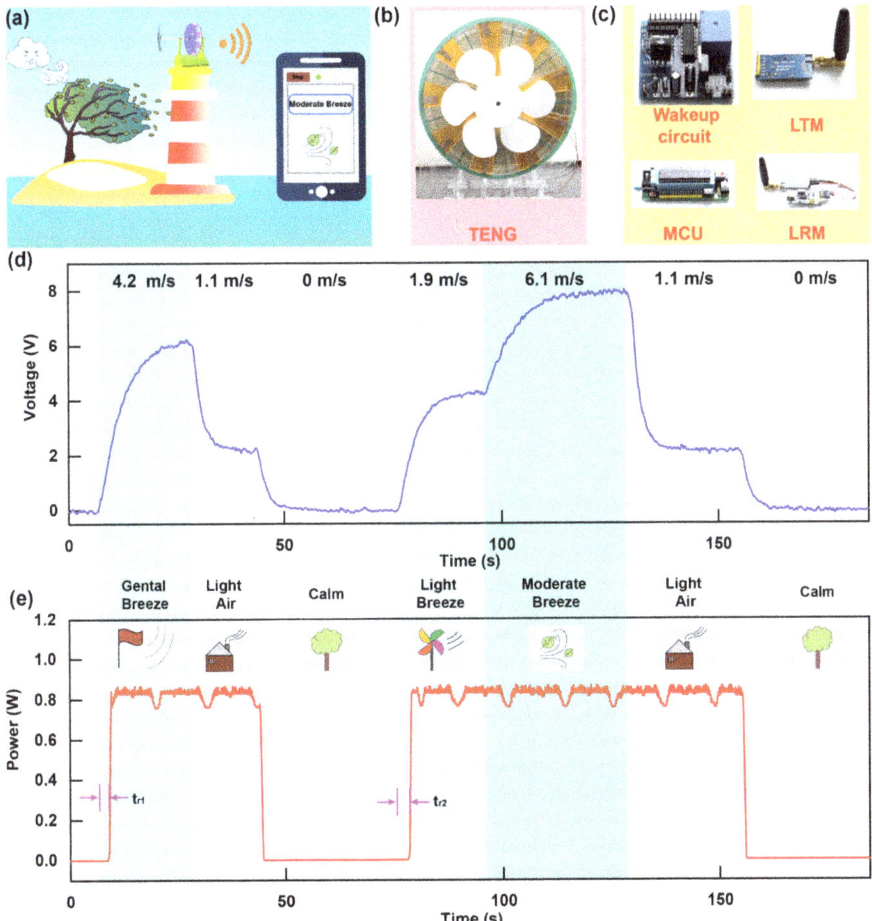

Figure 4. Demonstration of Beaufort scale of wind force monitoring. (**a**) Sketch of the NP-TWS for Beaufort scale of wind force monitoring. (**b**) The photo of the TENG. (**c**) The photos of the wake-up circuit, microprogrammed control unit (MCU), LoRa transmitting module (LTM), and LoRa receiving module (BRM). (**d**) The comparison of voltage of the system in a simulated wind environment. (**e**) The power consumption of the NP-TWS and the received data of the terminal in a simulated wind environment.

In the experiments, the system is automatically awakened up when the wind blows and the Beaufort scale of wind force is displayed on the terminal device. Even if the

Beaufort scale of wind force changes, the system can detect it immediately. When the wind stops blowing, the system will return to standby mode with ultralow static power consumption and wait for the next monitoring. The dynamic display of wake-up is shown in Video S1.

4. Experimental Section

The fabrication of rotary TENG. The two 280 mm diameter acrylic disks with a center hole were divided into 18 fan shapes with a 19 degree central angle. The 18 fan shapes of the stator disk were covered with copper, which both acted as the friction layer and the electrode. A fan-shaped area was left between every two film coated areas, and the nine fan shapes of the rotator were attached with fluorinated ethylene propylene (FEP), which acted as the friction layer. The stator and the rotor were connected through the shaft, and a fan blade was added in front of the rotor.

Characterization of electric output. The output voltage and current of TENG was measured by the electrometer (Keithley 6514). The current of the system was measured by the data acquisition (Keithley DAQ6510). Recording the realtime data and receiving the results in terminal were completed with the software platform developed based on LabVIEW.

5. Conclusions

In summary, an NP-TWS driven by wind energy was developed to monitor the Beaufort scale of wind force. Based on a rotary TENG for energy conversion and a half-wave rectifier for voltage regulation, the voltage of the storage capacitor can be a signal to wake up the system. The wake-up circuit is designed with a transistor, which can accurately control the wake-up of the system. When the capacitor voltage accumulates to the threshold voltage of MOSFET, it turns on as an electronic switch and the system wakes up. In the active mode, the SC can judge the Beaufort scale of wind force according to the comparison voltage and send out the signal by wireless. When the wind scale reaches or exceeds light air, the system can switch to active mode within one second and accurately judge the Beaufort scale of wind force within 10 s. Even if the Beaufort scale of wind force changes, the system can detect it immediately. In standby mode, because there is only the leakage current of the MOSFET in the system, the power consumption of the system is only 14 nW. Owing to the ultra-low static power consumption, the system can monitor in an unattended area for a long time. This work was tested in the laboratory to verify the feasibility of the system. However, for use outside, it is necessary to design the package for TENG. The stability and practicability of the system will be improved in future work, so that the system can cope with a harsh environment. This work has provided a triboelectric sensor-based wake-up system with ultralow static power consumption, which shows promise for unattended environmental monitoring, hurricane warning, and big data acquisition.

Supplementary Materials: The following are available online at https://www.mdpi.com/article/10.3390/nanoenergyadv1020006/s1, Figure S1: The wind speed ranges corresponding to the Beaufort scale of wind force at grade 1–4. Figure S2: The voltage of the C1 with different capacitance in light breeze. Figure S3: The output of the comparator with the Beaufort scale of wind force at grade 1–4. Figure S4: The flow chart of the NP-TWS. Figure S5: Durability test of TENG result. Figure S6: The output power of TENG. Video S1: The dynamic display of Beaufort scale of wind force monitoring.

Author Contributions: Conceptualization, T.T. and C.Z.; methodology, G.L.; validation, Y.L. and S.X.; data curation, Y.L.; writing—original draft preparation, T.T.; writing—review and editing, G.L.; visualization, Y.L.; supervision, C.Z.; funding acquisition, C.Z. All authors have read and agreed to the published version of the manuscript.

Funding: The authors thank the support of the National Natural Science Foundation of China (51922023, 61874011), Fundamental Research Funds for the Central Universities (E1EG6804), Tribology Science Fund of State Key Laboratory of Tribology (SKLTKF19B02), and Open Research Foundation of State Key Laboratory of Digital Manufacturing Equipment & Technology (DMETKF2020014).

Data Availability Statement: Data are contained within the article or Supplementary Material.

Conflicts of Interest: The authors declare no conflict of interest.

References

1. Kusiak, A.; Zhang, Z.; Verma, A. Prediction, operations, and condition monitoring in wind energy. *Energy* **2013**, *60*, 1–12. [CrossRef]
2. Olsson, R.H.; Bogoslovov, R.B.; Gordon, C. Event Driven Persistent Sensing: Overcoming the Energy and Lifetime Limitations in Unattended Wireless Sensors. In Proceedings of the 2016 IEEE SENSORS, Orlando, FL, USA, 30 Octocber–3 November 2016; pp. 1–3.
3. Bassirian, P.; Moody, J.; Bowers, S.M. Event-Driven Wakeup Receivers: Applications and Design Challenges. In Proceedings of the 2017 IEEE 60th International Midwest Symposium on Circuits and Systems (MWSCAS), Boston, UK, 6–9 August 2017; pp. 1324–1327.
4. Hassanalieragh, M.; Soyata, T.; Nadeau, A.; Sharma, G. UR-SolarCap: An Open Source Intelligent Auto-Wakeup Solar Energy Harvesting System for Supercapacitor-Based Energy Buffering. *IEEE Access* **2017**, *4*, 542–557. [CrossRef]
5. Sanchez, A.; Aguilar, J.; Blanc, S.; Serrano, J.J. RFID-based wake-up system for wireless sensor networks. *Proc. Spie* **2011**, *8067*, 8–12.
6. Kpuska, V.Z.; Eljhani, M.M.; Hight, B.H. Front-end of Wake-Up-Word Speech Recognition System Design on FPGA. *J. Telecommun. Syst. Manage.* **2013**, *2*, 1000108.
7. Yen, W.C.; Chen, H.W. Low power and fast system wakeup circuit. *IEE Proc. Circuits Devices Syst.* **2005**, *152*, 223–228. [CrossRef]
8. Reger, R.W.; Barney, B.; Yen, S.; Satches, M.; Griffin, B.A. Near-Zero Power Accelerometer Wakeup System. In Proceedings of the 2017 IEEE SENSORS, Glasgow, UK, 29 Octocber–1 November 2017; pp. 1–3.
9. Reger, R.W.; Clews, P.J.; Bryan, G.M.; Keane, C.A.; Henry, M.D.; Griffin, B.A. Aluminum Nitride Piezoelectric Microphones as Zero-Power Passive Acoustic Filters. In Proceedings of the 19th International Conference on Solid-State Sensors, Actuators and Microsystems (TRANSDUCERS), Kaohsiung, Taiwan, 18–22 June 2017; pp. 2207–2210.
10. Vivekananthan, V.; Chandrasekhar, A.; Alluri, N.R.; Purusothaman, Y.; Khandelwal, G.; Pandey, R.; Kima, S.-J. Fe_2O_3 magnetic particles derived triboelectric-electromagnetic hybrid generator for zero-power consuming seismic detection—ScienceDirect. *Nano Energy* **2019**, *64*, 103926. [CrossRef]
11. Vivekananthan, V.; Chandrasekhar, A.; Alluri, N.R.; Purusothaman, Y.; Kim, S.-J. A highly reliable, impervious and sustainable triboelectric nanogenerator as a zero-power consuming active pressure sensor. *Nanoscale Adv.* **2020**, *2*, 746–754. [CrossRef]
12. Zhang, C.; Dai, K.; Liu, D.; Yi, F.; You, Z. Ultralow Quiescent Power-Consumption Wake-Up Technology Based on the Bionic Triboelectric Nanogenerator. *Adv. Sci.* **2020**, *7*, 2000254. [CrossRef] [PubMed]
13. Wang, S.; Lin, L.; Wang, Z.L. Nanoscale triboelectric-effect-enabled energy conversion for sustainably powering portable electronics. *Nano Lett.* **2012**, *12*, 6339–6346. [CrossRef] [PubMed]
14. Wang, Z.L. On Maxwell's displacement current for energy and sensors: The origin of nanogenerators. *Mater. Today* **2017**, *20*, 74–82. [CrossRef]
15. Zhou, Y.S.; Wang, S.; Yang, Y.; Zhu, G.; Niu, S.; Lin, Z.H.; Liu, Y.; Wang, Z.L. Manipulating Nanoscale Contact Electrification by an Applied Electric Field. *Nano Lett.* **2014**, *14*, 1567–1572. [CrossRef]
16. Chen, J.; Huang, Y.; Zhang, N.; Zou, H.; Liu, R.; Tao, C.; Fan, X.; Wang, Z.L. Micro-cable structured textile for simultaneously harvesting solar and mechanical energy. *Nat. Energy* **2016**, *1*, 16138. [CrossRef]
17. Niu, S.; Ying, L.; Wang, S.; Long, L.; Yu, S.Z.; Hu, Y.; Zhong, L.W. Theoretical Investigation and Structural Optimization of Single-Electrode Triboelectric Nanogenerators. *Adv. Funct. Mater.* **2014**, *24*, 3332–3340. [CrossRef]
18. Zhu, G.; Chen, J.; Liu, Y.; Bai, P.; Zhou, Y.S.; Jing, Q.; Pan, C.; Wang, Z.L. Linear-Grating Triboelectric Generator Based on Sliding Electrification. *Nano Lett.* **2013**, *13*, 2282–2289. [CrossRef] [PubMed]
19. Guo, T.; Zhao, J.; Liu, W.; Liu, G.; Pang, Y.; Bu, T.; Xi, F.; Zhang, C. Self-Powered Hall Vehicle Sensors Based on Triboelectric Nanogenerators. *Adv. Mater. Technol.* **2018**, *3*, 1800140. [CrossRef]
20. Liu, D.; Chen, B.; An, J.; Li, C.; Liu, G.; Shao, J.; Tang, W.; Zhang, C.; Wang, Z.L. Wind-driven self-powered wireless environmental sensors for Internet of Things at long distance. *Nano Energy* **2020**, *73*, 104819. [CrossRef]
21. Wang, Z.L. Triboelectric Nanogenerators as New Energy Technology for Self-Powered Systems and as Active Mechanical and Chemical Sensors. *ACS Nano* **2013**, *7*, 9533–9557. [CrossRef]
22. Rong, X.; Zhao, J.; Guo, H.; Zhen, G.; Dong, G. Material Recognition Sensor Array by Electrostatic Induction and Triboelectric Effects. *Adv. Mater. Technol.* **2020**, *5*, 2000641. [CrossRef]
23. Liu, W.; Xu, L.; Liu, G.; Yang, H.; Bu, T.; Fu, X.; Xu, S.; Fang, C.; Zhang, C. Network topology optimization of triboelectric nanogenerators for effectively harvesting ocean wave energy. *Iscience* **2020**, *23*, 101848. [CrossRef] [PubMed]
24. Jiang, D.; Liu, G.; Li, W.; Bu, T.; Wang, Y.; Zhang, Z.; Pang, Y.; Xu, S. A leaf-shaped triboelectric nanogenerator for multiple ambient mechanical energy harvesting. *IEEE Trans. Power Electron.* **2019**, *35*, 25–32. [CrossRef]
25. Quan, Z.; Han, C.B.; Jiang, T.; Wang, Z.L. Wind Energy: Robust Thin Films-Based Triboelectric Nanogenerator Arrays for Harvesting Bidirectional Wind Energy. *Adv. Energy Mater.* **2016**, *6*, 1501799. [CrossRef]
26. Yang, Y.; Zhu, G.; Zhang, H.; Chen, J.; Zhong, X.; Lin, Z.H.; Su, Y.; Bai, P.; Wen, X.; Wang, Z.L. Triboelectric nanogenerator for harvesting wind energy and as self-powered wind vector sensor system. *Acs Nano* **2013**, *7*, 9461–9468. [CrossRef]
27. Bi, M.; Wu, Z.; Wang, S.; Cao, Z.; Ye, X. Optimization of structural parameters for rotary freestanding-electret generators and wind energy harvesting. *Nano Energy* **2020**, *75*, 104968. [CrossRef]

28. Wang, J.; Ding, W.; Pan, L.; Wu, C.; Yu, H.; Yang, L.; Liao, R.; Wang, Z.L. Self-Powered Wind Sensor System for Detecting Wind Speed and Direction Based on a Triboelectric Nanogenerator. *ACS Nano* **2018**, *12*, 3954–3963. [CrossRef] [PubMed]
29. Wang, P.; Lun, P.; Wang, J.; Xu, M.; Dai, G.; Zou, H.; Dong, K.; Wang, Z.L. An Ultra-Low-Friction Triboelectric-Electromagnetic Hybrid Nanogenerator for Rotation Energy Harvesting and Self-Powered Wind Speed Sensor. *ACS Nano* **2018**, *12*, 9433–9440. [CrossRef] [PubMed]
30. Shan, L.; Gao, L.; Chen, X.; Tong, D.; Yu, H. Simultaneous energy harvesting and signal sensing from a single triboelectric nanogenerator for intelligent self-powered wireless sensing systems. *Nano Energy* **2020**, *75*, 104813.
31. Fu, X.; Bu, T.; Li, C.; Liu, G.; Zhang, C. Overview of micro/nano wind energy harvesters and sensors. *Nanoscale* **2020**, *12*, 23929–23944. [CrossRef] [PubMed]

Review

Mechanical Conversion and Transmission Systems for Controlling Triboelectric Nanogenerators

Nghia Dinh Huynh and Dukhyun Choi *

Department of Mechanical Engineering, Sungkyunkwan University, Suwon 16419, Korea; morganskku@skku.edu
* Correspondence: bred96@skku.edu

Abstract: Triboelectric nanogenerators (TENGs) are a promising renewable energy technology. Many applications have been successfully demonstrated, such as self-powered Internet-of-Things sensors and many wearables, and those portable power source devices are useful in daily life due to their light weight, cost effectiveness, and high power conversion. To boost TENG performance, many researchers are working to modulate the surface morphology of the triboelectric layer through surface-engineering, surface modification, material selection, etc. Although triboelectric material can obtain a high charge density, achieving high output performance that is predictable and uniform requires mechanical energy conversion systems (MECSs), and their development remains a huge challenge. Many previous works did not provide an MECS or introduced only a simple mechanical system to support the TENG integration system device. However, these kinds of designs cannot boost the output performance or control the output frequency waveform. Currently, some MECS designs use transmission conversion components such as gear-trains, cam-noses, spiral springs, flywheels, or governors that can provide the step-up, controllable, predictable, and uniform output performance required for TENGs to be suitable for daily applications. In this review, we briefly introduce various MECS designs for regulating the output performance of TENGs. First, we provide an overview of simple machines that can be used when designing MECSs and introduce the basic working principles of TENGs. The following sections review MECSs with gear-based, cam-based, flywheel-based, and multiple-stage designs and show how the MECS structure can be used to regulate the input flow for the energy harvester. Last, we present a perspective and outline for a full system design protocol to correlate MECS designs with future TENG applications.

Keywords: mechanical conversion; mechanical transmission; triboelectric nanogenerators (TENGs); external mechanical system control; regulated output; uniform output

Citation: Huynh, N.D.; Choi, D. Mechanical Conversion and Transmission Systems for Controlling Triboelectric Nanogenerators. *Nanoenergy Adv.* **2022**, *2*, 29–51. https://doi.org/10.3390/nanoenergyadv2010002

Academic Editors: Ya Yang and Zhong Lin Wang

Received: 13 December 2021
Accepted: 18 January 2022
Published: 21 January 2022

Publisher's Note: MDPI stays neutral with regard to jurisdictional claims in published maps and institutional affiliations.

Copyright: © 2022 by the authors. Licensee MDPI, Basel, Switzerland. This article is an open access article distributed under the terms and conditions of the Creative Commons Attribution (CC BY) license (https://creativecommons.org/licenses/by/4.0/).

1. Introduction

The depletion of restricted resources such as crude oil, natural gas, coal, clean water, and minerals is having large effects on the world economy and global warming. Global economic growth has always increased simultaneously with the use of fossil energy. Many researchers have been working to develop renewable and green energy. However, abundant energies continue to be wasted, such as the wind [1–5], vibration [6–16], acoustic vibrations [17–19], human motion [12,20–22], solar radiation [23–25], thermodynamic forces [26–28], and ocean waves [13,29–32], and they could be harvested using numerous technologies. Among those energy sources, mechanical energy is popular because it can be found anytime and anywhere. Since triboelectric nanogenerators (TENGs) were invented in 2012 [33], many works have investigated various designs for energy harvesting devices [7,30,34–37], material development [34,38–41], and surface engineering [42–49] to modify the topography, enhance the contact area in TENGs, and thereby boost their electric performance. Other researchers have sought to make TENGs simple, cost effective, and sustainable while ensuring their high output performance [8,33,50–55].

In addition to studying methods to improve the output performance of TENGs, researchers have also proposed some simple mechanical systems that can be assembled with TENGs to harvest [22,28,31,56–60] the wind energy, blue energy, and biomechanical energy always present in daily life. These kinds of designs cannot step-up or control the output performance of a TENG in the way needed for practical applications. To obtain suitably controllable TENG output, the mechanical device used for conversion and transmission needs to be designed to reduce mechanical and circuit loss, enhance the input force and torque, extend the TENG operation time, boost and convert the low input frequency into a regulated high output frequency, and so on.

Some previous mechanical conversion and transmission systems, such as windmills and magnetic energy harvesters [61,62], are large, complicated, heavy, and expensive [63,64], which results in mechanical losses due to friction, vibration, noise, and so on. Therefore, those systems require a high input to operate. In addition, mismatches between the input frequency and free oscillation frequency of a system can reduce the output performance or cause zero power generation. A suitably resonant system design can help to optimize the motion velocity and magnitude, enabling high-power generation. Many researchers claim that their TENG system designs are cost-effective, lightweight, small, and easy to maintain in harsh environments. However, these studies still lack the mechanical energy conversion systems (MECSs) needed to control and enhance TENG output or regulate its performance [65]. Based on the basic parameters of contact force, velocity, and frequency, TENG output power must be controlled and regulated to a certain output target to be practical in useful applications.

Many system designs for mechanical energy harvesters (TENGs) in previous works use low input energy in the form of applied force, irregular mechanical energy, or variable frequencies, so they have a very low, uncontrollable output performance that is difficult to predict. Generally, mechanical energy harvesters convert irregular input energy into uncontrollable power output. However, in practical applications, the electrical output from energy harvesters needs to be uniform and predictable, despite the variety in the input energy. When MECSs are fully understood, these problems can be solved by applying mechanical forces. MECSs can regulate almost any mechanical motion by virtue of their kinematic design, i.e., the movement of points, links, and systems. With the help of a good MECS design, the working frequency can be enhanced from low input frequencies, and the motion direction can be adjusted (e.g., from linear motion to single/bi-directional rotational motion and vice versa). The most common and familiar mechanical system is the gear-train [66], i.e., two or more gears are mounted in a box to change the torque, rotational speed, and direction of the input. On the other hand, rotational motion can be turned into linear motion by using a cam-nosed design [37,67]. Many types of mechanical motion occur during daily life that can produce the non-continuous operation of an energy scavenger. To nonetheless offer continuous power output, a spiral spring and flywheel/governor can be applied to improve the use of intermittent excitation energies. By storing energy as continuous rotational motion in a flywheel/governor and potential motion in a spiral spring [35,68,69], an energy harvester can run continuously and possibly convert intermittent motion into rotational movement with the integral of ratchet and pawl in mechanical transmission structure [70], which would be an improvement in energy harvester design.

In this review, we briefly introduce various MECS designs for regulating TENG output performance. First, we provide an overview of simple machines that can be applied to MECSs and introduce the basic working principles of TENGs. The following sections review gear-based [36,66,68,69], cam-based [66,67,69], flywheel-based [35,69], and multiple-stage [68,69] MECS designs and explain how the different MECS structures regulate the input of flow energy harvesters [71–73]. Finally, we present an outline for a full system design protocol to coordinate MECS designs for future TENG applications.

2. Overview of Simple Machines and TENG Basics

2.1. Simple Machines: An Overview

Simple machine mechanisms can modify motion to control the direction or magnitude of a force; however, the work done by the effort is equal to the work done on the load if friction is neglected, as shown in Figure 1a–d. In MECSs, the working velocity is also controlled to meet the design requirements.

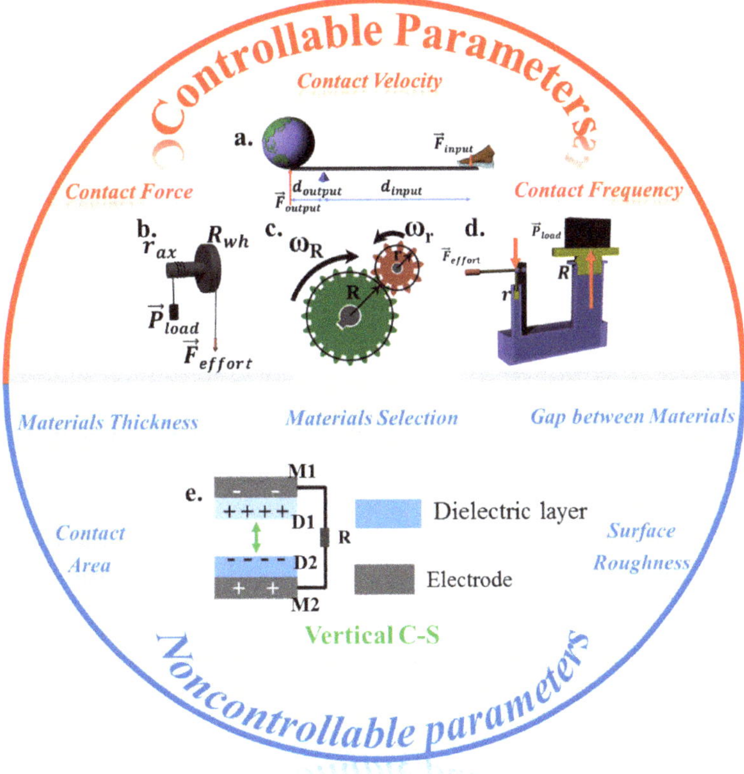

Figure 1. Controllable parameters and working modes of TENGs. (**a–d**) Schematic of simple machines that can govern the force, velocity, and frequency of input energy. (**e**) TENG with the vertical contact-separation working mode.

2.1.1. Force Controlled

The simplest mechanism uses leverage to increase force, as depicted in Figure 1a. It is a bar or beam resting on a fulcrum or pivot. A downward force applied on one end of the lever is transferred to the upward direction at the other end of the lever at a higher magnitude, which allows a small applied force to lift a heavy weight. The mechanical advantage (MA) of a lever depends on the length of the bar on either side of the fulcrum and can be determined by considering the balance of moments or torque, T, about the fulcrum, as shown below:

$$MA = \frac{F_{output}}{F_{input}} = \frac{d_{input}}{d_{output}} \tag{1}$$

As shown in Figure 1b, the structure of a wheel and axle mounted in the same axis of rotation is another simple machine. The wheel radius is larger than that of the axle. In addition to having the same MA as a lever, i.e., the ratio between F_{output} and F_{input}, the velocity ratio of a wheel and axle are another advantage. Wheels and axles allow people to easily undertake hard work quickly and for a long time. A wheel with a large radius can be turned easily but requires many turns. An axle rotates less but requires more force. So, a wheel can be used to create the MA of force, or high force or torque can be applied to an axle to rotate it and the attached wheel rapidly, as in a car. The velocity ratio is shown below:

$$v_{ratio} = \frac{R_{wh}}{r_{ax}} \tag{2}$$

2.1.2. Velocity-Frequency Controlled

Figure 1c shows a simple gear-train, two gears of different sizes, which is a promotion from the wheel and axle structure. One of the gears, named the driver gear, is mounted to a motor or crank; the other gear is rotated by driver gear and called the driven gear. In Figure 1c, if the larger gear is the driven gear or output gear (R), the gear train amplifies the input torque and reduces the rotation speed ($\omega_R < \omega_r$), meaning that this gear train is called a speed reducer. On the other hand, if the driven gear is the smaller one, the gear train increases the rotation speed and output frequency. The MA for a simple pair of meshing gears is as follows:

$$MA = \frac{R}{r} = \frac{\omega_r}{\omega_R} \tag{3}$$

Another simple machine can convert a small downward effort on a small piston (radius of r) into a greater force on a larger piston (radius of R), as shown in Figure 1d. This structure transmits liquid pressure on the small piston to the larger piston. As a result, a small force applied to the small piston produces a much greater force on the bigger piston. The velocity of the piston can also be modulated. The relation of velocity between the small and large pistons produces the MA shown below:

$$MA = \frac{F_{output}}{F_{input}} = v_{ratio} = \frac{R^2}{r^2} \tag{4}$$

In a flow-induced rotation TENG, wind flow is also used as a driver [55], or the airflow can be controlled by the Venturi effect [68].

2.2. Basic Working Principle of TENGs

2.2.1. Basic Principle of the Triboelectric Effect

The triboelectric effect is electrification caused by bringing two dissimilar materials into contact with each other. The charge exchange between the materials produces a positive charge in one material surface and a negative charge in the other. During contact, the transferred charges can be electrons or ions, depending on the interfacial energy matching between the two materials. Tribo-materials have different surface potentials, and when they are brought into contact, the surface energy state will change to reach equilibrium. The quantity of the transferred charges is determined by the surface charge density, polarization, and material selection. The signs and strengths of the charges transferred in a TENG depend on the relative polarity of the tribo-materials and the position of those materials in the triboelectric series. A TENG is a device that can convert mechanical energy into electrical energy by using the triboelectric effect. Since TENGs were invented [69] in 2012, many researchers have analyzed the principle of triboelectric nanogenerators.

2.2.2. Basic Working Modes of TENGs

TENGs can be operated in four fundamental modes: vertical contact-separation, lateral sliding, single-electrode, and free-standing triboelectric layers [43,74,75]. The most popular of those modes is vertical contact-separation, as shown in Figure 1e.

Most materials that people commonly use have triboelectrification effects, from metal to silk, plastic, polymers, wood, etc. Almost any material can be used to fabricate a

TENG device. The TENG structure uses two tribo-materials and two electrodes. Many parameters control TENG performance, such as material selection, material thickness, surface roughness, the gap between the materials, and the contact area, force, velocity, and frequency. However, the basic structure for a contact-separation TENG uses dielectric-to-dielectric forces, with tribo-material 1 (dielectric D1) and electrode 1 (metal M1) at the top and tribo-material 2 (dielectric D2) attached to electrode 2 (metal M2) at the bottom. In Figure 1e, surface charges form on the D1 and D2 surfaces after contact with an equal density of σ_{sc} because of contact electrification. Electrostatic induction then occurs right after tribo-materials 1 and 2 are separated. The positive and negative charges on both surfaces induce an electric field on both tribo-materials and lead to charge separation because of their electrostatic induction. As a result, a potential difference is established between the two tribo-materials. When electrodes M1 and M2 are connected by a conductive wire, the electrons repelled in the negative D2 tribo-material move to M1, which means that the TENG generates an alternating current when a periodic external mechanical force is applied. When the TENG reaches a balanced state, it returns to the initial state. As presented here, the theoretical TENG working mechanism can be understood to follow the Gauss law.

2.2.3. Governing Equations of TENG

TENG operation correlates with the Gauss theorem [43], which is derived from the Gauss law that describes the relationship of voltage–charge–motion (V–Q–X) as a function of time. Table 1 shows a theoretical model for a vertical contact-separation mode TENG and its governing equations for open-circuit voltage and short-circuit current.

Table 1. TENG governing equation and control parameters.

Vertical C-S Mode	Controllability by Mechanical Energy Conversion Systems							
	Material selection	Material thickness	Material gap	Surface roughness	Contact area	Contact force	Contact velocity	Contact frequency
	X	X	X	X	X	O	O	O

In the TENG schematic shown in the table, the two dielectric layers have thicknesses of d_1 and d_2 and relative dielectric constants of ε_{r1} and ε_{r2}, respectively. The two metal layers attached to the two dielectric layers serve as the electrodes. When an external force is applied to the TENG device, the distance (x) between the top and bottom parts can be changed. When the dielectric layers come into contact with each other as a result of the applied force, their inner surfaces become charged with opposite static charges (tribo-charges) with the density of σ due to contact electrification. During operation, the external force is released, and the two tribo-materials start to separate, which creates a potential difference (V) between the two electrodes. The number of charges transferred between the two electrodes as a result of the applied forced is called Q, and the transitory number of transferred charges at the two electrodes is Q and $-Q$, respectively.

This model consists of two electrodes (metal M1 and M2) and the two dielectrics D1 and D2. After repeated contact between the dielectrics, the surface charge density σ_{sc} forms on their surfaces. Based on the Gauss theorem, the voltage across the two electrodes is a summation of voltage across the air gap and dielectrics 1 and 2, as described in Equation (5).

$$V(t) = E_1 d_1 + E_2 d_2 + E_{air} x \tag{5}$$

To derive the V–Q–X relationship, substitute σ_{sc} into Equation (5), as shown in Equation (6).

$$V(t) = -\frac{Q}{S\varepsilon_0}\left(\frac{d_1}{\varepsilon_{r1}} + \frac{d_2}{\varepsilon_{r2}} + x(t)\right) + \frac{\sigma_{sc}}{\varepsilon_0}x(t) \tag{6}$$

Equation (6) yields the open circuit voltage (V_{OC}) and short circuit current (I_{SC}). In the open circuit condition, there is no moving charge in either electrode, so the current is equal to zero. The V_{OC} is defined in Equation (7).

$$V_{OC}(t) = \frac{\sigma_{sc}}{\varepsilon_0}x(t) \tag{7}$$

In the short circuit condition, the potential difference is zero. Thus, the I_{SC} derived with $V(t)$ is zero. Current represents the number of charges per unit time. I_{SC} can be expressed as indicated in Equation (8).

$$I_{SC} = \frac{dQ_{SC}}{dt} = \frac{d}{dt}\left[\frac{S\sigma_{sc}x(t)}{\left(\frac{d_1}{\varepsilon_{r1}} + \frac{d_2}{\varepsilon_{r2}} + x(t)\right)}\right] = \frac{S\sigma_{sc}\left(\frac{d_1}{\varepsilon_{r1}} + \frac{d_2}{\varepsilon_{r2}}\right)v(t)}{\left(\frac{d_1}{\varepsilon_{r1}} + \frac{d_2}{\varepsilon_{r2}} + x(t)\right)^2} \tag{8}$$

The target design parameters for controlling the triboelectric performance can be determined using Equations (7) and (8). Related to Equation (7), many researchers have tried to enhance the surface charge density by focusing on material selection and fabrication, various surface engineering techniques, and surface modification. Studies on MECSs that focus on regulating the contact force, frequency, and velocity are still needed. This review presents MECS designs that convert fixed, irregular, or intermittent input values for TENGs by using simple machines such as levers, pulleys, gear-trains, and hydraulic presses to control the input mechanical energy, as shown in Figure 1a–d.

The MECSs used in previous studies to control the external energy input to TENGs are here divided into six categories: (1) gear-based, (2) cam-based, (3) flywheel and governor-based, (4) gear- and cam-based combinations, (5) flywheel- and spiral-spring-based combinations, and (6) input-flow-based. As shown in Figure 2, these designs all enable otherwise wasted rotating mechanical energy and linear motion to be harvested for reliable electric power. MECSs can obtain predictable output performance even when the mechanical energy input is irregular, which allows the designed TENG circuit to be accurate and suitable for practical applications. This review focuses on the regulation of force parameters and exhibits the importance of MECS design in TENG development.

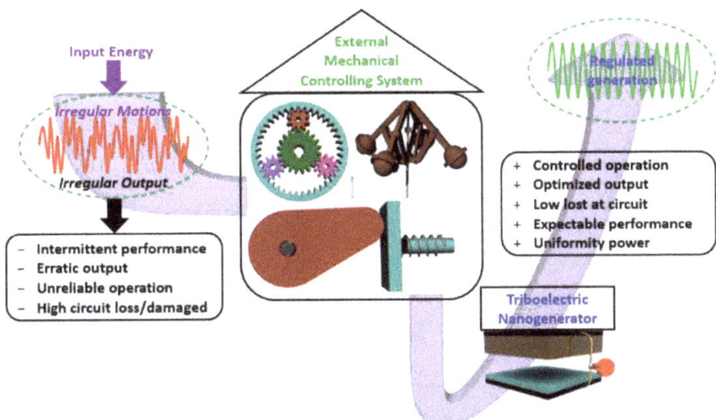

Figure 2. Flow chart for controlling TENG parameters.

3. Gear-Based Mechanical Control Systems

The output performance of rotary TENG systems can be improved by increasing the rotational velocity (i.e., angular velocity) [56,57,60,64,76–83]. The gear-train mechanism not only adjusts the torque (or force) but also modifies the rotation speed, the direction of rotation, and related motion. Specifically, the rotation speed can be accelerated or decelerated by changing the gear ratio in the system [64,66,68,69]. As shown in Equation (3), external input torque can be transmitted from a big gear (as a driving gear) to a smaller gear (as a driven gear) to boost the rotational speed of TENG devices. Many gear-train designs have been used in assembled TENG systems, as depicted in Figure 3. A TENG system with a simple straight-cut gear and a slider-crank mechanism is shown in Figure 3a$_1$ [66,69]. Three different gear ratios were designed to produce various angular speeds, and their effects on the TENG output power were compared. To boost the rotational speed, a large spur gear accepts the external input rotation while engaged with a smaller gear connected to the crank. At the slider-crank, the rotational motion is converted into periodical linear motion to produce a TENG with the vertical contact-separation mode. The working frequencies measured at input frequencies of 1.5, 3.0, and 4.5 Hz are shown in Figure 3a$_2$. When the gear ratio was 1.7, the working frequencies increased linearly to 2.4, 4.8, and 7.2 Hz, respectively, about 1.6 times higher than the input frequencies. With the higher gear ratio of 5, higher transmission losses were observed, and the increase in the working frequencies was only about 3.5 times. Figure 3a$_3$ shows the output voltages of the TENG as a function of the gear ratio when the input frequency was 4.5 Hz. At a gear ratio of 1.7 (f_w = 7.2 Hz), the output voltage improved to 290 V, higher than when the gear ratio was 1.0. When the gear ratio was 5.0 (f_w = 16 Hz), the output voltage improved to 340 V, a 1.7-fold enhancement. To clearly understand the output power of this gear-based TENG design, the researchers evaluated its capacitor charging behavior under a constant input frequency of 4.5 Hz at different gear-train ratios, as shown in Figure 3a$_4$. Surprisingly, at gear ratios of 1.0, 1.7, and 5.0, the voltage charge of the capacitor, V_c, became saturated at 5, 12, and 40 V, respectively. This performance represents a maximum eight-fold improvement, indicating that a gear-based TENG significantly increases the output power at a certain input energy.

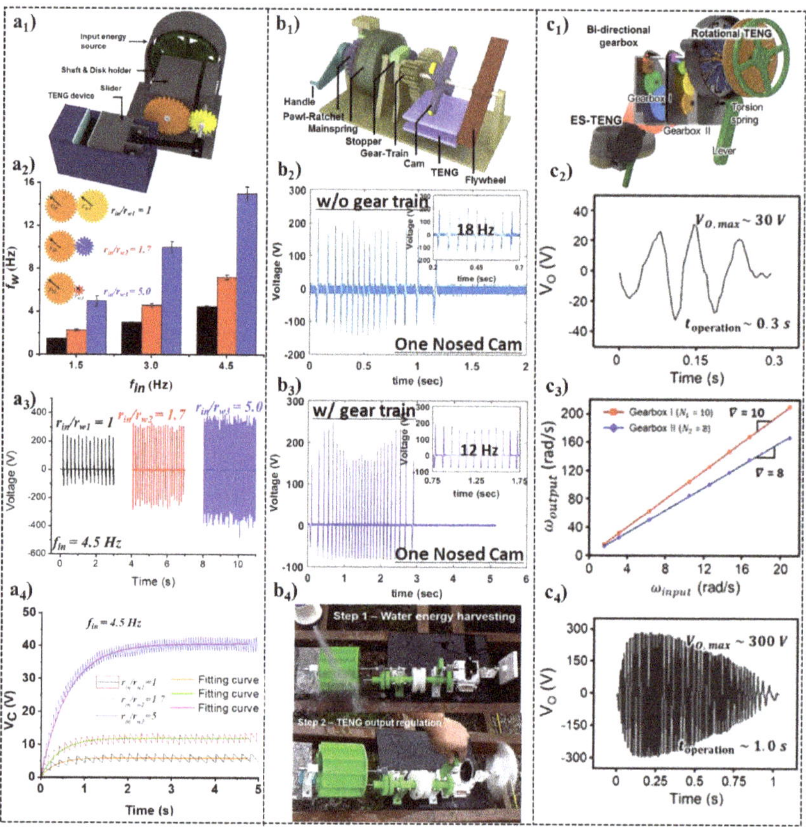

Figure 3. Gear-based mechanical system designs to control the input energy and regulate TENG performance. (**a₁**) Schematic of TENG system based on the gear and slider-crank mechanism. (**a₂**) Relationship between working frequencies and input frequencies at different gear ratios. (**a₃**) TENG output voltage at various ratios and an input frequency of 4.5 Hz. (**a₄**) Measured capacitor charging output voltage by TENG with a fixed frequency input of 4.5 Hz at different gear ratios. Reprinted with permission from ref. [66], Copyright 2016, Elsevier. (**b₁**) Full system schematic of the MFR-TENG and the controlled energy-release part of the MFR-TENG, The influence of a gear-train using a single flywheel and one-nosed cam on TENG output, as shown by comparing systems (**b₂**) without and (**b₃**) with a gear-train. (**b₄**) Demonstration of a waterwheel MFR-TENG system in a natural environmental application, with a gear-train that enables low input forces to turn a spiral spring. Reprinted with permission from ref. [69], Copyright 2018, Wiley. (**c₁**) Schematic illustration of the ES-TENG, which consists of a bi-directional gearbox and rotational TENG. The measured output voltage of (**c₂**) the rotational TENG without any transmission unit and (**c₃**) the rotational TENG integrated with the bi-directional gearbox. (**c₄**) The angular speed output of the rotational TENG at different angular speed inputs for gearboxes I and II. Reprinted with permission from ref. [36], Copyright 2021, Elsevier.

In other cases, the mechanical input energy can be irregular because it comes from different sources in the surrounding environment, which leads to an irregular electric output from the mechanical energy harvesting system. In such a case, the mechanical energy harvester should be regularly excited to regulate its output. Therefore, Bhatia et al. proposed a TENG design that fixed the input forces and input frequency and called it a mechanical frequency regulator TENG (MFR-TENG), as depicted in Figure 3b₁. Their system uses many elements for assembly, such as a pawl ratchet, mainspring, stopper, gear-train, and flywheel. The input energy turns the handle and winds up the spiral spring (i.e., mainspring). If the mechanical input energy is irregular or disappears, the mainspring releases the stored energy. A pawl-ratchet mechanism is installed between the handle and mainspring to prevent that from happening. The stored energy can only be released by the stopper. Following that energy transmission line, the mechanically stored energy rotates the gear-train, which causes rotation of the cam and a flywheel mounted on the shaft, so that torque can be transferred and stored as required. Here, we focus on the gear-train that supports the MFR-TENG system. The effect of a gear-train with one flywheel and a single cam-nose on the TENG output is shown in Figure $3b_2,b_3$. In this demonstration, the driving gear to driven gear ratio is 3:1, which decreases the torque of the driven gear and increases the influence of follower resistance, reducing the TENG output frequency. Thus, the use of this gear-train increases the output voltage and running time of the TENG system. In addition, the TENG output frequency obtained from using this gear-train is considerably more stable than the system without a gear-train. Lastly, a water wheel with a small gear radius is assembled on the input side, as shown in Figure $3b_4$, to demonstrate the applicability of the MFR-TENG system in a natural environment. The water wheel can transfer the water flow energy to a spiral spring for storage when the stopper is blocked and for frequency output regulation when the stopper is released.

To overcome the limitation that a low input frequency provides a low output frequency, the exo-shoe TENG (ES-TENG) with an integral transmission unit was proposed. The ES-TENG design consists of a rotational TENG to harvest biomechanical energy from footsteps, with a power transmission unit assembled outside of the shoe to improve the amplitude and frequency of the output, as shown in Figure $3c_1$. It uses gearboxes I and II to convert reciprocating footsteps with high force/torque, short displacement, and low frequency into bi-directional rotation that has a smaller force, bigger displacement, and higher frequency. To characterize the effects of gearboxes I and II, the output power of a rotational TENG without the transmission unit was evaluated as a reference. Figure $3c_2$ depicts the output voltage of the rotational TENG without any transmission unit. In that simple design, a reciprocating stroke with subsequent compress and release steps was generated at high input force. The resulting rotational TENG provides low output performance and a short duration of operation. The maximum output voltage was about 0.3 V, and the operation time was the same as the input duration (0.3 s). In the ES-TENG, the gear ratios of integrated gearboxes I and II were 10 and 8, respectively. Figure $3c_3$ shows the linear enhancement of the angular velocity output at different angular velocity inputs for each gearbox, and the enhancement exactly matches the gear ratio of each box. The higher gear ratio provides higher rotational speed, which enhances the output performance of the rotational TENG. In that work, the case 2 design, which used gearbox I, generated 10 times more power and elongated the output power (0.8 s) from the rotational TENG under an input motion identical to that in case 1 (not shown here). To further boost the output power, the bi-directional gearbox (case 3) was designed, as shown in Figure $3c_4$. Its output voltage was the same as that of case 2, but the operation time was longer (1 s) because gearbox II was in the bi-directional gearbox.

4. Cam-Based Mechanical Control Systems

In this section, we review cam-based TENG (C-TENG), MFR-assembled cam noses, and windmill-integrated cam noses that drive TENG to overcome the limitations of rotating energy harvesters—friction and high cost. Many rotational scavengers have short lifetimes and high costs due to friction between the moving parts. Cam-based TENGs that can transform rotational movement into linear motion are sustainable, high-performance harvesters based on contact-separation TENG. With the cam design in the mechanical system, the working frequency can be enhanced by changing the number of cams or the rotation speed of the cam-shaft during the energy conversion from rotation to linear motion of the TENG.

The rotational torque (i.e., force) and rotation speed affect the output of a rotational motion-based TENG, as shown below.

$$P(W) = \frac{\tau(N.m) \times 2\pi\left(\frac{rad}{rev}\right) \times n(rpm)}{60} \tag{9}$$

A cam system can be used to convert the rotational energy into the vertical contact-separation mode for a TENG, as depicted in Figure 4a$_1$. Basically, a simple C-TENG system contains a cam and a TENG device. The bumper springs between the bottom TENG plate and the frame are used to prevent sticking during cam encounters. The bumper springs enhance the smooth rotation and clear contact-separation of the TENG device. Without them, the TENG device can become stuck during the rotation of the cam. The ratio stiffnesses between the spacer and the bumper springs (k_s/k_b) and their influence on the output performance of the C-TENG can be used to understand this relationship. In addition, the ratio between the stiffness of the spacer and the bumper spring needs to be considered because it affects the repulsive force and contact-separation speed of the follower. According to Lee et al., the short-circuit current, I_{sc}, is determined as

$$I_{sc} = \frac{S\sigma\frac{d}{\varepsilon_r}}{\left(\frac{d}{\varepsilon_r} + x(t)\right)^2} \cdot r \cdot cos\theta(t) \cdot \omega \cdot \frac{1}{1 + k_s/k_b} \tag{10}$$

where S is the contact area; σ represents the static tribo-charges of the tribo-surfaces; ε_r and d are the relative dielectric constant and thickness of the negative tribo-layer, respectively; $x(t)$ is the distance between the negative tribo-layer and the bottom electrode; r and ω are the distance between the cam shaft and the nose and the angular speed, respectively; and k_s and k_b are the stiffness values of the spacer and bumper spring, respectively. The stiffness ratio between k_s and k_b directly affects the repulsive force of the TENG device when the cam comes into contact with it. Changing that stiffness ratio changes the force applied to the TENG device, which determines the TENG output performance. Here, however, the focus is on the effect that the cam noses have on the working frequency. As shown in Figure 4a$_2$, the working frequencies in the output of one TENG unit with single, dual, and quad cam noses increase linearly to 6.5, 13, and 26 Hz, respectively, at 400 rpm. Surprisingly, the peak voltage and current output are not changed by the various numbers of cam noses, but the output frequencies are. To clearly illustrate this behavior, the output peak waveforms for a single rotation with a single nose and quad noses are shown in Figure 4a$_3$,a$_4$. The separation time, Δt_s, between the two tribo-materials is the same for any number of cam noses, even though the number of voltage pulses increases along with the working frequency. This result affirms that the contact speed of the two tribo-materials is the same. Thus, the C-TENG can be operated with different numbers of cam noses but a constant angular speed of the cam, which causes the contact velocities of the tribo-materials to become the same, as shown in Equation (10). In other words, the output power of the C-TENG depends on the contact speed of the tribo-materials. Even the output peak amplitude is constant with various cam designs; however, the cam design does contribute to the performance of external connections to the C-TENG, such as transformers or rectifiers.

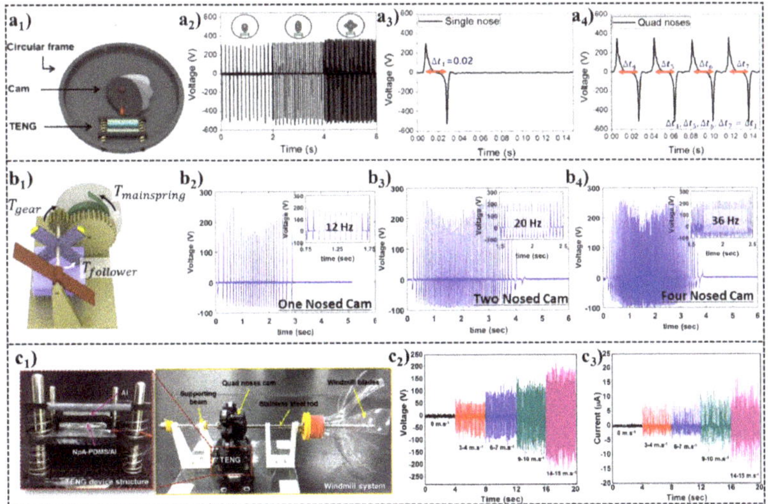

Figure 4. Cam-based mechanical system designs to control the input energy and regulate TENG performance. (a_1) Schematic diagram of a C-TENG. Output voltage measurement of (a_2) single, dual, and quad cam noses with a single TENG unit, the peak voltages for a single rotation with (a_3) a single cam nose and (a_4) quad cam noses. Reprinted with permission from ref. [37], Copyright 2017, Elsevier. (b_1) Schematic of the MFR-TENG and the influence the number of cam noses has on the TENG output in the presence of a single flywheel and gear-train assembled for a (b_2) one-nosed cam, (b_3) two-nosed cam, and (b_4) four-nosed cam. Reprinted with permission from ref. [69], Copyright 2018, Wiley. (c_1) Schematic of a windmill with an integrated four-nosed cam driving a TENG. The (c_2) output voltage and (c_3) current of the TENG windmill at different wind speeds. Reprinted with permission from ref. [42], Copyright 2017, Elsevier.

Another way to apply a cam in a TENG harvesting system is demonstrated by the MFR-TENG system described above and shown in Figure $4b_1$. An MFR-TENG with a gear-train and single flywheel of 320 ± 3 g was used to test how the number of cam noses affects TENG output. According to Equation (11), given by Bhatia et al., the TENG output frequency will increase with the number of cam noses.

$$f_{TENG} = \frac{N_c}{2\pi} \sqrt{\frac{2}{I_{total}} \left(\frac{1}{N_g} \int T_{mainspring}\, d\theta - \int T_{follower}\, d\theta \right)} \qquad (11)$$

The measured output voltage of the MFR-TENG clearly correlates with that design, as shown in Figure $4b_2$–b_4. The MFR-TENGs with one, two, and four cam noses had output frequencies of 12, 20, and 36 Hz, respectively. Thus, the output frequency is proportional to the number of cam noses. Rationally designing the gear-train, the number of cam noses, and the flywheel can produce a uniform TENG output and improve the running time.

An example TENG application is a windmill designed with four cam noses to convert wind flow energy into the vertical contact-separation working mode for the TENG. The windmill TENG photograph shown in Figure $4c_1$ shows some of the components: a stainless-steel rod, windmill blades made from recycled plastic bottles, a supporter beam, the quad-nosed cam, and a vertical contact-separation TENG device. Under wind speeds from 0 to 14–15 ms^{-1}, the windmill TENG produced the output voltages and currents depicted in Figure $4c_2$,c_3, respectively. As shown in the figure, the output voltage and current increased linearly with the wind speed. This broadband output performance can harvest breezes and strong winds in remote areas or moving vehicles.

As shown in Equation (13), the TENG output frequency is inversely proportional to the rotating arm radius and the square root of the governor mass. As shown in Equation (14), the TENG operation time is inversely proportional to the working frequency. The experimental data are well-matched with the theoretical calculation.

6. Gear- and Cam-Based Mechanical Control Systems

In this section, we review mechanical system designs that combine gears and cams to control the working time, the working frequency and output peak of TENG. The role of each component is discussed for the MFR and ASMFR systems described above. In addition, a gear-cam system design is used to achieve controlled high-frequency output from ionic triboelectric nanogenerators (iTENGs) with and without grounding.

Using the MFR system, as shown in Figure 6a$_1$, Divij et al. experimentally demonstrated the separate effects of gear and cam components. Because the spiral spring torque decreases as it unwinds, the TENG output frequency is shown to decrease gradually in all cases. A higher number of cam noses increases the TENG output frequency, as shown in Equation (11) and Figure 6a$_3$,a$_4$. Additionally, as described in Equation (11), the gear ratio is 3:1, which decreases the driven-gear torque and increases the influence of follower resistance to reduce the TENG output frequency. The experimental results match well with Equation (11). As shown in Figure 6a$_2$,a$_3$, the output frequency with a one-nosed cam decreased from ~22 to 15 Hz without and with a gear-train. In that comparison, the running time and voltage output increased. Thus, the effect of the spiral spring torque reduces the uniformity of TENG output frequency.

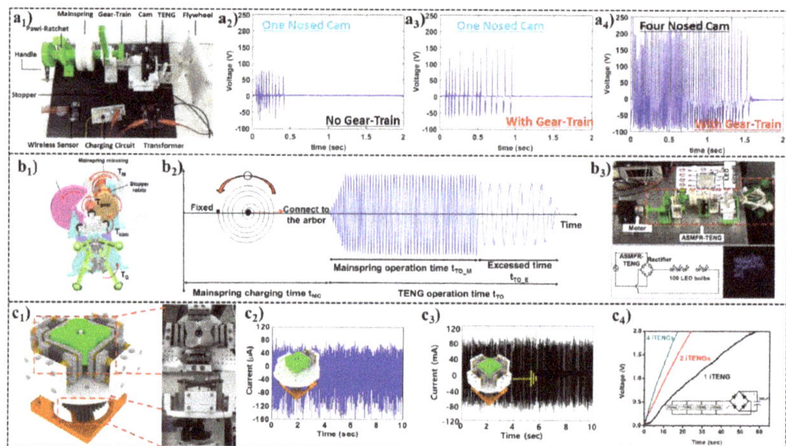

Figure 6. Effects of combining a gear-train and multi-nosed cam when regulating TENG input energy. (**a$_1$**) Photograph of a full MFR-TENG system. Influence of the gear-train on TENG output: a one-nosed cam (**a$_2$**) without and (**a$_3$**) with a gear-train. (**a$_4$**) Comparison of the running time and stability of the MFR-TENG output with different numbers of cam noses and an integrated gear-train. Reprinted with permission from ref. [69], Copyright 2018, Wiley. (**b$_1$**) ASMFR-TENG schematic showing the working mechanism on the output-side of TENG operation. (**b$_2$**) Automatic switching operation with $\frac{3}{4}$ blocker shape to enable continuous output regulation. (**b$_3$**) Photograph of 100 blue LEDs powered by the ASMFR-TENG and driven by a DC motor. Reprinted with permission from ref. [68], Copyright 2021, Elsevier. (**c$_1$**) Integration of iTENGs with gear- and cam-assembled kinematic systems. (**c$_2$**) Output currents for a single iTENG system with integrated gears and cam and (**c$_3$**) without grounding. (**c$_4$**) Output currents for four iTENGs in parallel connection with a grounding layer.Reprinted with permission from ref. [39], Copyright 2019, Wiley.

The ASMFR (Figure 6b$_1$) combines a gear-train, four cam noses, and a stopper and blocker mechanism for automatic switching operation, and after mainspring charging and operation, the TENG output frequency becomes uniform. Figure 6b$_2$ shows the uniform output frequency of the ASMFR-TENG driven by the mainspring. During spiral spring operation time, the output frequency is uniform. However, the excess TENG operation time obtained directly from the input energy transmitted through the four-nosed cam causes an irregular TENG output frequency, as also shown in Figure 6b$_2$. Figure 6b$_3$ shows the ASMFR-TENG in use as a power source driven by a DC motor to light 100 blue LEDs.

Hwang et al. designed a kinematic system integrating a gear-train and four cam noses into what they called iTENGs to achieve a high, controlled frequency of 60 Hz from an input rotation of 100 rpm (~1.67 Hz), as shown in Figure 6c$_1$. The inset is a photograph of the kinematic gear-cam system with a final gear ratio of 9 and four cam noses, which enhanced the output frequency by 36 times. A single iTENG without a grounded structure provided about 157 μA, as shown in Figure 6c$_2$. To further boost the output current, four iTENGs driven by the gear-cam system were connected in parallel with grounding to obtain a current of about 90 mA without transformation, as shown in Figure 6c$_3$. The kinematic frequency regulator helps to reduce the circuit loss based on the regulated output frequency of iTENGs in the gear-cam kinematic system. To demonstrate the practical applicability of the iTENG integrated gear-cam kinematic system, a 330 μF capacitor was charged by one, two, and four iTENGs, as depicted in Figure 6c$_4$. The designed iTENG system takes 58, 24, and 16 s to reach 2 V when using one, two, and four iTENGs, respectively. Thus, increasing of the number of iTENGs promotes the charging performance of the capacitor.

7. Gear-, Spiral-Spring- and Flywheel-Based Mechanical Systems

In this part, we review system designs that combine gear/rack systems, spiral springs, and flywheels with TENG (FSS-TENG) and an escapement mechanism-based TENG (EM-TENG). Those systems are intended to effectively harvest many types of intermittent mechanical motion using non-continuous operation. The gear can be used to enlarge the torque applied to the system, after which the energy is stored as potential in spiral spring and finally controlled by the flywheel for the running time and also amplitude of TENG. The FSS-TENG integrates a flywheel and spiral spring to improve the energy harvest from intermittent excitation. The transmission unit of the FSS-TENG is integrated on the box to convert intermittent motion into rotational motion, as shown in Figure 7a$_1$. Longer running time is obtained with the help of the flywheel. The transmission unit contains a rack and pinion that receives the intermittent mechanical force and converts it into rotational motion. The rotation of the shaft is driven by the gears. As the shaft rotates with one-way transmission (with the help of a one-way clutch), the spiral spring is compressed. The optimal spiral spring stiffness and flywheel masses provide a high load voltage and short-circuit current I_{sc}. The load voltage is almost constant, but the I_{sc} decreases as the mass increases, as shown in Figure 7a$_2$,a$_3$. The relationship between the running time and the flywheel mass at various spiral spring stiffnesses is shown in Figure 7a$_4$. As the flywheel mass increases, the running time of the FSS-TENG increases for each stiffness value. As a result, the FSS-TENG has the longest running time and maximum output at 29 N mm/rad because the higher stiffness of 42 N mm/rad produces unstable compression.

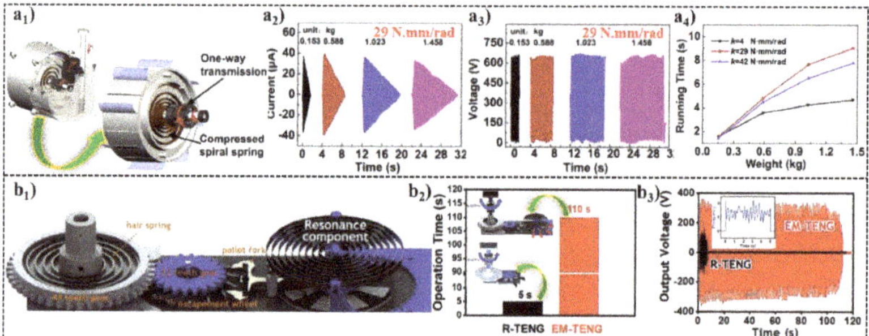

Figure 7. Mechanical design of TENG assembled with gears, a spiral spring, and a flywheel to control TENG output. (**a$_1$**) Schematic of a FSS-TENG system, (**a$_2$**) short-circuit current and (**a$_3$**) open circuit voltage output performance of the FSS-TENG with spiral spring stiffness of 29 N.mm/rad and different flywheel masses. (**a$_4$**) Dependence of running time on flywheel mass at different spiral spring stiffnesses. Reprinted with permission from ref. [35], Copyright 2019, Elsevier. (**b$_1$**) Schematic of EM-TENG, (**b$_2$**) controlling operation time and (**b$_3$**) open circuit voltage under irregular energy source of R-TENG and EM-TENG. Reprinted with permission from ref. [63], Copyright 2021, Wiley.

To harvest the extremely low input frequency (less than 0.1 Hz) from irregular motion, the EM-TENG is composed of a mechanical energy storage device (hair spring), an escapement mechanism, and a torsional resonator. A schematic of the full system of EM-TENG is provided in Figure 7b$_1$. With the very low input frequency of 0.067 Hz, the EM-TENG generated an open circuit voltage of 320 V and a short-circuit current of 0.59 mA.m^{-2}. Specifically, the EM-TENG provided a long-lasting and stable power output of 110 s with only 5 s of input energy, representing a 22-fold expansion of its operation time, as depicted in Figure 7b$_2$. The energy storage stage consists of a gear and a torsional hair spring. The hair spring can store the irregular and low input frequency, transfer it to the gear rotation and then transmit it to the escapement mechanism component. The escapement wheel then contacts the pallet fork, producing the rotational resonance of the hair spring and rotor. Then, a non-contact rotational TENG in freestanding mode is connected to the hair spring in the resonance component to produce electrical energy at a regulated high frequency from random motion. To prove the effectiveness of the EM-TENG with an extremely low input frequency, the EM-TENG was compared with a rotating TENG (R-TENG) without an escapement mechanism. Under the irregular input energy triggered by the average frequency of ocean waves in Korea for 5 s (inset figure in Figure 7b$_2$), the R-TENG generated an unstable output voltage of 150 V (maximum) for 5 s, the same as the input energy duration. Under the same conditions, the EM-TENG produced about 320 V for 110 s, as shown in Figure 7b$_3$. Thus, the EM-TENG can provide long-lasting and stable output from very low-frequency and irregular mechanical motion in the surrounding environment.

8. Mechanical Systems to Control the Input Flow

In this section, we review ways to control the input flow for TENG output regulation. As shown by parametric studies of the structure correlated with the flow behavior, i.e., turbulent, or laminar flow, the rotation speed or flow speed can be controlled.

N.D. Huynh et al. showed the design of a Tesla turbine for driving a rotational TENG disc and electromagnetic generator (T^3-E hybrid system). Figure 8a$_1$,a$_2$ shows a schematic of the T^3-E hybrid system design and a photograph of the Tesla turbine, respectively. In their turbine parametric study, the disc gap, disc thickness and disc diameter were experimentally manipulated. However, only the disc gap and disc thickness are presented here, because they had larger effects on the input flow behavior than the disc diameter. The output voltages of the Tesla turbine TENG with different turbine disc gaps and disc thicknesses to control the input rpm of the turbine are shown in Figure 8a$_3$,a$_4$, respectively.

The effect of the gap between the turbine discs was investigated by analyzing the output voltage as the gap was varied from 0.5 to 2 mm in 0.5 mm increments. The T1.1 model, with a gd = 0.5 mm, was the optimized turbine gap model, providing a peak output voltage of about 735 V. The larger gap models of 1, 1.5, and 2 mm reached peak output voltages of 725, 566, and 495 V, respectively. The rpm was a function of the flow input when the gap between discs was constant. However, an increase in the disc spacing caused a decrease in the rotational speed. In addition, the friction between the input flow and discs increased, leading to the formation of a boundary layer at the disc surface when the gap was optimized. The turbine disc thickness was another important factor in enhancing the turbine rotation velocity. Disc thicknesses from 0.5 to 2 mm, with a step size of 0.5 mm, were used for this test and were referred to as T2.1, T2.2, T2.3, and T2.4, respectively. As the turbine disc thickness increased from 0.5 to 1 mm, the output voltage of the T^3 doubled from 513.5 to 1005 V. However, turbine discs thicker than 1 mm showed reduced TENG performance. The T2.3 and T2.4 turbine models provided output voltages of 786 and 786 V, respectively. Apparently, airflow into the gap is not easy at the higher disc thicknesses, leading to the poor aerodynamic performance. On the other hand, thinner discs were less stiff than thicker ones, which increased the mechanical stress, vibration, and noise during turbine operation.

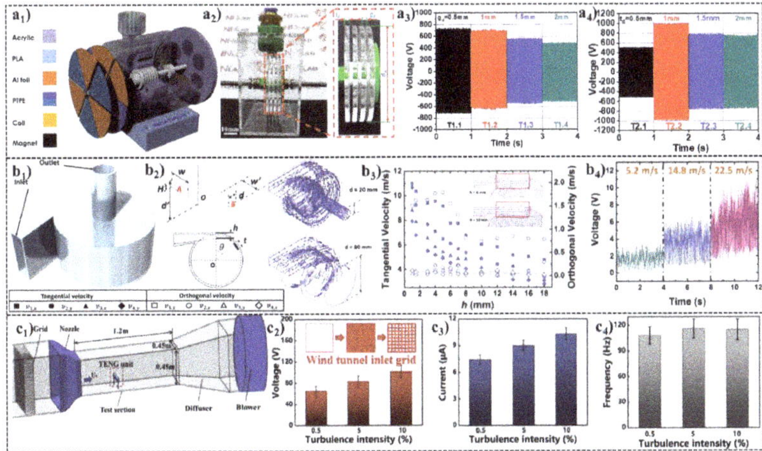

Figure 8. Mechanical design for controlling the input flow driving a TENG. (**a₁**) Schematic of the T^3-E hybrid system. (**a₂**) Photograph of a Tesla turbine to show its output voltage when using different (**a₃**) turbine disc gaps and (**a₄**) turbine disc thicknesses to control the input rpm of the turbine. Reprinted with permission from ref. [71], Copyright 2021, Elsevier. (**b₁**) A schematic of the wind-rolling triboelectric nanogenerator (WR-TENG) and (**b₂**) the computational fluid dynamics results from a 3D model used to simulate the influence of the outlet size on wind flow inside the whistle. (**b₃**) Tangential and orthogonal wind velocities calculated depending on the height of the throat. The inset figure is the CFD analysis of the nozzle shape throat effect. (**b₄**) The V_{OC} output of multiple WR-TENG patterns based on wind velocities. Reprinted from ref. [72]. (**c₁**) Schematic of the structural optimization experiments for the TENG unit in a wind tunnel, which shows the dependence of the power generation performance of the TENG unit on turbulence intensity with the (**c₂**) measured V_{OC}, (**c₃**) I_{SC}, and (**c₄**) membrane flutter frequency plotted versus turbulence intensity. Reprinted with permission from ref. [73], Copyright 2020, Elsevier.

To harvest wind energy, a wind-rolling triboelectric nanogenerator (WR-TENG) was proposed, as shown in Figure 8b₁. A lightweight dielectric sphere rolls along the whistle wall during vortex wind flow through the WR-TENG to generate electricity. A computation fluid dynamics (CFD) analysis simulated the wind flow movement to form a vortex flow.

For that purpose, the vortex whistle shape parameters were controlled to optimize and increase the kinetic energy conversion rate. Figure 8b$_2$ shows a 3D model used in the simulation with various optimization parameters and the outlet size of the wind flow inside the whistle simulated by CFD. As shown in Figure 8b$_3$, the tangential and orthogonal wind velocities were calculated depending on the height of the throat; the inset figures show the CFD analysis of the nozzle shape throat effect. A reduction in h increased the tangential velocity due to the Venturi effect during air flows through the nozzle, although the orthogonal velocity was almost unchanged for the points v_2, v_3, and v_4 (at the 3, 6, and 9 o'clock positions). On the other hand, at v_1 (in the 12 o'clock position near the nozzle), the orthogonal velocity was higher than at other points, especially until $h = 8$ mm. Figure 8b$_4$ shows the WR-TENG V_{oc} output following a change in wind velocity. The effective contact area was enhanced as the rotating ball velocity increased the centrifugal force that the expanded polystyrene sphere received in the cycle. As a result, the V_{oc} increased with the flow speed.

Lastly, a multi-functional wind barrier integrated with manifold TENG units is introduced. Figure 8c$_1$ shows a schematic of the experimental apparatus for the TENG unit in a wind tunnel. Those authors installed a $100 \times 10 \times 3$ mm^3 TENG unit in a wind tunnel to investigate the effects of specific wind conditions on the TENG performance. They found that the flutter characteristics of the fluorinated ethylene propylene membrane in the TENG depended on the turbulence intensity. To investigate that issue, the TENG power generation was plotted against the wind turbulence intensity in the wind tunnel using different wind tunnel inlet grid parameters. At a wind speed of 10 m/s and three typical turbulence intensities (0.5%, 5%, and 10%) at the entrance of the test section, the TENG output performances were measured in terms of the V_{oc}, I_{sc}, and flutter frequency. The results are displayed in Figure 8c$_2$–c$_4$. Higher turbulence intensity boosted the TENG output performance. The V_{oc} and I_{sc} increased by 56.2% and 33.3%, respectively, as the turbulence intensity increased from 0.5% to 10%. The flutter frequency, on the other hand, was independent of the turbulence intensity, as shown in Figure 8c$_4$. This implies that higher turbulence intensity might interrupt the vibration mode of the membrane and increase the local contact area. The flutter frequency mainly depends on the mean velocity instead of the turbulence intensity.

9. Summary and Perspective

Many researchers have attempted to control the output performance of TENGs by conducting material studies, such as various micro/nanoscale fabrication techniques, nanoparticle deposition techniques, or physical or chemical etching processes. However, the fabricated material itself cannot control the output performance; well-designed MECSs are needed to do that. Therefore, we have summarized previous studies that used MECSs to obtain a predictable, uniform output from TENG-based mechanical energy harvesters that receive low input frequencies from broadband energy sources that range from light breezes to high-speed winds. These kinds of wasted mechanical energies can be boosted, controlled, and predicted with the help of MECSs to further the industrialization and real application of TENGs. MECSs often integrate gears, cams, flywheels, and governors, individually and in combination structures.

In this review, we have described some system designs, such as MFR-TENG, ASMFR-TENG, T^3-E, WR-TENG, FSS-TENG, and C-TENG, that have been reported to achieve a predictable, uniform, controllable TENG output with low energy losses and extended operation times for energy scavenging from irregular mechanical forces. The regulators typically use a storage unit such as a spiral spring for the low-frequency mechanical input, a transmission unit based on some combination of gear-trains and cams, and flywheels to regulate the torque/force and accurately control the working frequency of energy release. We have also reviewed the theoretical bases of regulator dynamics studied using simulation processes. Overall, we affirm the inverse relationship between the frequency and operation time of TENG output. For an optimized process of TENG-based system design, including the consideration of mechanical, material, device, and circuit design factors, a closed-loop, total integration design is shown in Figure 9.

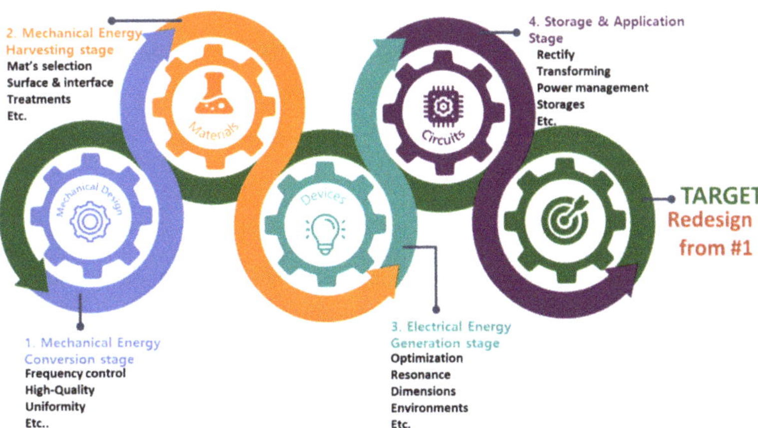

Figure 9. A closed-loop, total integration design for optimizing TENG-based systems.

In the view of mechanical structure design, controlling the input energy for a constant energy source is important in regulating TENG output for further applications. Currently, more studies are needed to improve MECSs for controllable input parameters and increase the efficiency of small input frequency conversions. To solve these problems, MECS designs for TENGs and hybrid systems still need to be developed, which will require research into the following issues:

(1) To obtain higher efficiency TENG systems, miniaturization, high adaptability, sustainability, endurance, and friction are big problems that need to be solved for the development of TENGs and hybrid system devices.

(2) Due to the irregular and intermittent external energy sources, TENGs and hybrid systems based on a rotary system design should be adapted for bi-directional rotation to easily harvest surrounding energy, such as wind, from multiple directions.

(3) For automatic and continuously operating TENG systems, MECSs should be designed to use unmanned device technology. TENGs integrated with such MECS designs could power Internet-of-Things systems and sensing networks automatically and precisely. We hope these problems can soon be solved and that TENGs can soon begin contributing to smart cities and other industrial applications.

Author Contributions: Conceptualization, D.C.; writing—original draft preparation, N.D.H.; writing—review and editing, D.C. and N.D.H. All authors have read and agreed to the published version of the manuscript.

Funding: This work was supported by the Mid-career Researcher Program (No. 2019R1A2C2083934) through the National Research Foundation (NRF) of Korea and the Technology Innovation Program (20013794, Center for Composite Materials and Concurrent Design) funded by the Ministry of Trade, Industry & Energy (MOTIE, Korea).

Conflicts of Interest: The authors declare no conflict of interest.

References

1. Chen, J.; Yang, J.; Guo, H.; Li, Z.; Zheng, L.; Su, Y.; Wen, Z.; Fan, X.; Wang, Z.L. Automatic Mode Transition Enabled Robust Triboelectric Nanogenerators. *ACS Nano* **2015**, *9*, 12334–12343. [CrossRef]
2. Yang, Y.; Zhu, G.; Zhang, H.; Chen, J.; Zhong, X.; Lin, Z.-H.; Su, Y.; Bai, P.; Wen, X.; Wang, Z.L. Triboelectric Nanogenerator for Harvesting Wind Energy and as Self-Powered Wind Vector Sensor System. *ACS Nano* **2013**, *7*, 9461–9468. [CrossRef]

3. Zhang, L.; Zhang, B.; Chen, J.; Jin, L.; Deng, W.; Tang, J.; Zhang, H.; Pan, H.; Zhu, M.; Yang, W.; et al. Lawn Structured Triboelectric Nanogenerators for Scavenging Sweeping Wind Energy on Rooftops. *Adv. Mater.* **2016**, *28*, 1650–1656. [CrossRef]

4. Zhang, B.; Chen, J.; Jin, L.; Deng, W.; Zhang, L.; Zhang, H.; Zhu, M.; Yang, W.; Wang, Z.L. Rotating-Disk-Based Hybridized Electromagnetic–Triboelectric Nanogenerator for Sustainably Powering Wireless Traffic Volume Sensors. *ACS Nano* **2016**, *10*, 6241–6247. [CrossRef]

5. Quan, Z.; Han, C.B.; Jiang, T.; Wang, Z.L. Robust Thin Films-Based Triboelectric Nanogenerator Arrays for Harvesting Bidirectional Wind Energy. *Adv. Energy Mater.* **2016**, *6*, 1501799. [CrossRef]

6. Bhatia, D.; Hwang, H.J.; Huynh, N.D.; Lee, S.; Lee, C.; Nam, Y.; Kim, J.-G.; Choi, D. Continuous scavenging of broadband vibrations via omnipotent tandem triboelectric nanogenerators with cascade impact structure. *Sci. Rep.* **2019**, *9*, 8223. [CrossRef]

7. Bhatia, D.; Kim, W.; Lee, S.; Kim, S.W.; Choi, D. Tandem triboelectric nanogenerators for optimally scavenging mechanical energy with broadband vibration frequencies. *Nano Energy* **2017**, *33*, 515–521. [CrossRef]

8. Wang, X.; Niu, S.; Yi, F.; Yin, Y.; Hao, C.; Dai, K.; Zhang, Y.; You, Z.; Wang, Z.L. Harvesting Ambient Vibration Energy over a Wide Frequency Range for Self-Powered Electronics. *ACS Nano.* **2017**, *11*, 1728–1735. Available online: https://pubs.acs.org/doi/full/10.1021/acsnano.6b07633 (accessed on 29 October 2021). [CrossRef]

9. Wang, L.; Yuan, F.G. Vibration energy harvesting by magnetostrictive material. *Smart Mater. Struct.* **2008**, *17*, 045009. [CrossRef]

10. Wang, S.; Niu, S.; Yang, J.; Lin, L.; Wang, Z.L. Quantitative measurements of vibration amplitude using a contact-mode freestanding triboelectric nanogenerator. *ACS Nano* **2014**, *8*, 12004–12013. [CrossRef]

11. Chen, J.; Zhu, G.; Yang, W.; Jing, Q.; Bai, P.; Yang, Y.; Hou, T.C.; Wang, Z.L. Harmonic-resonator-based triboelectric nanogenerator as a sustainable power source and a self-powered active vibration sensor. *Adv. Mater.* **2013**, *25*, 6094–6099. [CrossRef]

12. Yang, W.; Chen, J.; Zhu, G.; Yang, J.; Bai, P.; Su, Y.; Jing, Q.; Cao, X.; Wang, Z.L. Harvesting energy from the natural vibration of human walking. *ACS Nano* **2013**, *7*, 11317–11324. [CrossRef]

13. Wen, X.; Yang, W.; Jing, Q.; Wang, Z.L. Harvesting broadband kinetic impact energy from mechanical triggering/vibration and water waves. *ACS Nano* **2014**, *8*, 7405–7412. [CrossRef]

14. Yang, W.; Chen, J.; Zhu, G.; Wen, X.; Bai, P.; Su, Y.; Lin, Y.; Wang, Z. Harvesting vibration energy by a triple-cantilever based triboelectric nanogenerator. *Nano Res.* **2013**, *6*, 880–886. [CrossRef]

15. Yang, J.; Chen, J.; Yang, Y.; Zhang, H.; Yang, W.; Bai, P.; Su, Y.; Lin Wang, Z.; Yang, J.; Chen, J.; et al. Broadband Vibrational Energy Harvesting Based on a Triboelectric Nanogenerator. *Adv. Energy Mater.* **2014**, *4*, 1301322. [CrossRef]

16. Zhang, H.; Yang, Y.; Su, Y.; Chen, J.; Adams, K.; Lee, S.; Hu, C.; Wang, Z.L. Triboelectric Nanogenerator for Harvesting Vibration Energy in Full Space and as Self-Powered Acceleration Sensor. *Adv. Funct. Mater.* **2014**, *24*, 1401–1407. [CrossRef]

17. Liu, Y.; Zhu, Y.; Liu, J.; Zhang, Y.; Liu, J.; Zhai, J. Design of Bionic Cochlear Basilar Membrane Acoustic Sensor for Frequency Selectivity Based on Film Triboelectric Nanogenerator. *Nanoscale Res. Lett.* **2018**, *13*, 191. [CrossRef]

18. Yang, J.; Chen, J.; Liu, Y.; Yang, W.; Su, Y.; Wang, Z.L. Triboelectrification-based organic film nanogenerator for acoustic energy harvesting and self-powered active acoustic sensing. *ACS Nano* **2014**, *8*, 2649–2657. [CrossRef]

19. Fan, X.; Chen, J.; Yang, J.; Bai, P.; Li, Z.; Wang, Z.L. Ultrathin, rollable, paper-based triboelectric nanogenerator for acoustic energy harvesting and self-powered sound recording. *ACS Nano* **2015**, *9*, 4236–4243. [CrossRef]

20. Mitcheson, P.D.; Yeatman, E.M.; Rao, G.K.; Holmes, A.S.; Green, T.C. Energy harvesting from human and machine motion for wireless electronic devices. *Proc. IEEE* **2008**, *96*, 1457–1486. [CrossRef]

21. Wang, S.; Xie, Y.; Niu, S.; Lin, L.; Wang, Z.L. Freestanding triboelectric-layer-based nanogenerators for harvesting energy from a moving object or human motion in contact and non-contact modes. *Adv. Mater.* **2014**, *26*, 2818–2824. [CrossRef] [PubMed]

22. Bai, P.; Zhu, G.; Lin, Z.H.; Jing, Q.; Chen, J.; Zhang, G.; Ma, J.; Wang, Z.L. Integrated multilayered triboelectric nanogenerator for harvesting biomechanical energy from human motions. *ACS Nano* **2013**, *7*, 3713–3719. [CrossRef]

23. Chen, J.; Huang, Y.; Zhang, N.; Zou, H.; Liu, R.; Tao, C.; Fan, X.; Wang, Z.L. Micro-cable structured textile for simultaneously harvesting solar and mechanical energy. *Nat. Energy* **2016**, *1*, 16138. [CrossRef]

24. Zhang, N.; Chen, J.; Huang, Y.; Guo, W.; Yang, J.; Du, J.; Fan, X.; Tao, C.; Zhang, N.; Huang, Y.; et al. A Wearable All-Solid Photovoltaic Textile. *Adv. Mater.* **2016**, *28*, 263–269. [CrossRef]

25. Tian, B.; Kempa, T.J.; Lieber, C.M. Single nanowire photovoltaics. *Chem. Soc. Rev.* **2009**, *38*, 16–24. [CrossRef]

26. Niu, X.; Yu, J.; Wang, S. Experimental study on low-temperature waste heat thermoelectric generator. *J. Power Sources* **2009**, *188*, 621–626. [CrossRef]

27. Zi, Y.; Lin, L.; Wang, J.; Wang, S.; Chen, J.; Fan, X.; Yang, P.-K.; Yi, F.; Wang, Z.L. Triboelectric–Pyroelectric–Piezoelectric Hybrid Cell for High-Efficiency Energy-Harvesting and Self-Powered Sensing. *Adv. Mater.* **2015**, *27*, 2340–2347. [CrossRef]

28. Yang, Y.; Zhang, H.; Lin, Z.-H.; Liu, Y.; Chen, J.; Lin, Z.; Zhou, Y.S.; Wong, C.P.; Wang, Z.L. A hybrid energy cell for self-powered water splitting. *Energy Environ. Sci.* **2013**, *6*, 2429–2434. [CrossRef]

29. Wang, X.; Niu, S.; Yin, Y.; Yi, F.; You, Z.; Wang, Z.L. Triboelectric Nanogenerator Based on Fully Enclosed Rolling Spherical Structure for Harvesting Low-Frequency Water Wave Energy. *Adv. Energy Mater.* **2015**, *5*, 1501467. [CrossRef]

30. Liu, L.; Shi, Q.; Lee, C. A novel hybridized blue energy harvester aiming at all-weather IoT applications. *Nano Energy* **2020**, *76*, 105052. [CrossRef]

31. Shao, H.; Cheng, P.; Chen, R.; Xie, L.; Sun, N.; Shen, Q.; Chen, X.; Zhu, Q.; Zhang, Y.; Liu, Y.; et al. Triboelectric–Electromagnetic Hybrid Generator for Harvesting Blue Energy. *Nano-Micro Lett.* **2018**, *10*, 54. [CrossRef] [PubMed]

32. Zhang, L.M.; Han, C.B.; Jiang, T.; Zhou, T.; Li, X.H.; Zhang, C.; Wang, Z.L. Multilayer wavy-structured robust triboelectric nanogenerator for harvesting water wave energy. *Nano Energy* **2016**, *22*, 87–94. [CrossRef]
33. Fan, F.R.; Tian, Z.Q.; Lin Wang, Z. Flexible triboelectric generator. *Nano Energy* **2012**, *1*, 328–334. [CrossRef]
34. Cui, N.; Gu, L.; Lei, Y.; Liu, J.; Qin, Y.; Ma, X.; Hao, Y.; Wang, Z.L. Dynamic Behavior of the Triboelectric Charges and Structural Optimization of the Friction Layer for a Triboelectric Nanogenerator. *ACS Nano* **2016**, *10*, 6131–6138. [CrossRef] [PubMed]
35. Yang, W.; Wang, Y.; Li, Y.; Wang, J.; Cheng, T.; Wang, Z.L. Integrated flywheel and spiral spring triboelectric nanogenerator for improving energy harvesting of intermittent excitations/triggering. *Nano Energy* **2019**, *66*, 104104. [CrossRef]
36. Yun, Y.; Jang, S.; Cho, S.; Lee, S.H.; Hwang, H.J.; Choi, D. Exo-shoe triboelectric nanogenerator: Toward high-performance wearable biomechanical energy harvester. *Nano Energy* **2021**, *80*, 105505. [CrossRef]
37. Lee, Y.; Kim, W.; Bhatia, D.; Hwang, H.J.; Lee, S.; Choi, D. Cam-based sustainable triboelectric nanogenerators with a resolution-free 3D-printed system. *Nano Energy* **2017**, *38*, 326–334. [CrossRef]
38. Menge, H.G.; Huynh, N.D.; Hwang, H.J.; Han, S.; Choi, D.; Park, Y.T. Designable Skin-like Triboelectric Nanogenerators Using Layer-by-Layer Self-Assembled Polymeric Nanocomposites. *ACS Energy Lett.* **2021**, *6*, 2451–2459. [CrossRef]
39. Hwang, H.J.; Kim, J.S.; Kim, W.; Park, H.; Bhatia, D.; Jee, E.; Chung, Y.S.; Kim, D.H.; Choi, D. An Ultra-Mechanosensitive Visco-Poroelastic Polymer Ion Pump for Continuous Self-Powering Kinematic Triboelectric Nanogenerators. *Adv. Energy Mater.* **2019**, *9*, 1803786. [CrossRef]
40. Park, H.W.; Huynh, N.D.; Kim, W.; Hwang, H.J.; Hong, H.; Choi, K.H.; Song, A.; Chung, K.B.; Choi, D. Effects of embedded TiO_{2-x} nanoparticles on triboelectric nanogenerator performance. *Micromachines* **2018**, *9*, 407. [CrossRef]
41. Park, H.-W.; Huynh, N.D.; Kim, W.; Lee, C.; Nam, Y.; Lee, S.; Chung, K.-B.; Choi, D. Electron blocking layer-based interfacial design for highly-enhanced triboelectric nanogenerators. *Nano Energy* **2018**, *50*, 9–15. [CrossRef]
42. Dudem, B.; Huynh, N.D.; Kim, W.; Kim, D.H.; Hwang, H.J.; Choi, D.; Yu, J.S. Nanopillar-array architectured PDMS-based triboelectric nanogenerator integrated with a windmill model for effective wind energy harvesting. *Nano Energy* **2017**, *42*. [CrossRef]
43. Niu, S.; Wang, S.; Lin, L.; Liu, Y.; Zhou, Y.S.; Hu, Y.; Wang, Z.L. Theoretical study of contact-mode triboelectric nanogenerators as an effective power source. *Energy Environ. Sci.* **2013**, *6*, 3576–3583. [CrossRef]
44. Lee, S.; Lee, Y.; Kim, D.; Yang, Y.; Lin, L.; Lin, Z.H.; Hwang, W.; Wang, Z.L. Triboelectric nanogenerator for harvesting pendulum oscillation energy. *Nano Energy* **2013**, *2*, 1113–1120. [CrossRef]
45. Zhang, X.-S.; Han, M.-D.; Wang, R.-X.; Zhu, F.-Y.; Li, Z.-H.; Wang, W.; Zhang, H.-X. Frequency-Multiplication High-Output Triboelectric Nanogenerator for Sustainably Powering Biomedical Microsystems. *Nano Lett.* **2013**, *13*, 1168–1172. [CrossRef] [PubMed]
46. Fan, F.-R.; Lin, L.; Zhu, G.; Wu, W.; Zhang, R.; Wang, Z.L. Transparent Triboelectric Nanogenerators and Self-Powered Pressure Sensors Based on Micropatterned Plastic Films. *Nano Lett.* **2012**, *12*, 3109–3114. [CrossRef]
47. Muthu, M.; Pandey, R.; Wang, X.; Chandrasekhar, A.; Palani, I.A.; Singh, V. Enhancement of triboelectric nanogenerator output performance by laser 3D-Surface pattern method for energy harvesting application. *Nano Energy* **2020**, *78*, 105205. [CrossRef]
48. Chandrasekhar, A.; Vivekananthan, V.; Khandelwal, G.; Kim, W.J.; Kim, S.J. Green energy from working surfaces: A contact electrification–enabled data theft protection and monitoring smart table. *Mater. Today Energy* **2020**, *18*, 100544. [CrossRef]
49. Chandrasekhar, A.; Khandelwal, G.; Rao Alluri, N.; Vivekananthan, V.; Kim, S.-J. Battery-Free Electronic Smart Toys: A Step toward the Commercialization of Sustainable Triboelectric Nanogenerators. *ACS Sustain. Chem. Eng.* **2018**, *6*, 26. [CrossRef]
50. Dudem, B.; Ko, Y.H.; Leem, J.W.; Lee, S.H.; Yu, J.S. Highly Transparent and Flexible Triboelectric Nanogenerators with Subwavelength-Architectured Polydimethylsiloxane by a Nanoporous Anodic Aluminum Oxide Template. *ACS Appl. Mater. Interfaces* **2015**, *7*, 20520–20529. [CrossRef]
51. Wang, S.; Lin, L.; Wang, Z.L. Nanoscale Triboelectric-Effect-Enabled Energy Conversion for Sustainably Powering Portable Electronics. *Nano Lett.* **2012**, *12*, 6339–6346. [CrossRef] [PubMed]
52. Xie, Y.; Wang, S.; Niu, S.; Lin, L.; Jing, Q.; Yang, J.; Wu, Z.; Wang, Z.L. Grating-Structured Freestanding Triboelectric-Layer Nanogenerator for Harvesting Mechanical Energy at 85% Total Conversion Efficiency. *Adv. Mater.* **2014**, *26*, 6599–6607. [CrossRef] [PubMed]
53. Wang, S.; Lin, L.; Wang, Z.L. Triboelectric nanogenerators as self-powered active sensors. *Nano Energy* **2015**, *11*, 436–462. [CrossRef]
54. Chun, J.; Ye, B.U.; Lee, J.W.; Choi, D.; Kang, C.Y.; Kim, S.W.; Wang, Z.L.; Baik, J.M. Boosted output performance of triboelectric nanogenerator via electric double layer effect. *Nat. Commun.* **2016**, *7*, 12985. [CrossRef] [PubMed]
55. Yi, F.; Wang, X.; Niu, S.; Li, S.; Yin, Y.; Dai, K.; Zhang, G.; Lin, L.; Wen, Z.; Guo, H.; et al. A highly shape-adaptive, stretchable design based on conductive liquid for energy harvesting and self-powered biomechanical monitoring. *Sci. Adv.* **2016**, *2*, e1501624. [CrossRef] [PubMed]
56. Tang, Q.; Yeh, M.H.; Liu, G.; Li, S.; Chen, J.; Bai, Y.; Feng, L.; Lai, M.; Ho, K.C.; Guo, H.; et al. Whirligig-inspired triboelectric nanogenerator with ultrahigh specific output as reliable portable instant power supply for personal health monitoring devices. *Nano Energy* **2018**, *47*, 74–80. [CrossRef]
57. Wang, P.; Pan, L.; Wang, J.; Xu, M.; Dai, G.; Zou, H.; Dong, K.; Wang, Z.L. An Ultra-Low-Friction Triboelectric-Electromagnetic Hybrid Nanogenerator for Rotation Energy Harvesting and Self-Powered Wind Speed Sensor. *ACS Nano* **2018**, *12*, 9433–9440. [CrossRef] [PubMed]

58. Wang, S.; Mu, X.; Yang, Y.; Sun, C.; Gu, A.Y.; Wang, Z.L. Flow-Driven Triboelectric Generator for Directly Powering a Wireless Sensor Node. *Adv. Mater.* **2015**, *27*, 240–248. [CrossRef]

59. Park, M.; Cho, S.; Yun, Y.; La, M.; Park, S.J.; Choi, D. A highly sensitive magnetic configuration-based triboelectric nanogenerator for multidirectional vibration energy harvesting and self-powered environmental monitoring. *Int. J. Energy Res.* **2021**, *45*, 18262–18274. [CrossRef]

60. Lin, Z.; Zhang, B.; Zou, H.; Wu, Z.; Guo, H.; Zhang, Y.; Yang, J.; Wang, Z.L. Rationally designed rotation triboelectric nanogenerators with much extended lifetime and durability. *Nano Energy* **2020**, *68*, 104378. [CrossRef]

61. Ahmed, R.; Kim, Y.; Mehmood, M.U.; Shaislamov, U.; Chun, W. Power generation by a thermomagnetic engine by hybrid operation of an electromagnetic generator and a triboelectric nanogenerator. *Int. J. Energy Res.* **2019**, *43*, 5852–5863. [CrossRef]

62. Liu, S.; Li, X.; Wang, Y.; Yang, Y.; Meng, L.; Cheng, T.; Wang, Z.L. Magnetic switch structured triboelectric nanogenerator for continuous and regular harvesting of wind energy. *Nano Energy* **2021**, *83*, 105851. [CrossRef]

63. Han, K.W.; Kim, J.N.; Rajabi-Abhari, A.; Bui, V.T.; Kim, J.S.; Choi, D.; Oh, I.K. Long-Lasting and Steady Triboelectric Energy Harvesting from Low-Frequency Irregular Motions Using Escapement Mechanism. *Adv. Energy Mater.* **2021**, *11*, 2002929. [CrossRef]

64. Tcho, I.W.; Jeon, S.B.; Park, S.J.; Kim, W.G.; Jin, I.K.; Han, J.K.; Kim, D.; Choi, Y.K. Disk-based triboelectric nanogenerator operated by rotational force converted from linear force by a gear system. *Nano Energy* **2018**, *50*, 489–496. [CrossRef]

65. Kim, W.; Bhatia, D.; Jeong, S.; Choi, D. Mechanical energy conversion systems for triboelectric nanogenerators: Kinematic and vibrational designs. *Nano Energy* **2019**, *56*, 307–321. [CrossRef]

66. Kim, W.; Hwang, H.J.; Bhatia, D.; Lee, Y.; Baik, J.M.; Choi, D. Kinematic design for high performance triboelectric nanogenerators with enhanced working frequency. *Nano Energy* **2016**, *21*, 19–25. [CrossRef]

67. Kim, H.; Hwang, H.J.; Huynh, N.D.; Pham, K.D.; Choi, K.; Ahn, D.; Choi, D. Magnetic Force Enhanced Sustainability and Power of Cam-Based Triboelectric Nanogenerator. *Research* **2021**, *2021*, 6426130. [CrossRef] [PubMed]

68. Pham, K.D.; Bhatia, D.; Huynh, N.D.; Kim, H.; Baik, J.M.; Lin, Z.H.; Choi, D. Automatically switchable mechanical frequency regulator for continuous mechanical energy harvesting via a triboelectric nanogenerator. *Nano Energy* **2021**, *89*, 106350. [CrossRef]

69. Bhatia, D.; Lee, J.; Hwang, H.J.; Baik, J.M.; Kim, S.; Choi, D. Design of Mechanical Frequency Regulator for Predictable Uniform Power from Triboelectric Nanogenerators. *Adv. Energy Mater.* **2018**, *8*, 1702667. [CrossRef]

70. Jiang, D.; Ouyang, H.; Shi, B.; Zou, Y.; Tan, P.; Qu, X.; Chao, S.; Xi, Y.; Zhao, C.; Fan, Y.; et al. A wearable noncontact free-rotating hybrid nanogenerator for self-powered electronics. *InfoMat* **2020**, *2*, 1191–1200. [CrossRef]

71. Huynh, N.D.; Lin, Z.H.; Choi, D. Dynamic balanced hybridization of TENG and EMG via Tesla turbine for effectively harvesting broadband mechanical pressure. *Nano Energy* **2021**, *85*, 105983. [CrossRef]

72. Yong, H.; Chung, J.; Choi, D.; Jung, D.; Cho, M.; Lee, S. Highly reliable wind-rolling triboelectric nanogenerator operating in a wide wind speed range. *Sci. Rep.* **2016**, *6*, 33977. [CrossRef]

73. Wang, Y.; Wang, J.; Xiao, X.; Wang, S.; Kien, P.T.; Dong, J.; Mi, J.; Pan, X.; Wang, H.; Xu, M. Multi-functional wind barrier based on triboelectric nanogenerator for power generation, self-powered wind speed sensing and highly efficient windshield. *Nano Energy* **2020**, *73*, 104736. [CrossRef]

74. Jiang, T.; Chen, X.; Han, C.B.; Tang, W.; Wang, Z.L. Theoretical Study of Rotary Freestanding Triboelectric Nanogenerators. *Adv. Funct. Mater.* **2015**, *25*, 2928–2938. [CrossRef]

75. Zhang, C.; Tang, W.; Han, C.; Fan, F.; Wang, Z.L. Theoretical comparison, equivalent transformation, and conjunction operations of electromagnetic induction generator and triboelectric nanogenerator for harvesting mechanical energy. *Adv. Mater.* **2014**, *26*, 3580–3591. [CrossRef]

76. Lin, L.; Wang, S.; Niu, S.; Liu, C.; Xie, Y.; Wang, Z.L. Noncontact free-rotating disk triboelectric nanogenerator as a sustainable energy harvester and self-powered mechanical sensor. *ACS Appl. Mater. Interfaces* **2014**, *6*, 3031–3038. [CrossRef]

77. Guo, T.; Liu, G.; Pang, Y.; Wu, B.; Xi, F.; Zhao, J.; Bu, T.; Fu, X.; Li, X.; Zhang, C.; et al. Compressible hexagonal-structured triboelectric nanogenerators for harvesting tire rotation energy. *Extrem. Mech. Lett.* **2018**, *18*, 1–8. [CrossRef]

78. Cao, R.; Zhou, T.; Wang, B.; Yin, Y.; Yuan, Z.; Li, C.; Wang, Z.L. Rotating-Sleeve Triboelectric-Electromagnetic Hybrid Nanogenerator for High Efficiency of Harvesting Mechanical Energy. *ACS Nano* **2017**, *11*, 8370–8378. [CrossRef]

79. Roh, H.; Yu, J.; Kim, I.; Chae, Y.; Kim, D. Dynamic Analysis to Enhance the Performance of a Rotating-Disk-Based Triboelectric Nanogenerator by Injected Gas. *ACS Appl. Mater. Interfaces* **2019**, *11*, 25170–25178. [CrossRef] [PubMed]

80. Zhong, X.; Yang, Y.; Wang, X.; Wang, Z.L. Rotating-disk-based hybridized electromagnetic-triboelectric nanogenerator for scavenging biomechanical energy as a mobile power source. *Nano Energy* **2015**, *13*, 771–780. [CrossRef]

81. Ren, X.; Fan, H.; Wang, C.; Ma, J.; Li, H.; Zhang, M.; Lei, S.; Wang, W. Wind energy harvester based on coaxial rotatory freestanding triboelectric nanogenerators for self-powered water splitting. *Nano Energy* **2018**, *50*, 562–570. [CrossRef]

82. Lin, L.; Wang, S.; Xie, Y.; Jing, Q.; Niu, S.; Hu, Y.; Wang, Z.L. Segmentally structured disk triboelectric nanogenerator for harvesting rotational mechanical energy. *Nano Lett.* **2013**, *13*, 2916–2923. [CrossRef] [PubMed]

83. He, J.; Cao, S.; Zhang, H. Cylinder-based hybrid rotary nanogenerator for harvesting rotational energy from axles and self-powered tire pressure monitoring. *Energy Sci. Eng.* **2020**, *8*, 291–299. [CrossRef]

84. Yang, Y.; Yu, X.; Meng, L.; Li, X.; Xu, Y.; Cheng, T.; Liu, S.; Wang, Z.L. Triboelectric nanogenerator with double rocker structure design for ultra-low-frequency wave full-stroke energy harvesting. *Extrem. Mech. Lett.* **2021**, *46*, 101338. [CrossRef]

85. Yin, M.; Lu, X.; Qiao, G.; Xu, Y.; Wang, Y.; Cheng, T.; Wang, Z.L. Mechanical Regulation Triboelectric Nanogenerator with Controllable Output Performance for Random Energy Harvesting. *Adv. Energy Mater.* **2020**, *10*, 2000627. [CrossRef]
86. Bhatia, D.; Jo, S.H.; Ryu, Y.; Kim, Y.; Kim, D.H.; Park, H.S. Wearable triboelectric nanogenerator based exercise system for upper limb rehabilitation post neurological injuries. *Nano Energy* **2021**, *80*, 105508. [CrossRef]

 *nanoenergy advances*

Article

Ultrathin Stretchable All-Fiber Electronic Skin for Highly Sensitive Self-Powered Human Motion Monitoring

Yapeng Shi [1,2], Tianyi Ding [3], Zhihao Yuan [1], Ruonan Li [1], Baocheng Wang [1,2] and Zhiyi Wu [1,2,4,*]

[1] Beijing Institute of Nanoenergy and Nanosystems, Chinese Academy of Sciences, Beijing 101400, China; shiyapeng@binn.cas.cn (Y.S.); 1907301106@st.gxu.edu.cn (Z.Y.); 1504280108@st.gxu.edu.cn (R.L.); wangbaocheng@binn.cas.cn (B.W.)
[2] School of Nanoscience and Technology, University of Chinese Academy of Sciences, Beijing 100049, China
[3] State Key Laboratory of Physical Chemistry of Solid Surfaces, iChEM, College of Chemistry and Chemical Engineering, Xiamen University, Xiamen 361005, China; tyding@stu.xmu.edu.cn
[4] CUSTech Institute of Technology, Wenzhou 325024, China
* Correspondence: wuzhiyi@binn.cas.cn

Abstract: Advances in the technology of wearable electronic devices have necessitated much research to meet their requirements, such as stretchability, sustainability, and maintenance-free functioning. In this study, we developed an ultrathin all-fiber triboelectric nanogenerator (TENG)-based electronic skin (TE-skin) with high stretchability, using electrospinning and spraying, whereby the silver nanowire (Ag NW) electrode layer is deposited between two electrospinning thermoplastic polyurethane (TPU) fibrous layers. Due to its extraordinary stretchability and prominent Ag NW conductive networks, the TE-skin exhibits a high sensitivity of 0.1539 kPa^{-1} in terms of pressure, superior mechanical property with a low-resistance electrode of 257.3 Ω at a strain of 150%, great deformation recovery ability, and exceptional working stability with no obvious fluctuation in electrical output before and after stretching. Based on the outstanding performances of the TE-skin, an intelligent electronic glove was fabricated to detect multifarious hand gestures. Moreover, the TE-skin has the potential to record human motion for real-time physiological signal monitoring, which provides promising applications in the fields of flexible robots, human-machine interaction, and multidimensional sports monitoring in next-generation electronics.

Keywords: stretchable electronic skin; triboelectric nanogenerator; self-powered sensing; human motion monitoring; thermoplastic polyurethane fibers

Citation: Shi, Y.; Ding, T.; Yuan, Z.; Li, R.; Wang, B.; Wu, Z. Ultrathin Stretchable All-Fiber Electronic Skin for Highly Sensitive Self-Powered Human Motion Monitoring. *Nanoenergy Adv.* **2022**, 2, 52–63. https://doi.org/10.3390/nanoenergyadv2010003

Academic Editors: Yogendra Kumar Mishra, Ya Yang and Zhong Lin Wang

Received: 29 November 2021
Accepted: 28 January 2022
Published: 30 January 2022
Corrected: 13 July 2022

Publisher's Note: MDPI stays neutral with regard to jurisdictional claims in published maps and institutional affiliations.

Copyright: © 2022 by the authors. Licensee MDPI, Basel, Switzerland. This article is an open access article distributed under the terms and conditions of the Creative Commons Attribution (CC BY) license (https://creativecommons.org/licenses/by/4.0/).

1. Introduction

With the current rapid progress in the fields of the Internet of Things [1] and artificial intelligence [2,3], various kinds of wearable and portable electronics have recently attracted intensive research, with different potential applications in human healthcare monitoring [4–6], smart robots [7,8], human motion sensing [9,10] and human–machine interaction [11–13], etc. Pang et al. [14] realized the real-time monitoring of wound temperature by an integrated sensor to release antibiotics from a hydrogel, with on-demand infection treatment by in situ UV irradiation. Li et al. [15] developed a closed-loop system that is fully integrated with a microneedle platform and wearable electronics, which could concurrently monitor and treat diabetes in situ. These devices were generally powered by a sustainable and reliable power supply; however, this power comes from sources such as batteries with the disadvantages of a limited lifespan, high recharging costs and increasingly serious environmental problems. Meanwhile, a great number of wearable electronic devices without integrating batteries have been reported in the same period. Liu et al. [16] proposed a simple and low-cost method to fabricate skin-like electronics, using a material composed of ternary piezoelectric rubber composite, including polydimethylsiloxane

(PDMS), graphene, and lead zirconate titanate (PZT). Due to the customized device geometries and the optimized integration of rubbery electronics with the human body, the developed device could be successfully mounted on human joints to realize excellent output and working stability during a sequence of large deformations, demonstrating the application of skin-integrated electronics for energy harvesting and mechanical sensing. Kim et al. [17] introduced a novel skin-like sensor that could detect the minute signals from small skin deformations through unique laser-induced crack structures. By integrating with a deep neural network, a deep-learned skin-like sensor system was developed to obtain data from different areas of the wrist; the system was also applied to the pelvis to generate dynamic gait motions in real-time, realizing the measurement of human motion remotely. Although battery-free electronic devices have shown great progress in a variety of cutting-edge applications, such as advanced human–machine interfaces for clinical diagnostics [18] and evolvable skin electronics with active accommodation for human physiological measurements and skin condition tracking [19], it is still urgently necessary and highly desirable to develop sustainable, environmentally friendly, and maintenance-free sensing technology.

In recent years, electronic skin, as a significant form of multifunctional wearable electronics, has dramatically facilitated modern life. Among different electromechanical sensing principles, such as capacitance [20,21], piezoresistivity [22,23], piezoelectricity [24,25], and triboelectricity [26,27], triboelectric nanogenerator (TENG)-based electronic skins (TE-skins) as automatic sensors offer various advantages, including pollution-free manufacturing technology, low cost, and diverse material selection. Based on contact electrification and the electrostatic induction coupling effect, TE-skins have been widely applied as self-powered sensors to realize real-time tactile sensing [27,28], sports sensing [29], respiratory sensing [30,31], motion monitoring [32,33], etc. Generally, the developed TE-skins need to be elastic and flexible so that they could be attached to human skin and adapt to all sorts of deformations. Recently, a variety of stretchable TENGs were fabricated, based on thermoplastic polyurethane (TPU), demonstrating great potential for energy harvesting and self-powered sensing [27–29,34]. Unfortunately, it is still a challenge to construct such devices with high stretchability. Jeong et al. [35] prepared a hyper-stretchable and elastic-composite generator (SEG) with extraordinary output performance, strain capacity, and mechanical stability; this is composed of rubber-based piezoelectric elastic composite (PEC) and very long nanowire percolation (VLNP) electrodes. Yu et al. [36] proposed a triboelectric–piezoelectric hybrid nanogenerator (TPHNG) using piezoelectric-polyvinylidene fluoride (PVDF), polydimethylsiloxane (PDMS), and a network of welded silver nanowires (AgNWs), achieving mechanical energy harvesting and physiological signal monitoring. Although a series of stretchable energy devices have been developed [37], the multidimensional conductive networks in these devices can easily be damaged in the stretching process. Therefore, it is crucial to effectively reinforce the adhesion of the conductive networks on the substrate.

Considering the above-mentioned issues, a highly stretchable TENG-based all-fiber electronic skin was fabricated using the industrial manufacturing technologies of electrospinning and spraying. Initially, silver nanowire (Ag NW) is distributed on the surface of thermoplastic polyurethane (TPU) fibers, forming a layer-by-layer network structure by spraying. Subsequently, the Ag NW networks are tightly wrapped by double-layered TPU fibers, effectively enhancing the adhesion of the Ag NW conductive networks on the TPU substrate. The resulting TE-skin, with high stretchability, has excellent mechanical properties, with a low-resistance electrode of 257.3 Ω at a strain of 150%, and shows extraordinary deformation recovery ability, working stability and durability. Due to a sensitivity of 0.1539 kPa^{-1} in terms of pressure, an intelligent electronic glove was constructed to detect different kinds of gestures, including the "ok" sign, the "victory" sign, and the "good" sign. More specifically, the TE-skin was further applied as a wearable self-powered sensor in sports to achieve real-time motion monitoring of different parts of the human body. Consequently, the outstanding performance of the developed TE-skin, including substantial stretchability, light weight, large-scale production feasibility, and high sensitivity promises future applications in human–machine interaction, multidimensional sports monitoring, and flexible robots.

2. Experimental Section

2.1. Electrospinning of TPU Fibers

First, N,N-dimethylformamide (DMF, $\geq$99.8%, ACS reagent) and tetrahydrofuran (THF, $\geq$99.5%, ACS reagent) were mixed at a volume ratio of 1:1. After stirring for 10 min, the TPU pellets (Shanghai BASF Polyurethane Specialties Co., Ltd., Shanghai, China) were added to the resulting solution at a concentration of 20 wt%, followed by continuous stirring for 12 h, until it is completely dissolved to form a homogeneous electrospinning solution. In the electrospinning process, the needle-collector distance and the electrospinning voltage of the electrospinning device (DP30, Tianjin Yunfan Technology Co., Ltd., Tianjin, China) were 10 cm and 9 kV, respectively. The manufactured TPU substrate was dried overnight in a vacuum environment at 50 °C for further use.

2.2. Fabrication of the TE-Skin

After electrospinning of TPU fibrous substrate, a spray gun (PS289, 0.3 mm, Mr. HOBBY, Shanghai Henghui Model Co., Ltd., Shanghai, China) connected with a portable air pump (601G, USTAR, Shanghai Henghui Model Co., Ltd., Shanghai, China) was used to implement the spraying of the Ag NW electrode. Specifically, the Ag NW solution (2 mg mL^{-1}) was spray-coated onto the TPU substrate under a discharge speed of 80 μL/min, and the distance between the mouth of the spray gun and the TPU substrate was set as 10 cm. After the spraying of the Ag NW electrode, the TPU substrate, coated with Ag NW conductive networks, was wrapped by another TPU sensing layer that was fabricated by electrospinning under the same experimental conditions as the TPU substrate.

2.3. Characterization and Measurement

The micromorphologies of TPU fibers and Ag NW were observed using field emission scanning electron microscopy (ZEISS Gemini 300, Carl Zeiss optics (Guangzhou) Co., Ltd., Guangzhou, China) and a transmission electron microscope (JEM-2100plus, JEOL (BEIJING) Co., Ltd., Beijing, China), respectively. A tensile testing machine (ESM303, Shenzhen Chenyi Technology Co., Ltd., Shenzhen, China) was utilized to evaluate the mechanical properties of the TPU and the TE-skin. The hydrophobicity of the TPU film was characterized using a contact angle meter (XG-CAM, Shanghai Xuanyichuangxi Industrial Equipment Co., Ltd., Shanghai, China). The resistances of the stretchable conductive electrodes were measured by a flexible electronic tester (Prtronic FT2000, Shanghai Mifang Electronic Technology Co., Ltd., Shanghai, China). Through the periodic contact and separation movements provided by a commercial linear mechanical motor (Linmot E1100, Shenzhen Nuoxide Trading Co., Ltd., Shenzhen, China), the electrical signals of the TE-skin were acquired using a programmable electrometer (Keithley model 6514, Guangzhou Meidake Data Technology Co., Ltd., Guangzhou, China).

3. Results and Discussion

3.1. Fabrication and Structural Design of the TE-Skin

As illustrated in Figure 1a, the TE-skin was manufactured using electrospinning and spraying technique; details of the fabrication process are described in the Experimental Section. Considering that the properties of the triboelectric layer are crucial to the sensing capability of the TE-skin, a commercial TPU that is widely used for clothing fabric, with good wearability, biocompatibility, and nontoxicity was selected for the top encapsulation layer and the bottom sensing layer [38]. The TE-skin forms a skin contact interface and is composed of three functional layers, with a fiber network and three-dimensional layered porous structure in which the Ag NW layer (Figure 1c) is sandwiched between two TPU fiber layers (Figure 1b). After optimization of the construction parameters, it is worth mentioning that the thickness and the weight of the whole device are only 0.12 mm and

0.0415 g, respectively (3 × 3 cm^2, Figure 1d and Supplementary Figure S1), showing a relatively light and thin all-fiber structure that can cope with various kinds of complex mechanical deformations, such as stretching, twisting, and bending, indicating its excellent mechanical properties (Figure 1e). Due to the TE-skin's all-fiber structure with superior breathing ability, there was no discomfort, including itching and inflammation or diseases of the skin, even after the TE-skin had adhered to the arm for 12 h (Supplementary Figure S2). Above all, TPU fibers with hydrophobicity (contact angle, 125° (Supplementary Figure S3)) can guarantee the outstanding waterproof performance of the TE-skin. Therefore, the TE-skin, as a human-friendly skin-interfaced biosensor, can be attached to different parts of the human body to continuously monitor real-time signals during sports activity.

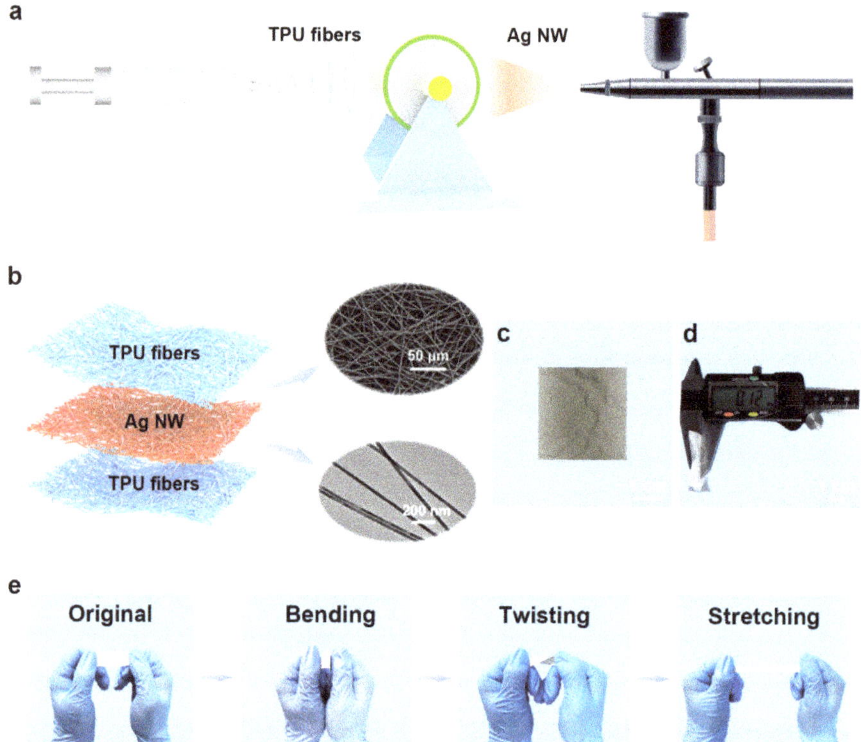

Figure 1. Fabrication and characterization of the TE-skin. (**a**) Schematic diagram of the experimental process for fabricating the TE-skin. (**b**) The sandwich structure of the TE-skin. (**c**) Photograph of the Ag NW electrode layer that is deposited on the TPU fiber substrate. (**d**) Photograph of the TE-skin with a total thickness of 120 µm. (**e**) Photographs of the TE-skin in the original, bent, twisted, and stretched states.

3.2. Characterization and Working Mechanism of the TE-Skin

By using an electrospinning technique, the two TPU layers present a typical stacked-fiber structure (Figure 2a,c), with a corresponding average fiber diameter of 1.9 µm (Supplementary Figure S4). In addition, the Ag NW electrode layer is uniformly distributed over the surface of the TPU fibers by spraying (Figure 2b), which process constructs excellent conductive networks. The energy-dispersive spectrometer (EDS) mapping images (Supplementary Figure S5) further confirm this pivotal result. In addition, the TE-skin has numerous micro/nanopores that are formed by the multilayer interconnected compact fiber networks, ensuring good breathability. To explain the working mechanism of the TE-skin in detail, polytetrafluoroethylene (PTFE) film was adopted in order to have a contact-separation process with the TE-skin in single-electrode mode (Figure 2d), wherein the PTFE is electronegative in comparison with the TPU, due to the diverse triboelectric polarities. Once the TPU is in contact with the PTFE, electrification will occur at their interface, and an equal number of charges with opposite polarities will be generated on the surface of the TPU and PTFE. Due to its ability to attract more electrons, the PTFE will be negatively charged, whereas the TPU will be positively charged in this period (Figure 2d, ii). As the PTFE begins to separate from the TPU, a potential difference is formed. Owing to the electrostatic induction effect, the negative charges will be transferred from the ground to the Ag NW conductive networks (Figure 2d, iii). The electron flow continues until the separation between the PTFE and the TPU is complete; the positive and negative charges are neutralized during this process (Figure 2d, iv). Notably, the accumulated charges will be retained for a sufficiently long time due to the inherent characteristic of the insulator, rather than being lost immediately. In contrast, if the PTFE is close to the TPU, the electrons will flow back from the Ag NW conductive networks to the ground through the external load, compensating for the electrical potential differences (Figure 2d, v) until the whole system reverts to its original state. As a result, an alternating potential and current are produced in the contact–separation process occurring between the PTFE and the TE-skin. To assess the electricity-generating process quantitatively, the COMSOL was utilized to simulate the potential distribution result of the PTFE and the TE-skin in different contact and separation processes. Figure 2e shows that the potential difference will grow moderately from 0 to 80 V with an increase in the separation distance between the PTFE and the TE-skin from 0 to 5 mm.

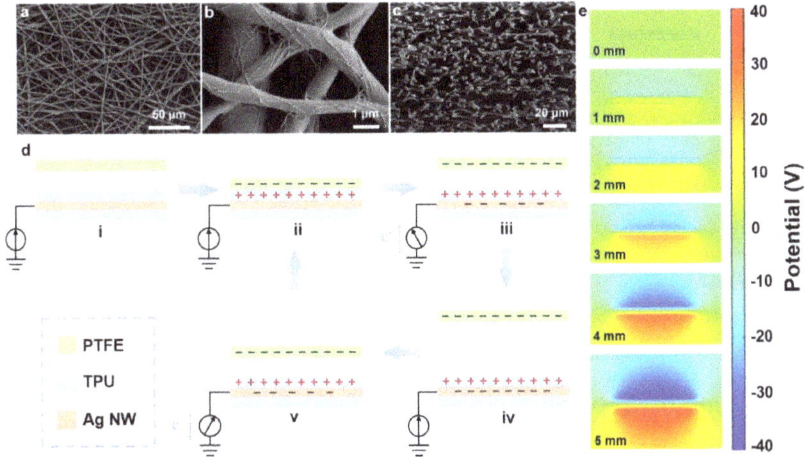

Figure 2. The micromorphologies and working mechanism of the TE-skin. (**a**) The SEM image of the TPU layer. (**b**) The SEM image of the Ag NW that is deposited on the TPU fibers. (**c**) The cross-section SEM image of the TE-skin. (**d**) The working mechanism of the TE-skin upon contact with skin. (**e**) The potential simulation by COMSOL to elucidate the working principle of the TE-skin.

3.3. Electrical Output Performance of the TE-Skin

To quantitatively evaluate the electrical output performance, a mechanical linear motor was used to drive the TE-skin at different frequencies (1–5 Hz). It turned out that the open-circuit voltage (V_{OC}) and short-circuit transferred charge (Q_{SC}) remained almost steady (108 V and 35 nC, respectively), while the short-circuit current (I_{SC}) increased gradually (from 0.3 to 3.5 µA) as the frequencies increased from 1 to 5 Hz (Figure 3a–c, at a fixed load of 1 N). Since V_{OC} and Q_{SC} are independent of speed, the increase in movement frequency does not affect them, whereas the I_{SC} is determined by the relative movement speed, and so presented the characteristics of gradual growth with an increase in speed. Moreover, the power output performance of the TE-skin was systematically investigated by loading different external resistances at an applied frequency of 1 Hz. As depicted in Figure 3d, the output current decreased dramatically, while the output voltage showed the opposite trend as the resistance increased from 10 kΩ to 1000 GΩ. The output power density (P) of the TE-skin was further calculated using the following formula:

$$p = \frac{U^2}{RA} \tag{1}$$

where R represents the loading resistance; U is the output voltage; A stands for the contact area of the TE-skin. The power density achieved a peak of 28.8 mW m^{-2} with an external load resistance of 0.1 GΩ (Figure 3e), which is greater than that shown in some previous works (Supplementary Table S1). To demonstrate the working stability and durability of the TE-skin, the V_{OC} is shown in Figure 3f (the fixed load is 1 N, with applied frequencies of 1 Hz). Accordingly, there was no obvious reduction in V_{OC} after long-term cycles for 10 h, revealing the impressive working stability and durability of the TE-skin. In addition, the charging ability of the TE-skin with various types of commercial capacitors was also analyzed and compared. The results showed that the voltage value will decrease gradually and the charging speed will become slower at the same time as the capacitances of the capacitors increase (Figure 3g). After being rectified, this delivered output performance could power some small electronics like LEDs (Figure 3h), indicating great potential regarding energy supply for wearable electronics.

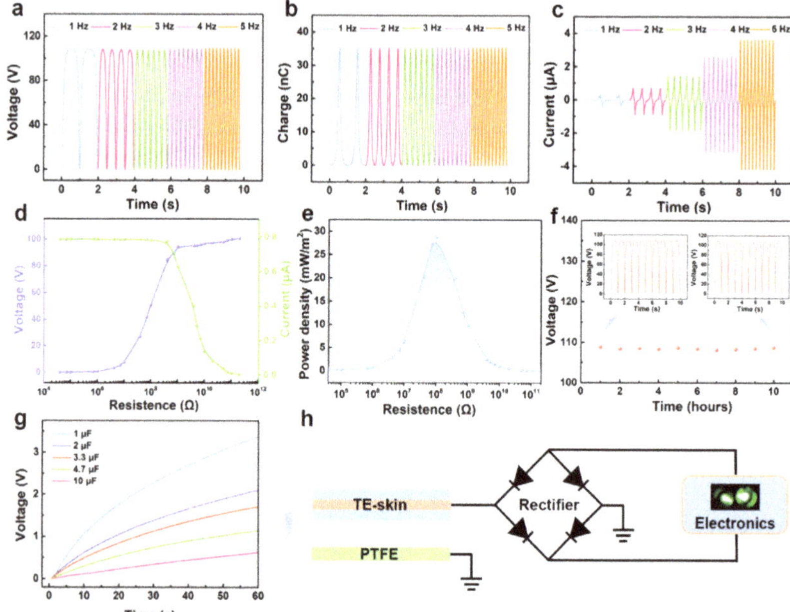

Figure 3. The electrical output performance of the TE-skin. (**a–c**) The electrical outputs of the TE-skin under the frequencies from 1 Hz to 5 Hz. (**d,e**) The output voltage, current, and power density of the TE-skin when loading different external resistances from 10 kΩ to 1000 GΩ. (**f**) The long-term cyclic test of the TE-skin for 10 h. The insets demonstrated in left and right are the corresponding output voltages after testing for 1 h and 10 h. (**g**) The charging performance of the TE-skin under different capacitors from 1 μF to 10 μF. (**h**) Schematic diagram of powering LEDs with a rectifier circuit by the contact–separation process between the TE-skin and the PTFE.

The pressure sensitivity of the TE-skin (S) as a function of the loading force was also explored, which is calculated as follows:

$$S = \frac{d(\Delta V/V_s)}{dP_A} \tag{2}$$

where ΔV and V_s are the relative change of voltage and the saturated voltage, respectively, and P_A represents the fixed pressure. As diagrammed in Figure 4b, the relationship between normalized voltage and the applied pressure is a nearly linear feature within the pressure range of 3.5 kPa (Figure 4b), and the resulting sensitivity regarding pressure (0.1539 kPa^{-1}) is higher than those of other TENG-based pressure sensors worked in single-electrode mode [39–41]. Due to the high biocompatibility, skin compatibility, and ultrathin structure of the TE-skin, a piece of pigskin (3 cm × 3 cm) was placed on the linear mechanical motor to achieve a contact–separation process with the TE-skin, further validating the electrical performance characteristics of the TE-skin (Figure 4c). Considering that the human motion frequency range is 1.1–3.8 Hz [42], the applied frequencies were set at 1.5, 2, 2.5, 3, and 3.5 Hz, respectively. Interestingly, the V_{OC} and Q_{SC} were relatively stable, while the I_{SC} gradually increased with the acceleration in frequency (Figure 4d–f), which was consistent with the previous results shown in Figure 4a–c. It is undeniable that the TE-skin has the potential to be driven by human skin to work as a self-powered sensor for human motion monitoring. Given that the human body will sweat during sport, the V_{OC} of the TE-skin in different relative humidity (RH) conditions was also evaluated. Accordingly, although the V_{OC} decreased gradually from 9.28 to 1.18 V with the increase in RH from 30% to 80%, the electrical output at 80% RH was sufficient to ensure that the TE-skin can work well for human motion-sensing in such a high humidity environment (Supplementary Figure S6).

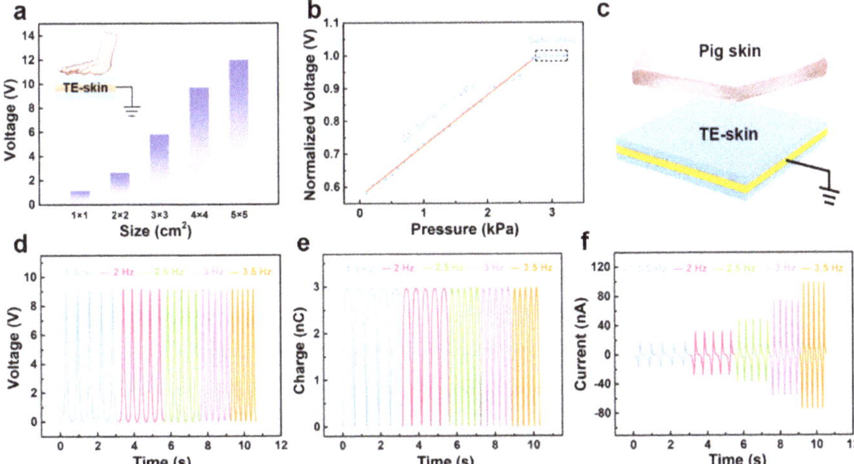

Figure 4. The electrical outputs of the skin-interfaced TE-skin. (**a**) The V_{OC} of the TE-skin from hand slapping, with the respective increased size. (**b**) The relationship between the corresponding normalized voltage and a range of applied pressures (0–3.5 kPa). (**c**) Schematic illustration of the experimental setup for the skin-interfaced electrical output test. (**d–f**) Skin-interfaced electrical outputs of the TE-skin under frequencies from 1.5 Hz to 3.5 Hz.

3.4. Stretchability Characterization

As a wearable self-powered sensor for human motion monitoring, the TE-skin needs to generate electrical signals under a progression of extreme deformation conditions. Figure 5a shows that the TE-skin was capable of being stretched to a strain of 500% in comparison with the original length, presenting outstanding mechanical properties. Compared with the failure strain and the ultimate stress of pure TPU film (546%, 1.73 MPa), the TE-skin exhibited enhanced failure strain and ultimate stress, which were 1019% and 2.64 MPa, respectively (Figure 5b). It is worth explicitly noting that the improvement in the tensile strength and toughness of the TE-skin can be attributed to the interface adhesion of the three layers in the process of electrospinning and spraying. More significantly, the cyclic stress-strain curves of the TE-skin under different strains were also investigated to further verify its excellent mechanical properties. Whether it was stretched by 30%, 50%, or 100%, the TE-skin always retained great deformation recovery ability after 5 cycles (Figure 5c and Supplementary Figure S7), which is conducive to the sensing capacity of the TE-skin needed for human motion monitoring. Figure 5d revealed the resistance of the Ag NW electrode under different strains. Although the resistance of the Ag NW electrode grew moderately as the TE-skin was uniaxially stretched, the TE-skin still possessed a great capacity for electron transportation, due to the reserved Ag NW conductive networks during the stretching process. Notably, the V_{OC} of the TE-skin was equivalent to the initial sample after the strain was completely released (Figure 5e), showing extraordinary stretchability and the possibility for numerous applications in the field of flexible and wearable electronics. The V_{OC} of the TE-skin under different strains was tested further. The V_{OC} decreased gradually with the increase in strain from 0% to 100% (Figure 5f), which resulted from the extension of the electron transfer path in Ag NW conductive networks. However, there was still a certain electrical output even at a strain of 100%, and the V_{OC} under each strain was relatively stable throughout 50 cycles. Undoubtedly, the differences of V_{OC} under different strains could be applied in the field of deformation sensing, which further verified the promising prospect of the TE-skin in human motion monitoring.

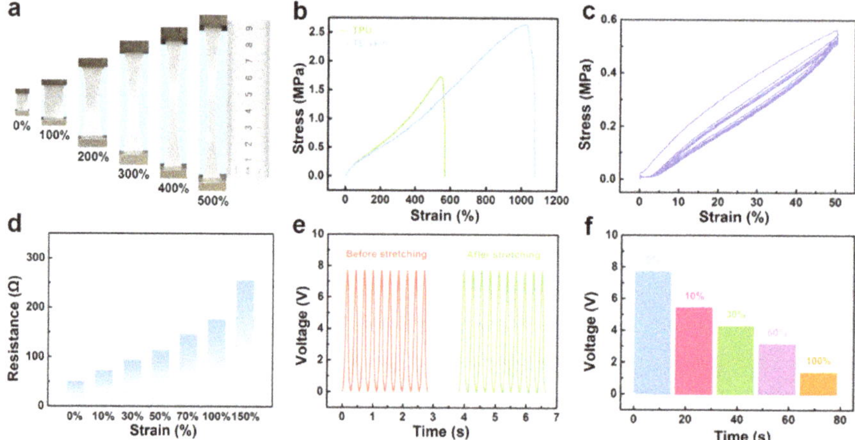

Figure 5. Stretchability characterization of the TE-skin. (**a**) Photographs of the TE-skin, stretched at different strains. (**b**) The strain-stress of the TPU film and the TE-skin. (**c**) The successive loading–unloading stress-strain of the TE-skin, stretched by a strain of 50% for 5 cycles. (**d**) The resistance of the Ag NW electrode layer under different strains. (**e**) The V_{OC} of the TE-skin before and after being stretched by a strain of 50%. (**f**) The V_{OC} of the TE-skin under different strains for 50 cycles (3.5 Hz and 1 N).

3.5. Applications of the TE-Skin in Human Motion Monitoring

A range of experiments was implemented to better investigate the practical applications of the TE-skin in wearable electronics technologies. As shown in Figure 6a, an intelligent electronic glove was developed to monitor the different motions of human hands, in which five TE-skins were attached to the five digits. Furthermore, a data acquisition system integrated with five channels was customized, based on this intelligent electronic glove, for detecting the electrical signals generated by different hand gestures. When a hand gesture was made, the output signal of the bending finger in the corresponding channel could be acquired significantly, whereas the output signals of the other fingers in the rest of the channels were maintained in the initial state. Supplementary Figure S8 illustrates that the intelligent electronic glove has the potential to record various gestures, including the "ok" sign, the "victory" sign, and the "good" sign. The corresponding detailed electrical signals in each channel of the mentioned gestures are shown in the bottom half of Supplementary Figure S8. Through machine learning and self-training, the recognition of more complicated gestures might be achievable based on the tested TE-skin, with promising applications in the field of multifunctional prostheses or flexible robots in the future. In addition, sports monitoring is a complex multi-dimensional analytical process that needs a process of quantification and the analysis of relevant physiological indicators. Due to the preeminent stretchability and sensing performance of the material under test, the TE-skin can also be used to detect and record human motion for physiological signal monitoring, which is a significant feature in wearable devices. Figure 6b demonstrates that the TE-skin was attached to the elbow (i), finger (ii), wrist (iii), and leg (iv) to realize the real-time monitoring of different human body parts. Different pressures obtained from different parts of the human body will reflect the individual voltage signals of these TE-skins during exercise. Undoubtedly, the coordinated operation of each TE-skin will provide a leap forward in terms of improvement in the performance of the whole human motion monitoring system.

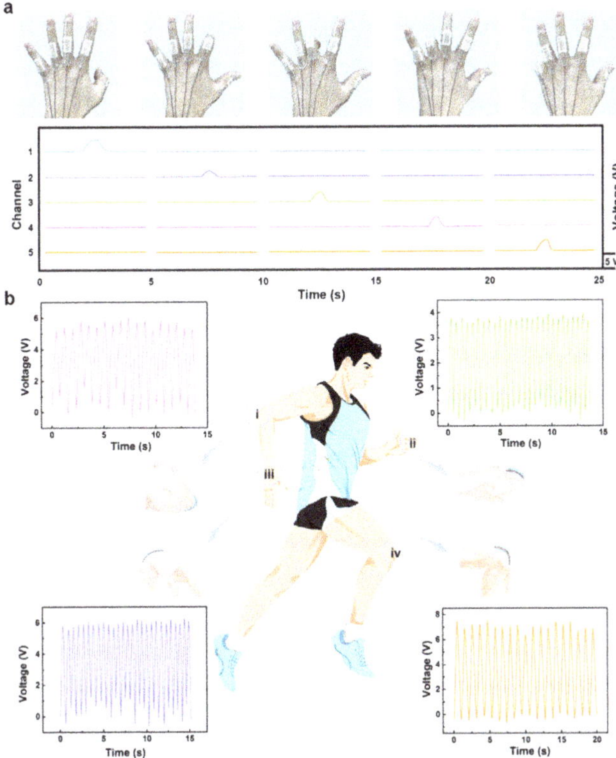

Figure 6. The application of the TE-skin regarding human motion monitoring. (**a**) The corresponding voltage signals of the five TE-skins that are fixed on 5 digits. Insets are the corresponding hand gestures. (**b**) The corresponding voltage signals of bending elbow (i), finger (ii), wrist (iii), and leg (iv) in the different states of motion.

4. Conclusions

In summary, an ultrathin stretchable all-fiber TE-skin was created via an electrospinning and spraying technique and is presented in this work. Owing to the Ag NW conductive networks designed in a layer-by-layer structure, the adhesion between the Ag NW and TPU layer is reinforced, and the high stretchability of the TE-skin is achieved with a low-resistance electrode of 257.3 Ω at a strain of 150%. In addition, the TE-skin possesses a high pressure sensitivity of 0.1539 kPa^{-1} and outstanding deformation recovery ability. Based on the excellent sensing performances, an intelligent electronic glove has been developed to detect several human gestures, exhibiting its potential in artificial intelligence applications. Moreover, several TE-skins were utilized to achieve the real-time monitoring of different human body parts during sports. It is envisaged that the fabricated TE-skin presented in this paper can be utilized in the field of next-generation electronics for the applications of artificial intelligence, flexible robots, and multidimensional sports monitoring.

Supplementary Materials: The following supporting information can be downloaded at: https://www.mdpi.com/article/10.3390/nanoenergyadv2010003/s1, Figure S1: Photograph of the TE-skin with a total weight of 41.5 mg (3×3 cm^2). Figure S2: The biocompatibility test of the TE-skin. Figure S3: The contact angle of the TPU film. Figure S4: The diameter distribution of TPU fibers. Figure S5: Energy dispersive spectrometer (EDS) analysis of Ag NW deposited on the TPU fibers. (a) SEM image. (b) element overlay. (c) Ag element mapping. (d) C element mapping. (e) N element mapping. (f) O element mapping. Figure S6: The open-circuit voltage of the TE-skin under different relative humidity. Figure S7: The successive loading-unloading stress-strain of the TE-skin stretched to the strains of (a) 30% and (b) 100% for 5 cycles. Figure S8: Different hand gestures, including (i) the "ok" sign, (ii) the "victory" sign, and (iii) the "good" sign, with the corresponding electrical signals. Table S1. Comparison of the power output with the findings of previous works. Reference [43] is cited in the Supplementary Materials.

Author Contributions: Y.S., T.D. and Z.W. conceived the project, designed the research, and prepared the manuscript. Y.S. fabricated the electronic skin and performed all the experiments. Z.Y., R.L. and B.W. performed the electrical experiment. All authors participated in the analysis and discussions of the results and agreed to publish the manuscript in the current version. All authors have read and agreed to the published version of the manuscript.

Funding: This work was financially supported by the National Natural Science Foundation of China (61503051).

Data Availability Statement: Data are contained within the article and Supplementary Materials.

Conflicts of Interest: The authors declare no conflict of interest.

References

1. Yuan, M.; Li, C.; Liu, H.; Xu, Q.; Xie, Y. A 3D-printed acoustic triboelectric nanogenerator for quarter-wavelength acoustic energy harvesting and self-powered edge sensing. *Nano Energy* **2021**, 105962. [CrossRef]
2. Liao, X.; Song, W.; Zhang, X.; Yan, C.; Li, T.; Ren, H.; Liu, C.; Wang, Y.; Zheng, Y. A bioinspired analogous nerve towards artificial intelligence. *Nat. Commun.* **2020**, *11*, 268. [CrossRef] [PubMed]
3. Araromi, O.A.; Graule, M.A.; Dorsey, K.L.; Castellanos, S.; Foster, J.R.; Hsu, W.-H.; Passy, A.E.; Vlassak, J.J.; Weaver, J.C.; Walsh, C.J.; et al. Ultra-sensitive and resilient compliant strain gauges for soft machines. *Nature* **2020**, *587*, 219–224. [CrossRef] [PubMed]
4. Wu, M.; Yao, K.; Li, D.; Huang, X.; Liu, Y.; Wang, L.; Song, E.; Yu, J.; Yu, X. Self-powered skin electronics for energy harvesting and healthcare monitoring. *Mater. Today Energy* **2021**, *21*, 100786. [CrossRef]
5. Wu, H.; Yang, G.; Zhu, K.; Liu, S.; Guo, W.; Jiang, Z.; Li, Z. Materials, devices, and systems of on-skin electrodes for electrophysiological monitoring and human-machine interfaces. *Adv. Sci.* **2021**, *8*, 2001938. [CrossRef]
6. Zhang, S.; Li, S.; Xia, Z.; Cai, K. A review of electronic skin: Soft electronics and sensors for human health. *J. Mater. Chem. B* **2020**, *8*, 852–862. [CrossRef]
7. Liu, Y.-Q.; Chen, Z.-D.; Han, D.-D.; Mao, J.-W.; Ma, J.-N.; Zhang, Y.-L.; Sun, H.-B. Bioinspired soft robots based on the moisture-responsive graphene oxide. *Adv. Sci.* **2021**, *8*, 2002464. [CrossRef]
8. Fu, C.; Xia, Z.; Hurren, C.; Nilghaz, A.; Wang, X. Textiles in soft robots: Current progress and future trends. *Biosens. Bioelectron.* **2022**, *196*, 113690. [CrossRef]

9. Ning, C.; Dong, K.; Cheng, R.; Yi, J.; Ye, C.; Peng, X.; Sheng, F.; Jiang, Y.; Wang, Z.L. Flexible and stretchable fiber-shaped triboelectric nanogenerators for biomechanical monitoring and human-interactive sensing. *Adv. Funct. Mater.* **2021**, *31*, 2006679. [CrossRef]
10. He, M.; Du, W.; Feng, Y.; Li, S.; Wang, W.; Zhang, X.; Yu, A.; Wan, L.; Zhai, J. Flexible and stretchable triboelectric nanogenerator fabric for biomechanical energy harvesting and self-powered dual-mode human motion monitoring. *Nano Energy* **2021**, *86*, 106058. [CrossRef]
11. Sim, K.; Rao, Z.; Zou, Z.; Ershad, F.; Lei, J.; Thukral, A.; Chen, J.; Huang, Q.-A.; Xiao, J.; Yu, C. Metal oxide semiconductor nanomembrane-based soft unnoticeable multifunctional electronics for wearable human-machine interfaces. *Sci. Adv.* **2019**, *5*, eaav9653. [CrossRef] [PubMed]
12. Dong, B.; Yang, Y.; Shi, Q.; Xu, S.; Sun, Z.; Zhu, S.; Zhang, Z.; Kwong, D.-L.; Zhou, G.; Ang, K.-W. Wearable triboelectric-human-machine interface (THMI) using robust nanophotonic readout. *ACS Nano* **2020**, *14*, 8915–8930. [CrossRef] [PubMed]
13. Luo, Y.; Li, Y.; Sharma, P.; Shou, W.; Wu, K.; Foshey, M.; Li, B.; Palacios, T.; Torralba, A.; Matusik, W. Learning human-environment interactions using conformal tactile textiles. *Nat. Electron.* **2021**, *4*, 193–201. [CrossRef]
14. Pang, Q.; Lou, D.; Li, S.; Wang, G.; Qiao, B.; Dong, S.; Ma, L.; Gao, C.; Wu, Z. Smart flexible electronics-integrated wound dressing for real-time monitoring and on-demand treatment of infected wounds. *Adv. Sci.* **2020**, *7*, 1902673. [CrossRef]
15. Li, X.; Huang, X.; Mo, J.; Wang, H.; Huang, Q.; Yang, C.; Zhang, T.; Chen, H.-J.; Hang, T.; Liu, F.; et al. A fully integrated closed-loop system based on mesoporous microneedles-iontophoresis for diabetes treatment. *Adv. Sci.* **2021**, *8*, 2100827. [CrossRef] [PubMed]
16. Liu, Y.; Zhao, L.; Wang, L.; Zheng, H.; Li, D.; Avila, R.; Lai, K.W.C.; Wang, Z.; Xie, Z.; Zi, Y.; et al. Skin-Integrated graphene-embedded lead zirconate titanate rubber for energy harvesting and mechanical sensing. *Adv. Mater. Technol.* **2019**, 1900744. [CrossRef]
17. Kim, K.K.; Ha, I.; Kim, M.; Choi, J.; Won, P.; Jo, S.; Ko, S.H. A deep-learned skin sensor decoding the epicentral human motions. *Nat. Commun.* **2020**, *11*, 2149. [CrossRef]
18. Kim, K.K.; Choi, J.; Ko, S.H. Energy harvesting untethered soft electronic devices. *Adv. Healthcare Mater.* **2021**, *10*, 2002286. [CrossRef]
19. Kim, K.K.; Choi, J.; Kim, J.-H.; Nam, S.; Ko, S.H. Evolvable skin electronics by in situ and in operando adaptation. *Adv. Funct. Mater.* **2021**, *32*, 2106329. [CrossRef]
20. Lee, S.; Franklin, S.; Hassani, F.A.; Yokota, T.; Nayeem, M.O.G.; Wang, Y.; Leib, R.; Cheng, G.; Franklin, D.W.; Someya, T. Nanomesh pressure sensor for monitoring finger manipulation without sensory interference. *Science* **2020**, *370*, 966–970. [CrossRef]
21. Núñez, C.G.; Navaraj, W.T.; Polat, E.O.; Dahiya, R. Energy-autonomous, flexible, and transparent tactile skin. *Adv. Funct. Mater.* **2017**, *27*, 1606287. [CrossRef]
22. Bae, G.Y.; Pak, S.W.; Kim, D.; Lee, G.; Kim, D.H.; Chung, Y.; Cho, K. Linearly and highly pressure-sensitive electronic skin based on a bioinspired hierarchical structural array. *Adv. Mater.* **2016**, *28*, 5300–5306. [CrossRef] [PubMed]
23. Yang, T.; Deng, W.; Chu, X.; Wang, X.; Hu, Y.; Fan, X.; Song, J.; Gao, Y.; Zhang, B.; Tian, G.; et al. Hierarchically microstructure-bioinspired flexible piezoresistive bioelectronics. *ACS Nano* **2021**, *15*, 11555–11563. [CrossRef] [PubMed]
24. He, H.; Fu, Y.; Zang, W.; Wang, Q.; Xing, L.; Zhang, Y.; Xue, X. A flexible self-powered T-ZnO/PVDF/fabric electronic-skin with multi-functions of tactile-perception, atmosphere-detection and self-clean. *Nano Energy* **2017**, *31*, 37–48. [CrossRef]
25. Yuan, X.; Gao, X.; Shen, X.; Yang, J.; Li, Z.; Dong, S. A 3D-printed, alternatively tilt-polarized PVDF-TrFE polymer with enhanced piezoelectric effect for self-powered sensor application. *Nano Energy* **2021**, *85*, 105985. [CrossRef]
26. Lan, L.; Yin, T.; Jiang, C.; Li, X.; Yao, Y.; Wang, Z.; Qu, S.; Ye, Z.; Ping, J.; Ying, Y. Highly conductive 1D-2D composite film for skin-mountable strain sensor and stretchable triboelectric nanogenerator. *Nano Energy* **2019**, *62*, 319–328. [CrossRef]
27. Zhou, K.; Zhao, Y.; Sun, X.; Yuan, Z.; Zheng, G.; Dai, K.; Mi, L.; Pan, C.; Liu, C.; Shen, C. Ultra-stretchable triboelectric nanogenerator as high-sensitive and self-powered electronic skins for energy harvesting and tactile sensing. *Nano Energy* **2020**, *70*, 104546. [CrossRef]
28. Jiang, Y.; Dong, K.; Li, X.; An, J.; Wu, D.; Peng, X.; Yi, J.; Ning, C.; Cheng, R.; Yu, P.; et al. Stretchable, washable, and ultrathin triboelectric nanogenerators as skin-like highly sensitive self-powered haptic sensors. *Adv. Funct. Mater.* **2021**, *31*, 2005584. [CrossRef]
29. Shi, Y.; Wei, X.; Wang, K.; He, D.; Yuan, Z.; Xu, J.; Wu, Z.; Wang, Z.L. Integrated all-fiber electronic skin toward self-powered sensing sports systems. *ACS Appl. Mater. Interfaces* **2021**, *13*, 50329–50337. [CrossRef]
30. Peng, X.; Dong, K.; Ning, C.; Cheng, R.; Yi, J.; Zhang, Y.; Sheng, F.; Wu, Z.; Wang, Z.L. All-nanofiber self-powered skin-interfaced real-time respiratory monitoring system for obstructive sleep apnea-hypopnea syndrome diagnosing. *Adv. Funct. Mater.* **2021**, *31*, 2103559. [CrossRef]
31. Zhu, M.; Lou, M.; Yu, J.; Li, Z.; Ding, B. Energy autonomous hybrid electronic skin with multi-modal sensing capabilities. *Nano Energy* **2020**, *78*, 105208. [CrossRef]
32. Jiang, Y.; Dong, K.; An, J.; Liang, F.; Yi, J.; Peng, X.; Ning, C.; Ye, C.; Wang, Z.L. UV-protective, self-cleaning, and antibacterial nanofiber-based triboelectric nanogenerators for self-powered human motion monitoring. *ACS Appl. Mater. Interfaces* **2021**, *13*, 11205–11214. [CrossRef] [PubMed]
33. Ganesh, R.S.; Yoon, H.-J.; Kim, S.-W. Recent trends of biocompatible triboelectric nanogenerators toward self-powered e-skin. *EcoMat* **2020**, *2*, e12065. [CrossRef]

34. Zhang, W.; Liu, Q.; Chao, S.; Liu, R.; Cui, X.; Sun, Y.; Ouyang, H.; Li, Z. Ultrathin stretchable triboelectric nanogenerators improved by postcharging electrode material. *ACS Appl. Mater. Interfaces* **2021**, *13*, 42966–42976. [CrossRef]

35. Jeong, C.K.; Lee, J.; Han, S.; Ryu, J.; Hwang, G.-T.; Park, D.Y.; Park, J.H.; Lee, S.S.; Byun, M.; Ko, S.H.; et al. A hyper-stretchable elastic-composite energy harvester. *Adv. Mater.* **2015**, *27*, 2866–2875. [CrossRef]

36. Yu, X.; Liang, X.; Krishnamoorthy, R.; Jiang, W.; Zhang, L.; Ma, L.; Zhu, P.; Hu, Y.; Sun, R.; Wong, C.-P. Transparent and flexible hybrid nanogenerator with welded silver nanowire networks as the electrodes for mechanical energy harvesting and physiological signal monitoring. *Smart Mater. Struct.* **2020**, *29*, 045040. [CrossRef]

37. Jung, J.; Cho, H.; Yuksel, R.; Kim, D.; Lee, H.; Kwon, J.; Lee, P.; Yeo, J.; Hong, S.; Unalan, H.E.; et al. Stretchable/flexible silver nanowire electrodes for energy device applications. *Nanoscale* **2019**, *11*, 20356–20378. [CrossRef]

38. Tan, C.; Dong, Z.; Li, Y.; Zhao, H.; Huang, X.; Zhou, Z.; Jiang, J.-W.; Long, Y.-Z.; Jiang, P.; Zhang, T.-Y.; et al. A high performance wearable strain sensor with advanced thermal management for motion monitoring. *Nat. Commun.* **2020**, *11*, 3530. [CrossRef]

39. Gogurla, N.; Kim, S. Self-powered and imperceptible electronic tattoos based on silk protein nanofiber and carbon nanotubes for human–machine interfaces. *Adv. Energy Mater.* **2021**, *11*, 2100801. [CrossRef]

40. Wang, X.; Zhang, H.; Dong, L.; Han, X.; Du, W.; Zhai, J.; Pan, C.; Wang, Z.L. Self-powered high-resolution and pressure-sensitive triboelectric sensor matrix for real-time tactile mapping. *Adv. Mater.* **2016**, *28*, 2896–2903. [CrossRef]

41. Peng, X.; Dong, K.; Ye, C.; Jiang, Y.; Zhai, S.; Cheng, R.; Liu, D.; Gao, X.; Wang, J.; Wang, Z.L. A breathable, biodegradable, antibacterial, and self-powered electronic skin based on all-nanofiber triboelectric nanogenerators. *Sci. Adv.* **2020**, *6*, eaba9624. [CrossRef] [PubMed]

42. Li, R.; Wei, X.; Xu, J.; Chen, J.; Li, B.; Wu, Z.; Wang, Z.L. Smart wearable sensors based on triboelectric nanogenerator for personal healthcare monitoring. *Micromachines* **2021**, *12*, 352. [CrossRef] [PubMed]

43. Du, W.; Nie, J.; Ren, Z.; Jiang, T.; Xu, L.; Dong, S.; Zheng, L.; Chen, X.; Li, H. Inflammation-free and gas-permeable on-skin triboelectric nanogenerator using soluble nanofibers. *Nano Energy* **2018**, *51*, 260–269. [CrossRef]

 *nanoenergy advances*

Review

Recent Advances on Hybrid Piezo-Triboelectric Bio-Nanogenerators: Materials, Architectures and Circuitry

Massimo Mariello

Laboratory for Processing of Advanced Composites (LPAC), École Polytechnique Fédérale de Lausanne (EPFL), 1015 Lausanne, Switzerland; massimo.mariello@epfl.ch or massimomariello@gmail.com

Abstract: Nanogenerators, based on piezoelectric or triboelectric materials, have emerged in the recent years as an attractive cost-effective technology for harvesting energy from renewable and clean energy sources, but also for human sensing and biomedical wearable/implantable applications. Advances in materials engineering have enlightened new opportunities for the creation and use of novel biocompatible soft materials as well as micro/nano-structured or chemically-functionalized interfaces. Hybridization is a key concept that can be used to enhance the performances of the single devices, by coupling more transducing mechanisms in a single-integrated micro-system. It has attracted plenty of research interest due to the promising effects of signal enhancement and simultaneous adaptability to different operating conditions. This review covers and classifies the main types of hybridization of piezo-triboelectric bio-nanogenerators and it also provides an overview of the most recent advances in terms of material synthesis, engineering applications, power-management circuits and technical issues for the development of reliable implantable devices. State-of-the-art applications in the fields of energy harvesting, in vitro/in vivo biomedical sensing, implantable bioelectronics are outlined and presented. The applicative perspectives and challenges are finally discussed, with the aim to suggest improvements in the design and implementation of next-generation hybrid bio-nanogenerators and biosensors.

Keywords: nanogenerators; piezoelectricity; triboelectricity; biosensors; hybridization

Citation: Mariello, M. Recent Advances on Hybrid Piezo-Triboelectric Bio-Nanogenerators: Materials, Architectures and Circuitry. *Nanoenergy Adv.* **2022**, *2*, 64–109. https://doi.org/10.3390/nanoenergyadv2010004

Academic Editors: Ya Yang and Zhong Lin Wang

Received: 13 December 2021
Accepted: 31 January 2022
Published: 10 February 2022

Publisher's Note: MDPI stays neutral with regard to jurisdictional claims in published maps and institutional affiliations.

Copyright: © 2022 by the author. Licensee MDPI, Basel, Switzerland. This article is an open access article distributed under the terms and conditions of the Creative Commons Attribution (CC BY) license (https://creativecommons.org/licenses/by/4.0/).

1. Introduction

Nanogenerators have emerged in the last years as a novel useful technology for several applications, i.e., for scavenging energy from the environments, for sensing human motion, for monitoring physiological parameters, for supplying self-powered sensor systems, for providing energy to miniaturized storage devices. The word nanogenerator refer to whichever micro-device that is able to convert energy from the available sources (mechanical, thermal, biochemical, etc.) into electrical energy, at the micro/nano scale [1]. These types of devices can act as power generators, sensors or actuators and represent a revolutionary alternative to standard bulky systems for energy generation or sensing [2–5].

The major issue addressed by the employment of nanogenerators is the global energy problem. A today's grand challenge is to exploit more massively natural resources of renewable and clean energy in order to reduce and avoid in the end the use of polluting and expensive fossil fuels [1,2,6–8]. Solar energy is the most widely present form of natural energy and research on solar/photovoltaic cells based on new materials and architecture (e.g., perovskites [9]) has received a remarkable boost in the recent years. Other energy sources include wind flows, ocean waves/currents, rainfalls, environmental vibrations, human body motions (voluntary movements, heartbeat, breathing, blood circulation, etc.).

These are more convenient to be exploited in places where solar energy is not really accessible because of unstable sunshine and weather conditions [10–12]: mechanical energy is ubiquitous, abundant and manifold [13] and scavenging this energy through distributed networks of nanogenerators is a promising solution to pursue in parallel to standard current power plants [4,6,14,15].

The main purpose of harvesting energy is to overcome issues related to chemical batteries (limited lifetime, need of periodic recharging, environmental impact, unsuitability for positioning in unusual and hostile places [16–20]), thus to supply storage devices or to satisfy the increasing demand of highly efficient self-powered (battery-less) energy or sensing systems. These are often flexible, lightweight and wearable/implantable smart sensors based on nanogenerators, and they find application in diverse fields: real-time monitoring of life parameters [21–23], sensing of human motion [24–26], sports and athletics [27–29], smart grids, robotics [30–32] and prosthetics [33,34], biomedical active implants [22,35,36]. In the context of the Internet of Things (IoT) [37,38] and Internet of Bodies (IoB) [39], intelligent devices need to be adaptable to different operating conditions and places, even in remote locations, thus they require to be supplied with energy harvested locally [13]. Hence, nanogenerators are growingly based on cost-friendly, downscaling and widely-implementable flexible electronics with softness, conformability, biocompatibility and specific designs/architectures. Research and industry efforts have been moving towards the development of novel materials [14,40,41] and new environmental or biomedical applications [21–29,42–44].

The most widely employed nanogenerators are based on piezoelectricity and triboelectricity [4,40,45]. These two transducing mechanisms rely on charge generation due to mechanical deformations or contact friction and their practical realization is compatible with the required properties of advanced smart devices, such as flexibility, adaptability, multifunctionality, conformability and biocompatibility [4,46,47]. Their intrinsic capability of generating charges as well as their light mass and structural simplicity, make them advantageous with respect to other energy harvesting devices (e.g., electromagnetic, electrostatic, etc.). In terms of sensing properties, they are low-cost and affordable alternatives to wearable systems based on strain gauges [48], capacitive devices [49] or iontronics [50,51].

Additionally, their hybrid coupling has recently shown to obtain enhanced performances in terms of output power or sensitivity, beyond the linear summation of the single components [52]. The hybridization also allows to adapt the overall device to different operating conditions, without any constraints imposed by the single working mechanisms. Hybrid systems relying on the combination of piezoelectric and triboelectric devices, based on biocompatible materials and aimed for clean energy harvesting from the environment or for biomedical applications are hereafter called hybrid bio-nanogenerators (HBNGs). Figure 1 illustrates a qualitative comparison between HBNG, piezoelectric nanogenerators (PNGs) and triboelectric nanogenerators (TNGs), in terms of flexibility, output voltage, output current, output power density, lifetime and reliability, ease of miniaturization, low-frequency operation, high-frequency operation, biocompatibility. The most usual fields of applications for HBNGs are also displayed, i.e., energy harvesting, wearable bioelectronics and implantable bioelectronics, with some specific examples: tactile sensors, electronic skins, biosensors, biomedical implants, actuation, stimulation, robotics, human-machine interfaces, self-powered sensors and renewable energy sources.

In this review the main types of hybridization of piezo-triboelectric bio-nanogenerators are described and classified with a general overview of the most recent examples of HBNGs for energy harvesting, wearable or implantable bioelectronics. Section 2 provides a description of the working principles of piezoelectric and triboelectric nanogenerators, in terms of basic governing equations and most used materials. The different kinds of hybridization of the two mechanisms are presented in Section 3, where a classification is established according to the physical domains involved in the hybridization process, as well as the relative position of the piezoelectric and triboelectric components. State-of-the-art applications in the fields of energy harvesting, in vitro/in vivo biomedical sensing, implantable bioelectronics are outlined and presented in Section 4. A specific focus is reserved in Section 5 for the electronic interfaces of these devices and the power management circuits. In order to design and fabricate properly these circuitry, different issues have to be addressed, including the attachment of electrical connections which must be suitable for the employed materials, and the analytical and equivalent-circuit models of each type of transducer. Finally, the applicative perspectives and future challenges are discussed, with the aim to suggest improvements in the design and implementation of next-generation hybrid bio-nanogenerators and biosensors. Hybrid devices able to sense, monitor, interact, recognize, learn or stimulate, represent the building blocks for all the low-cost and accessible technologies of the current fast-evolving IoT revolution.

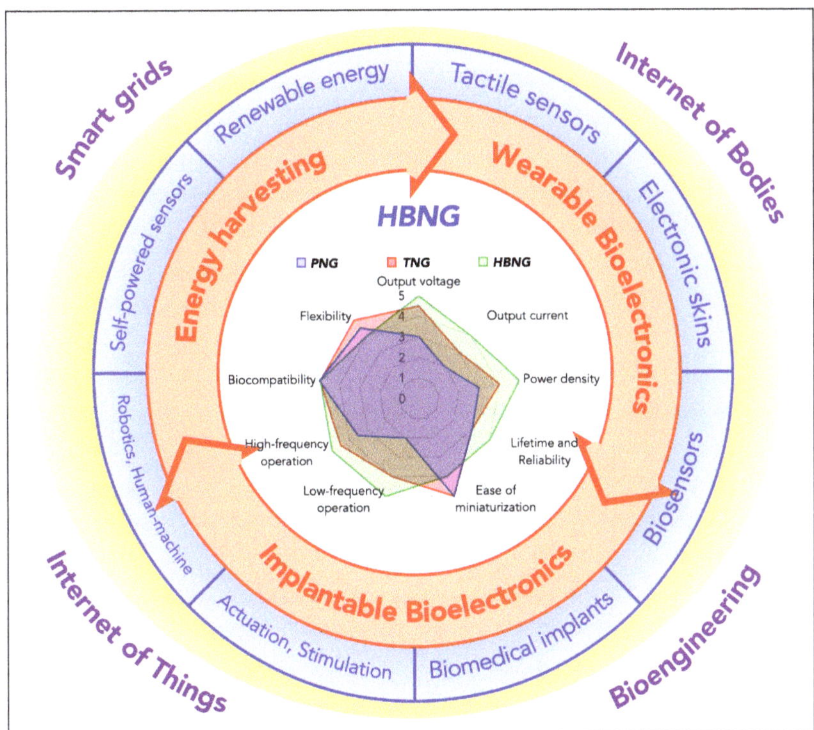

Figure 1. Illustrations of the potential applications of hybrid bio-nanogenerators (HBNGs). The radar chart in the center shows qualitatively the properties of piezoelectric nanogenerators (PNGs), tribo-electric nanogenerators (TNGs) and hybrid bio-nanogenerators (HBNGs): flexibility, output voltage, output current, output power density, lifetime and reliability, ease of miniaturization, low-frequency operation, high-frequency operation, biocompatibility. The surrounding scheme shows the most usual fields of applications for HBNGs, i.e., energy harvesting, wearable bioelectronics and implantable bioelectronics, with some specific examples: tactile sensors, electronic skins, biosensors, biomedical implants, actuation, stimulation, robotics, human-machine interfaces, self-powered sensors and renewable energy sources. The words outside the circle indicate the main technological contexts for the aforementioned applications: smart grids and cities, Internet of Things (IoT), Internet of Bodies (IoB) and Bioengineering.

2. Piezo-Triboelectric Transducing Mechanisms

In this section the governing equations of the analytical model for piezoelectric and triboelectric nanogenerators are outlined because they can be used as useful guidelines for the study and analysis of HBNGs. The main features of the materials usually adopted for exploiting piezoelectricity and triboelectricity are also briefly described.

2.1. Piezoelectric Materials and Mechanisms

Piezoelectric materials directly convert applied strain energy into electrical energy and exhibit high output power densities [46,53,54]: they are generally used in the form of nanostructured materials [55–57] or as thin films in layered structures [58–62]. The common feature is the non-centrosymmetric microstructure [63] which is responsible of the intrinsic capability of converting even ultra-weak mechanical stimuli into electrical energy, without external powering [64] (see Figure 2a). There are several choices among piezoelectric materials: they include piezo-ceramics or polymers [63,64] and they can be deposited on soft substrates through low-cost techniques for the fabrication of flexible and bendable piezoelectric nanogenerators and sensors [47]. In addition to lead zirconate titanate ($PbZr_{1-x}Ti_xO_3$, PZT) and its derivatives, which are highly performing but also less environmentally friendly, other valuable ceramic lead-free alternatives are barium titanate ($BaTiO_3$) [65], potassium sodium niobite (KNN) [66], lithium niobite ($LiNbO_3$) [67], zinc oxide (ZnO) [58,68] or aluminum nitride (AlN) [69,70]. AlN [5,71–74] has emerged as a promising material in the form of transparent μm-thick films deposited on flexible substrates [13,69,75–77] and with high-temperature and humidity resistance [78], mechanical and chemical stability [79,80], and biocompatibility [47,59]. $LiNbO_3$ is another cheap and easily accessible epitaxially grown piezoelectric material exploited for acoustic and optical devices [81]. It is chemically inert, heat-resistant, with lower dielectric constant than other piezo-ceramics [82], and its processability as thin films has been investigated and demonstrated [67]. Poly(vinylidene fluoride) (PVDF) and its co-polymers [83,84] are widely used for flexible polymeric biocompatible piezoelectric devices. Although it has a high dissipation factor, high dielectric constants, poor heat resistance, damped and worse performances than inorganic piezo-materials [85], it is very lightweight, soft and suitable for being employed in fluid flows or for wearable/implantable patches [84,86,87].

A piezoelectric nanogenerator (PNG) basically consists of a piezoelectric element sandwiched between two electrodes, which is subjected to a given mechanical load, depending on the specific application, thus generating electric power (Figure 2a(I,II)). In order to have a unified model of a PNG, in the context of mechanical vibrations and harmonic excitations, it may be considered as an oscillator vibrating along some directions, in the simplest case only one (therefore with a single degree of freedom, sDOF (Figure 2b(I)). This model can be used for rigid beams and as an approximated model for flexible oscillating beams. The dynamic behaviour of a piezoelectric beam/cantilever is generally described by the equations of mechanical motion:

$$\frac{\partial^2 \mathcal{M}(s,t)}{\partial s^2} + c_s \mathcal{I} \frac{\partial^5 w(s,t)}{\partial s^4 \partial t} + c_a \frac{\partial w(s,t)}{\partial t} + m \frac{\partial^2 w(s,t)}{\partial t^2} = -m \frac{\partial^2 w_b(s,t)}{\partial t^2} \tag{1}$$

where m is the equivalent mass per unit length; s is the longitudinal spatial coordinate; t is time; $\mathcal{M}(s,t)$ is the internal bending moment; c_s, c_a are the strain rate damping coefficient and viscous air damping coefficient, respectively; $\mathcal{I}$ is the moment of inertia of the cross section; $w(s,t)$ is the relative displacement, related to the base displacement $w_b(s,t)$ by the expression $\mathcal{W}(s,t) = w(s,t) + w_b(s,t)$, with $\mathcal{W}(s,t)$ the absolute displacement.

To explicit (1) for piezoelectric materials, the e-form piezoelectric constitutive equations may be applied:

$$\begin{cases} T_i = \sum_{j=1}^6 c_{ij}^E S_j - \sum_{\kappa=1}^3 e_{\kappa i} E_\kappa, & i = 1, \ldots, 6 \\ D_i = \sum_{j=1}^6 e_{ij} S_j + \sum_{\kappa=1}^3 \epsilon_{\kappa\kappa}^S E_\kappa, & i = 1, \ldots, 3 \end{cases} \tag{2}$$

where T_i, S_i are the stress and strain components along direction "i"; c_{ij}^E are the elastic stiffness matrix components at constant electric field; D_i, E_i are the electric displacement and electric field along direction "i"; e_{ij} are the piezoelectric coefficients; ϵ_{ii}^S are the electric permittivities at constant strain.

Using a simplified notation, the piezoelectric polarization charge density ρ_P and the electric potential can be deduced from the following expressions:

$$\rho_P = d_P T, \ \nabla E = -\frac{\rho_P}{\epsilon} \tag{3}$$

where d_P is the piezoelectric coefficient, T is the applied stress and ∇E is the divergence of the electric field. Equations (2) and (3) hold true for any type of piezoelectric material: although most of them are dielectrics, piezoelectric semiconductors have been also synthesised and studied such as ZnO or GaN fibers, nanotubes, belts and films. In this case, the basic behaviour of these materials with semiconductive properties relies on the equations of linear piezoelectricity and the conservation of charge for electron and holes, electrostatic equations, current-density equations and continuity equations [88–94], which are developed in the theories of piezotronics and piezo-phototronics [95–99].

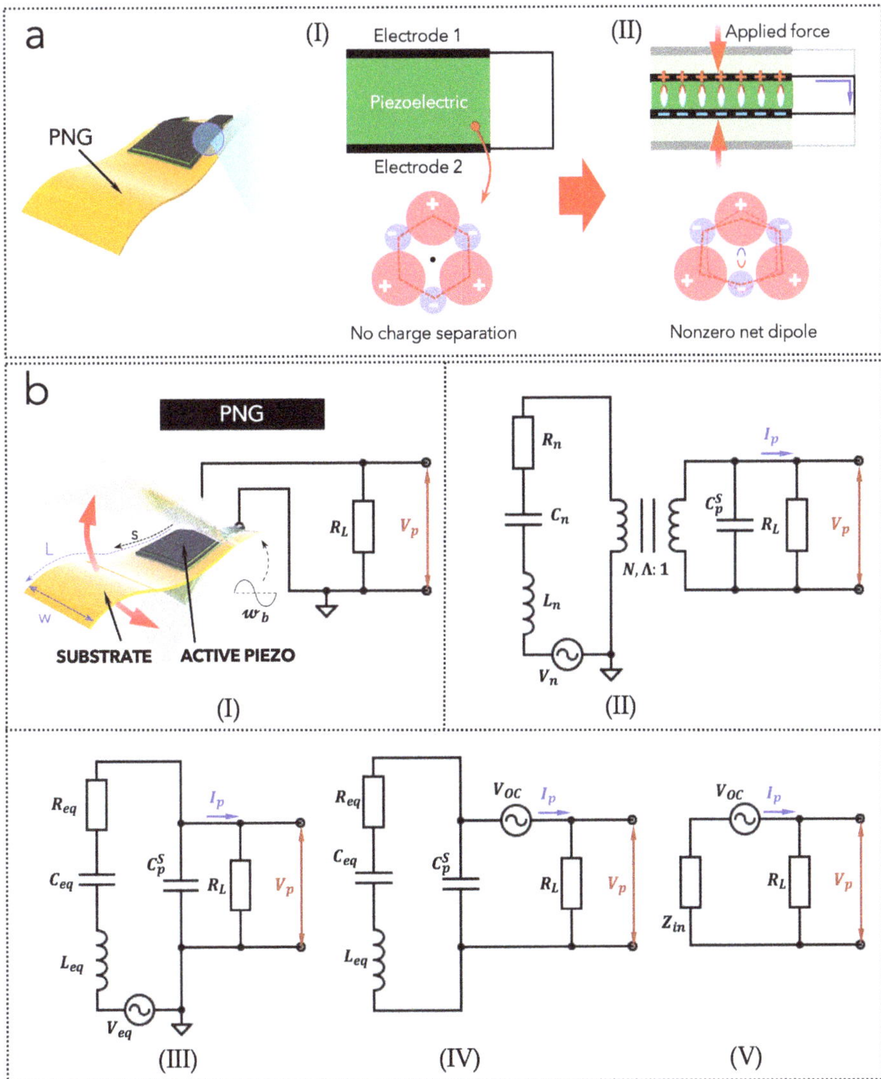

Figure 2. (**a**) Piezoelectric mechanism: (**I**) unloaded state and no charge separation; (**II**) loaded state with an applied force or strain and a non-zero net electrical dipole. (**b**(**I–V**)) Equivalent circuit model of a PNG. (**I**) PNG connected to a resistive load (R_L) and subjected to a base excitation (w_b). (**II–IV**) Full equivalent circuit model and simplified versions. (**V**) Final equivalent circuit model of a PNG, consisting of an internal impedance (Z_{in}) and an alternate voltage source (V_{OC}).

For a sDOF PNG, the following assumptions hold true: (i) Euler-Bernoulli beam hypothesis; (ii) negligible external excitation from air damping; (iii) strain rate damping and viscous air damping are proportional to the bending stiffness and mass per length of the beam; (iv) uniform electric field over the piezoelectric thickness; (v) small deformations of the PNG [100]. It is worth noticing that these assumptions do not account for non-linear structural behaviour or for very large deformations (e.g., very flexible PNG in turbulent flows).

Therefore, as discussed in [100–103], the piezoelectric constitutive equations can be simplified and together with Equation (1) yield:

$$
\begin{cases}
Y\mathcal{I}\frac{\partial^4 w}{\partial s^4} + c_s\mathcal{I}\frac{\partial^5 w}{\partial s^4 \partial t} + c_a\frac{\partial w}{\partial t} + m\frac{\partial^2 w}{\partial t^2} + KV_p(t)\left[\frac{d\delta(s)}{ds} - \frac{d\delta(s-L)}{ds}\right] = -m\frac{\partial^2 w_b}{\partial t^2} \\
\frac{V_p(t)}{R_L} + C_p^S\frac{dV(t)}{dt} + \int_0^L e_{31}h_{pn}w\frac{\partial^3 w}{\partial^2 s\partial t}ds = 0
\end{cases}
\tag{4}
$$

where $Y\mathcal{I}$ is the average bending stiffness; L, w are the length and width of the beam; K is the electromechanical coupling coefficient; $\delta(s)$ is the Dirac delta function; $V_p(t)$ is the output voltage; R_L is a generic resistive load connected to the PNG; h_{pn} is the distance of the center of the piezoelectric layer from the beam's neutral axis.

The mechanical motion of a sDOF PNG subjected to harmonic vibrations, fluid/pressure-induced oscillations may be represented as $w(s,t) = \sum_n \lambda_n(s)u_n(t)$, where $\lambda_n(s)$ and $u_n(t)$ are, respectively, the mass normalized eigenfunction and the modal coordinate for the nth mode of the beam harmonic motion.

Hence, the modal electromechanical equations for the PNG are:

$$
\begin{cases}
\ddot{u}_n + 2\zeta_n\omega_n\,\dot{u}_n + \omega_n^2 u + \Lambda_n V_p = -f_n\ddot{u}_b \\
I_p + C_p^S\dot{V}_p - \sum_n \Lambda_n\,\dot{u}_n = 0
\end{cases}
\tag{5}
$$

where $I_p(t) = V_p(t)/R_L$ is the output current; $\omega_n, \zeta_n, \Lambda_n, f_n$ are the natural frequency, the damping ratio, the modal electromechanical coupling coefficient (or force-voltage factor) and the modal mechanical forcing function for the n-th mode, respectively, with $\Lambda_n = \int_0^L K\frac{d^2\lambda_n(s)}{ds^2}ds$ and $f_n = \int_0^L m\lambda_n(s)ds$.

The analytical model takes into account only a resistive load connected to the PNG and the maximum output power consumption on the load is yielded by the optimal load resistance [101]. Accordingly, a load resistor is usually taken into account for during the transducer design and optimisation stage. However, this approach is not exhaustive, in fact, the circuit downstream of the PNG is often more complex in order to transfer efficiently the power to an energy storage, as discussed in [104]. On the other hand, the approach of electromechanical analogy, i.e., between electrical and mechanical domains, is a useful tool for solving this issue. Each element in the mechanical motion of the PNG may be translated to a homologous electric counterpart, therefore the PNG can be replaced by its equivalent circuit model (ECM) [100–102].

For the *n*th mode, the base excitation $-f_n\ddot{u}_b$ is equivalent to an AC voltage source V_n; the electromechanical coupling may be considered as a force-voltage ideal transformer with ratio $N_n = \Lambda_n$; the mass m (per unit length), the compliance $1/\omega_n^2$ and the mechanical damping $2\zeta_n\omega_n$ correspond, respectively, to an inductance L_n, a capacitance C_n and a resistance R_n. The electrical counterparts of the modal coordinate and velocity are the generated charge q_n and flowing current I_n. Figure 2b(II–IV) reports the variations of the ECM for a PNG, based on the electromechanical analogy and circuital simplifications. In the simplest case a PNG may be replaced by a current source connected in parallel or a voltage source connected in series to the free capacitor of the piezoelectric material (Figure 2b(V)). The RC matching method asserts that the optimal resistance is the impedance of the free capacitor [105]. However, other ECMs based on lumped parameters have been proposed [106–108], which are useful for designing the power management circuits and for accurately account for the power output across the whole frequency range [102,109,110]. In particular, a pure resistive load that matches the internal impedance of the transducer only takes into account the case of an existing zero-phase impedance matching, which yields a power peak in the power/resistance curve [108,111].

2.2. Triboelectric Materials and Mechanisms

Triboelectricity is a phenomenon through which the relative contact between two dissimilar materials induces the generation of superficial charges [112–114]. Electrostatic induction is responsible of the redistribution of charges from these materials to some conductive or metal electrodes which collect and transfer them to an external circuit to generate electrical power [115–117]. Surface charge density is a key parameter and it is determined by the surface triboelectric polarities of materials, i.e., the capability of attaining electrons or positive charges on the surface. Triboelectric series list most of materials in terms of their triboelectric capability, depending on the electron affinity and the work function [115,118,119]. The charge transfer during contact electrification is basically a surface process so it can be favored or enhanced by different nanostructuring/patterning techniques, by acting on the surface roughness, the porosity and the wettability properties to increase the effective friction area [120–128].

Triboelectric nanogenerators (TNGs) present several advantages against other electromagnetic generators, e.g., light mass, low density, cost-effectiveness, higher scalability and no need of bulky coils and magnets [14,24,113,129–134]. Flexible devices rely on soft substrates and polymers as triboelectric layers, due to their flexibility and deformability. Common examples are polytetrafluoroethylene (PTFE), Polyimide (PI), polyethyleneterephtalate (PET), polydimethyl siloxane (PDMS), etc. Parylene C (poly(para-xylylene)) is also another promising material for TNG because of several attractive properties [135,136], e.g., chemical inertness, conformality, FDA-approved biocompatibility, barrier properties, anti-corrosive/antibiofouling behavior and high flexibility [137–142]. It can be easily deposited through an efficient, room-temperature, controlled process (Gorham route) [143] and it can also be variously surface-treated, e.g., under oxygen plasma, ozonized UV, or by wet and dry etching processes [5,76,137,144,145].

A TNG can be basically regarded as a pair of materials separated by a gap, which come into contact owing to an external mechanical load, generating an electrical potential and a current in an external circuit that serve to re-establish the electrostatic equilibrium broken by the contact electrification (Figure 3a(I–III)). Several operating modes are possible for a TNG (vertical contact-separation mode, sliding mode, free-standing mode, single-electrode mode) [5], but the theoretical foundation of the triboelectrification phenomenon is the Maxwell's displacement current, given by:

$$J_D = \frac{\partial D}{\partial t} = \epsilon \frac{\partial E}{\partial t} + \frac{\partial P_S}{\partial t} \tag{6}$$

where D is the displacement field, E is the electric field and P_S is the polarization due to the presence of surface polarization charges.

The triboelectric potential and the triboelectric current may be defined by the following equations [146]:

$$\Delta V = -\frac{\rho_T d}{\epsilon}, \quad I = C_T \frac{\partial \Delta V}{\partial t} + \Delta V \frac{\partial C_T}{\partial t} \tag{7}$$

where ρ_T is the generated triboelectric charge density, d is the gap distance between the materials, C_T is the capacitance of the overall system.

The analytical model of a TNG can be derived from a Lagrangian formulation, as reported in [147]. In particular, for a sDOF generic dual dielectric contact-mode triboelectric generator, the electromechanical equations of motions can be written as:

$$\begin{cases} m\ddot{x} + c_{eq}\dot{x} + k_{eq}x - \frac{S}{2\epsilon_2}\left(\sigma_2 + \frac{q}{S}\right)^2 = f + mg \\ R\dot{q} + \frac{q}{A}\left(\frac{d_1}{\epsilon_1} + \frac{d_2}{\epsilon_2} + \frac{d_3}{\epsilon_3} - \frac{x}{\epsilon_2}\right) + \frac{\sigma_2}{\epsilon_2}(d_2 - x) + \frac{d_1}{\epsilon_1}(\sigma_1 + \sigma_2) = 0 \end{cases} \tag{8}$$

where x is the displacement of the top electrode, q is the electrical displacement of the bottom dielectric, k_{eq} is the equivalent stiffness of the system, S is the surface active area, c_{eq} is the equivalent damping coefficient of the system; d_1, d_2, d_3 are the thicknesses of the upper dielectric, the air gap and the bottom dielectric; σ_1, σ_2 are the surface charge density on the top and bottom dielectrics; f is a mechanical load; m is the mass of the system.

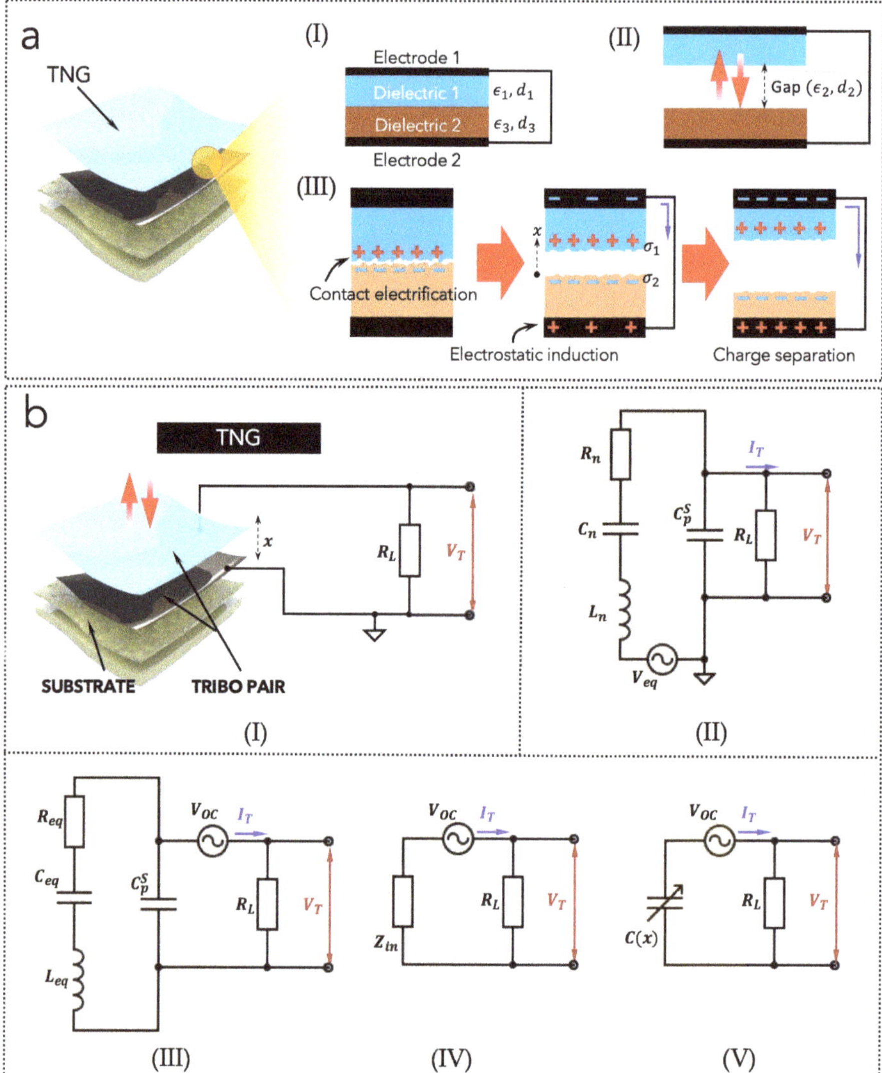

Figure 3. (**a**) Pressed state of a TNG with two dielectrics in contact. (**I**) Released state of the TNG. (**II,III**) Contact electrification and process of triboelectric charge separation. (**b**) Equivalent circuit model of a TNG. (**I**) Schematic of a TNG connected to a resistive load (R_L) and subjected to a periodic vertical contact-separation motion (x). (**II–IV**) Full equivalent circuit model and simplified versions. (**V**) Final equivalent circuit model of a TNG, made of an internal variable capacitance ($C_{(x)}$) and an alternate voltage source (V_{OC}).

In the assumption of small oscillations, the equations of motions may be simplified into the following:

$$\begin{cases} \ddot{x} + 2\xi\omega_n\,\dot{x} + \omega_n^2 x - \frac{\alpha}{m}q - \frac{\beta}{m}q^2 = \frac{f}{m} \\ \dot{q} + \frac{1}{C(x)R}q - \frac{\alpha}{R}x = 0 \end{cases} \tag{9}$$

where $C(x)$ is the variable equivalent capacitance; $\xi = c_{eq}/2\sqrt{mk_{eq}}$ is the damping ratio, $\omega_n = \sqrt{k_{eq}/m}$ is the natural frequency; R is a load resistance; α and β are two lumped parameters defined as $\alpha = 1/\epsilon_2 \cdot (\sigma_2 + Q_0/S)$ and $\beta = 1/2\epsilon_2 S$, with Q_0 the initial total charge on the dielectrics.

As for the PNGs, from the electromechanical equations it is possible to deduce the electromechanical analogy and pass from the mechanical to the electrical domain. The applied force is equivalent to an alternating voltage source V_n; the mass m, the compliance $1/\omega_n^2$ and the mechanical damping $2\zeta\omega_n$ correspond, respectively, to an inductance L_n, a capacitance C_n and a resistance R_n [148].

Figure 3b(I–V) reports the variations of the ECM for a TNG, based on the electromechanical analogy and circuital simplifications. The simplest ECM of a TNG can be represented as an alternating voltage source in series with a capacitor with varying capacitance [148,149], and according to the RC matching method, the optimal resistance is the impedance of this capacitor [105].

3. Piezo-Triboelectric Hybridization

Hybrid bio-nanogenerators (HBNGs) are aimed at constituting sustainable power systems that integrate different kinds of transducing mechanisms, e.g., mechanical, photovoltaic, thermal, exploiting electromagnetic induction, triboelectrification, piezoelectricity, solar light, temperature gradients [52]. The objective of this innovative systems is to enhance the performances of the single components, to harness simultaneously several renewable energy sources and to improve the sensitivity range of the devices.

In fact, PNGs and TNGs often require specific operating conditions and present some disadvantages. In particular, the type of output voltages and currents originating from the piezoelectric and triboelectric effects are different: the piezoelectric signal is very effective with impulsive loads and rapid mechanical deformations, the triboelectric signal is more perceptible with continuous slow friction. Additionally, their impedances are different implying that the output voltage/current is lower/higher for piezoelectrics and higher/lower for triboelectrics. Therefore, the hybridization of the two mechanisms allows to overcome these constraints, to integrate their functionalities in the same device, enhancing the overall sensitivity and widening the measurement ranges [120,150–154].

HBNGs can be considered multi-source, multi-function and multi-mechanism devices. In this section, the hybridization of PNGs and TNGs is taken into consideration. Depending on the way these two components are connected and coupled, different structural schemes can be distinguished, and the circuit model and the power management circuits necessary to extract signals are different.

The first classification regards the physical domains involved in the energy conversion process: if they belong to the same species and are ruled by the same physical processes, then the hybridization is called *"intra-domain"*. This holds true in most cases for piezoelectric-triboelectric devices because the common domain is the mechanical one. In all the other cases, when solar, thermal or biochemical energy is involved for energy/signal generation, the hybridization is called *"inter-domain"*.

The second classification is related to the way the materials of the piezoelectric or triboelectric components are physically coupled. If they are not permanently in contact with each other and belong to distinct single devices, then the hybridization occurs *"with physical separation"*. On the other hand, if there is a physical connection or interpenetration between the active materials responsible for the two transducing mechanisms, then the hybridization is *"with physical integration"*. This difference is also crucial for the number and positioning of the electrodes within the hybrid device.

The third classification concerns the number of mechanisms coupled together in the same hybrid device. A "coupling number" (CN) can be defined in this respect as the number of processes separately responsible for charge generation. For instance, most of piezo-triboelectric hybrid devices have a CN equal to 2, a piezo-triboelectric-electromagnetic hybridized device has a CN equal to 3. In case of wearable patches, the triboelectrification due to the contact with human skin can generate further charges [76] and this would count as an additional process, increasing the CN. Hence, the CN is correlated with the structural complexity of the whole system. A higher CN implies more electrodes, connections and processes involved in the transduction mechanism.

Han et al. [155] have investigated and compared (theoretically and experimentally) different piezoelectric-triboelectric nanogenerators, taking into account the number of terminals and contacted materials, the polarization direction and electrode connection (Figure 4a). They classified six different configurations (three in conductor-to-dielectric mode and three in dielectric-to-dielectric mode), as well as one, three and nine possible configurations for two-terminal, three-terminal and four-terminal devices, respectively. Additionally, for each configuration, two possibilities can be distinguished with forward or reverse polarization of the piezoelectric material. Thus, 52 configurations have been considered in total. In general, if n is the number of terminals, the following formula can be proposed for the number of possible configurations:

$$\begin{cases} 4 \cdot \prod\limits_{\substack{k=0 \\ k \le \frac{n}{2}-1}} \begin{pmatrix} n-2k \\ 2 \end{pmatrix}, & n \le 3 \\[2em] 4 \cdot \prod\limits_{\substack{k=0 \\ k \le \frac{n}{2}-1}} \begin{pmatrix} n-2k \\ 2 \end{pmatrix} \cdot \frac{3}{2}, & n > 3 \end{cases} \tag{10}$$

where the factor 4 accounts for the two conductor/dielectric modes and for the two forward/reverse polarization modes, whereas the factor 3/2 is for avoiding repetitions.

The structure, polarization direction and electrode connection strongly affect the output performance of the overall hybrid device. In particular, it can be enhanced or reduced almost to zero just modifying the electrode connection: this can be an advantage because with the same polarization direction and configuration, the performances can be tuned to maximize the output power (in case of an energy harvester) or to minimize the electrostatic charges (in case of preventing charging in electronic devices).

Another key point is that with the same parameters and configurations, the performances can be modulated by acting on the mechanical design, regardless the piezoelectric coefficients or the triboelectric surface charge densities. In this respect, the piezoelectric component can be designed to obtain specific required stress variations during the contact-separation cycle of the triboelectric component, so that there is a simultaneous coupling of the two mechanisms.

The combination of piezoelectric and triboelectric effects for a specific configuration can be correlated with an output state characterized by different multi-level electric outputs, depending on the electrode connections. This can be exploited for logic communication devices, SC/OC detection systems or active impedance-based sensors [155].

For an HBNG, deriving and applying the electromechanical equations is very complex, especially when the motions of the two components are closely related. A general system of equations that govern an HBNG is given by the combinations of the piezoelectric and triboelectric equations: in particular, the Maxwell's law of displacement current, i.e., (6 [156], holds true but it is diversified within the piezoelectric or triboelectric domain, thus the analytical model can be represented in general as follows (assuming a polarization and a contact-separation cycle along the z-axis, and no external applied electric field):

$$\begin{cases} J_D^{piezo}\Big|_x = \frac{\partial D_x}{\partial t} = \frac{\partial P_x}{\partial t} = (e)_{ij}\left(\frac{\partial S}{\partial t}\right)_j\Big|_x = \frac{\partial \sigma_P(x)}{\partial t} \\[1em] J_D^{tribo}\Big|_x = \frac{\partial D_x}{\partial t} = \frac{\partial \sigma_I(x,t)}{\partial x} = \sigma_c \frac{dx}{dt} \frac{d_1\epsilon_0/\epsilon_1 + d_2\epsilon_0/\epsilon_2}{[d_1\epsilon_0/\epsilon_1 + d_2\epsilon_0/\epsilon_2 + x]^2} \\[1em] \mathscr{F}\left(\Delta V_{piezo}, \Delta V_{tribo}, J_D^{piezo}, J_D^{tribo}\right) = 0. \end{cases} \tag{11}$$

where x is the separation distance between the triboelectric layers or the distance inside the piezoelectric material; $\sigma_P(x)$ is the surface polarization charge density on the piezoelectric material; $\sigma_I(x,t)$ is the surface density of free electrons accumulated in the electrodes of the triboelectric material; d_i, ϵ_i are the thicknesses and dielectric constants of two dielectrics of the triboelectric pair; $\Delta V_{piezo}, \Delta V_{tribo}$ are the output voltage downstream of the piezoelectric and the triboelectric device; $\mathscr{F}$ is a function that involves the output voltages and currents of the piezoelectric and triboelectric components and which depends on the specific configuration and electrode connections.

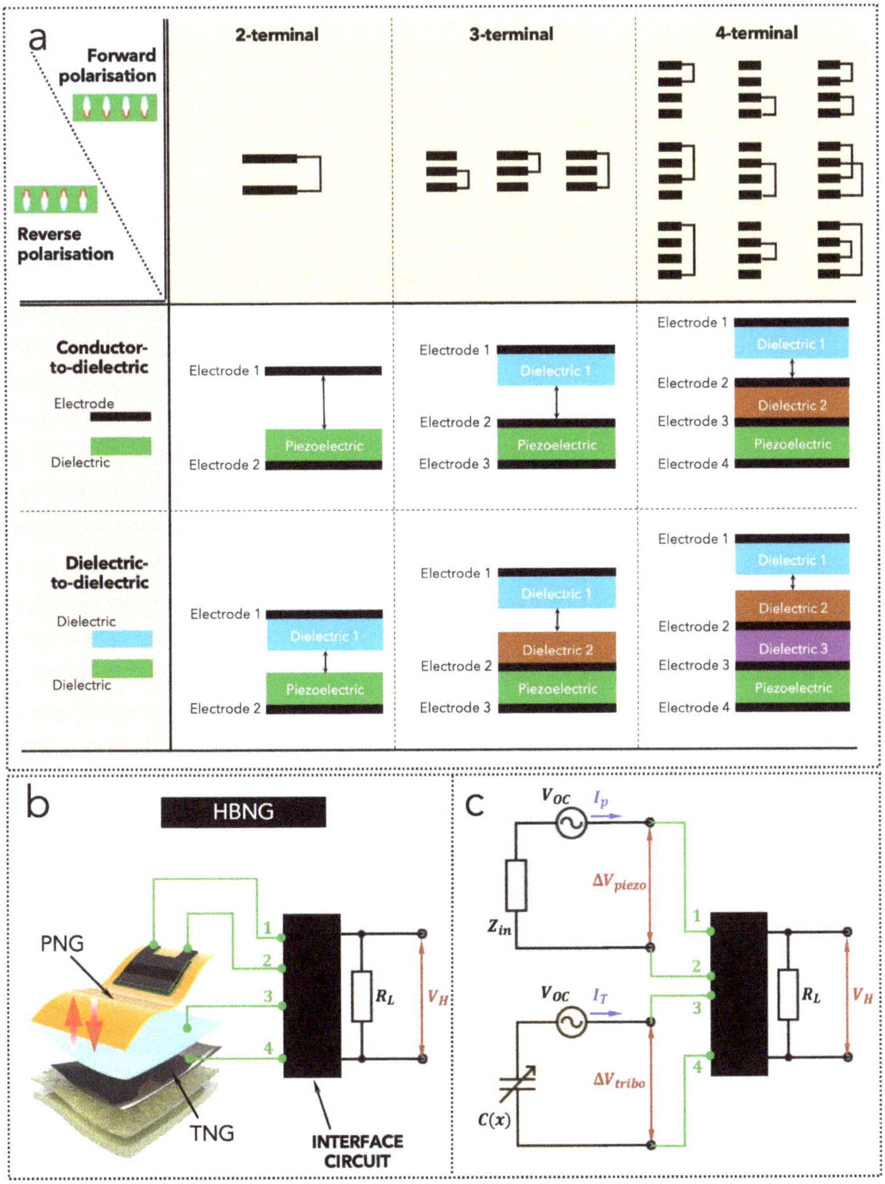

Figure 4. (**a**) Classification of the HBNGs with respect to the number of electrodes, the contact mode (conductor-to-dielectric or dielectric-to-dielectric) and the polarization direction of the piezoelectric material. (**b**,**c**) Equivalent circuit model of a HBNG. (**b**) Schematic of a HBNG connected to a resistive load (R_L) and subjected to a periodic vertical contact-separation motion. (**c**) Simplified equivalent circuit model of a HBNG.

Chen et al. [157] proposed a first example of theoretical analysis of the contact-mode hybrid piezoelectric-triboelectric nanogenerator, describing the relationships among transfer charges, voltage, current and average output power in terms of material properties and operating conditions. This model can be used for a general 3-electrode HBNG with a single-electrode TNG: assuming that the triboelectric charges are distributed uniformly on the triboelectric layer with a small decay, the contact area is much larger than the gap and only vertical components of the electric fields exist in the air gap, the charge balance holds true:

$$\sigma_1 + \sigma_2 + \sigma_3 - \sigma = 0 \tag{12}$$

where σ is the triboelectric charge density and $\sigma_1, \sigma_2, \sigma_3$ the surface charge densities on the electrodes. The electric fields of the piezoelectric, triboelectric and air gap. Then, for a closed loop in a circuit, it is possible to write:

$$V_a + V_T + V_1(t) = 0, \quad V_P + V_2(t) = 0 \tag{13}$$

where V_a, V_T, V_P are the voltages across the air gap, the triboelectric layer and the piezoelectric layer; V_1, V_2 are the voltages across two external resistors connected to the PNG and TNG. Combining the previous equations with the V-Q-x relationship for the TNG unit and the charge transfer equation for the PNG unit, it is possible to obtain the current and voltage expressions for the full hybrid device [157].

Therefore, since it is challenging to predict theoretically the function $\mathscr{F}$, the ECM is necessary in this case for determining the electromechanical behaviour of the device (Figure 4b). Here a simple ECM is proposed for a piezoelectric and triboelectric HBNG based on contact-mode (Figure 4c). The HBNG can be considered as a set of two separate ECMs for the PNG and TNG, respectively, and then their distinct outputs ($\Delta V_{piezo}, \Delta V_{tribo}$) are used as inputs of an interface circuit that provides the output voltage (V_H) of the HBNG, as a two-terminal device. The electrodes deriving from the PNG and TNG are in total four but they can vary depending on the configuration.

The next sections describe the main types of hybridization that can occur between PNGs and TNGs, and examples of HBNGs from previous works are comprehensively summarized in Table 1.

3.1. Intra-Domain Hybridization with Physical Separation

This category includes the devices where piezoelectric and triboelectric nanogenerators are coupled separately (Figure 5a). This means there are at least two electrodes for the PNG and one electrode for the TNG. The 3-terminal and 4-terminal configurations illustrated in Figure 5a are included in this group. The physical separation between the two devices allows to extract separately the signal generated by each component and perform an additive sum through rectifying circuits downstream of the hybrid device. Thus, the hybridization occurs mostly within the electronic interface according to the combination of output voltages and currents: it can be considered an *"external hybridization"*. A slight contribution is only provided by the materials interaction because the two devices work as if they were alone: this is especially true for a PNG and a two-electrode TNG (four-terminal HBNG). When the TNG has only one single electrode (three-terminal HBNG), although it is still separated from the PNG, the contact electrification occurs between the triboelectric layer and one of the layers of the PNG (Figure 5b): in this case the interaction between the two devices plays a role in the hybridization [13]. In particular, the piezoelectric polarization direction and the triboelectric charge polarity on the common electrode is determining for the enhancement or reduction of the hybrid signal [13].

Table 1. Hybrid bio-nanogenerators (HBNGs) classified in terms of hybridization approach, materials, performances and applicative context. The green, yellow, blue and pink colors are, respectively, for: 2-mechanism hybrid piezo-triboelectric, 2-mechanism hybrid with piezoelectric or with triboelectric, intra-domain 3-mechanism hybrid with piezo-triboelectric, 2/3-mechanism inter-domain hybrid. The coupling number (CN) represents the total number of transducing mechanisms present in the device.

Ref.	Coupled Mechanisms	CN	Hybridization Type	Materials	Thickness /Size	Output Performances	Application Field
Chen et al. [158]	Piezo-Tribo	2	intraD-PI	PVDF NMFs, Cu-PCB	~500 μm	15 V; 0.0783 μW/cm^2	MEH
Chen et al. [159]	Piezo-Tribo	2	intraD-PI	P(VDF-TrFE) NFs, Kapton, Cu, PET/ITO, PDMS	-	96 V; 0.887 μW/cm^2(piezo) 8 V; 1.14 μW/cm^2(tribo1) 16 V; 17.6 μW/cm^2 (tribo2)	MEH
Zhao et al. [160]	Piezo-Tribo	2	intraD-PS	P(VDF-TrFE), Au, PET, Al, PTFE	650 μm	210 V; 6040 μW/cm^2	MEH
Xia et al. [161]	Piezo-Tribo	2	intraD-PI	PVDF, Al, PDMS	~600 μm	5.36 V; 1 μW/cm^2 (Piezo) 5.84; 1.68 μW/cm^2 (Tribo)	MEH
Suo et al. [162]	Piezo-Tribo	2	intraD-PI	PET/ITO, BTO/PDMS, Cu	200 μm	3.3 V	MEH
Singh et al. [68]	Piezo-Tribo	2	intraD-PS	Au, ZnO/PVDF, PTFE, Al	~245 μm	~97 V; 24.5 μW/cm^2	MEH
Guo et al. [163]	Piezo-Tribo	2	intraD-PI	Conductive mat, Silk FBs, PVDF FBs	145.3 μm	~500 V; 310 μW/cm^2	MEH, WBS
Wang et al. [164]	Piezo-Tribo	2	intraD-PI	Al, ZnO, P(VDF-TrFE), PDMS	~200 μm	751.1 mW/m^2	MEH, WBS
Huang et al. [165]	Piezo-Tribo	2	intraD-PI	PVDF FBs, conductive mat	-	210 V; 70 μW/cm^2	MEH
Chen et al. [166]	Piezo-Tribo	2	intraD-PS	P(VDF-TrFE), PDMS, PU/CNTs/AgNWs	-	57.1 V; 0.11 μW/cm^2 (Piezo) 183 V; 84 μW/cm^2 (Tribo)	MEH, WBS
Wang et al. [150]	Piezo-Tribo	2	intraD-PS	Al, ZnO, P(VDF-TrFE), PDMS/MWCNTs, Al	130 μm	2.5 V; 0.689 mW/cm^3 (Piezo) 25 V; 1.98 mW/cm^3 (Tribo)	MEH
Mariello et al. [13,76]	Piezo-Tribo	2	intraD-PS	Mo, AlN/Kapton, pC, Ti/Au, PDMS/Ecoflex	~86 μm	16 V; 0.8 W/m^2	MEH, WBS
Jung et al. [167]	Piezo-Tribo	2	intraD-PS	Polyimide, Al, PVDF, PTFE, Au	~200 μm	370 V, 12 μA/cm^2, 4.44 mW/cm^2	MEH
Li et al. [168]	Piezo-Tribo	2	intraD-PS	Nylon, ZnO, PDMS, Cu	3 mm	10.2 mW/m^2 (piezo) 42.6 mW/m^2 (tribo)	MEH
Mariello et al. [84]	Piezo-Tribo	2	intraD-PI	PVDF NFs, Cardanol, PDMS, Kapton	~120 μm	~0.5 V; ~0.85 10^{-2} μW/cm^2	MEH, WBS
Zhu et al. [169]	Piezo-Tribo	2	intraD-PS	PDMS, skin, PVDF-TrFE NFs	~500 μm	25.8 V. 6.15 μW/g (6 mm amplitude)	WBS
Wang et al. [170]	Piezo-Tribo	2	intraD-PS	Ecoflex, nickel fabric, PZT	2.9 cm	15 V/g (0–1.5 g) (g: gravity acc.)	MEH, IoT
Yu et al. [153]	Piezo-Tribo	2	intraD-PI	PDMS, PZT-PVDF-TrFE NFs	~200 μm	15.43 V/kPa (0–100 kPa) 18.96 V/kPa (100–800 kPa)	WBS
Yang et al. [171]	Piezo-Tribo	2	intraD-PS	Acrylic, PVDF-TiO2, PDMS-GQD, ITO, PET, Al, Cu	-	88 V, 19.5 μA343.5 μW (Tribo), 43.2 μW (Piezo)	MEH

Table 1. *Cont.*

Ref.	Coupled Mechanisms	CN	Hybridization Type	Materials	Thickness /Size	Output Performances	Application Field
Wang et al. [172]	Tribo-ElectroM	2	intraD-PS	PTFE, Kapton, Cu, Magnet	2 cm	8.8 mW/g (Tribo) 0.3 mW/g (ElectroM)	MEH
Yang et al. [173]	Tribo-ElectroM	2	intraD-PS	PTFE, PDMS, NaNbO3, acrylic	mm-cm	15.5 V/cm^2; 0.08 mW/cm^2	MEH
Wang et al. [174]	Tribo-ElectroM	2	intraD-PS	Acrylic, Polymer films, Cu, Magnet	4 cm	65 V; 438.9 mW/kg (Tribo) 7 V; 181 mW/kg (ElectroM)	MEH
Hu et al. [175]	Tribo-ElectroM	2	intraD-PS	Acrylic, Kapton, Cu	-	7.7 mW (Tribo) 1.9 mW (ElectroM)	MEH
Quan et al. [176]	Tribo-ElectroM	2	intraD-PS	Acrylic, Kapton, PVB/PDMS, Cu	3 cm	0.1 mW (Tribo); 2.8 mW (ElectroM)	MEH
Zhang et al. [177]	Tribo-ElectroM	2	intraD-PS	Acrylic, PTFE, Al, magnet	-	17.5 mW, 55.7 W/m^3	MEH
Ran et al. [178]	Tribo-ElectroM	2	intraD-PS	PLA, FEP, Cu, magnet	~10 cm	48 V, 1 mA, 13 mW	MEH
Zhao et al. [179]	Piezo-ElectroM	2	intraD-PS	PZT, metal, Copper, Magnet	mm-cm	11 µW/cm^2 (@11 m/s)	MEH
Hamid and Yuce [180]	Piezo-ElectroM	2	intraD-PS	PZT, magnets, coils	37 mm	1.8 V, 3.3 V; 0.550 mW	MEH, WBS
Fan et al. [181]	Piezo-ElectroM	2	intraD-PS	PMMA, Nd-Fe-B-N35, Phosphor bronze Brass, PZT	14 mm	0.13 mW (Piezo); 2.94 mW (ElectroM)	MEH
Toyabur et al. [182]	Piezo-ElectroM	2	intraD-PS	Acrylate, Copper, NdFeB magnet, PZT, Al	mm-cm	250.23 µW (Piezo) 244.17 µW (ElectroM)	MEH
Gong et al. [182]	Piezo-ElectroS	2	intraD-PI	PLLA, PLA, Au	~500 µm	0.31 mW; 35 V, 1 µA	MEH, WBS
Lagomarsini et al. [182]	Piezo-ElectroS	2	intraD-PS	VHB, PZT, PVDF	500 µm	85 V, 17 µJ (PZT); 15 V, 0.45 µJ (PVDF)	MEH, WBS
He et al. [183]	Piezo-Tribo-ElectroM	3	intraD-PS	Acrylic, PVDF, Cu, PVS, Steel, Magnet	37 mm	12.6 V; 41.0 µW (Piezo) 362.1 mV; 66.5 µW (ElectroM) 13.3 V; 4.6 µW (Tribo)	MEH
Rodrigues et al. [184]	Piezo-Tribo-ElectroM	3	intraD-PS	Au, ZnO, PMMA, ITO, PET, Magnet, Copper, Nylon, Kapton, PTFE	0.1 mm	34 V (430 N)	MEH
He et al. [152]	Piezo-Tribo-ElectroM	3	intraD-PS	Silicone, CNT, Copper, PZT, Magnet, ABS	2.6 cm	78.4 µW (Tribo) 36 mW, 38.4 mW (ElectroM) 122 mW, 105 mW (Piezo)	MEH
Singh et al. [185]	Piezo-Tribo-ElectroM	3	intraD-PS	Acrylic, Copper, Gold, PTFE, ZnO-PVDF	6 cm	192 V, 2.78 mA	MEH
Wang et al. [186]	Piezo-Tribo-Pyro	3	interD	Kapton, Al, Cu, PVDF, PTFE	~200 µm	5.12 µW (Piezo-Tribo) 6.05 µW (Pyro)	MEH, TEH

Table 1. *Cont.*

Ref.	Coupled Mechanisms	CN	Hybridization Type	Materials	Thickness /Size	Output Performances	Application Field
Zheng et al. [187]	Piezo-Tribo-Pyro	3	interD	FEP, PVDF, Kapton, Cu	156 µm	350 V; 4.74 mW (Tribo) 20 V; 184.32 µW (Piezo-Pyro)	MEH, TEH
Xu et al. [188]	Piezo-Solar	2	interD	GaN, Si, ZnO	~500 µm	0.424 V; 4.5 µA/cm^2; 1.908 µW/cm^2	MEH, SEH
Xu et al. [189]	Piezo-Solar	2	interD	GaN, Glass, ZnO NWs	~10 µm	0.433 V; 252 µA/cm^2; 34.5 µW/cm^2	MEH, SEH
Yoon et al. [190]	Piezo-Solar	2	interD	PEN, Glass, ZnO, P3HT:PC$_{60}$BM	~50 µm	0.71 V; 10.17 mA/cm^2	MEH, SEH
Liu et al. [191]	Piezo-Solar	2	interD	PEN, Si, ZnO NWs	~126 µm	3V; 280 µA	MEH, SEH
Yang et al. [192]	Piezo-Pyro-Solar	3	interD	PET, PVDF, ZnO-P3HT	~500 nm	~5 V, 50 nA (Piezo-Pyro) 0.41 V, 31 µA/cm^2 (Solar)	MEH, SEH, TEH
Pan et al. [193]	Piezo-Biofuel	2	interD	ZnO, Au, CNT	~10 µm	3.1 V, 300 nA	MEH, BEH
Hansen et al. [194]	Piezo-Biofuel	2	interD	Kapton, PVDF, PBS	530 µm	50–95 mV	MEH, BEH
Wu et al. [195]	Tribo-Solar-Chemical	3	interD	Acrylic, PTFE, Al, Si, NaCl	3 mm	60 V, 500 µA	MEH, SHE, BEH
Li et al. [196]	Tribo-Solar	2	interD	PET, PTFE, Si	~600 µm	30 V, 4.2 mA/cm^2 (tribo) 0.6 V; 350 A/m^2 (solar)	MEH, SEH
Yang et al. [197]	Tribo-Solar	2	interD	PET, Al, PDMS, ITO, Si	>300µm	12 V; 17.4 mA	MEH, SEH
Dudem et al. [198]	Tribo-Solar	2	interD	PET, Glass, PDMS, ZnO NWs, Dye, TiO$_2$	0.6–2 µm	2–10 V	MEH, SEH
Wang et al. [199]	Tribo-Solar	2	interD	Acrylic, FEP, Cu	~12.5 mm	12 mA	MEH, SEH
Wang et al. [200]	Tribo-Thermal-ElectroM	3	interD	Acrylic, PTFE, Polyamide, magnet	~2–3 cm	5 V, 160 mA	MEH, TEH
Kim et al. [201]	Tribo-Thermal	2	interD	Buffer, PTFE, Al	3.4 mm	14.98 mW/cm^2	MEH, TEH
Mariello et al. [74]	Piezo-Optical	2	interD	SiO$_2$ optical fiber, AlN	~220 µm	~35 mV (buckling tests)	IBE

intraD-PS: intra-domain hybridization with physical separation; intraD-PI: intra-domain with physical integration; interD: inter-domain hybridization. Piezo: piezoelectric; Tribo: triboelectric; ElectroM: electromagnetic. ElectroS: electrostatic, electret. MEH: mechanical energy harvesting; SEH: solar energy harvesting; BEH: biochemical energy harvesting; TEH: thermal energy harvesting; WBS: wearable biosensing; IBE: implantable bioelectronics, IoT: Internet of Things.

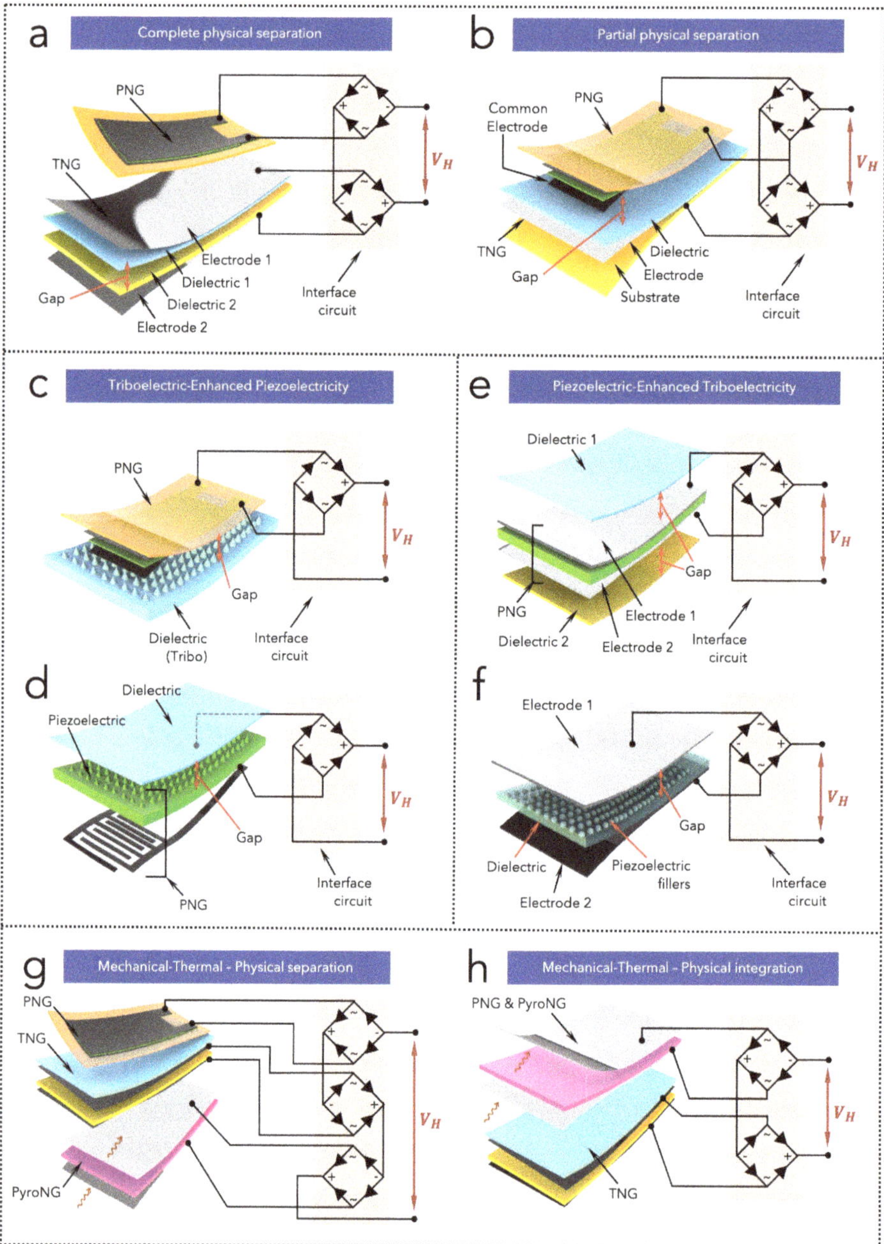

Figure 5. (**a**,**b**) Classification of intra-domain hybridization with complete (**a**) or partial (**b**) physical separation. (**b**–**f**) Classification of intra-domain hybridization with physical integration, as Triboelectric Enhanced Piezoelectricity, TeP (**c**,**d**) and Piezoelectric-Enhanced Triboelectricity, PeT, (**e**,**f**). The classification is based on the surface micro-/nano-patterning of the triboelectric (**c**) or piezoelectric (**d**) layer, on the interposition of piezoelectric between triboelectric layers (**e**), or on the incorporation of piezoelectric fillers into a triboelectric layer. (**g**,**h**) Classification of inter-domain hybridization with (**g**) physical separation or (**h**) physical integration, using as example a piezoelectric-triboelectric-pyroelectric nanogenerator.

3.2. Intra-Domain Hybridization with Physical Integration

In HBNGs based on physical integration, the hybridization occurs before the signal extraction through the electronic interface. This happens when the materials used for the PNG and TNG are not physically separated thus the same deformation or friction induces simultaneously the piezoelectric and triboelectric effects: this can be considered an *"internal hybridization"* (Figure 5c–f). The physical integration can be performed by incorporating piezoelectric nanofillers into polymeric matrices, such as lead-free inorganic compounds (BaTiO$_3$, KNN, etc.), or non-piezoelectric nanofillers into piezoelectric polymers, such as metallic particles, carbon nanofillers or ionic salts. These nanofillers are effective nucleating fillers to induce polar electroactive phases, with no need of electrical poling [202–208]. In addition, they play a role in the triboelectrification because they enhance the stretchability, flexibility, robustness, charge transport and breakdown fields [209]. Common strategies to enhance the performances consist of aligning conductive fillers along the perpendicular direction of the electric field, or of electrically poling the whole matrix to orient properly the nanofillers with the right electroactive phase. Hence, these fillers embedded in a polymer matrix represent a source for the enhancement of the piezo-triboelectric hybridization, which occurs before the actual extraction of electronic signals through the terminals into an external circuit [210,211]. Thus, the physical integration can provide two types of hybridization modes: the piezoelectric-enhanced triboelectricity (PeT, Figure 5c,d) and the triboelectric-enhanced piezoelectricity (TeP, Figure 5e,f). The key point is that the main transducing mechanism is enhanced by the other mechanism owing to the physical integration and without adding more electrodes to the overall system. Figure 5c shows a PeT-HBNG based on a PNG (two-electrodes) faced on a micro-patterned dielectric layer which acts as an additional source of charges through triboelectrification, whenever this layer comes into contact with the PNG. Figure 5d shows that it would be possible to micro-pattern the surface of the piezoelectric material and use it for amplify the piezoelectric charge generation through the contact with another dielectric layer; in this case the PNG includes planar electrodes, e.g., interdigital electrodes. Figure 5e illustrates a PNG interposed between two dielectric layers: in this case the triboelectric charge transfer is enhanced by the piezoelectric charge separation during the contact, but there are still two electrodes for the signal extraction. Figure 5f corresponds to the embedding of piezoelectric fillers into a dielectric layer: the presence of additional piezoelectric materials enhances the triboelectric generation [212].

3.3. Inter-Domain Hybridization

This category of HBNGs includes the combination of piezo-triboelectric devices with systems belonging to a non-mechanical physical domain. This can be a solar/photovoltaic cell (light domain), a biofuel cell (biochemical domain), a thermoelectric/pyroelectric nanogenerator (thermal domain), etc. For instance, Figure 5g,h show the integration of a piezo-triboelectric device with a pyroelectric device to exploit simultaneously the mechanical and thermal domain. The inter-domain hybridization requires the presence of more than one energy source, thus it is suitable or necessary for more complex operating conditions. It can occur at the electronic interface where the signals of the two coupled devices are added through some rectifying circuits (Figure 5g), or in the devices themselves, in case one of the active materials is used for both systems (Figure 5h). In the latter case, a multifunctional material is generally deployed: ferroelectric materials, for instance, can be used to exploit simultaneously the piezoelectric, pyroelectric and photoelectric effects [213]. This internal hybridization reduced the total number of electrodes in the coupled hybrid device and allows to magnify the charge quantity and electric power from various energy resources, making the HBNG useful when they are available either individually or simultaneously [214]. A side type of inter-domain internal hybridization regards the use of one domain as trigger source and the other domain as a detecting source. Piezoelectric photoacoustic probes are an example: the light injected through the optical component induces a photoelastic effect in the targeted tissue, resulting in thermo-mechanical vibrations that can be detected by a piezoelectric transducer [215–217]. The hybridization occurs when the optical stimulator and the electromechanical transducers are integrated in the same device [74,218].

4. Hybrid Piezo-Triboelectric Bio-Nanogenerators (HBNGs)

4.1. HBNGs for Energy Harvesting

The most investigated application field of HBNGs is energy harvesting, especially from mechanical energy sources [5], such as wind flows, water waves, raindrops, environmental vibrations. In most cases, the HBNG is manufactured into the shape of a flapping-foil device, for instance for wind energy harvesting. As a renewable energy source, wind and water have a strong potential to meet the growing global energy demand: onshore wind is considered the cheapest form of new power generation in Europe, covering 15% of Europe's electricity demand in 2019 [5,219]. 320 GW of wind energy capacity is expected to be installed in the EU by 2030, 254 GW of onshore wind and 66 GW of offshore wind, with a resulting production of 778 TWh of electricity (24.4% of the EU's electricity demand) [219]. Water also represents another promising choice: it is plentiful and widely accessible in different forms, e.g., precipitations, slow flows, buoying waves and ocean currents. The blue energy is that conveyed by water, typically in five forms, i.e., wave energy, current energy, thermal energy, tidal energy and osmotic energy. Wave energy, in particular, represents the most widely exploitable and is estimated to be more than 2 TW globally, around the coast [220,221].

HBNGs represent a complementary solution to the bulky standard technologies for wind/water energy harvesting, based on turbines and power plants [2–4].

Wind energy harvesting through HBNGs is based on flow-induced structural vibrations and deflections of an oscillating element that is generally made of piezoelectric material and during the flapping it touches double frames at each side triggering the triboelectric effect [5]. Since wind is intermittently available due to highly variable weather and working conditions [222–229], the performances of HBNGs are often increased by applying concurrent vibration sources [224,225] or by optimizing the aerodynamic shape to exploit fluid vortexes [225,230–234]. Piezo-triboelectric nanogenerators are most commonly combined in a hybrid unit [150,158–160,235–237] and their charge generation mechanisms do not overlap [162]. In order to exploit efficiently vortex-induced vibrations (VIV), nonlinear restoring forces, e.g., magnetic or external axial loading forces, can be introduced and integrated in the device to widen the range of possible resonant frequencies and thus increase the efficiency [4,238–241]. These restoring forces can be used also to exploit buckling instability of the piezoelectric cantilevers, enhancing the performance level of harvested power [242]. For instance, Yang et al. [240] demonstrated a magnet-induced nonlinearity for a double-beam harvester made of two piezoelectric beams with two magnets attached which allow the reduce the critical wind speed needed to activate the galloping vibrations up to about 42%. Huang et al. [243] presented a hybrid magnetic-assisted noncontact TNG combined with a magnetic responsive composite (PDMS with magnetic Fe-Co-Ni powder): the key mechanism is that the wind forces are converted into the contact-separation action for the triboelectricity, enhanced by the magnetic forces.

Jung et al. [167] presented a HBNG based on PVDF and poly(tetrafluoroethylene) (PTFE) and demonstrated the interaction between piezoelectricity and triboelectricity (Figure 6a(I,II)). At the full-contact state, in fact, the piezoelectric material generates a potential between the electrodes due to the application of an external force and the triboelectric charge transfer is maximum, achieving rectified peak output voltage, current density and power density of 370 V, 12 μAcm^{-2} and 4400 μWcm^{-2}, respectively. Xia et al. [161] demonstrated a HBNG based on a flapping-leaf mechanism using two PVDF leaves-shape cantilevers symmetrically assembled on top and bottom of micro-patterned PDMS films. Driven by the wind flow, the cantilevers undergo bending, generating piezoelectric charges, and become in intermittent contact with the PDMS frame, triggering the triboelectrification. This HBNG can generate, at 4.5 m/s speed, peak voltages and peak powers of 5.36 V, 5.84 V and 1 μWcm^{-2}, 1.68 μWcm^{-2}) for the piezoelectric and triboelectric parts, respectively. Chen et al. [158] proposed a HBNG based on piezoelectric PVDF nano-/micro-fibers and a triboelectric nano-textured Cu–PCB, which can obtain an output voltage of 15 V and a peak power density of 0.0783 μWcm^{-2}, at 3.4–15 m/s wind speed. Zhao et al. [160] demonstrated a HBNG by integrating a bimorph-based PNG into a one-end fixed TNG: exploiting a rotor-stator mechanism the device can achieve output voltage, output current and power density of 150 V, 150 μA, 6040 μWcm^{-2}, at 14 m/s wind speed (Figure 6b(I,II)). Chen et al. [159] adopted electrospinning to prepare flexible piezoelectric P(VDF-TrFE) nanofiber film and they interposed the PNG between two TNGs based on double-side metallized Kapton film. The device exhibits an enhancement of the performances due to the piezoelectric-triboelectric interaction, of 49.7%, 10% an 4.4% with respect to the PNG, the upper and lower TNG, respectively. Singh and Khare [68] fabricated a hybrid energy harvester with power density of 24.5 $\mu W/cm2$, for mechanical vibrations in which the piezoelectric film (ZnO nanorods-embedded PVDF) is also one triboelectric layer paired with PTFE (Figure 6c(I,II)).

Water energy harvesting can be similarly based on flow-induced oscillations but several other designs and structures can be exploited, as reviewed by Wang et al. [221]: rolling ball structure, multilayer structure, grating structure, pendulum structure, mass-spring structure, spacing structure, water-solid contact structure. Most of these structures harness one or more TNGs, but HBNGs have recently been explored as well. Su et al. [244] presented a hybrid TNG for water wave energy harvesting, composed of an interfacial electrification enabled TNG and an impact-TNG. This is an example of "broad-sense" hybrid device because it is based on two transducers with the same working mechanism (triboelectricity). The first component is made of FEP with electrodes to scavenge electrostatic energy at water-solid interface; the second one is made of nanostructured PTFE to scavenge impact energy of water waves. The device is able to produce, at wave-propagation speed of 0.5 m/s, short-circuit currents of 5.1 μA and 4.3 μA for the first and second components, respectively. Mariello et al. [13] reported on the fabrication of a multifunctional flexible, biocompatible low-thickness (~86 μm) HBNG for water energy harvesting (Figure 6d(I–VIII)). The PNG is based on a 1 μm-thick piezoelectric AlN sandwiched between Molybdenum (Mo) thin electrodes, all sputter-deposited on a flexible substrate. The TNG is instead composed of a Ti/Au metallized porous elastomeric patch. The hybrid device exploits the piezoelectricity and triboelectricity to harvest water-conveyed energy from different mechanical sources, i.e., impacts/breakwaters, raindrops and sea waves, achieving power densities of 0.8 W/m^2, 9 mW/m^2, 3.2 mW/m^2, respectively.

TNGs are often integrated with electromagnetic generators (EMGs) to increase the current generation, thus many examples of hybrid TNG-EMG devices have been proposed for wind or water energy harvesting [245–251]. Wang et al. [172] proposed a electromagnetic−triboelectric nanogenerator to scavenge airflow kinetic energy, based on two TNGs and two EMGs, able to generate 1.8 mW and 3.5 mW at 18 m/s wind speed, respectively. Wang et al. [174] demonstrated another similar hybrid system with a rotary-blade TNG and a rotary EMG, for harvesting rotation energy: under a wind speed of 5.7 m/s, the device can generate 438.9 mW/kg and 181 mW/kg for the TNG and EMG, respectively. Three-mechanism HBNGs have been also proposed, based on the hybridization between piezoelectricity, triboelectricity and another transducing mechanism. In the context of mechanical energy, the EMG is often used as a third component, exhibiting important advantages, i.e., superior output performances in terms of power density and current. He et al. [183] described, for instance, a PNG-TNG-EMG device with four PVDF L-shaped vibrating beams with NdFeB permanent magnets and four beams with patterned PDMS and PET. The PNG, TNG and EMG units can generate 12.6 V (41.0 μW), 13.3 V (4.6 μW) and 362.1 mV (66.5 μW) at ~20 Hz and an acceleration of 0.5 g.

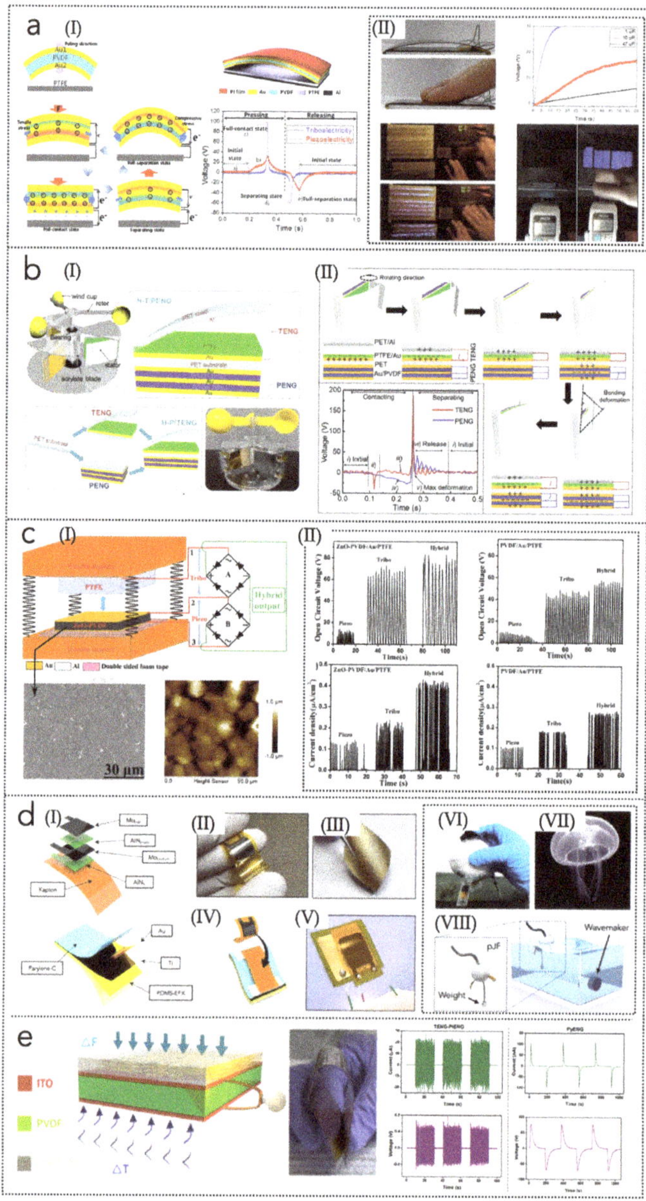

Figure 6. HBNGs for energy harvesting. (**a(I,II)**) High-output HBNG based on PVDF and PTFE. Reprinted with permission from ref. [167], 2015, Copyright Springer Nature. (**b(I,II)**) HBNG for efficient and stable rotation energy harvesting. Reprinted with permission from ref. [160], 2018, Copyright Elsevier. (**c(I,II)**) Flexible ZnO-PVDF/PTFE HBNG for energy harvesting. The incorporation of ZnO enhances the piezoelectric and triboelectric properties of PVDF. Reprinted with permission from ref. [68], 2018, Copyright Elsevier. (**d(I–V)**) HBNG based on AlN and PDMS-parylene for water wave energy harvesting. The system piezoJellyFish (pJF) is designed to have three devices connected simultaneously for scavenging energy from buoying waves (**VI–VIII**). Reprinted with permission from ref. [13], 2021, Copyright Elsevier. (**e**) Piezoelectric-Triboelectric-Pyroelectric HBNG for harvesting mechanical and thermal energy. Reprinted with permission from ref. [187], 2016, Copyright Wiley-VCH Verlag GmbH and Co. KGaA.

The inter-domain hybridization has been often used for combining mechanical and thermal energy, exploiting the piezo-triboelectric and the pyroelectric effects [213]. Wang et al. [186] characterized a TNG-PNG-PyroNG device based on PVDF nanowires-PDMS composite film as a triboelectric layer and a polarized PVDF film, sandwiched between ITO electrodes, as piezoelectric/pyroelectric layer. The device can produce 5.12 μW and 6.05 μW with the PNG-TNG and PyroNG mechanisms, respectively, and it is suitable for harvesting energy from wind and temperature variations simultaneously. Zheng et al. [187] presented a wind-driven TNG-PNG-PyroNG system based on fluorinated ethylene propylene (FEP) and PVDF films as triboelectric and piezoelectric layers, on a Kapton substrate (Figure 6e). The outputs of this device are 184.32 μW and 4.74 mW at 18 m/s wind speed, for the PNG-PyroNG and the TNG, respectively.

Hence, HBNGs have been demonstrated as suitable tools to harvest mechanical (kinetic) energy (rotational or oscillatory) from different sources, e.g., wind and water waves. The contribution of the HBNGs to the energy harvesting field must be ascribed to two aspects. First, the use of them as renewable energy sources employed in complementary way with the large energy platforms (power plants, wind turbines, hydraulic turbines, hydrovoltaic systems, etc.), in order to mitigate the energy crisis. Second, the deployment of these devices for supplying energy to small self-powered IoT sensors used for healthcare, sport, medicine, automotive, robotics, etc. The use of energy harvesting HBNG devices as wearable or implantable systems is discussed in the next sections.

4.2. HBNGs for Wearable Bioelectronics

Wearable bioelectronics (WBE) includes all the systems that can be applied on the human body and that can be employed for applications in health monitoring, robotics, sports, rehabilitation. The main requirements for WBE are flexibility, conformality, shape-adaptability, lightweight and high sensitivity. PNGs and TNGs have been widely used as wearable sensors [27,125,252–256] and their integration in a single hybrid device allows to increase the measurement range and the sensitivity. The hybridization can occur by coupling separately the two devices, or by using the same flexible ferroelectric material as piezoelectric and triboelectric layer (e.g., PVDF or inorganic perovskites such as $BeFiO_3$ [85]). PNGs and TNGs can thus be coupled and applied to monitor cardiovascular, respiratory or neurological disorders [257].

Yu et al. [153] reported on a tactile PeT-HBNG based on a PZT-PDMS composite film (Figure 7a), demonstrating enhancements in terms of sensitivity, good skin-conformality, good linearity, fast response and high stability. To avoid the use of PZT, a similar PeT-HBNG was presented by Guo et al. [163]: the device is based on electrospun silk fibroin and PVDF nanofibers, exhibiting high energy harvesting capability, good air permeability, possibility of identifying various types of body motion. Figure 7b(I,II) shows the structure and materials of the HBNG, its working principle and the simulated/experimental piezo-triboelectric interaction effects in accordant and opposite state (depending on the piezoelectric polarization). Wang et al. [150] proposed a flexible wearable HBNG made of P(VDF-TrFE) electrospun nanofibers and multiwall carbon nanotubes (MWCNT)-doped PDMS composite membrane (Figure 7c). The device exhibits, as a power source, a peak-to-peak voltage and power density, respectively, of 25 V, 1.98 mW/cm^3 for PNG, and 2.5 V, 0.689 mW/cm^3 for TNG. Electrospinning is a useful technology to produce films of piezoelectric materials without the need of electrical poling, even though the process parameters are characterized by a wide variability.

Tang et al. [258] described a self-powered HBNG used for the simultaneous detection of multi-parameter sensory information of finger mechanical operations, for applications in human-machine interfaces. The TNG is made of a PTFE friction layer, two PET films, a nylon grid, whereas the PNG is composed of an Al-sandwiched PVDF. Chen et al. [166] reported on a conformable HBNG for harvesting touch energies and real-time monitoring human physiological signals. The PNG is made of a P(VDF-TrFE) nanofiber mat sandwiched between two electrospun PU films coated with CNTs and AgNWs; the TNG is a single-electrode device based on micro-patterned PDMS. Suo et al. [162] developed a HBNG based on $BaTiO_3$ NPs/PDMS composite film, demonstrating the mutual enhancing effect of piezoelectricity and triboelectricity in a single material component (hybridization with physical integration). Mariello et al. [76] proposed a flexible, ultra-thin biocompatible HBNG as wearable conformal sensor on the human body. The PNG is based on thin-film AlN sputter deposited on a polyimide and sandwiched between thin-film Mo electrodes, and the TNG is made of an ultra-soft PDMS-Ecoflex patch encapsulated in a parylene C friction film, surface-treated with UV/ozone and oxygen plasma. The HBNG is an inorganic/elastomeric 3-electrode device for multi-site and multifunctional conformal sensing: it works as a skin-adaptable sensor for gait walking, gestures recognition and monitoring of joints movements, yielding repeatable and real-time rectified signals. The architecture of the device is based on a bridge configuration, i.e., only the ends of the device are attached on the skin, so that an additional triboelectric mechanism is exploited between the skin and the bottom triboelectric layer (skin-contact actuation), besides the piezo-tribo hybrid contact (Figure 7d(I,II)). This allows to widen the measurement range for tiny and irregular human motions, combining together the impulsiveness of piezoelectricity and the slowness and sensitivity of friction-based triboelectricity.

Definitely, the main practical issues related to wearable HBNGs are: high thicknesses, poor layer adhesion and delamination, non-optimal deformability and conformability, complex design, unproper choice of materials [76].

4.3. HBNGs for Implantable Bioelectronics

Implantable systems need more stringent requirements in terms of materials, design, mechanical response and reliability, since they have to be employed inside the human body or of the selected animal models, and they must be in direct contact with tissues, organic and biofluids. This implies the need of flexibility, stretchability, conformality and durability. These devices often require barrier encapsulations to avoid water permeation inside the electronic components, and the barriers are usually made of ultrathin coatings deposited from the vapor phase (e.g., SiO_2 [259,260], SiN_x [261], Al_2O_3 [262], $TiSi_2$ [263], SnO_2 [264], etc.).

HBNGs are still not massively adopted for implantable applications but the use of PNGs and TNGs as implants has been recently explored [265–269]. However, the friction effects with the surrounding tissues are always present due to the sliding movements of the devices, thus the resulting signal produced by the system can be a combination with triboelectric signals, therefore an internal hybridization is possible to occur. Dagdeviren et al. [270] reported on biocompatible, flexible PZT-based energy harvesters monolithically integrated with rectifiers and millimeter-scale batteries for simultaneous power generation and storage, including high-power, multilayer designs. The devices are evaluated in live animal models, on various locations/orientations on different internal organs (heart, lung, diaphragm), as well as inside the body of a bovine model through thoracotomy experiments. Output voltages of ∼4 V, ∼4 V, and ∼2 V, respectively, can be generated from contraction and relaxation of the right ventriculus, lung, and diaphragm, whereas a stack of five devices can harvest 1.2 $\mu W/cm^2$ power density. The same authors [271] presented ultrathin, stretchable networks of mechanical actuators (seven) and sensors (six) constructed with PZT nanoribbons, that allows for in vivo measurements of viscoelasticity in the near-surface regions of the epidermis as well as on ex vivo freshly explanted, unpreserved bovine organs (heart and lung). Zhang et al. [272] fabricated a PNG based on 200 μm-thick Al/PVDF wrapped around the ascending aorta of a pig, for harvesting energy from the variations in diameter of the carotid artery between the systolic and diastolic states (Figure 8a(I,II)). The resulting peak output voltage is 1.5 V at a heart rate and blood pressure of 120 bpm and 160/105 mmHg, respectively.

Yang et al. [273] demonstrated a single-wire PNG consisting of a ZnO nanowire affixed laterally at its two ends on a flexible substrate, for scavenging biomechanical energy (e.g., the movement of a human finger and the body motion of a live hamster) into electrical power (Figure 8b(I,II)). Four series-connected devices located on the back of a running hamster yield an output voltage of ~0.1–0.15 V. With the same principle, Li et al. [274] investigated the use of a single-wire PNG for energy harvesting under in vivo conditions, in a live rate, generating ~1 mV and 3 V output voltages from breathing (diaphragm) and heart beating, respectively (Figure 8c). Hwang et al. [275] presented a thin-film PMN-PT PNG to harvest energy from the motion of a rat's heart (with an output voltage of 8.2 V at a 0.36% strain and 2.3%/s strain rate) and also to act as an artificial heart stimulator.

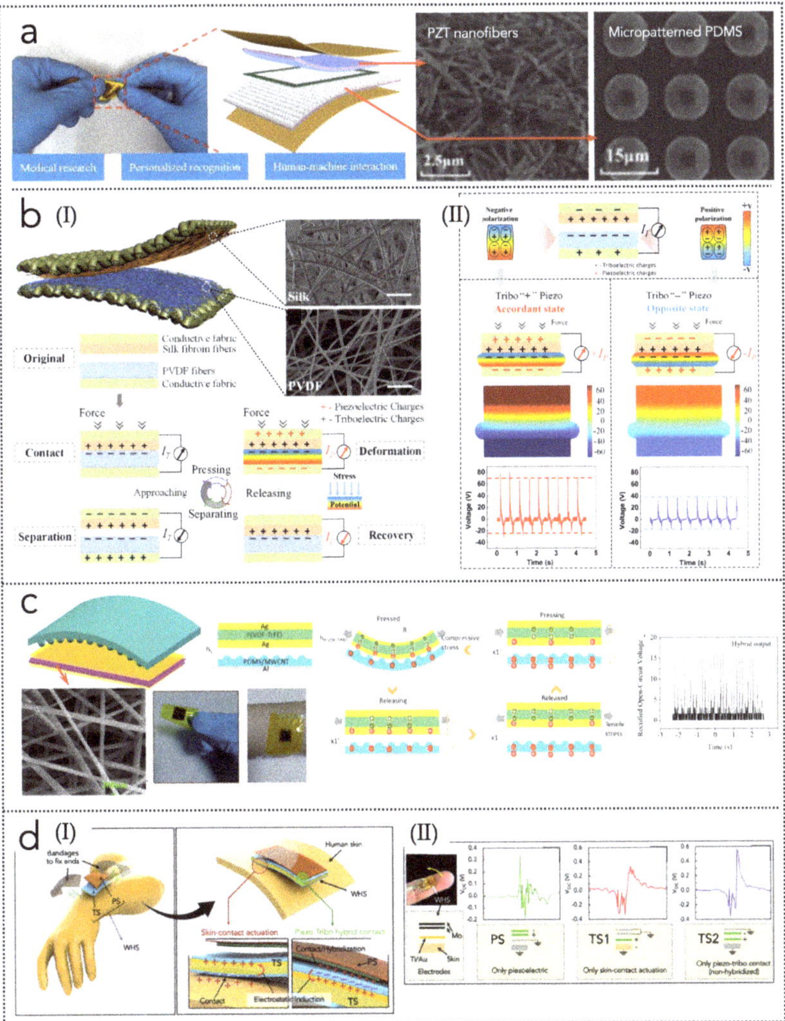

Figure 7. HBNGs for wearable bioelectronics. (**a**) Tactile PeT-HBNG based on a PZT-PDMS composite film. Reprinted with permission from ref. [153], 2019, Copyright Elsevier. (**b(I,II)**) PeT-HBNG based on electrospun silk fibroin and PVDF nanofibers. Reprinted with permission from ref. [163], 2018, Copyright, Elsevier. (**c**) Flexible wearable HBNG made of P(VDF-TrFE) electrospun nanofibers and multiwall carbon nanotubes (MWCNT)-doped PDMS composite membrane. Reprinted with permission from ref. [150], 2016, Copyright, Springer Nature. (**d(I,II)**) Inorganic/elastomeric 3-electrode skin-contact-actuation HBNG for multi-site and multifunctional conformal sensing. Reprinted from ref. [76], 2021, Copyright Wiley-VCH GmbH, Creative Commons CC-BY.

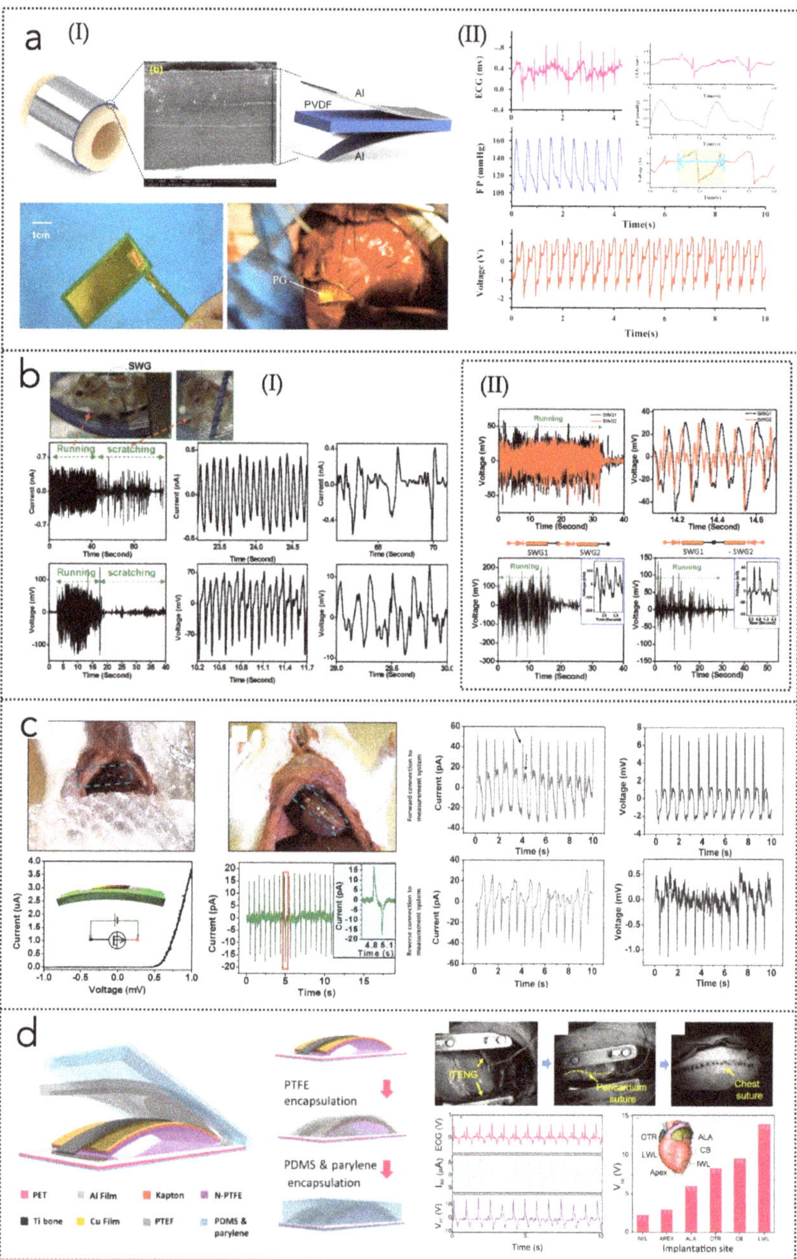

Figure 8. HBNGs for implantable bioelectronics. (**a**(**I,II**)) PNG based on 200 μm-thick Al/PVDF wrapped around the ascending aorta of a pig. Reprinted with permission from ref. [272], 2015, Copyright Elsevier. (**b**(**I,II**)) Single-wire PNG for scavenging biomechanical energy from a live hamster. Reprinted with permission from ref. [273], 2009, Copyright American Chemical Society. (**c**) Single-wire PNG for energy harvesting from breathing and heartbeat in a live rate. Reprinted with permission from ref. [274], 2010, Copyright Wiley-VCH Verlag GmbH and Co. KGaA. (**d**) Implantable TNG for in vivo biomechanical energy harvesting from the heartbeat of adult swine. Reprinted with permission from ref. [22], 2016, Copyright American Chemical Society.

Many examples of implantable TNGs can be found in literature [276–279]. Ryu et al. [280] reported on a commercial coin battery-sized high-performance inertia-driven TNG based on body motion and gravity, able to produce 4.9 μW/cm^3. The TNG has been also integrated with a cardiac pacemaker, demonstrating the successful operation mode of a self-rechargeable pacemaker system. Zheng et al. [22,281] proposed an innovative implantable TNG for in vivo biomechanical energy harvesting from the heartbeat of adult swine (Figure 8d). The structure is designed properly to achieve a leak-proof performance in vivo and it is composed of a multilayered core/shell/shell structure ("keel structure"), with nanostructured PTFE as triboelectric layer on Au-metallized Kapton as flexible substrate. A highly resilient titanium strip is introduced as the keel structure to strengthen the overall mechanical property and guarantee the contact-separation in vivo. The device can also be integrated in a self-powered wireless transmission system for real-time wireless cardiac monitoring.

The existing approaches that have been demonstrated for energy harvesting from the bodies of living subjects (animals and humans) for self-powered electronics are summarized in a review by Dagdeviren et al. [282] who described the material choices, device layouts and working principles (especially biofuel cells, thermoelectricity, triboelectricity and piezoelectricity) with a specific focus on in vivo applications, with future perspectives in disease diagnostics, treatment and prevention. Zheng et al. [209] and Zhang et al. [209] also reviewed the recent progresses of piezo-triboelectric flexible sensors and energy harvesters, providing examples of in vivo energy harvesting and direct stimulation of living cells, tissues and organs.

Energy harvesting, wearable and implantable bioelectronics represent the main application fields of HBNGs and while the first one has been widely explored, the others are promising in terms of scientific and commercial perspectives. Additionally, cutting-edge applications such as HBNGs for tissue engineering and nanomedicine may require in the future new research efforts.

5. Electronic Interfaces and Power Management Circuits for HBNGs

Interface electronics allows to transfer the output electrical signal generated by a transducer to a supplier or an energy storage downstream of the device. Generally speaking, integrated circuits (ICs) and discrete components are often used, and the demand of higher performances in terms of speed, low power, reduced parasitic capacitances, low noise-to-signal ratio and low cost is progressively increasing [283]. The recent needs of high multi-functionality, adaptability and diversification of transducers, especially with the introduction of hybridized systems, require in many cases the combination of interface electronics with high-performance packaging or their monolithic integration in the same system. In this context, the approaches of system-on-chip (SoC) and System-in-package (SiP) [284] with on-board systems for signal-processing and wireless communication [283,285] represent the most adopted methods for the fabrication of novel micro-systems, but are challenging when dealing with flexible and soft materials and devices. In fact, stretchable and conformal substrates and patches for flexible electronics require a multidisciplinary approach for the connection and integration with interface electronics, i.e., including fields of electrical circuits, material science, nanotechnology and packaging. This is further a more complex issue when the systems must be employed in contact with the human body or for the fabrication of implantable biomedical devices (IBDs): in this case requirements of conformability, biocompatibility and even biodegradability are essential, as well as durability of materials and reliability of electrical connections [286,287]. Moreover, the implementation of interface electronics strongly depends on the type of transducer and the ensemble of characteristics of the final micro-system, in terms of response, accuracy, cost and power consumption, is unavoidably determined by both the device and the connected electronics [283,284]. For instance, mechanical transducers, especially used for energy harvesting, require different types of power management circuits depending on the applied mechanical loads, exciting frequency and working mechanism: low-frequency mechanical source, such as water waves or weak breezes, require very low-threshold rectifying circuits or high-yield amplification [13,14,212].

In addition, PNGs and TNGs made of soft and flexible materials present different characteristics in terms of internal impedance, output power generation capability, frequency range of utilization and working principles, therefore they need specific circuits to convey the harvested power or sensory signals [100,115]. Since their outputs are usually composed of noncontinuous pulses with irregular magnitude, they cannot directly drive electronic devices that need a constant DC voltage, so to act as a battery, the PNG/TNG needs a self-charging power unit (SCPU) with at least an energy storage device. Therefore, the circuit attached to a nanogenerator is more complicated than a resistor for current regulation and power management, thus the challenge remains to establish a more general modelling approach to account for both complex operating conditions and practical energy harvesting circuit. This is further complicated when the two components are coupled to make HBNGs because signal rectification is necessary when they are out-of-phase. Additionally, their employment in harsh environments such as water, the human body, strong airflows or heavy gases, is a further issue to be addressed for the design and package of circuitry.

5.1. Circuits for Piezoelectricity

The simplest method to extract signal from a PNG is to connect it to a load resistance and detect the corresponding voltage. However, especially in the field of energy harvesting, this is not enough to power portable electronic devices and the AC signal must be converted into a DC signal. The most common approach is the rectification which can be implemented according to different models. The first is the Direct Energy Transfer (DET) approach which relies on a direct connection of the PNG with a bridge rectifier, a DC-DC converter and a storage capacitor (Figure 9a). The main issue of this approach is the risk that the energy stored in the capacitor flows back to the PNG when its instantaneous power is negative (energy return phenomenon [288]). Additionally, there is a problem of impedance matching with the internal resistance and capacitance of the PNG. A traditional interface used to match the PNG output impedance is described by Li et al. [289] and it is based on two coils with the same resonance frequency (i.e., a transformer, Figure 9b) which, however, needs a high value of inductance to have a proper impedance matching. A development of this interface consists of using a frequency up-conversion technique to shift the working operational frequency to higher values [290] (Figure 9c).

The second approach is based on resonant rectifiers (RR), i.e., the PNG is intermittently connected to the resonant electronic interface for very short time intervals and switched magnetic components (inductors or transformers, directly connected to the PNG or after a rectification stage) are responsible of the resonant effect (Figure 9d). This way, the impedance matching can be obtained whenever the RR is connected to the PNG, behaving similar to an LC oscillator, with a resulting enhancement of the maximum generated voltage with respect to an open-circuit condition [291,292]. Several RR have been introduced in previous works and they are comprehensively reviewed by Dicken et al. [293] and Dell'Anna et al. [294]: (1) series-synchronized switch harvesting on inductor (S-SSHI) [295] (Figure 9e); (2) parallel-synchronized switch harvesting on inductor (P-SSHI) [296] (Figure 9f); (3) synchronized switching and discharging to a storage capacitor through an inductor (SSDCI) [297] (Figure 9g); (4) synchronous electric charge extraction (SECE) [298] (Figure 9h); (5) phase shift synchronous electric charge extraction (PS-SECE) [299]; (6) synchronized switch harvesting on inductor magnetic rectifier (MR-SSHI) [300] (Figure 9i); (7) Hybrid MR/P-SSHI (Figure 9l); (8) Double synchronized switch harvesting (DSSH) [301] (Figure 9m); (9) enhanced synchronized switch harvesting (ESSH) [302]; (10) adaptive synchronized switch harvesting (ASSH) [302]; (11) energy injection technique [303] (Figure 9n). All the SSHI approaches depend strongly on the load, thus an additional DC-DC converter is usually required in practical implementations. The other methods (SECE, DSSH, ESSH) are more independent on the connected load, but they are affected by a linear decrease of the rectified voltage. The SSDCI approach allows a constant power extraction at certain loads and the EI technique provides higher output power peak performances. Another rectification approach relies on employing Maximum Power Point Tracking (MPPT) algorithms to control the impedance seen by the PNG, coupled with full-bridge (FB), half-bridge (HB) rectifiers or voltage/charge doublers (VD, CD) [304].

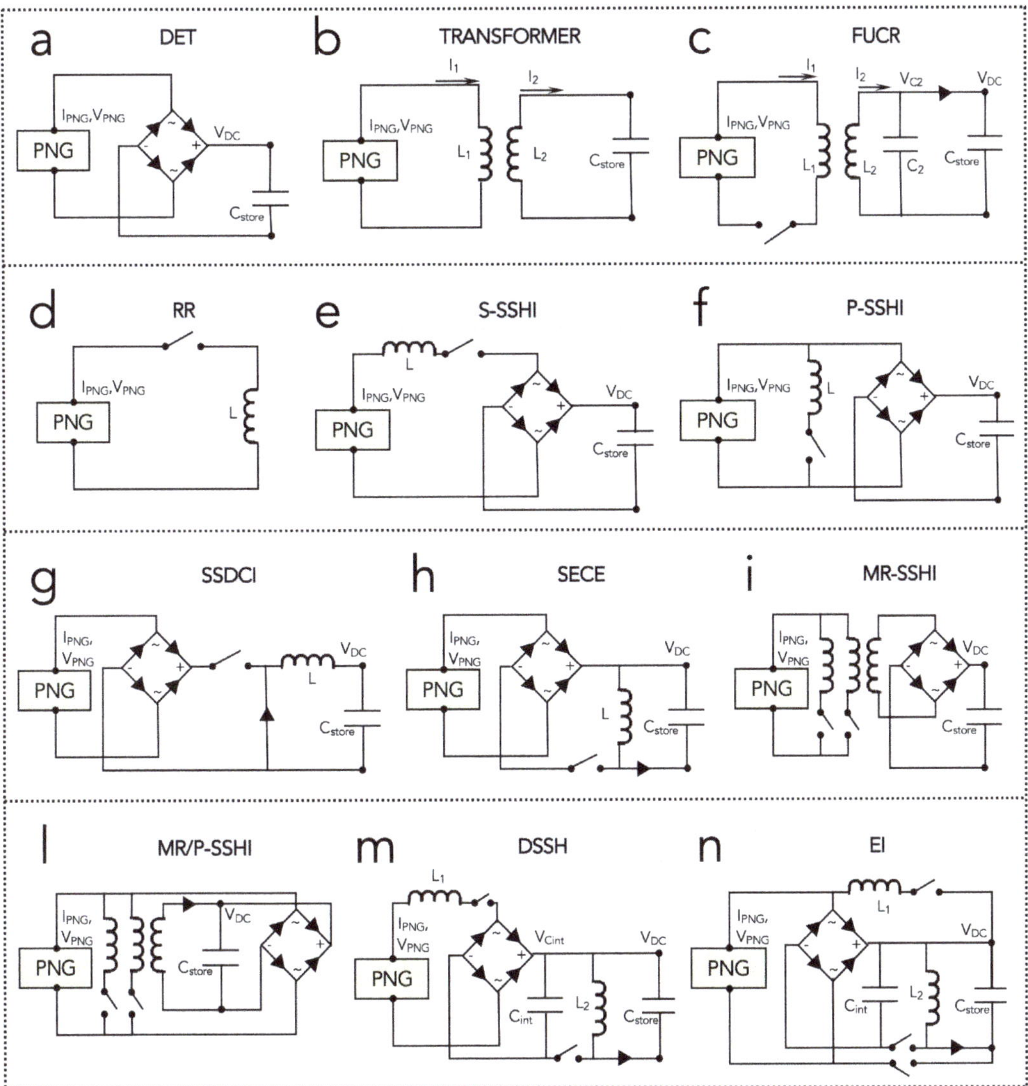

Figure 9. Electronic interfaces and power management circuits for PNGs. (**a**) Direct energy transfer (DET) approach. (**b**) Transformer-based approach. (**c**) Frequency up-conversion rectifier (FUCR). (**d**) Resonant rectifier (RR). (**e**) Series-Synchronized Switch Harvesting on Inductor (S-SSHI). (**f**) Parallel-Synchronized Switch Harvesting on Inductor (P-SSHI). (**g**) Synchronized Switching and Discharging to a storage Capacitor through an Inductor (SSDCI). (**h**) Synchronous Electric Charge Extraction (SECE). (**i**) Synchronized Switch Harvesting on Inductor Magnetic Rectifier (MR-SSHI). (**l**) Hybrid SSHI, made of MR-SSHI and P-SSHI. (**m**) Double Synchronized Switch Harvesting (DSSH). (**n**) Energy Injection (EI) technique.

5.2. Circuits for Triboelectricity

There are several issues related to the power management of TNGs. The first is that their high internal impedance (>10 MΩ) does not match with that of most standard electronic interfaces, thus the power transfer and the energy storage occurs with low efficiency [221,305]. The power management circuit plays a crucial role in the TNG-based energy harvesting, for efficiently storing the AC energy and overcome the huge impedance mismatch. The second issue is about the outputs of the TNG, i.e., high voltage and low current: this limits the practical applications of TNGs unless another system is used for the conversion to high currents. This could in principles be performed by a transformer but this has a high working frequency and a low-input-load resistance, which are the opposite of the TNG characteristics (low frequency and high-input-load resistance). The third issue regards the need of maximizing the extracted energy per cycle: generally, a switch approach between open-circuit and short-circuit conditions is needed [115,116,306]. The fourth aspect is the networking of several TNGs for distributed arrays of devices.

In order to reduce the TNG efficiency loss due to the impedance unmatching, Hu et al. [307] proposed an adaptable interface conditioning circuit, composed of an impedance matching circuit, a synchronous rectifier bridge, a control circuit and an energy storage device (Figure 10a). In particular, a novel bi-directional switch control mechanism is adopted to increase the frequency and reduce the energy loss of coupling impedance. Effectively, a TNG exhibits an increase of efficiency by 50% with the proposed interface.

The switch approach is usually adopted to maximize the extracted energy from the TNG. The common technique is to combine an electric switch and an inductance transducer and efficiencies up to 80% were reached in previous works [308,309], although plenty of energy is absorbed by the electric switch. Bao et al. [310] presented a new power management circuit for TNGs, that includes a rectification storage circuit and a DC-DC management circuit (Figure 10b). The first part (with a diode rectifier bridge and an optimal intermediate capacitor) serves to maximize the energy storage efficiency. The second circuit includes an automatic switch (transistor controlled by a low-power logic control circuit), an inductor and a diode in a buck-boost static converter; it works to obtain impedance match with the storage device or load. A rotating TNG equipped with the proposed circuit exhibits a storage efficiency of 50%.

Wang et al. [305] designed an SCPU for an arch-shaped TNG by integrating it with a flexible Li-ion battery (Figure 10c). It was employed to power a ZnO-nanowire-based UV sensor: the hybridization of the mechanical energy harvesting and energy storage in the SPCU allows the sensor to be self-powered continuously and sustainably, without any drop in the output current and voltage. Xi et al. [311,312] designed and patented a universal, efficient and autonomous power management module (PMM) made of a tribotronic energy extractor and a DC-DC buck converter (Figure 10d). The first serves to transfer energy efficiently from the TNG, and the second generates the DC output on the load. With the implemented PMM, about 85% energy can be autonomously extracted from the TNG as DC output voltage, and the matched impedance is reduced from 35 MΩ to 1 MΩ at 80% efficiency. Liang et al. [313] employed the PMM with a hexagonal TNG network consisting of spherical TNG units with spring-assisted multi-layered structure for water wave energy harvesting (Figure 10e): the units connected in series can produce a steady and continuous DC voltage on the load resistance and the PMM helps improving the energy storing by a factor 96.

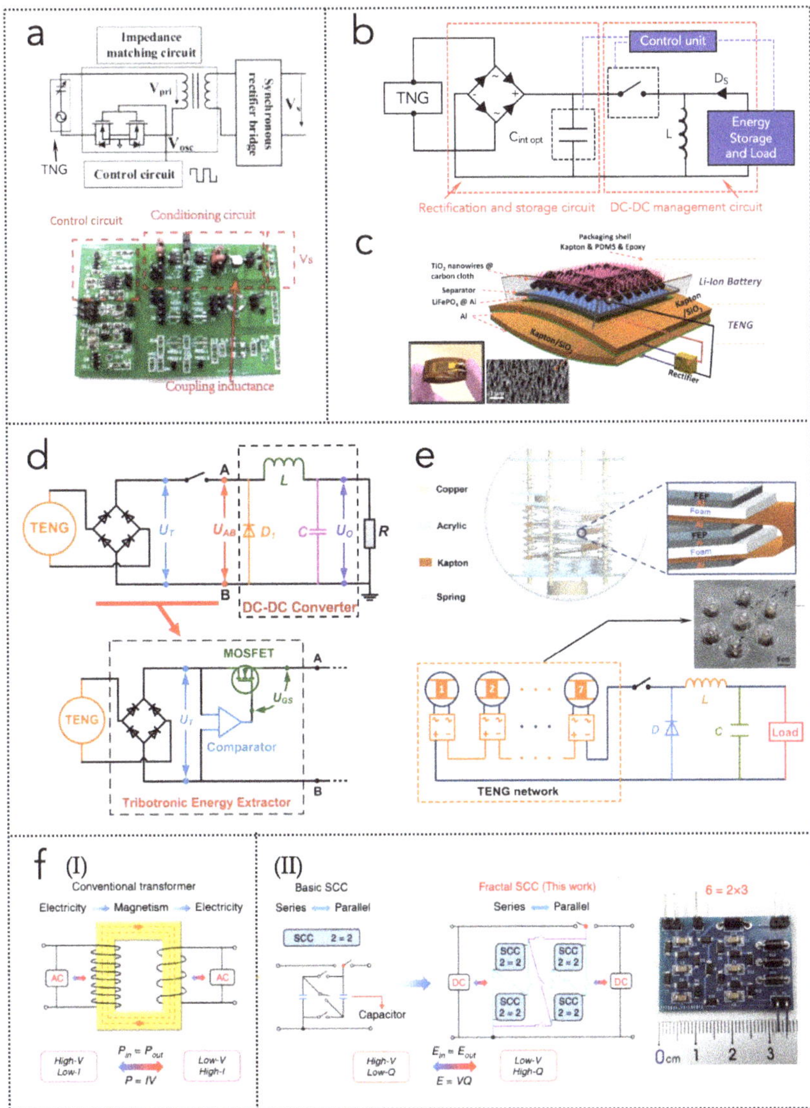

Figure 10. Electronic interfaces and power management circuits for TNGs. (**a**) Diagram and real photo of a power management circuit based on an impedance matching circuit and a synchronous rectifier bridge. Reprinted from ref. [307]. (**b**) Diagram of a power management circuit made of a rectification circuit and a DC-DC management circuit with an automatic switch. (**c**) Self-charging power unit (SCPU) of a TNG integrated with a Li-ion battery. Reprinted with permission from ref. [305], 2013, Copyright America Chemical Society. (**d**) Circuit diagram of PMM made of tribotronic energy extractor and DC-DC buck converter. Reprinted with permission from refs. [311,312], Copyright 2017, Elsevier. (**e**) PMM implemented with a hexagonal network of TNGs. Reprinted with permission from ref. [313], 2019, Copyright, WILEY-VCH Verlag GmbH and Co. KGaA. (**f(I,II)**) Fractal-design-based switched-capacitor-convertor compared to a conventional transformer. Reprinted with permission from ref. [314], 2020 Copyright Springer Nature, Creative Commons CC BY.

To avoid the energy loss due to the electric switches, another strategy has been proposed, based on an inductor-free mechanical switch between serial and parallel connections of capacitors [315,316]. Liu et al. [314] report on a fractal-design-based switched-capacitor-convertor with high conversion efficiency, minimum output impedance and electrostatic voltage applicability (Figure 10f(I,II)). The convertor is magnet-free, light-weight, easily integrated on printed circuit boards and if coupled to a TNG, it can enhance the charge transfer with a 67-fold factor, reaching a power density of 954 Wm^{-2} in pulse mode at 1 Hz, with a >94% total energy transfer efficiency. Xu et al. [188] proposed a multilayered-electrode-based TNG with the aim of lowering the output voltage and preserving the total power (to overcome the problem of high-voltage/low-current). With this design, the current flow is controlled by a mechanical switch, yielding an increasing total charge transport. The concept of multilayer design is also adopted by Xu et al. [317] who fabricated the switches by attaching Al foils on the air chambers of a TNG: the switches close when the corresponding foils acquire contact and current flow by the drive of the established voltage due to charge separation. The switch circuit reduces the duration of the charge transfer thus enhancing the output current and power.

5.3. Circuits for HBNGs

Since an optimal universal solution for power management of PNGs and TNGs is still lacking due to the wide range of operating conditions and of technical issues to overcome, it is even harder and more challenging to provide an established approach for the electronic interfaces of HBNGs. In fact, the hybridization occurring inside the system is not necessarily additive in terms of voltages and currents, since this depends basically on the electrodes' configuration and the polarization direction of the piezoelectric material, as discussed in Section 3. In the simplest case, each elemental device (PNG or TNG) inside a HBNG needs a rectifying circuit to extract the signal, unless the HBNG is based on a 2-terminal hybridization with physical integration. Therefore, the most common electronic interfaces for a HBNG consists of some full-wave bridge rectifiers (FWBRs) connected in parallel (Figure 11a(i)) or in series (Figure 11a(ii)) and a storage capacitor [318]. Li et al. [319] reported on a HBNG to harvest energy from low-frequency ambient vibrations, made of a PNG patch, a TNG patch, a spring-mass system and an amplitude limiter, introduced to achieve the desired frequency up-conversion effect. The circuit for signal extraction and power management contains two rectifiers to invert the negative current signals into positive ones and they are connected independently to the PNG and TNG in parallel to a storage capacitor. A similar type of connection is used for the HBNGs presented by Wang et al. [150], whereas only one full-wave rectifier is adopted by Li et al. [320]. Mariello et al. [13,76] employed an electronic interface with two full-wave rectifiers in series or in parallel to extract the hybrid signal from the 3-electrode HBNGs, whereas to achieve the single voltage and current waveforms they used two differential high-impedance (~200 MΩ–1 GΩ) buffer circuits, supplied by 9 V batteries, to decouple the internal impedance of the measuring oscilloscope and separate the different signals (two for PNG, two for TNG) (Figure 11b). Singh and Khare [185] fabricated a piezoelectric-triboelectric-electromagnetic HBNG, based on ZnO-PVDF, to harvest mechanical vibrations. The electrical connections used to extract simultaneously the power outputs from the three transducing mechanisms include five terminals connected to three full-wave low-leakage bridge rectifier ICs (DF10M): all of the outputs from the rectifiers are connected in parallel. Intermediate transformers are often used between the nanogenerator and the bridge rectifier, in order to match the impedance between the different parts, reduce the power consumption in the power source and enhance the power transfer capability [175,177].

Different approaches can also be adopted and they can consist of exploiting a transducing component of the HBNG as an active part of the power management circuit. For instance, Lallart and Lombardi [321] presented a hybrid nonlinear interface combining piezoelectric and electromagnetic effects for energy harvesting purposes (Figure 11c(i,ii)). Inspired to the SSHI interface mentioned in Section 5.1, the new approach relies on the replacement of the passive inductance with an electromagnetic transducer, thus it can be ascribed in the power management circuits for HBNGs. It is called synchronized switch harvesting on electromagnetic system (SSH-EM) and provides an increase of the output voltage and harvested power, compared to the SSHI or the standard DET approaches.

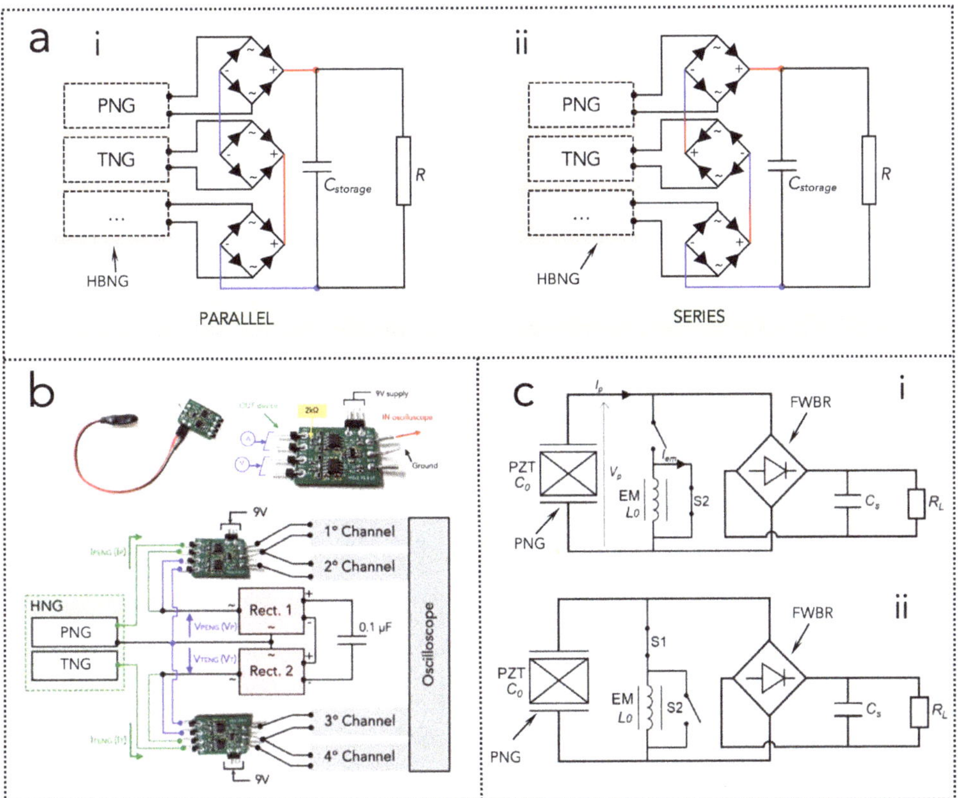

Figure 11. Power management circuits for HBNGs. (**a**) Electrical interfaces for HBNGs with full-wave bridge rectifiers connected in parallel (**i**) or in series (**ii**). (**b**) Differential high-impedance buffer circuit used to measure simultaneously voltage and current signals for the PNG and TNG, functioning as a voltmeter and an ammeter. In the bottom, circuit employed to obtain synchronized voltage/current waveforms for PNG and TNG. Two buffer circuits are used and each of them provides two pairs of outputs, detected as four separate channels by the oscilloscope. Reprinted with permission from ref. [13], Copyright 2021, Elsevier. (**c**) SSH-EM circuit configuration at default state (**i**) and switching state (**ii**). Reprinted with permission from ref. [321], Copyright 2019, Elsevier.

In conclusion, a unified solution for the power management of HBNGs with more than two terminals is still lacking and needs further investigation and analysis.

6. Conclusions and Future Challenges

Nowadays, the global energy problem implies a high demand of novel environmentally friendly, efficient and cost-effective technologies to scavenge energy from the environment and from humans. Nanogenerators represent a valid complementary option to the standard bulky systems such as power plants or turbines. On the other hand, in the context of the spreading Internet of Things revolution, there is a high demand of easy-to-use, rapid, biocompatible and non-invasive method for monitoring human life parameters, for sensing human body motion, for stimulating chemical reactions or for implantable applications. Nanogenerators have also emerged as powerful tools for these human-centered applications and many research efforts have been made for designing new architectures, for developing and using novel materials, for increasing the device's performances. Against most current devices based on single-mechanism transducers (nanogenerators or sensors), the future direction of these technologies is to exploit multi mechanisms simultaneously. Hybridization brings some key advantages such as a further miniaturization of the devices and enhanced signal generation. The technology of the hybrid bio-nanogenerators (HBNGs) is promising and in continuous growth.

In this paper, the recent advances of hybrid nanogenerators and sensors based on piezoelectric and triboelectric transduction mechanisms are reviewed. These two mechanisms are the most widely adopted and their advantages to other systems are discussed: when combined together, both mechanical deformation and contact friction of materials can be exploited in the same hybrid device to scavenge or convert mechanical energy.

Different possibilities for the design of hybridized devices are described, based on the physical domains involved in the energy conversion. In this respect, intra-domain or inter-domain hybridization can be exploited. The piezoelectric and triboelectric transducers can be used individually or together, in combination with other transducing principles (electromagnetic, solar, photovoltaic, etc.). Regarding the relative position and physical interpenetration of the piezoelectric and triboelectric components the hybridization can occur with physical separation or with physical integration.

Power management of piezoelectric, triboelectric and hybrid nanogenerators is also discussed, highlighting the higher complexity that hybrid devices bring for signal extraction. The three main applications of HBNGs are presented: energy harvesting, wearable bioelectronics and implantable bioelectronics. The first category is so far the most widely explored with plenty of examples of nanogenerators that scavenge energy from environmental clean-energy sources (wind, water etc.). The last two categories are less investigated and robust commercializable solutions are still lacking because of additional challenges regarding the skin adhesion, conformality, adaptability, miniaturization and implantation.

As similarly described in [294,322] for piezoelectric energy harvesting systems, there are many criteria to evaluate the performances of a HBNG: (1) efficiency, (2) standalone operation, (3) circuit complexity, (4) adaptivity to the environment, (5) micro-scale compatibility, (6) start-up operation, (7) minimum operating voltage, (8) internal impedance, (9) flexibility and conformality. Future research work is urgently needed for several key aspects and challenges, that are pointed out below and illustrated in Figure 12.

The first aspect to focus on is the *output generation performances* of the HBNGs, which depends on the performances of the single components and also the mutual combination and interaction. To enhance the performances, novel materials with higher capability of generation of piezo-potentials or surface charges are needed, in particular polymers with lower dielectric permittivity, electrostatic breakdown rigidity and mechanical robustness [221,323]. On the other hand, acting on the design and architecture of the hybridized device plays an important role in enhancing the overall performances. This holds true both for increasing the power density (for energy harvesting applications) and for increasing the sensitivity of sensing devices (for biomedical applications). The optimization of the design can be performed on two sides, i.e., the structural design or the electronic circuitry. General principles for the enhancement of triboelectrification are (i) multilayer structures, (ii) grating structures, (iii) softness to increase contact areas, (iv) charge pumping or surface treatments/functionalization strategies to increase surface charge densities [212,221].

The second aspect that needs further investigation and urgent solutions regards the *reliability* of HBNGs. This includes the durability of the materials involved in the fabrication of the devices: since they are supposed to undergo mechanical deformations or friction/contact, the stability of all the interfaces is necessary for a long-term operation. A second factor that affects the reliability, especially for devices aimed at being employed in water or implanted in the human body, is the package and external embodiment of the HBNGs. Hermetic packages or ultra-high barrier encapsulations are necessary for protecting the devices against biofouling, humidity or water permeation that can induce internal short-circuits, corrosion or delaminations. Water affects negatively both triboelectricity, by forming thin layers on the tribo-surfaces, and piezoelectricity, by reducing the dielectric strength of the active material. Acting on the wettability (hydrophobicity) of the involved layered materials in HBNGs can be a solution for impeding the water absorption [137]. Other harsh agents of device failure are dryness, high temperatures (which cause the loss of piezoelectricity beyond the Curie temperature and a reduction of triboelectricity through electron thermion emission), UV radiations (which cause the desorption of oxygen molecules and formation of free radicals that influence the charge generation), the presence of chemical substances, contaminants, interstitial atomic hydrogen and oxygen vacancies in ceramic materials [209].

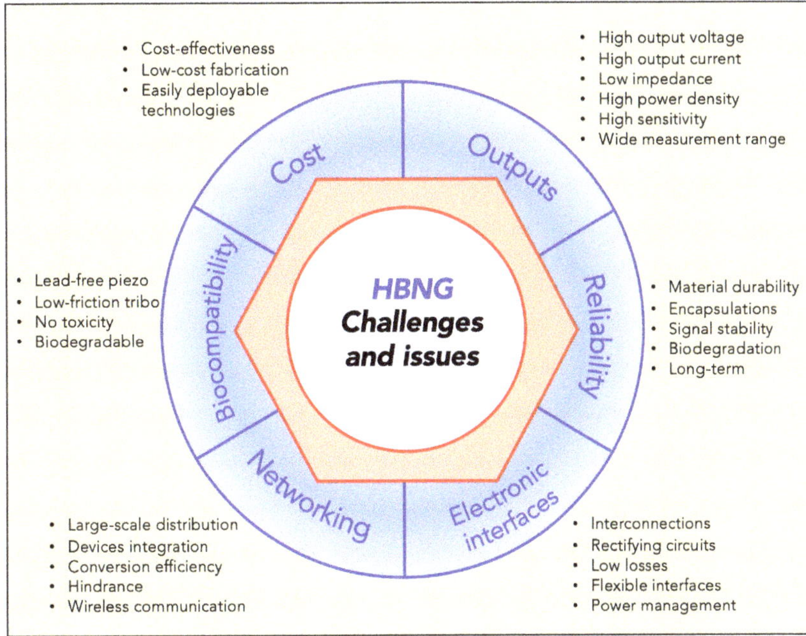

Figure 12. Illustrations of challenges and issues to be solved for the future employment of HBNGs.

The third aspect concerns the power management circuits and *electronic interfaces* used to extract signals from the HBNGs. They need to be optimized to minimize the losses, maximize the signal-to-noise ratio, increase the efficiency, especially for energy harvesting applications, or lower the sensitivity limit of the devices, for sensing applications. The sensitivity, as well as linearity, is crucial for physiological monitoring of heartbeat, intestinal movement or respiratory rate and hybrid devices can improve it to provide broad sensing ranges.

In case of implantable or wearable bioelectronic devices, the electronic interfaces can be incorporated in the package and they need to be carried by flexible or even stretchable substrates. This means that new methods of fabricating the electronic circuitry are required, based on soft substrates and reliable deposition of the metal interconnections. Additionally, piezoelectricity produces a low output voltage and higher output current while triboelectricity produces a high output voltage but with low output current. Leakage due to arc between the triboelectric materials can occur in the same device leading to degraded performances. Hence, proper circuits for handling the high triboelectric impedance must be designed.

The fourth aspect is the *networking*, i.e., the possibility to have distributed arrays of HBNGs for specific applications. An example is related to the blue energy harvesting for which coupling together different units of HBNGs can be exploited to enhance the overall efficiency of power scavenging. Distributed electrodes or devices are increasingly demanded for implantable applications, especially for stimulating simultaneously different regions of the human brain or other organs. This poses further challenges for the electronic systems that need to handle the signal of multiple devices simultaneously, thus multiplexing instrumentations are required. In addition, infrastructures for big data, artificial intelligence (AI) and deep learning are growingly required to connect large arrays of distributed self-powered sensors and to improve the systems for intelligent signal acquisition [323].

The fifth aspect is related to the *biocompatibility* of the materials used for the realization of HBNGs. The demand of environmentally friendly or biodegradable materials is of utmost important in order to develop green technologies both for power generation and for wearable/implantable sensing/actuation.

The sixth aspect is the *cost* of these novel technologies. Although they are really more cost-effective than the standard counterparts (power plants, turbines, etc.), the deposition techniques and the fabrication are still based on processes and equipment that are not affordable at very low price. These machines and processes typically are performed with cleanroom facilities and need frequent maintenance, thus lowering the price of these technologies can favor the spread and wide utilization of HBNGs, even on a large scale.

Author Contributions: M.M. wrote and edited the manuscript. All authors have read and agreed to the published version of the manuscript.

Funding: This research received no external funding.

Data Availability Statement: Not applicable.

Conflicts of Interest: The authors declare no conflict of interest.

References

1. Wang, Z.L.; Song, J. Piezoelectric Nanogenerators Based on Zinc Oxide Nanowire Arrays. *Science* **2006**, *312*, 242–246. [CrossRef] [PubMed]
2. Zhang, Z.; Li, X.; Yin, J.; Xu, Y.; Fei, W.; Xue, M.; Wang, Q.; Zhou, J.; Guo, W. Emerging Hydrovoltaic Technology. *Nat. Nanotechnol.* **2018**, *13*, 1109–1119. [CrossRef] [PubMed]
3. Helseth, L.E.; Guo, X.D. Contact Electrification and Energy Harvesting Using Periodically Contacted and Squeezed Water Droplets. *Langmuir* **2015**, *31*, 3269–3276. [CrossRef] [PubMed]
4. Mariello, M.; Guido, F.; Mastronardi, V.M.; Todaro, M.T.; Desmaële, D.; De Vittorio, M. Nanogenerators for Harvesting Mechanical Energy Conveyed by Liquids. *Nano Energy* **2019**, *57*, 141–156. [CrossRef]
5. Mariello, M.; Guido, F.; Mastronardi, V.M.; Madaro, F.; Mehdipour, I.; Todaro, M.T.; Rizzi, F.; De Vittorio, M. Chapter 13—Micro- and nanodevices for wind energy harvesting. In *Nano Tools and Devices for Enhanced Renewable Energy*; Devasahayam, S., Hussain, C.M., Eds.; Micro and Nano Technologies; Elsevier: Amsterdam, The Netherlands, 2021; pp. 291–374. ISBN 978-0-12-821709-2.
6. Wang, Z.L. Catch Wave Power in Floating Nets. *Nat. News* **2017**, *542*, 159. [CrossRef] [PubMed]
7. Qin, Y.; Wang, X.; Wang, Z.L. Microfibre–Nanowire Hybrid Structure for Energy Scavenging. *Nature* **2008**, *451*, 809–813. [CrossRef] [PubMed]
8. Lai, Y.-C.; Hsiao, Y.-C.; Wu, H.-M.; Wang, Z.L. Waterproof Fabric-Based Multifunctional Triboelectric Nanogenerator for Universally Harvesting Energy from Raindrops, Wind, and Human Motions and as Self-Powered Sensors. *Adv. Sci.* **2019**, *6*, 1801883. [CrossRef]
9. Ippili, S.; Jella, V.; Thomas, A.M.; Yoon, S.-G. The Recent Progress on Halide Perovskite-Based Self-Powered Sensors Enabled by Piezoelectric and Triboelectric Effects. *Nanoenergy Adv.* **2021**, *1*, 3–31. [CrossRef]
10. Chen, J.; Huang, Y.; Zhang, N.; Zou, H.; Liu, R.; Tao, C.; Fan, X.; Wang, Z.L. Micro-Cable Structured Textile for Simultaneously Harvesting Solar and Mechanical Energy. *Nat. Energy* **2016**, *1*, 16138. [CrossRef]
11. Ren, Z.; Zheng, Q.; Wang, H.; Guo, H.; Miao, L.; Wan, J.; Xu, C.; Cheng, S.; Zhang, H. Wearable and Self-Cleaning Hybrid Energy Harvesting System Based on Micro/Nanostructured Haze Film. *Nano Energy* **2020**, *67*, 104243. [CrossRef]
12. Cho, Y.; Lee, S.; Hong, J.; Pak, S.; Hou, B.; Lee, Y.-W.; Jang, J.E.; Im, H.; Sohn, J.I.; Cha, S.; et al. Sustainable Hybrid Energy Harvester Based on Air Stable Quantum Dot Solar Cells and Triboelectric Nanogenerator. *J. Mater. Chem. A* **2018**, *6*, 12440–12446. [CrossRef]

13. Mariello, M.; Fachechi, L.; Guido, F.; De Vittorio, M. Multifunctional Sub-100 μm Thickness Flexible Piezo/Triboelectric Hybrid Water Energy Harvester Based on Biocompatible AlN and Soft Parylene C-PDMS-EcoflexTM. *Nano Energy* **2021**, *83*, 105811. [CrossRef]

14. Khan, U.; Kim, S.-W. Triboelectric Nanogenerators for Blue Energy Harvesting. *ACS Nano* **2016**, *10*, 6429–6432. [CrossRef] [PubMed]

15. Viet, N.V.; Wu, N.; Wang, Q. A Review on Energy Harvesting from Ocean Waves by Piezoelectric Technology. *J. Model. Mech. Mater.* **2017**, *1*. [CrossRef]

16. Wang, Z.L. Self-Powered Nanosensors and Nanosystems. *Adv. Mater.* **2012**, *24*, 280–285. [CrossRef]

17. Wu, Z.; Cheng, T.; Wang, Z.L. Self-Powered Sensors and Systems Based on Nanogenerators. *Sensors* **2020**, *20*, 2925. [CrossRef]

18. Wang, Z.L. Energy Harvesting for Self-Powered Nanosystems. *Nano Res.* **2008**, *1*, 1–8. [CrossRef]

19. Pu, X.; Li, L.; Liu, M.; Jiang, C.; Du, C.; Zhao, Z.; Hu, W.; Wang, Z.L. Wearable Self-Charging Power Textile Based on Flexible Yarn Supercapacitors and Fabric Nanogenerators. *Adv. Mater.* **2016**, *28*, 98–105. [CrossRef]

20. Sim, H.J.; Choi, C.; Kim, S.H.; Kim, K.M.; Lee, C.J.; Kim, Y.T.; Lepró, X.; Baughman, R.H.; Kim, S.J. Stretchable Triboelectric Fiber for Self-Powered Kinematic Sensing Textile. *Sci. Rep.* **2016**, *6*, 35153. [CrossRef]

21. Chen, X.; Shao, J.; An, N.; Li, X.; Tian, H.; Xu, C.; Ding, Y. Self-Powered Flexible Pressure Sensors with Vertically Well-Aligned Piezoelectric Nanowire Arrays for Monitoring Vital Signs. *J. Mater. Chem. C* **2015**, *3*, 11806–11814. [CrossRef]

22. Zheng, Q.; Zhang, H.; Shi, B.; Xue, X.; Liu, Z.; Jin, Y.; Ma, Y.; Zou, Y.; Wang, X.; An, Z.; et al. In Vivo Self-Powered Wireless Cardiac Monitoring via Implantable Triboelectric Nanogenerator. *ACS Nano* **2016**, *10*, 6510–6518. [CrossRef] [PubMed]

23. Gu, Y.; Zhang, T.; Chen, H.; Wang, F.; Pu, Y.; Gao, C.; Li, S. Mini Review on Flexible and Wearable Electronics for Monitoring Human Health Information. *Nanoscale Res. Lett.* **2019**, *14*, 263. [CrossRef] [PubMed]

24. Zhang, Z.; Du, K.; Chen, X.; Xue, C.; Wang, K. An Air-Cushion Triboelectric Nanogenerator Integrated with Stretchable Electrode for Human-Motion Energy Harvesting and Monitoring. *Nano Energy* **2018**, *53*, 108–115. [CrossRef]

25. Mitcheson, P.D.; Yeatman, E.M.; Rao, G.K.; Holmes, A.S.; Green, T.C. Energy Harvesting from Human and Machine Motion for Wireless Electronic Devices. *Proc. IEEE* **2008**, *96*, 1457–1486. [CrossRef]

26. Guido, F.; Qualtieri, A.; Algieri, L.; Lemma, E.D.; De Vittorio, M.; Todaro, M.T. AlN-Based Flexible Piezoelectric Skin for Energy Harvesting from Human Motion. *Microelectron. Eng.* **2016**, *159*, 174–178. [CrossRef]

27. Yang, Y.; Zhang, H.; Lin, Z.-H.; Zhou, Y.S.; Jing, Q.; Su, Y.; Yang, J.; Chen, J.; Hu, C.; Wang, Z.L. Human Skin Based Triboelectric Nanogenerators for Harvesting Biomechanical Energy and as Self-Powered Active Tactile Sensor System. *ACS Nano* **2013**, *7*, 9213–9222. [CrossRef]

28. Shi, Q.; Wang, H.; Wu, H.; Lee, C. Self-Powered Triboelectric Nanogenerator Buoy Ball for Applications Ranging from Environment Monitoring to Water Wave Energy Farm. *Nano Energy* **2017**, *40*, 203–213. [CrossRef]

29. Olsen, M.; Zhang, R.; Örtegren, J.; Andersson, H.; Yang, Y.; Olin, H. Frequency and Voltage Response of a Wind-Driven Fluttering Triboelectric Nanogenerator. *Sci. Rep.* **2019**, *9*, 5543. [CrossRef]

30. Shi, Q.; Zhang, Z.; Chen, T.; Lee, C. Minimalist and Multi-Functional Human Machine Interface (HMI) Using a Flexible Wearable Triboelectric Patch. *Nano Energy* **2019**, *62*, 355–366. [CrossRef]

31. O'Flynn, B.; Sanchez-Torres, J.; Tedesco, S.; Walsh, M. Challenges in the development of wearable human machine interface systems. In Proceedings of the 2019 IEEE International Electron Devices Meeting (IEDM), San Francisco, CA, USA, 7–11 December 2019; pp. 10.4.1–10.4.4.

32. Sun, Z.; Zhu, M.; Lee, C. Progress in the Triboelectric Human–Machine Interfaces (HMIs)-Moving from Smart Gloves to AI/Haptic Enabled HMI in the 5G/IoT Era. *Nanoenergy Adv.* **2021**, *1*, 81–120. [CrossRef]

33. Gerratt, A.P.; Michaud, H.O.; Lacour, S.P. Elastomeric Electronic Skin for Prosthetic Tactile Sensation. *Adv. Funct. Mater.* **2015**, *25*, 2287–2295. [CrossRef]

34. Brunelli, D.; Tadesse, A.M.; Vodermayer, B.; Nowak, M.; Castellini, C. Low-cost wearable multichannel surface EMG acquisition for prosthetic hand control. In Proceedings of the 6th International Workshop on Advances in Sensors and Interfaces (IWASI), Gallipoli, Italy, 18–19 June 2015; pp. 94–99.

35. Kim, B.-Y.; Lee, W.-H.; Hwang, H.-G.; Kim, D.-H.; Kim, J.-H.; Lee, S.-H.; Nahm, S. Resistive Switching Memory Integrated with Nanogenerator for Self-Powered Bioimplantable Devices. *Adv. Funct. Mater.* **2016**, *26*, 5211–5221. [CrossRef]

36. Zhang, X.-S.; Han, M.-D.; Wang, R.-X.; Zhu, F.-Y.; Li, Z.-H.; Wang, W.; Zhang, H.-X. Frequency-Multiplication High-Output Triboelectric Nanogenerator for Sustainably Powering Biomedical Microsystems. *Nano Lett.* **2013**, *13*, 1168–1172. [CrossRef] [PubMed]

37. Mukhopadhyay, S.C.; Suryadevara, N.K. *Internet of Things: Challenges and Opportunities*; Mukhopadhyay, S.C., Ed.; Smart Sensors, Measurement and Instrumentation; Springer International Publishing: Cham, Switzerland, 2014; pp. 1–17. ISBN 978-3-319-04223-7.

38. Yin, Y.; Zeng, Y.; Chen, X.; Fan, Y. The Internet of Things in Healthcare: An Overview. *J. Ind. Inf. Integr.* **2016**, *1*, 3–13. [CrossRef]

39. Lee, M.; Boudreaux, B.; Chaturvedi, R.; Romanosky, S.; Downing, B. *The Internet of Bodies: Opportunities, Risks, and Governance*; RAND Corporation: Santa Monica, CA, USA, 2020.

40. Fan, F.R.; Tang, W.; Wang, Z.L. Flexible Nanogenerators for Energy Harvesting and Self-Powered Electronics. *Adv. Mater.* **2016**, *28*, 4283–4305. [CrossRef]

41. Lee, J.-H.; Lee, K.Y.; Kumar, B.; Tien, N.T.; Lee, N.-E.; Kim, S.-W. Highly Sensitive Stretchable Transparent Piezoelectric Nanogenerators. *Energy Environ. Sci.* **2012**, *6*, 169–175. [CrossRef]
42. Akaydin, H.D.; Elvin, N.; Andreopoulos, Y. Energy Harvesting from Highly Unsteady Fluid Flows Using Piezoelectric Materials. *J. Intell. Mater. Syst. Struct.* **2010**, *21*, 1263–1278. [CrossRef]
43. Dagdeviren, C.; Joe, P.; Tuzman, O.L.; Park, K.-I.; Lee, K.J.; Shi, Y.; Huang, Y.; Rogers, J.A. Recent Progress in Flexible and Stretchable Piezoelectric Devices for Mechanical Energy Harvesting, Sensing and Actuation. *Extreme Mech. Lett.* **2016**, *9*, 269–281. [CrossRef]
44. Choi, A.Y.; Lee, C.J.; Park, J.; Kim, D.; Kim, Y.T. Corrugated Textile Based Triboelectric Generator for Wearable Energy Harvesting. *Sci. Rep.* **2017**, *7*, 45583. [CrossRef]
45. Wang, A.C.; Wu, C.; Pisignano, D.; Wang, Z.L.; Persano, L. Polymer Nanogenerators: Opportunities and Challenges for Large-Scale Applications. *J. Appl. Polym. Sci.* **2018**, *135*, 45674. [CrossRef]
46. Todaro, M.T.; Guido, F.; Mastronardi, V.; Desmaele, D.; Epifani, G.; Algieri, L.; De Vittorio, M. Piezoelectric MEMS Vibrational Energy Harvesters: Advances and Outlook. *Microelectron. Eng.* **2017**, *183–184*, 23–36. [CrossRef]
47. Todaro, M.T.; Guido, F.; Algieri, L.; Mastronardi, V.M.; Desmaële, D.; Epifani, G.; Vittorio, M.D. Biocompatible, Flexible, and Compliant Energy Harvesters Based on Piezoelectric Thin Films. *IEEE Trans. Nanotechnol.* **2018**, *17*, 220–230. [CrossRef]
48. Lu, N.; Lu, C.; Yang, S.; Rogers, J. Highly Sensitive Skin-Mountable Strain Gauges Based Entirely on Elastomers. *Adv. Funct. Mater.* **2012**, *22*, 4044–4050. [CrossRef]
49. Zheng, Y.-N.; Yu, Z.; Mao, G.; Li, Y.; Pravarthana, D.; Asghar, W.; Liu, Y.; Qu, S.; Shang, J.; Li, R.-W. A Wearable Capacitive Sensor Based on Ring/Disk-Shaped Electrode and Porous Dielectric for Noncontact Healthcare Monitoring. *Glob. Chall.* **2020**, *4*, 1900079. [CrossRef] [PubMed]
50. Lei, Z.; Wu, P. A Supramolecular Biomimetic Skin Combining a Wide Spectrum of Mechanical Properties and Multiple Sensory Capabilities. *Nat. Commun.* **2018**, *9*, 1134. [CrossRef] [PubMed]
51. Zhang, X.; Sheng, N.; Wang, L.; Tan, Y.; Liu, C.; Xia, Y.; Nie, Z.; Sui, K. Supramolecular Nanofibrillar Hydrogels as Highly Stretchable, Elastic and Sensitive Ionic Sensors. *Mater. Horiz.* **2019**, *6*, 326–333. [CrossRef]
52. Ryu, H.; Yoon, H.-J.; Kim, S.-W. Hybrid Energy Harvesters: Toward Sustainable Energy Harvesting. *Adv. Mater.* **2019**, *31*, 1802898. [CrossRef] [PubMed]
53. Kim, S.-G.; Priya, S.; Kanno, I. Piezoelectric MEMS for Energy Harvesting. *MRS Bull.* **2012**, *37*, 1039–1050. [CrossRef]
54. Li, P.; Ryu, J.; Hong, S. *Piezoelectric/Triboelectric Nanogenerators for Biomedical Applications*; IntechOpen: London, UK, 2019; ISBN 978-1-83881-060-3.
55. Fang, J.; Wang, X.; Lin, T. Electrical Power Generator from Randomly Oriented Electrospun Poly(Vinylidene Fluoride) Nanofibre Membranes. *J. Mater. Chem.* **2011**, *21*, 11088–11091. [CrossRef]
56. Chen, X.; Xu, S.; Yao, N.; Shi, Y. 1.6 V Nanogenerator for Mechanical Energy Harvesting Using PZT Nanofibers. *Nano Lett.* **2010**, *10*, 2133–2137. [CrossRef]
57. Bai, S.; Xu, Q.; Gu, L.; Ma, F.; Qin, Y.; Wang, Z.L. Single Crystalline Lead Zirconate Titanate (PZT) Nano/Micro-Wire Based Self-Powered UV Sensor. *Nano Energy* **2012**, *1*, 789–795. [CrossRef]
58. Wang, P.; Du, H. ZnO Thin Film Piezoelectric MEMS Vibration Energy Harvesters with Two Piezoelectric Elements for Higher Output Performance. *Rev. Sci. Instrum.* **2015**, *86*, 075002. [CrossRef] [PubMed]
59. Algieri, L.; Todaro, M.T.; Guido, F.; Mastronardi, V.; Desmaële, D.; Qualtieri, A.; Giannini, C.; Sibillano, T.; De Vittorio, M. Flexible Piezoelectric Energy-Harvesting Exploiting Biocompatible AlN Thin Films Grown onto Spin-Coated Polyimide Layers. *ACS Appl. Energy Mater.* **2018**, *1*, 5203–5210. [CrossRef]
60. Eom, C.-B.; Trolier-McKinstry, S. Thin-Film Piezoelectric MEMS. *MRS Bull.* **2012**, *37*, 1007–1017. [CrossRef]
61. Wasa, K.; Matsushima, T.; Adachi, H.; Kanno, I.; Kotera, H. Thin-Film Piezoelectric Materials for a Better Energy Harvesting MEMS. *J. Microelectromech. Syst.* **2012**, *21*, 451–457. [CrossRef]
62. Won, S.S.; Sheldon, M.; Mostovych, N.; Kwak, J.; Chang, B.-S.; Ahn, C.; Kingon, A.; Won Kim, I.; Kim, S.-H. Piezoelectric Poly(Vinylidene Fluoride Trifluoroethylene) Thin Film-Based Power Generators Using Paper Substrates for Wearable Device Applications. *Appl. Phys. Lett.* **2015**, *107*, 202901. [CrossRef]
63. Curie, J.; Curie, P. Phénomènes Électriques Des Cristaux Hémièdres à Faces Inclinées. *J. Phys. Theor. Appl.* **1882**, *1*, 245–251. [CrossRef]
64. Kim, H.S.; Kim, J.-H.; Kim, J. A Review of Piezoelectric Energy Harvesting Based on Vibration. *Int. J. Precis. Eng. Manuf.* **2011**, *12*, 1129–1141. [CrossRef]
65. Wang, Z.J.; Narita, F. Corona Poling Conditions for Barium Titanate/Epoxy Composites and Their Unsteady Wind Energy Harvesting Potential. *Adv. Eng. Mater.* **2019**, *21*, 1900169. [CrossRef]
66. Kosec, M.; Malič, B.; Benčan, A.; Rojac, T. KNN-based piezoelectric ceramics. In *Piezoelectric and Acoustic Materials for Transducer Applications*; Safari, A., Akdoğan, E.K., Eds.; Springer: Boston, MA, USA, 2008; pp. 81–102. ISBN 978-0-387-76540-2.
67. Clementi, G.; Lombardi, G.; Margueron, S.; Suarez, M.A.; Lebrasseur, E.; Ballandras, S.; Imbaud, J.; Lardet-Vieudrin, F.; Gauthier-Manuel, L.; Dulmet, B.; et al. LiNbO3 Films—A Low-Cost Alternative Lead-Free Piezoelectric Material for Vibrational Energy Harvesters. *Mech. Syst. Signal Process. MSSP* **2021**, *149*, 107171. [CrossRef]
68. Singh, H.H.; Khare, N. Flexible ZnO-PVDF/PTFE Based Piezo-Tribo Hybrid Nanogenerator. *Nano Energy* **2018**, *51*, 216–222. [CrossRef]

69. Qualtieri, A.; Rizzi, F.; Todaro, M.T.; Passaseo, A.; Cingolani, R.; De Vittorio, M. Stress-Driven AlN Cantilever-Based Flow Sensor for Fish Lateral Line System. *Microelectron. Eng.* **2011**, *88*, 2376–2378. [CrossRef]

70. Caro, M.A.; Zhang, S.; Riekkinen, T.; Ylilammi, M.; Moram, M.A.; Lopez-Acevedo, O.; Molarius, J.; Laurila, T. Piezoelectric Coefficients and Spontaneous Polarization of ScAlN. *J. Phys. Condens. Matter* **2015**, *27*, 245901. [CrossRef]

71. Koblmueller, G.; Averbeck, R.; Geelhaar, L.; Riechert, H.; Hösler, W.; Pongratz, P. Growth Diagram and Morphologies of AlN Thin Films Grown by Molecular Beam Epitaxy. *J. Appl. Phys.* **2003**, *93*, 9591–9596. [CrossRef]

72. Chen, D.; Colas, J.; Mercier, F.; Boichot, R.; Charpentier, L.; Escape, C.; Balat-Pichelin, M.; Pons, M. High Temperature Properties of AlN Coatings Deposited by Chemical Vapor Deposition for Solar Central Receivers. *Surf. Coat. Technol.* **2019**, *377*, 124872. [CrossRef]

73. Mariello, M.; Blad, T.W.A.; Mastronardi, V.M.; Madaro, F.; Guido, F.; Staufer, U.; Tolou, N.; De Vittorio, M. Flexible Piezoelectric AlN Transducers Buckled through Package-Induced Preloading for Mechanical Energy Harvesting. *Nano Energy* **2021**, *85*, 105986. [CrossRef]

74. Mariello, M.; Guido, F.; Algieri, L.; Mastronardi, V.M.; Qualtieri, A.; Pisanello, F.; De Vittorio, M. Microstructure and Electrical Properties of Novel Piezo-Optrodes Based on Thin-Film Piezoelectric Aluminium Nitride for Sensing. *IEEE Trans. Nanotechnol.* **2021**, *20*, 10–19. [CrossRef]

75. Akiyama, M.; Morofuji, Y.; Kamohara, T.; Nishikubo, K.; Tsubai, M.; Fukuda, O.; Ueno, N. Flexible Piezoelectric Pressure Sensors Using Oriented Aluminum Nitride Thin Films Prepared on Polyethylene Terephthalate Films. *J. Appl. Phys.* **2006**, *100*, 114318. [CrossRef]

76. Mariello, M.; Fachechi, L.; Guido, F.; Vittorio, M.D. Conformal, Ultra-Thin Skin-Contact-Actuated Hybrid Piezo/Triboelectric Wearable Sensor Based on AlN and Parylene-Encapsulated Elastomeric Blend. *Adv. Funct. Mater.* **2021**, *31*, 2101047. [CrossRef]

77. Signore, M.A.; Rescio, G.; De Pascali, C.; Iacovacci, V.; Dario, P.; Leone, A.; Quaranta, F.; Taurino, A.; Siciliano, P.; Francioso, L. Fabrication and Characterization of AlN-Based Flexible Piezoelectric Pressure Sensor Integrated into an Implantable Artificial Pancreas. *Sci. Rep.* **2019**, *9*, 17130. [CrossRef]

78. Jackson, N.; Keeney, L.; Mathewson, A. Flexible-CMOS and Biocompatible Piezoelectric AlN Material for MEMS Applications. *Smart Mater. Struct.* **2013**, *22*, 115033. [CrossRef]

79. Hu, X.; Yang, F.; Guo, M.; Pei, J.; Zhao, H.; Wang, Y. Fabrication of Polyimide Microfluidic Devices by Laser Ablation Based Additive Manufacturing. *Microsyst. Technol.* **2020**, *26*, 1573–1583. [CrossRef]

80. Sun, Y.; Lacour, S.; Brooks, R.; Rushton, N.; Fawcett, J.; Cameron, R.E. Assessment of the Biocompatibility of Photosensitive Polyimide for Implantable Medical Device Use. *J. Biomed. Mater. Res. A* **2009**, *90*, 648–655. [CrossRef]

81. Bartasyte, A.; Margueron, S.; Baron, T.; Oliveri, S.; Boulet, P. Toward High-Quality Epitaxial LiNbO3 and LiTaO3 Thin Films for Acoustic and Optical Applications. *Adv. Mater. Interfaces* **2017**, *4*, 1600998. [CrossRef]

82. Kovacs, G.; Anhorn, M.; Engan, H.E.; Visintini, G.; Ruppel, C.C.W. Improved material constants for LiNbO/Sub 3/and LiTaO/Sub 3/. In Proceedings of the IEEE Symposium on Ultrasonics, Honolulu, HI, USA, 4–7 December 1990; Volume 1, pp. 435–438.

83. Pi, Z.; Zhang, J.; Wen, C.; Zhang, Z.; Wu, D. Flexible Piezoelectric Nanogenerator Made of Poly(Vinylidenefluoride-Co-Trifluoroethylene) (PVDF-TrFE) Thin Film. *Nano Energy* **2014**, *7*, 33–41. [CrossRef]

84. Mariello, M.; Qualtieri, A.; Mele, G.; De Vittorio, M. Metal-Free Multilayer Hybrid PENG Based on Soft Electrospun/-Sprayed Membranes with Cardanol Additive for Harvesting Energy from Surgical Face Masks. *ACS Appl. Mater. Interfaces* **2021**, *13*, 20606–20621. [CrossRef]

85. Liu, J.; Yu, D.; Zheng, Z.; Huangfu, G.; Guo, Y. Lead-Free BiFeO3 Film on Glass Fiber Fabric: Wearable Hybrid Piezoelectric-Triboelectric Nanogenerator. *Ceram. Int.* **2021**, *47*, 3573–3579. [CrossRef]

86. Chiu, Y.-Y.; Lin, W.-Y.; Wang, H.-Y.; Huang, S.-B.; Wu, M.-H. Development of a Piezoelectric Polyvinylidene Fluoride (PVDF) Polymer-Based Sensor Patch for Simultaneous Heartbeat and Respiration Monitoring. *Sens. Actuators Phys.* **2013**, *189*, 328–334. [CrossRef]

87. Bae, J.-H.; Chang, S.-H. PVDF-Based Ferroelectric Polymers and Dielectric Elastomers for Sensor and Actuator Applications: A Review. *Funct. Compos. Struct.* **2019**, *1*, 012003. [CrossRef]

88. Hutson, A.R.; White, D.L. Elastic Wave Propagation in Piezoelectric Semiconductors. *J. Appl. Phys.* **1962**, *33*, 40–47. [CrossRef]

89. Auld, B.A. *Acoustic Fields and Waves in Solids*; John Wiley & Sons: New York, NY, USA, 1973; Volume 1, ISBN 978-0-471-03700-2.

90. Pierret, R. *Semiconductor Fundamentals: Volume I*; Pearson: Reading, MA, USA, 1988; ISBN 978-0-201-12295-4.

91. Asthana, A.; Asayesh-Ardakani, H.; Yap, Y.K.; Yassar, R. Real Time Observation of Mechanically Triggered Piezoelectric Current in Individual ZnO Nanobelts. *J. Mater. Chem. C* **2014**, *2*, 3995–4004. [CrossRef]

92. Yang, G.; Du, J.; Wang, J.; Yang, J. Electromechanical Fields in a Nonuniform Piezoelectric Semiconductor Rod. *J. Mech. Mater. Struct.* **2018**, *13*, 103–120. [CrossRef]

93. Luo, Y.; Zhang, C.; Chen, W.; Yang, J. An Analysis of PN Junctions in Piezoelectric Semiconductors. *J. Appl. Phys.* **2017**, *122*, 204502. [CrossRef]

94. Huang, H.; Qian, Z.; Yang, J. I-V Characteristics of a Piezoelectric Semiconductor Nanofiber under Local Tensile/Compressive Stress. *J. Appl. Phys.* **2019**, *126*, 164902. [CrossRef]

95. Wang, Z. Piezopotential Gated Nanowire Devices: Piezotronics and Piezo-Phototronics. *Nano Today* **2010**, *5*, 540–552. [CrossRef]

96. Liu, Y.; Zhang, Y.; Yang, Q.; Niu, S.; Wang, Z.L. Fundamental Theories of Piezotronics and Piezo-Phototronics. *Nano Energy* **2015**, *14*, 257–275. [CrossRef]

97. Zhang, Y.; Leng, Y.; Willatzen, M.; Huang, B. Theory of Piezotronics and Piezo-Phototronics. *MRS Bull.* **2018**, *43*, 928–935. [CrossRef]

98. Zhang, Y.; Liu, Y.; Wang, Z.L. Fundamental Theory of Piezotronics. *Adv. Mater.* **2011**, *23*, 3004–3013. [CrossRef] [PubMed]

99. Wang, Z.L.; Wu, W. Piezotronics and Piezo-Phototronics: Fundamentals and Applications. *Natl. Sci. Rev.* **2014**, *1*, 62–90. [CrossRef]

100. Yang, Y.; Tang, L. Equivalent Circuit Modeling of Piezoelectric Energy Harvesters. *J. Intell. Mater. Syst. Struct.* **2009**, *20*, 2223–2235. [CrossRef]

101. Kuang, Y.; Chew, Z.J.; Zhu, M. Strongly Coupled Piezoelectric Energy Harvesters: Finite Element Modelling and Experimental Validation. *Energy Convers. Manag.* **2020**, *213*, 112855. [CrossRef]

102. Liang, J.; Liao, W.-H. Impedance Modeling and Analysis for Piezoelectric Energy Harvesting Systems. *IEEE/ASME Trans. Mechatron.* **2012**, *17*, 1145–1157. [CrossRef]

103. Erturk, A.; Inman, D.J. Issues in Mathematical Modeling of Piezoelectric Energy Harvesters. *Smart Mater. Struct.* **2008**, *17*, 065016. [CrossRef]

104. Kuang, Y.; Ruan, T.; Chew, Z.J.; Zhu, M. Energy Harvesting during Human Walking to Power a Wireless Sensor Node. *Sens. Actuators Phys.* **2017**, *254*, 69–77. [CrossRef]

105. Elvin, N.G.; Elvin, A.A.; Spector, M. A Self-Powered Mechanical Strain Energy Sensor. *Smart Mater. Struct.* **2001**, *10*, 293–299. [CrossRef]

106. Elvin, N.G.; Elvin, A.A. A General Equivalent Circuit Model for Piezoelectric Generators. *J. Intell. Mater. Syst. Struct.* **2008**, *20*, 3–9. [CrossRef]

107. Pozo, B.; Garate, J.I.; Araujo, J.Á.; Ferreiro, S. Energy Harvesting Technologies and Equivalent Electronic Structural Models—Review. *Electronics* **2019**, *8*, 486. [CrossRef]

108. Kong, N.; Ha, D.S.; Erturk, A.; Inman, D.J. Resistive Impedance Matching Circuit for Piezoelectric Energy Harvesting. *J. Intell. Mater. Syst. Struct.* **2010**, *21*, 1293–1302. [CrossRef]

109. Qi, Y.; Jafferis, N.T.; Lyons, K.; Lee, C.M.; Ahmad, H.; McAlpine, M.C. Piezoelectric Ribbons Printed onto Rubber for Flexible Energy Conversion. *Nano Lett.* **2010**, *10*, 524–528. [CrossRef] [PubMed]

110. Cheng, C.; Chen, Z.; Shi, H.; Liu, Z.; Xiong, Y. System-Level Coupled Modeling of Piezoelectric Vibration Energy Harvesting Systems by Joint Finite Element and Circuit Analysis. Available online: https://www.hindawi.com/journals/sv/2016/2413578/ (accessed on 12 October 2020).

111. Lei, A.; Xu, R.; Borregaard, L.; Guizzetti, M.; Hansen, O.; Thomsen, E. Impedance Based Characterization of a High-Coupled Screen Printed PZT Thick Film Unimorph Energy Harvester. *J. Microelectromech. Syst.* **2014**, *23*, 842–854. [CrossRef]

112. Fan, F.-R.; Tian, Z.-Q.; Lin Wang, Z. Flexible Triboelectric Generator. *Nano Energy* **2012**, *1*, 328–334. [CrossRef]

113. Wang, Z.; Lin, L.; Chen, J.; Niu, S.; Zi, Y. *Triboelectric Nanogenerators*; Green Energy and Technology; Springer International Publishing: Berlin/Heidelberg, Germany, 2016; ISBN 978-3-319-40038-9.

114. Luo, J.; Wang, Z.L. Recent Progress of Triboelectric Nanogenerators: From Fundamental Theory to Practical Applications. *EcoMat* **2020**, *2*, e12059. [CrossRef]

115. Niu, S.; Wang, Z.L. Theoretical Systems of Triboelectric Nanogenerators. *Nano Energy* **2015**, *14*, 161–192. [CrossRef]

116. Niu, S.; Liu, Y.; Wang, S.; Lin, L.; Zhou, Y.S.; Hu, Y.; Wang, Z.L. Theoretical Investigation and Structural Optimization of Single-Electrode Triboelectric Nanogenerators. *Adv. Funct. Mater.* **2014**, *24*, 3332–3340. [CrossRef]

117. Henniker, J. Triboelectricity in Polymers. *Nature* **1962**, *196*, 474. [CrossRef]

118. Zou, H.; Zhang, Y.; Guo, L.; Wang, P.; He, X.; Dai, G.; Zheng, H.; Chen, C.; Wang, A.; Xu, C.; et al. Quantifying the Triboelectric Series. *Nat. Commun.* **2019**, *10*, 1427. [CrossRef] [PubMed]

119. Ibrahim, M.; Jiang, J.; Wen, Z.; Sun, X. Surface Engineering for Enhanced Triboelectric Nanogenerator. *Nanoenergy Adv.* **2021**, *1*, 58–80. [CrossRef]

120. Arcot Narasimulu, A.; Zhao, P.; Soin, N.; Kovur, P.; Ding, P.; Chen, J.; Dong, S.; Chen, L.; Zhou, E.; Montemagno, C.; et al. Significant Triboelectric Enhancement Using Interfacial Piezoelectric ZnO Nanosheet Layer. *Nano Energy* **2017**, *40*, 471–480. [CrossRef]

121. Ravichandran, A.N.; Ramuz, M.; Blayac, S. Increasing Surface Charge Density by Effective Charge Accumulation Layer Inclusion for High-Performance Triboelectric Nanogenerators. *MRS Commun.* **2019**, *9*, 682–689. [CrossRef]

122. Kim, H.-J.; Yim, E.-C.; Kim, J.-H.; Kim, S.-J.; Park, J.-Y.; Oh, I.-K. Bacterial Nano-Cellulose Triboelectric Nanogenerator. *Nano Energy* **2017**, *33*, 130–137. [CrossRef]

123. Wen, R.; Guo, J.; Yu, A.; Zhang, K.; Kou, J.; Zhu, Y.; Zhang, Y.; Li, B.-W.; Zhai, J. Remarkably Enhanced Triboelectric Nanogenerator Based on Flexible and Transparent Monolayer Titania Nanocomposite. *Nano Energy* **2018**, *50*, 140–147. [CrossRef]

124. Kil Yun, B.; Soo Kim, H.; Joon Ko, Y.; Murillo, G.; Hoon Jung, J. Interdigital Electrode Based Triboelectric Nanogenerator for Effective Energy Harvesting from Water. *Nano Energy* **2017**, *36*, 233–240. [CrossRef]

125. Mariello, M.; Scarpa, E.; Algieri, L.; Guido, F.; Mastronardi, V.M.; Qualtieri, A.; De Vittorio, M. Novel Flexible Triboelectric Nanogenerator Based on Metallized Porous PDMS and Parylene C. *Energies* **2020**, *13*, 1625. [CrossRef]

126. Qian, Y.; Lyu, Z.; Kim, D.-H.; Kang, D.J. Enhancing the Output Power Density of Polydimethylsiloxane-Based Flexible Triboelectric Nanogenerators with Ultrathin Nickel Telluride Nanobelts as a Co-Triboelectric Layer. *Nano Energy* **2021**, *90*, 106536. [CrossRef]

127. Qian, Y.; Kang, D.J. Large-Area High-Quality AB-Stacked Bilayer Graphene on h-BN/Pt Foil by Chemical Vapor Deposition. *ACS Appl. Mater. Interfaces* **2018**, *10*, 29069–29075. [CrossRef]
128. Harnchana, V.; Ngoc, H.V.; He, W.; Rasheed, A.; Park, H.; Amornkitbamrung, V.; Kang, D.J. Enhanced Power Output of a Triboelectric Nanogenerator Using Poly(Dimethylsiloxane) Modified with Graphene Oxide and Sodium Dodecyl Sulfate. *ACS Appl. Mater. Interfaces* **2018**, *10*, 25263–25272. [CrossRef]
129. Kim, D.Y.; Kim, H.S.; Kong, D.S.; Choi, M.; Kim, H.B.; Lee, J.-H.; Murillo, G.; Lee, M.; Kim, S.S.; Jung, J.H. Floating Buoy-Based Triboelectric Nanogenerator for an Effective Vibrational Energy Harvesting from Irregular and Random Water Waves in Wild Sea. *Nano Energy* **2018**, *45*, 247–254. [CrossRef]
130. Chen, J.; Yang, J.; Li, Z.; Fan, X.; Zi, Y.; Jing, Q.; Guo, H.; Wen, Z.; Pradel, K.C.; Niu, S.; et al. Networks of Triboelectric Nanogenerators for Harvesting Water Wave Energy: A Potential Approach toward Blue Energy. *ACS Nano* **2015**, *9*, 3324–3331. [CrossRef]
131. Wang, Z.L.; Jiang, T.; Xu, L. Toward the Blue Energy Dream by Triboelectric Nanogenerator Networks. *Nano Energy* **2017**, *39*, 9–23. [CrossRef]
132. Xiong, J.; Lin, M.-F.; Wang, J.; Gaw, S.L.; Parida, K.; Lee, P.S. Wearable All-Fabric-Based Triboelectric Generator for Water Energy Harvesting. *Adv. Energy Mater.* **2017**, *7*, 1701243. [CrossRef]
133. Luo, J.; Tang, W.; Fan, F.R.; Liu, C.; Pang, Y.; Cao, G.; Wang, Z.L. Transparent and Flexible Self-Charging Power Film and Its Application in a Sliding Unlock System in Touchpad Technology. *ACS Nano* **2016**, *10*, 8078–8086. [CrossRef] [PubMed]
134. Luo, J.; Wang, Z.; Xu, L.; Wang, A.C.; Han, K.; Jiang, T.; Lai, Q.; Bai, Y.; Tang, W.; Fan, F.R.; et al. Flexible and Durable Wood-Based Triboelectric Nanogenerators for Self-Powered Sensing in Athletic Big Data Analytics. *Nat. Commun.* **2019**, *10*, 5147. [CrossRef] [PubMed]
135. Genter, S.; Langhof, T.; Paul, O. Electret-Based Out-Of-Plane Micro Energy Harvester with Parylene-C Serving as the Electret and Spring Material. *Procedia Eng.* **2015**, *120*, 341–344. [CrossRef]
136. Lo, H.; Tai, Y.-C. Parylene-Based Electret Power Generators. *J. Micromech. Microeng.* **2008**, *18*, 104006. [CrossRef]
137. Mariello, M.; Guido, F.; Mastronardi, V.M.; De Donato, F.; Salbini, M.; Brunetti, V.; Qualtieri, A.; Rizzi, F.; De Vittorio, M. Captive-Air-Bubble Aerophobicity Measurements of Antibiofouling Coatings for Underwater MEMS Devices. *Nanomater. Nanotechnol.* **2019**, *9*, 1847980419862075. [CrossRef]
138. Mariello, M.; Guido, F.; Mastronardi, V.M.; Giannuzzi, R.; Algieri, L.; Qualteri, A.; Maffezzoli, A.; De Vittorio, M. Reliability of Protective Coatings for Flexible Piezoelectric Transducers in Aqueous Environments. *Micromachines* **2019**, *10*, 739. [CrossRef]
139. Golda-Cepa, M.; Brzychczy-Wloch, M.; Engvall, K.; Aminlashgari, N.; Hakkarainen, M.; Kotarba, A. Microbiological Investigations of Oxygen Plasma Treated Parylene C Surfaces for Metal Implant Coating. *Mater. Sci. Eng. C* **2015**, *52*, 273–281. [CrossRef]
140. Chen, T.-N.; Wuu, D.-S.; Wu, C.-C.; Chiang, C.-C.; Chen, Y.-P.; Horng, R.-H. Improvements of Permeation Barrier Coatings Using Encapsulated Parylene Interlayers for Flexible Electronic Applications. *Plasma Process. Polym.* **2007**, *4*, 180–185. [CrossRef]
141. Applerot, G.; Abu-Mukh, R.; Irzh, A.; Charmet, J.; Keppner, H.; Laux, E.; Guibert, G.; Gedanken, A. Decorating Parylene-Coated Glass with ZnO Nanoparticles for Antibacterial Applications: A Comparative Study of Sonochemical, Microwave, and Microwave-Plasma Coating Routes. *ACS Appl. Mater. Interfaces* **2010**, *2*, 1052–1059. [CrossRef]
142. Cieślik, M.; Engvall, K.; Pan, J.; Kotarba, A. Silane–Parylene Coating for Improving Corrosion Resistance of Stainless Steel 316L Implant Material. *Corros. Sci.* **2011**, *53*, 296–301. [CrossRef]
143. Gorham, W.F. A New, General Synthetic Method for the Preparation of Linear Poly-p-Xylylenes. *J. Polym. Sci. A1* **1966**, *4*, 3027–3039. [CrossRef]
144. Calcagnile, P.; Blasi, L.; Rizzi, F.; Qualtieri, A.; Athanassiou, A.; Gogolides, E.; De Vittorio, M. Parylene C Surface Functionalization and Patterning with PH-Responsive Microgels. ACS Appl. Mater. *Interfaces* **2014**, *6*, 15708–15715. [CrossRef]
145. Scarpa, E.; Dattoma, T.; Calcagnile, P.; Blasi, L.; Qualtieri, A.; Rizzi, F.; Vittorio, M.D. Surface-tension-confined fluidics on Parylene C coated paper substrate. In Proceedings of the IEEE 17th International Conference on Nanotechnology, Pittsburgh, PA, USA, 25–27 July 2017; pp. 259–262.
146. Wu, C.; Wang, A.C.; Ding, W.; Guo, H.; Wang, Z.L. Triboelectric Nanogenerator: A Foundation of the Energy for the New Era. *Adv. Energy Mater.* **2019**, *9*, 1802906. [CrossRef]
147. Gauntt, S.; Batt, G.; Gibert, J. Dynamic modeling of triboelectric generators using lagrange's equation. In *Smart Materials, Adaptive Structures and Intelligent Systems*; American Society of Mechanical Engineers Digital Collection: New York, NY, USA, 2017.
148. Fang, C.; Tong, T.; Bu, T.; Cao, Y.; Xu, S.; Qi, Y.; Zhang, C. Overview of Power Management for Triboelectric Nanogenerators. *Adv. Intell. Syst.* **2020**, *2*, 1900129. [CrossRef]
149. Shao, J.; Willatzen, M.; Wang, Z.L. Theoretical Modeling of Triboelectric Nanogenerators (TENGs). *J. Appl. Phys.* **2020**, *128*, 111101. [CrossRef]
150. Wang, X.; Yang, B.; Liu, J.; Zhu, Y.; Yang, C.; He, Q. A Flexible Triboelectric-Piezoelectric Hybrid Nanogenerator Based on P(VDF-TrFE) Nanofibers and PDMS/MWCNT for Wearable Devices. *Sci. Rep.* **2016**, *6*, 36409. [CrossRef]
151. Nguyen, V.; Kelly, S.; Yang, R. Piezoelectric Peptide-Based Nanogenerator Enhanced by Single-Electrode Triboelectric Nanogenerator. *APL Mater.* **2017**, *5*, 074108. [CrossRef]

152. He, J.; Wen, T.; Qian, S.; Zhang, Z.; Tian, Z.; Zhu, J.; Mu, J.; Hou, X.; Geng, W.; Cho, J.; et al. Triboelectric-Piezoelectric-Electromagnetic Hybrid Nanogenerator for High-Efficient Vibration Energy Harvesting and Self-Powered Wireless Monitoring System. *Nano Energy* **2017**, *43*, 326–339. [CrossRef]

153. Yu, J.; Hou, X.; Cui, M.; Zhang, S.; He, J.; Geng, W.; Mu, J.; Chou, X. Highly Skin-Conformal Wearable Tactile Sensor Based on Piezoelectric-Enhanced Triboelectric Nanogenerator. *Nano Energy* **2019**, *64*, 103923. [CrossRef]

154. Han, Z.; Jiao, P.; Zhu, Z. Combination of Piezoelectric and Triboelectric Devices for Robotic Self-Powered Sensors. *Micromachines* **2021**, *12*, 813. [CrossRef]

155. Han, M.; Chen, X.; Yu, B.; Zhang, H. Coupling of Piezoelectric and Triboelectric Effects: From Theoretical Analysis to Experimental Verification. *Adv. Electron. Mater.* **2015**, *1*, 1500187. [CrossRef]

156. Wang, Z.L. On Maxwell's Displacement Current for Energy and Sensors: The Origin of Nanogenerators. *Mater. Today* **2017**, *20*, 74–82. [CrossRef]

157. Chen, S.; Tao, X.; Zeng, W.; Yang, B.; Shang, S. Quantifying Energy Harvested from Contact-Mode Hybrid Nanogenerators with Cascaded Piezoelectric and Triboelectric Units. *Adv. Energy Mater.* **2017**, *7*, 1601569. [CrossRef]

158. Chen, C.Y.; Tsai, C.Y.; Xu, M.H.; Wu, C.T.; Huang, C.Y.; Lee, T.H.; Fuh, Y.K. A Fully Encapsulated Piezoelectric–Triboelectric Hybrid Nanogenerator for Energy Harvesting from Biomechanical and Environmental Sources. *Express Polym. Lett.* **2019**, *13*, 533–542. [CrossRef]

159. Chen, X.; Han, M.; Chen, H.; Cheng, X.; Song, Y.; Su, Z.; Jiang, Y.; Zhang, H. A Wave-Shaped Hybrid Piezoelectric and Triboelectric Nanogenerator Based on P(VDF-TrFE) Nanofibers. *Nanoscale* **2017**, *9*, 1263–1270. [CrossRef] [PubMed]

160. Zhao, C.; Zhang, Q.; Zhang, W.; Du, X.; Zhang, Y.; Gong, S.; Ren, K.; Sun, Q.; Wang, Z.L. Hybrid Piezo/Triboelectric Nanogenerator for Highly Efficient and Stable Rotation Energy Harvesting. *Nano Energy* **2019**, *57*, 440–449. [CrossRef]

161. Xia, Y.; Zhou, J.; Chen, T.; Liu, H.; Liu, W.; Yang, Z.; Wang, P.; Sun, L. A hybrid flapping-leaf microgenerator for harvesting wind-flow energy. In Proceedings of the 29th International Conference on Micro Electro Mechanical Systems (MEMS), Shanghai, China, 24–28 January 2016; pp. 1224–1227.

162. Suo, G.; Yu, Y.; Zhang, Z.; Wang, S.; Zhao, P.; Li, J.; Wang, X. Piezoelectric and Triboelectric Dual Effects in Mechanical-Energy Harvesting Using BaTiO3/Polydimethylsiloxane Composite Film. *ACS Appl. Mater. Interfaces* **2016**, *8*, 34335–34341. [CrossRef] [PubMed]

163. Guo, Y.; Zhang, X.-S.; Wang, Y.; Gong, W.; Zhang, Q.; Wang, H.; Brugger, J. All-Fiber Hybrid Piezoelectric-Enhanced Triboelectric Nanogenerator for Wearable Gesture Monitoring. *Nano Energy* **2018**, *48*, 152–160. [CrossRef]

164. Wang, X.; Yang, B.; Liu, J.; Yang, C. A Transparent and Biocompatible Single-Friction-Surface Triboelectric and Piezoelectric Generator and Body Movement Sensor. *J. Mater. Chem. A* **2017**, *5*, 1176–1183. [CrossRef]

165. Huang, T.; Wang, C.; Yu, H.; Wang, H.; Zhang, Q.; Zhu, M. Human Walking-Driven Wearable All-Fiber Triboelectric Nanogenerator Containing Electrospun Polyvinylidene Fluoride Piezoelectric Nanofibers. *Nano Energy* **2015**, *14*, 226–235. [CrossRef]

166. Chen, X.; Song, Y.; Su, Z.; Chen, H.; Cheng, X.; Zhang, J.; Han, M.; Zhang, H. Flexible Fiber-Based Hybrid Nanogenerator for Biomechanical Energy Harvesting and Physiological Monitoring. *Nano Energy* **2017**, *38*, 43–50. [CrossRef]

167. Jung, W.-S.; Kang, M.-G.; Moon, H.G.; Baek, S.-H.; Yoon, S.-J.; Wang, Z.-L.; Kim, S.-W.; Kang, C.-Y. High Output Piezo/Triboelectric Hybrid Generator. *Sci. Rep.* **2015**, *5*, 9309. [CrossRef] [PubMed]

168. Li, X.; Lin, Z.-H.; Cheng, G.; Wen, X.; Liu, Y.; Niu, S.; Wang, Z.L. 3D Fiber-Based Hybrid Nanogenerator for Energy Harvesting and as a Self-Powered Pressure Sensor. *ACS Nano* **2014**, *8*, 10674–10681. [CrossRef]

169. Zhu, J.; Hou, X.; Niu, X.; Guo, X.; Zhang, J.; He, J.; Guo, T.; Chou, X.; Xue, C.; Zhang, W. The D-Arched Piezoelectric-Triboelectric Hybrid Nanogenerator as a Self-Powered Vibration Sensor. *Sens. Actuators Phys.* **2017**, *263*, 317–325. [CrossRef]

170. Wang, L.; He, T.; Zhang, Z.; Zhao, L.; Lee, C.; Luo, G.; Mao, Q.; Yang, P.; Lin, Q.; Li, X.; et al. Self-Sustained Autonomous Wireless Sensing Based on a Hybridized TENG and PEG Vibration Mechanism. *Nano Energy* **2021**, *80*, 105555. [CrossRef]

171. Yang, X.; Li, P.; Wu, B.; Li, H.; Zhou, G. A Flexible Piezoelectric-Triboelectric Hybrid Nanogenerator in One Structure with Dual Doping Enhancement Effects. *Curr. Appl. Phys.* **2021**, *32*, 50–58. [CrossRef]

172. Wang, X.; Wang, S.; Yang, Y.; Wang, Z.L. Hybridized Electromagnetic–Triboelectric Nanogenerator for Scavenging Air-Flow Energy to Sustainably Power Temperature Sensors. *ACS Nano* **2015**, *9*, 4553–4562. [CrossRef]

173. Yang, H.; Yang, H.; Lai, M.; Xi, Y.; Guan, Y.; Liu, W.; Zeng, Q.; Lu, J.; Hu, C.; Wang, Z.L. Triboelectric and Electromagnetic Hybrid Nanogenerator Based on a Crankshaft Piston System as a Multifunctional Energy Harvesting Device. *Adv. Mater. Technol.* **2020**, *4*, 1800278. [CrossRef]

174. Wang, P.; Pan, L.; Wang, J.; Xu, M.; Dai, G.; Zou, H.; Dong, K.; Wang, Z.L. An Ultra-Low-Friction Triboelectric–Electromagnetic Hybrid Nanogenerator for Rotation Energy Harvesting and Self-Powered Wind Speed Sensor. *ACS Nano* **2018**, *12*, 9433–9440. [CrossRef]

175. Hu, Y.; Yang, J.; Niu, S.; Wu, W.; Wang, Z.L. Hybridizing Triboelectrification and Electromagnetic Induction Effects for High-Efficient Mechanical Energy Harvesting. *ACS Nano* **2014**, *8*, 7442–7450. [CrossRef]

176. Quan, T.; Wang, X.; Wang, Z.L.; Yang, Y. Hybridized Electromagnetic–Triboelectric Nanogenerator for a Self-Powered Electronic Watch. *ACS Nano* **2015**, *9*, 12301–12310. [CrossRef] [PubMed]

177. Zhang, B.; Chen, J.; Jin, L.; Deng, W.; Zhang, L.; Zhang, H.; Zhu, M.; Yang, W.; Wang, Z.L. Rotating-Disk-Based Hybridized Electromagnetic-Triboelectric Nanogenerator for Sustainably Powering Wireless Traffic Volume Sensors. *ACS Nano* **2016**, *10*, 6241–6247. [CrossRef]

178. Cao, R.; Zhou, T.; Wang, B.; Yin, Y.; Yuan, Z.; Li, C.; Wang, Z.L. Rotating-Sleeve Triboelectric–Electromagnetic Hybrid Nanogenerator for High Efficiency of Harvesting Mechanical Energy. *ACS Nano* **2017**, *11*, 8370–8378. [CrossRef]

179. Zhao, L.-C.; Zou, H.-X.; Yan, G.; Liu, F.-R.; Tan, T.; Zhang, W.-M.; Peng, Z.-K.; Meng, G. A Water-Proof Magnetically Coupled Piezoelectric-Electromagnetic Hybrid Wind Energy Harvester. *Appl. Energy* **2019**, *239*, 735–746. [CrossRef]

180. Hamid, R.; Yuce, M.R. A Wearable Energy Harvester Unit Using Piezoelectric–Electromagnetic Hybrid Technique. *Sens. Actuators Phys.* **2017**, *257*, 198–207. [CrossRef]

181. Fan, K.; Tan, Q.; Liu, H.; Zhu, Y.; Wang, W.; Zhang, D. Hybrid Piezoelectric-Electromagnetic Energy Harvester for Scavenging Energy from Low-Frequency Excitations. *Smart Mater. Struct.* **2018**, *27*, 085001. [CrossRef]

182. Toyabur, R.M.; Salauddin, M.; Cho, H.; Park, J.Y. A Multimodal Hybrid Energy Harvester Based on Piezoelectric-Electromagnetic Mechanisms for Low-Frequency Ambient Vibrations. *Energy Convers. Manag.* **2018**, *168*, 454–466. [CrossRef]

183. He, X.; Wen, Q.; Sun, Y.; Wen, Z. A Low-Frequency Piezoelectric-Electromagnetic-Triboelectric Hybrid Broadband Vibration Energy Harvester. *Nano Energy* **2017**, *40*, 300–307. [CrossRef]

184. Rodrigues, C.; Gomes, A.; Ghosh, A.; Pereira, A.; Ventura, J. Power-Generating Footwear Based on a Triboelectric-Electromagnetic-Piezoelectric Hybrid Nanogenerator. *Nano Energy* **2019**, *62*, 660–666. [CrossRef]

185. Hemojit Singh, H.; Kumar, D.; Khare, N. A Synchronous Piezoelectric–Triboelectric–Electromagnetic Hybrid Generator for Harvesting Vibration Energy. Sustain. *Energy Fuels* **2021**, *5*, 212–218. [CrossRef]

186. Wang, S.; Wang, Z.L.; Yang, Y. A One-Structure-Based Hybridized Nanogenerator for Scavenging Mechanical and Thermal Energies by Triboelectric-Piezoelectric-Pyroelectric Effects. *Adv. Mater. Deerfield Beach Fla* **2016**, *28*, 2881–2887. [CrossRef]

187. Zheng, H.; Zi, Y.; He, X.; Guo, H.; Lai, Y.-C.; Wang, J.; Zhang, S.L.; Wu, C.; Cheng, G.; Wang, Z.L. Concurrent Harvesting of Ambient Energy by Hybrid Nanogenerators for Wearable Self-Powered Systems and Active Remote Sensing. *ACS Appl. Mater. Interfaces* **2018**, *10*, 14708–14715. [CrossRef]

188. Xu, C.; Wang, X.; Wang, Z.L. Nanowire Structured Hybrid Cell for Concurrently Scavenging Solar and Mechanical Energies. *J. Am. Chem. Soc.* **2009**, *131*, 5866–5872. [CrossRef]

189. Xu, C.; Wang, Z.L. Compact Hybrid Cell Based on a Convoluted Nanowire Structure for Harvesting Solar and Mechanical Energy. *Adv. Mater.* **2011**, *23*, 873–877. [CrossRef]

190. Yoon, G.C.; Shin, K.-S.; Gupta, M.K.; Lee, K.Y.; Lee, J.-H.; Wang, Z.L.; Kim, S.-W. High-Performance Hybrid Cell Based on an Organic Photovoltaic Device and a Direct Current Piezoelectric Nanogenerator. *Nano Energy* **2015**, *12*, 547–555. [CrossRef]

191. Liu, G.; Mrad, N.; Abdel-Rahman, E.; Ban, D. Cascade-Type Hybrid Energy Cells for Driving Wireless Sensors. *Nano Energy* **2016**, *26*, 641–647. [CrossRef]

192. Yang, Y.; Zhang, H.; Zhu, G.; Lee, S.; Lin, Z.-H.; Wang, Z.L. Flexible Hybrid Energy Cell for Simultaneously Harvesting Thermal, Mechanical, and Solar Energies. *ACS Nano* **2013**, *7*, 785–790. [CrossRef] [PubMed]

193. Pan, C.; Li, Z.; Guo, W.; Zhu, J.; Wang, Z.L. Fiber-Based Hybrid Nanogenerators for/as Self-Powered Systems in Biological Liquid. *Angew. Chem. Int. Ed. Engl.* **2011**, *50*, 11192–11196. [CrossRef] [PubMed]

194. Hansen, B.J.; Liu, Y.; Yang, R.; Wang, Z.L. Hybrid Nanogenerator for Concurrently Harvesting Biomechanical and Biochemical Energy. *ACS Nano* **2010**, *4*, 3647–3652. [CrossRef] [PubMed]

195. Wu, Y.; Zhong, X.; Wang, X.; Yang, Y.; Wang, Z.L. Hybrid Energy Cell for Simultaneously Harvesting Wind, Solar, and Chemical Energies. *Nano Res.* **2014**, *7*, 1631–1639. [CrossRef]

196. Zheng, L.; Lin, Z.-H.; Cheng, G.; Wu, W.; Wen, X.; Lee, S.; Wang, Z. Silicon-Based Hybrid Cell for Harvesting Solar Energy and Raindrop Electrostatic Energy. *Nano Energy* **2014**, *9*, 291–300. [CrossRef]

197. Yang, Y.; Zhang, H.; Liu, Y.; Lin, Z.-H.; Lee, S.; Lin, Z.; Wong, C.P.; Wang, Z.L. Silicon-Based Hybrid Energy Cell for Self-Powered Electrodegradation and Personal Electronics. *ACS Nano* **2013**, *7*, 2808–2813. [CrossRef] [PubMed]

198. Dudem, B.; Ko, Y.H.; Leem, J.W.; Lim, J.H.; Yu, J.S. Hybrid Energy Cell with Hierarchical Nano/Micro-Architectured Polymer Film to Harvest Mechanical, Solar, and Wind Energies Individually/Simultaneously. *ACS Appl. Mater. Interfaces* **2016**, *8*, 30165–30175. [CrossRef] [PubMed]

199. Wang, S.; Wang, X.; Wang, Z.L.; Yang, Y. Efficient Scavenging of Solar and Wind Energies in a Smart City. *ACS Nano* **2016**, *10*, 5696–5700. [CrossRef] [PubMed]

200. Wang, X.; Wang, Z.L.; Yang, Y. Hybridized Nanogenerator for Simultaneously Scavenging Mechanical and Thermal Energies by Electromagnetic-Triboelectric-Thermoelectric Effects. *Nano Energy* **2016**, *26*, 164–171. [CrossRef]

201. Kim, M.-K.; Kim, M.-S.; Jo, S.-E.; Kim, Y.-J. Triboelectric–Thermoelectric Hybrid Nanogenerator for Harvesting Frictional Energy. *Smart Mater. Struct.* **2016**, *25*, 125007. [CrossRef]

202. Kim, H.; Sohn, C.; Hwang, G.-T.; Park, K.-I.; Jeong, C.K. (K,Na)NbO3-LiNbO3 Nanocube-Based Flexible and Lead-Free Piezoelectric Nanocomposite Energy Harvesters. *J. Korean Ceram. Soc.* **2020**, *57*, 401–408. [CrossRef]

203. Mimura, K.; Liu, Z.; Itasaka, H.; Lin, Y.-C.; Suenaga, K.; Kato, K. One-Step Synthesis of BaTiO3/CaTiO3 Core-Shell Nanocubes by Hydrothermal Reaction. *J. Asian Ceram. Soc.* **2021**, *9*, 359–365. [CrossRef]

204. Mimura, K.; Kato, K.; Imai, H.; Wada, S.; Haneda, H.; Kuwabara, M. Piezoresponse Properties of Orderly Assemblies of BaTiO3 and SrTiO3 Nanocube Single Crystals. *Appl. Phys. Lett.* **2012**, *101*, 012901. [CrossRef]

205. Rovisco, A.; dos Santos, A.; Cramer, T.; Martins, J.; Branquinho, R.; Águas, H.; Fraboni, B.; Fortunato, E.; Martins, R.; Igreja, R.; et al. Piezoelectricity Enhancement of Nanogenerators Based on PDMS and ZnSnO3 Nanowires through Microstructuration. *ACS Appl. Mater. Interfaces* **2020**, *12*, 18421–18430. [CrossRef]

206. Qin, W.; Zhou, P.; Qi, Y.; Zhang, T. Lead-Free Bi3.15Nd0.85Ti3O12 Nanoplates Filler-Elastomeric Polymer Composite Films for Flexible Piezoelectric Energy Harvesting. *Micromachines* **2020**, *11*, 966. [CrossRef] [PubMed]
207. Ma, S.W.; Fan, Y.J.; Li, H.Y.; Su, L.; Wang, Z.L.; Zhu, G. Flexible Porous Polydimethylsiloxane/Lead Zirconate Titanate-Based Nanogenerator Enabled by the Dual Effect of Ferroelectricity and Piezoelectricity. *ACS Appl. Mater. Interfaces* **2018**, *10*, 33105–33111. [CrossRef]
208. Jeronimo, K.; Koutsos, V.; Cheung, R.; Mastropaolo, E. PDMS-ZnO Piezoelectric Nanocomposites for Pressure Sensors. *Sensors* **2021**, *21*, 5873. [CrossRef] [PubMed]
209. Zhang, J.; He, Y.; Boyer, C.; Kalantar-Zadeh, K.; Peng, S.; Chu, D.; Wang, C.H. Recent Developments of Hybrid Piezo–Triboelectric Nanogenerators for Flexible Sensors and Energy Harvesters. *Nanoscale Adv.* **2021**, *3*, 5465–5486. [CrossRef]
210. Hajra, S.; Padhan, A.M.; Sahu, M.; Alagarsamy, P.; Lee, K.; Kim, H.J. Lead-Free Flexible Bismuth Titanate-PDMS Composites: A Multifunctional Colossal Dielectric Material for Hybrid Piezo-Triboelectric Nanogenerator to Sustainably Power Portable Electronics. *Nano Energy* **2021**, *89*, 106316. [CrossRef]
211. He, W.; Qian, Y.; Lee, B.S.; Zhang, F.; Rasheed, A.; Jung, J.-E.; Kang, D.J. Ultrahigh Output Piezoelectric and Triboelectric Hybrid Nanogenerators Based on ZnO Nanoflakes/Polydimethylsiloxane Composite Films. *ACS Appl. Mater. Interfaces* **2018**, *10*, 44415–44420. [CrossRef] [PubMed]
212. Pongampai, S.; Charoonsuk, T.; Pinpru, N.; Pulphol, P.; Vittayakorn, W.; Pakawanit, P.; Vittayakorn, N. Triboelectric-Piezoelectric Hybrid Nanogenerator Based on BaTiO3-Nanorods/Chitosan Enhanced Output Performance with Self-Charge-Pumping System. *Compos. Part B Eng.* **2021**, *208*, 108602. [CrossRef]
213. Zi, Y.; Lin, L.; Wang, J.; Wang, S.; Chen, J.; Fan, X.; Yang, P.-K.; Yi, F.; Wang, Z.L. Triboelectric–Pyroelectric–Piezoelectric Hybrid Cell for High-Efficiency Energy-Harvesting and Self-Powered Sensing. *Adv. Mater.* **2020**, *27*, 2340–2347. [CrossRef] [PubMed]
214. Minhas, J.Z.; Hasan, M.A.M.; Yang, Y. Ferroelectric Materials Based Coupled Nanogenerators. *Nanoenergy Adv.* **2021**, *1*, 131–180. [CrossRef]
215. Wang, H.; Ma, Y.; Yang, H.; Jiang, H.; Ding, Y.; Xie, H. MEMS Ultrasound Transducers for Endoscopic Photoacoustic Imaging Applications. *Micromachines* **2020**, *11*, 928. [CrossRef] [PubMed]
216. Dangi, A.; Agrawal, S.; Kothapalli, S.-R. Lithium Niobate-Based Transparent Ultrasound Transducers for Photoacoustic Imaging. *Opt. Lett.* **2019**, *44*, 5326–5329. [CrossRef] [PubMed]
217. Qiu, Y.; Gigliotti, J.V.; Wallace, M.; Griggio, F.; Demore, C.E.M.; Cochran, S.; Trolier-McKinstry, S. Piezoelectric Micromachined Ultrasound Transducer (PMUT) Arrays for Integrated Sensing, Actuation and Imaging. *Sensors* **2015**, *15*, 8020–8041. [CrossRef] [PubMed]
218. Dangi, A.; Cheng, C.Y.; Agrawal, S.; Tiwari, S.; Datta, G.R.; Benoit, R.R.; Pratap, R.; Trolier-Mckinstry, S.; Kothapalli, S.-R. A Photoacoustic Imaging Device Using Piezoelectric Micromachined Ultrasound Transducers (PMUTs). *IEEE Trans. Ultrason. Ferroelectr. Freq. Control* **2020**, *67*, 801–809. [CrossRef] [PubMed]
219. WindEurope Homepage. Available online: https://windeurope.org/ (accessed on 12 December 2021).
220. Energy Harvesting: Solar, Wind, and Ocean Energy Conversion Systems. Available online: https://www.routledge.com/Energy-Harvesting-Solar-Wind-and-Ocean-Energy-Conversion-Systems/Khaligh-Onar/p/book/9781439815083 (accessed on 4 December 2021).
221. Wang, H.; Xu, L.; Wang, Z. Advances of High-Performance Triboelectric Nanogenerators for Blue Energy Harvesting. *Nanoenergy Adv.* **2021**, *1*, 32–57. [CrossRef]
222. Zhao, L.; Yang, Y. Toward Small-Scale Wind Energy Harvesting: Design, Enhancement, Performance Comparison, and Applicability. Available online: https://www.hindawi.com/journals/sv/2017/3585972/ (accessed on 23 May 2020).
223. Xu, C.; Pan, C.; Liu, Y.; Wang, Z.L. Hybrid Cells for Simultaneously Harvesting Multi-Type Energies for Self-Powered Micro/Nanosystems. *Nano Energy* **2012**, *1*, 259–272. [CrossRef]
224. Lai, Z.H.; Wang, J.L.; Zhang, C.L.; Zhang, G.Q.; Yurchenko, D. Harvest Wind Energy from a Vibro-Impact DEG Embedded into a Bluff Body. *Energy Convers. Manag.* **2019**, *199*, 111993. [CrossRef]
225. Zhao, L.-C.; Zou, H.-X.; Yan, G.; Liu, F.-R.; Tan, T.; Wei, K.-X.; Zhang, W.-M. Magnetic Coupling and Flextensional Amplification Mechanisms for High-Robustness Ambient Wind Energy Harvesting. *Energy Convers. Manag.* **2019**, *201*, 112166. [CrossRef]
226. Wang, X.; Pan, C.L.; Liu, Y.B.; Feng, Z.H. Electromagnetic Resonant Cavity Wind Energy Harvester with Optimized Reed Design and Effective Magnetic Loop. *Sens. Actuators Phys.* **2014**, *205*, 63–71. [CrossRef]
227. Emirates Aviation University; Aljadiri, R.T.; Taha, L.Y.; Windsor University; Ivey, P. Birmingham City University Electrostatic Energy Harvesting Systems: A Better Understanding of Their SustainabilityElectrostatic Energy Harvesting Systems: A Better Understanding of Their Sustainability. *J. Clean Energy Technol.* **2017**, *5*, 409–416. [CrossRef]
228. Bryant, M.; Garcia, E. Modeling and Testing of a Novel Aeroelastic Flutter Energy Harvester. *J. Vib. Acoust.* **2011**, *133*, 011010. [CrossRef]
229. Dai, H.L.; Abdelmoula, H.; Abdelkefi, A.; Wang, L. Towards Control of Cross-Flow-Induced Vibrations Based on Energy Harvesting. *Nonlinear Dyn.* **2017**, *88*, 2329–2346. [CrossRef]
230. Hu, G.; Tse, K.T.; Wei, M.; Naseer, R.; Abdelkefi, A.; Kwok, K.C.S. Experimental Investigation on the Efficiency of Circular Cylinder-Based Wind Energy Harvester with Different Rod-Shaped Attachments. *Appl. Energy* **2018**, *226*, 682–689. [CrossRef]
231. Usman, M.; Hanif, A.; Kim, I.-H.; Jung, H.-J. Experimental Validation of a Novel Piezoelectric Energy Harvesting System Employing Wake Galloping Phenomenon for a Broad Wind Spectrum. *Energy* **2018**, *153*, 882–889. [CrossRef]

232. Wang, J.; Tang, L.; Zhao, L.; Zhang, Z. Efficiency Investigation on Energy Harvesting from Airflows in HVAC System Based on Galloping of Isosceles Triangle Sectioned Bluff Bodies. *Energy* **2019**, *172*, 1066–1078. [CrossRef]
233. Wang, J.; Zhou, S.; Zhang, Z.; Yurchenko, D. High-Performance Piezoelectric Wind Energy Harvester with Y-Shaped At-tachments. *Energy Convers. Manag.* **2019**, *181*, 645–652. [CrossRef]
234. Zhou, S.; Wang, J. Erratum: "Dual Serial Vortex-Induced Energy Harvesting System for Enhanced Energy Harvesting". *AIP Adv.* **2018**, *8*, 089901. [CrossRef]
235. Jirayupat, C.; Wongwiriyapan, W.; Kasamechonchung, P.; Wutikhun, T.; Tantisantisom, K.; Rayanasukha, Y.; Jiemsakul, T.; Tansarawiput, C.; Liangruksa, M.; Khanchaitit, P.; et al. Piezoelectric-Induced Triboelectric Hybrid Nanogenerators Based on the ZnO Nanowire Layer Decorated on the Au/Polydimethylsiloxane–Al Structure for Enhanced Triboelectric Performance. *ACS Appl. Mater. Interfaces* **2018**, *10*, 6433–6440. [CrossRef]
236. Han, M.; Zhang, X.-S.; Meng, B.; Liu, W.; Tang, W.; Sun, X.; Wang, W.; Zhang, H. R-Shaped Hybrid Nanogenerator with Enhanced Piezoelectricity. *ACS Nano* **2013**, *7*, 8554–8560. [CrossRef]
237. Dai, H.L.; Abdelkefi, A.; Wang, L. Theoretical Modeling and Nonlinear Analysis of Piezoelectric Energy Harvesting from Vortex-Induced Vibrations. *J. Intell. Mater. Syst. Struct.* **2020**, *25*, 1861–1874. [CrossRef]
238. Abdelkefi, A. Aeroelastic Energy Harvesting: A Review. *Int. J. Eng. Sci.* **2016**, *100*, 112–135. [CrossRef]
239. Yang, K.; Wang, J.; Yurchenko, D. A Double-Beam Piezo-Magneto-Elastic Wind Energy Harvester for Improving the Galloping-Based Energy Harvesting. *Appl. Phys. Lett.* **2020**, *115*, 193901. [CrossRef]
240. Challa, V.R.; Prasad, M.G.; Fisher, F.T. A Coupled Piezoelectric–Electromagnetic Energy Harvesting Technique for Achieving Increased Power Output through Damping Matching. *Smart Mater. Struct.* **2009**, *18*, 095029. [CrossRef]
241. Naseer, R.; Dai, H.L.; Abdelkefi, A.; Wang, L. Piezomagnetoelastic Energy Harvesting from Vortex-Induced Vibrations Using Monostable Characteristics. *Appl. Energy* **2017**, *203*, 142–153. [CrossRef]
242. Huang, L.; Xu, W.; Bai, G.; Wong, M.-C.; Yang, Z.; Hao, J. Wind Energy and Blue Energy Harvesting Based on Magnetic-Assisted Noncontact Triboelectric Nanogenerator. *Nano Energy* **2016**, *30*, 36–42. [CrossRef]
243. Su, Y.; Wen, X.; Zhu, G.; Yang, J.; Bai, P.; Wu, Z.; Jiang, Y.; Wang, Z. Hybrid Triboelectric Nanogenerator for Harvesting Water Wave Energy and as a Self-Powered Distress Signal Emitter. *Nano Energy* **2014**, *9*, 186–195. [CrossRef]
244. Wang, X.; Wen, Z.; Guo, H.; Wu, C.; He, X.; Lin, L.; Cao, X.; Wang, Z.L. Fully Packaged Blue Energy Harvester by Hybridizing a Rolling Triboelectric Nanogenerator and an Electromagnetic Generator. *ACS Nano* **2016**, *10*, 11369–11376. [CrossRef]
245. Wen, Z.; Guo, H.; Zi, Y.; Yeh, M.-H.; Wang, X.; Deng, J.; Wang, J.; Li, S.; Hu, C.; Zhu, L.; et al. Harvesting Broad Frequency Band Blue Energy by a Triboelectric–Electromagnetic Hybrid Nanogenerator. *ACS Nano* **2016**, *10*, 6526–6534. [CrossRef]
246. Shao, H.; Wen, Z.; Cheng, P.; Sun, N.; Shen, Q.; Zhou, C.; Peng, M.; Yang, Y.; Xie, X.; Sun, X. Multifunctional Power Unit by Hybridizing Contact-Separate Triboelectric Nanogenerator, Electromagnetic Generator and Solar Cell for Harvesting Blue Energy. *Nano Energy* **2017**, *39*, 608–615. [CrossRef]
247. Zhu, J.; Wang, A.; Hu, H.; Zhu, H. Hybrid Electromagnetic and Triboelectric Nanogenerators with Multi-Impact for Wideband Frequency Energy Harvesting. *Energies* **2017**, *10*, 2024. [CrossRef]
248. Shao, H.; Cheng, P.; Chen, R.; Xie, L.; Sun, N.; Shen, Q.; Chen, X.; Zhu, Q.; Zhang, Y.; Liu, Y.; et al. Triboelectric–Electromagnetic Hybrid Generator for Harvesting Blue Energy. *Nano Micro Lett.* **2018**, *10*, 54. [CrossRef]
249. Wang, H.; Zhu, Q.; Ding, Z.; Li, Z.; Zheng, H.; Fu, J.; Diao, C.; Zhang, X.; Tian, J.; Zi, Y. A Fully-Packaged Ship-Shaped Hybrid Nanogenerator for Blue Energy Harvesting toward Seawater Self-Desalination and Self-Powered Positioning. *Nano Energy* **2019**, *57*, 616–624. [CrossRef]
250. Guo, H.; Wen, Z.; Zi, Y.; Yeh, M.-H.; Wang, J.; Zhu, L.; Hu, C.; Wang, Z.L. A Water-Proof Triboelectric–Electromagnetic Hybrid Generator for Energy Harvesting in Harsh Environments. *Adv. Energy Mater.* **2020**, *6*, 1501593. [CrossRef]
251. Sun, T.; Tasnim, F.; McIntosh, R.T.; Amiri, N.; Solav, D.; Anbarani, M.T.; Sadat, D.; Zhang, L.; Gu, Y.; Karami, M.A.; et al. Decoding of Facial Strains via Conformable Piezoelectric Interfaces. *Nat. Biomed. Eng.* **2020**, *4*, 954–972. [CrossRef] [PubMed]
252. Fernandez, S.V.; Cai, F.; Chen, S.; Suh, E.; Tiepelt, J.; McIntosh, R.; Marcus, C.; Acosta, D.; Mejorado, D.; Dagdeviren, C. On-Body Piezoelectric Energy Harvesters through Innovative Designs and Conformable Structures. *ACS Biomater. Sci. Eng.* **2021**; *in press*. [CrossRef] [PubMed]
253. Dhakar, L.; Pitchappa, P.; Tay, F.E.H.; Lee, C. An Intelligent Skin Based Self-Powered Finger Motion Sensor Integrated with Triboelectric Nanogenerator. *Nano Energy* **2016**, *19*, 532–540. [CrossRef]
254. Zhang, N.; Tao, C.; Fan, X.; Chen, J. Progress in Triboelectric Nanogenerators as Self-Powered Smart Sensors. *J. Mater. Res.* **2017**, *32*, 1628–1646. [CrossRef]
255. Zhang, H.; Yang, Y.; Hou, T.-C.; Su, Y.; Hu, C.; Wang, Z.L. Triboelectric Nanogenerator Built inside Clothes for Self-Powered Glucose Biosensors. *Nano Energy* **2013**, *2*, 1019–1024. [CrossRef]
256. Shawon, S.M.A.Z.; Sun, A.X.; Vega, V.S.; Chowdhury, B.D.; Tran, P.; Carballo, Z.D.; Tolentino, J.A.; Li, J.; Rafaqut, M.S.; Danti, S.; et al. Piezo-Tribo Dual Effect Hybrid Nanogenerators for Health Monitoring. *Nano Energy* **2021**, *82*, 105691. [CrossRef]
257. Tang, G.; Shi, Q.; Zhang, Z.; He, T.; Sun, Z.; Lee, C. Hybridized Wearable Patch as a Multi-Parameter and Multi-Functional Human-Machine Interface. *Nano Energy* **2021**, *81*, 105582. [CrossRef]
258. Fang, H.; Zhao, J.; Yu, K.J.; Song, E.; Farimani, A.B.; Chiang, C.-H.; Jin, X.; Xue, Y.; Xu, D.; Du, W.; et al. Ultrathin, Transferred Layers of Thermally Grown Silicon Dioxide as Biofluid Barriers for Biointegrated Flexible Electronic Systems. *Proc. Natl. Acad. Sci. USA* **2016**, *113*, 11682–11687. [CrossRef]

259. Wu, C.-Y.; Liao, R.-M.; Lai, L.-W.; Jeng, M.-S.; Liu, D.-S. Organosilicon/Silicon Oxide Gas Barrier Structure Encapsulated Flexible Plastic Substrate by Using Plasma-Enhanced Chemical Vapor Deposition. *Surf. Coat. Technol.* **2012**, *206*, 4685–4691. [CrossRef]

260. Song, E.; Fang, H.; Jin, X.; Zhao, J.; Jiang, C.; Yu, K.J.; Zhong, Y.; Xu, D.; Li, J.; Fang, G.; et al. Thin, Transferred Layers of Silicon Dioxide and Silicon Nitride as Water and Ion Barriers for Implantable Flexible Electronic Systems. *Adv. Electron. Mater.* **2017**, *3*, 1700077. [CrossRef]

261. Carcia, P.F.; McLean, R.S.; Reilly, M.H.; Groner, M.D.; George, S.M. Ca Test of Al2O3 Gas Diffusion Barriers Grown by Atomic Layer Deposition on Polymers. *Appl. Phys. Lett.* **2006**, *89*, 031915. [CrossRef]

262. Li, J.; Li, R.; Du, H.; Zhong, Y.; Chen, Y.; Nan, K.; Won, S.M.; Zhang, J.; Huang, Y.; Rogers, J.A. Ultrathin, Transferred Layers of Metal Silicide as Faradaic Electrical Interfaces and Biofluid Barriers for Flexible Bioelectronic Implants. *ACS Nano* **2019**, *13*, 660–670. [CrossRef] [PubMed]

263. Behrendt, A.; Friedenberger, C.; Gahlmann, T.; Trost, S.; Becker, T.; Zilberberg, K.; Polywka, A.; Görrn, P.; Riedl, T. Highly Robust Transparent and Conductive Gas Diffusion Barriers Based on Tin Oxide. *Adv. Mater.* **2015**, *27*, 5961–5967. [CrossRef]

264. Zheng, Q.; Shi, B.; Li, Z.; Wang, Z.L. Recent Progress on Piezoelectric and Triboelectric Energy Harvesters in Biomedical Systems. *Adv. Sci.* **2017**, *4*, 1700029. [CrossRef] [PubMed]

265. Shi, B.; Li, Z.; Fan, Y. Implantable Energy-Harvesting Devices. *Adv. Mater.* **2018**, *30*, 1801511. [CrossRef] [PubMed]

266. Han, S.A.; Lee, J.-H.; Seung, W.; Lee, J.; Kim, S.-W.; Kim, J.H. Patchable and Implantable 2D Nanogenerator. *Small* **2021**, *17*, 1903519. [CrossRef] [PubMed]

267. Dong, L.; Jin, C.; Closson, A.B.; Trase, I.; Richards, H.C.; Chen, Z.; Zhang, J.X.J. Cardiac Energy Harvesting and Sensing Based on Piezoelectric and Triboelectric Designs. *Nano Energy* **2020**, *76*, 105076. [CrossRef]

268. Zhang, G.; Li, M.; Li, H.; Wang, Q.; Jiang, S. Harvesting Energy from Human Activity: Ferroelectric Energy Harvesters for Portable, Implantable, and Biomedical Electronics. *Energy Technol.* **2018**, *6*, 791–812. [CrossRef]

269. Dagdeviren, C.; Yang, B.D.; Su, Y.; Tran, P.L.; Joe, P.; Anderson, E.; Xia, J.; Doraiswamy, V.; Dehdashti, B.; Feng, X.; et al. Conformal Piezoelectric Energy Harvesting and Storage from Motions of the Heart, Lung, and Diaphragm. *Proc. Natl. Acad. Sci. USA* **2014**, *111*, 1927–1932. [CrossRef] [PubMed]

270. Dagdeviren, C.; Shi, Y.; Joe, P.; Ghaffari, R.; Balooch, G.; Usgaonkar, K.; Gur, O.; Tran, P.L.; Crosby, J.R.; Meyer, M.; et al. Conformal Piezoelectric Systems for Clinical and Experimental Characterization of Soft Tissue Biomechanics. *Nat. Mater.* **2015**, *14*, 728–736. [CrossRef] [PubMed]

271. Zhang, H.; Zhang, X.-S.; Cheng, X.; Liu, Y.; Han, M.; Xue, X.; Wang, S.; Yang, F.; Smitha, S.A.; Zhang, H.; et al. A Flexible and Implantable Piezoelectric Generator Harvesting Energy from the Pulsation of Ascending Aorta: In Vitro and In Vivo Studies. *Nano Energy* **2015**, *12*, 296–304. [CrossRef]

272. Yang, R.; Qin, Y.; Li, C.; Zhu, G.; Wang, Z.L. Converting Biomechanical Energy into Electricity by a Muscle-Movement-Driven Nanogenerator. *Nano Lett.* **2009**, *9*, 1201–1205. [CrossRef] [PubMed]

273. Li, Z.; Zhu, G.; Yang, R.; Wang, A.C.; Wang, Z.L. Muscle-Driven in Vivo Nanogenerator. *Adv. Mater. Deerfield Beach Fla* **2010**, *22*, 2534–2537. [CrossRef] [PubMed]

274. Hwang, G.-T.; Park, H.; Lee, J.-H.; Oh, S.; Park, K.-I.; Byun, M.; Park, H.; Ahn, G.; Jeong, C.K.; No, K.; et al. Self-Powered Cardiac Pacemaker Enabled by Flexible Single Crystalline PMN-PT Piezoelectric Energy Harvester. *Adv. Mater.* **2014**, *26*, 4880–4887. [CrossRef]

275. Liu, Z.; Li, H.; Shi, B.; Fan, Y.; Wang, Z.L.; Li, Z. Wearable and Implantable Triboelectric Nanogenerators. *Adv. Funct. Mater.* **2019**, *29*, 1808820. [CrossRef]

276. Ouyang, H.; Liu, Z.; Li, N.; Shi, B.; Zou, Y.; Xie, F.; Ma, Y.; Li, Z.; Li, H.; Zheng, Q.; et al. Symbiotic Cardiac Pacemaker. *Nat. Commun.* **2019**, *10*, 1821. [CrossRef] [PubMed]

277. Shi, Q.; He, T.; Lee, C. More than Energy Harvesting—Combining Triboelectric Nanogenerator and Flexible Electronics Technology for Enabling Novel Micro-/Nano-Systems. *Nano Energy* **2019**, *57*, 851–871. [CrossRef]

278. Maiti, S.; Karan, S.K.; Kim, J.K.; Khatua, B.B. Nature Driven Bio-Piezoelectric/Triboelectric Nanogenerator as Next-Generation Green Energy Harvester for Smart and Pollution Free Society. *Adv. Energy Mater.* **2019**, *9*, 1803027. [CrossRef]

279. Ryu, H.; Park, H.-M.; Kim, M.-K.; Kim, B.; Myoung, H.S.; Kim, T.Y.; Yoon, H.-J.; Kwak, S.S.; Kim, J.; Hwang, T.H.; et al. Self-Rechargeable Cardiac Pacemaker System with Triboelectric Nanogenerators. *Nat. Commun.* **2021**, *12*, 4374. [CrossRef] [PubMed]

280. Zheng, Q.; Shi, B.; Fan, F.; Wang, X.; Yan, L.; Yuan, W.; Wang, S.; Liu, H.; Li, Z.; Wang, Z.L. In Vivo Powering of Pacemaker by Breathing-Driven Implanted Triboelectric Nanogenerator. *Adv. Mater.* **2014**, *26*, 5851–5856. [CrossRef] [PubMed]

281. Dagdeviren, C.; Li, Z.; Wang, Z.L. Energy Harvesting from the Animal/Human Body for Self-Powered Electronics. *Annu. Rev. Biomed. Eng.* **2017**, *19*, 85–108. [CrossRef] [PubMed]

282. Falconi, C.; Mandal, S. Interface Electronics: State-of-the-Art, Opportunities and Needs. *Sens. Actuators Phys.* **2019**, *296*, 24–30. [CrossRef]

283. Falconi, C.; Martinelli, E.; Di Natale, C.; D'Amico, A.; Maloberti, F.; Malcovati, P.; Baschirotto, A.; Stornelli, V.; Ferri, G. Electronic Interfaces. *Sens. Actuators B Chem.* **2007**, *121*, 295–329. [CrossRef]

284. Ponmozhi, J.; Frias, C.; Marques, T.; Frazão, O. Smart Sensors/Actuators for Biomedical Applications: Review. *Measurement* **2012**, *45*, 1675–1688. [CrossRef]

285. Schiavone, G.; Fallegger, F.; Kang, X.; Barra, B.; Vachicouras, N.; Roussinova, E.; Furfaro, I.; Jiguet, S.; Seáñez, I.; Borgognon, S.; et al. Soft, Implantable Bioelectronic Interfaces for Translational Research. *Adv. Mater.* **2020**, *32*, 1906512. [CrossRef]
286. Fallegger, F.; Schiavone, G.; Lacour, S. Conformable Hybrid Systems for Implantable Bioelectronic Interfaces. *Adv. Mater.* **2019**, *32*, 1903904. [CrossRef]
287. Liang, J.; Liao, W.-H. Energy Flow in Piezoelectric Energy Harvesting Systems. *Smart Mater. Struct.* **2010**, *20*, 015005. [CrossRef]
288. Li, P.; Wen, Y.; Jia, C.; Li, X. A Magnetoelectric Composite Energy Harvester and Power Management Circuit. *IEEE Trans. Ind. Electron.* **2011**, *58*, 2944–2951. [CrossRef]
289. Li, P.; Wen, Y.; Yin, W.; Wu, H. An Upconversion Management Circuit for Low-Frequency Vibrating Energy Harvesting. *IEEE Trans. Ind. Electron.* **2014**, *61*, 3349–3358. [CrossRef]
290. Guyomar, D.; Lallart, M. Recent Progress in Piezoelectric Conversion and Energy Harvesting Using Nonlinear Electronic Interfaces and Issues in Small Scale Implementation. *Micromachines* **2011**, *2*, 274–294. [CrossRef]
291. Szarka, G.D.; Stark, B.H.; Burrow, S.G. Review of Power Conditioning for Kinetic Energy Harvesting Systems. *IEEE Trans. Power Electron.* **2012**, *27*, 803–815. [CrossRef]
292. Dicken, J.; Mitcheson, P.D.; Stoianov, I.; Yeatman, E.M. Power-Extraction Circuits for Piezoelectric Energy Harvesters in Miniature and Low-Power Applications. *IEEE Trans. Power Electron.* **2012**, *27*, 4514–4529. [CrossRef]
293. Dell'Anna, F.; Dong, T.; Li, P.; Wen, Y.; Yang, Z.; Casu, M.R.; Azadmehr, M.; Berg, Y. State-of-the-Art Power Management Circuits for Piezoelectric Energy Harvesters. *IEEE Circuits Syst. Mag.* **2018**, *18*, 27–48. [CrossRef]
294. Lefeuvre, E.; Badel, A.; Richard, C.; Petit, L.; Guyomar, D. A Comparison between Several Vibration-Powered Piezoelectric Generators for Standalone Systems. *Sens. Actuators Phys.* **2006**, *126*, 405–416. [CrossRef]
295. Krihely, N.; Ben-Yaakov, S. Self-Contained Resonant Rectifier for Piezoelectric Sources Under Variable Mechanical Excitation. *IEEE Trans. Power Electron.* **2011**, *26*, 612–621. [CrossRef]
296. Long, Z.; Wang, X.; Li, P.; Wang, B.; Zhang, X.; Chung, H.S.-H.; Yang, Z. Self-Powered SSDCI Array Interface for Multiple Piezoelectric Energy Harvesters. *IEEE Trans. Power Electron.* **2021**, *36*, 9093–9104. [CrossRef]
297. Tang, L.; Yang, Y. Analysis of Synchronized Charge Extraction for Piezoelectric Energy Harvesting. *Smart Mater. Struct.* **2011**, *20*, 085022. [CrossRef]
298. Lefeuvre, E.; Badel, A.; Brenes, A.; Seok, S.; Yoo, C.-S. Power and Frequency Bandwidth Improvement of Piezoelectric Energy Harvesting Devices Using Phase-Shifted Synchronous Electric Charge Extraction Interface Circuit. *J. Intell. Mater. Syst. Struct.* **2017**, *28*, 2988–2995. [CrossRef]
299. Garbuio, L.; Lallart, M.; Guyomar, D.; Richard, C.; Audigier, D. Mechanical Energy Harvester with Ultralow Threshold Rectification Based on SSHI Nonlinear Technique. *IEEE Trans. Ind. Electron.* **2009**, *56*, 1048–1056. [CrossRef]
300. Lallart, M.; Garbuio, L.; Petit, L.; Richard, C.; Guyomar, D. Double Synchronized Switch Harvesting (DSSH): A New Energy Harvesting Scheme for Efficient Energy Extraction. *IEEE Trans. Ultrason. Ferroelectr. Freq. Control* **2008**, *55*, 2119–2130. [CrossRef]
301. Shen, H.; Qiu, J.; Ji, H.; Zhu, K.; Balsi, M. Enhanced Synchronized Switch Harvesting: A New Energy Harvesting Scheme for Efficient Energy Extraction. *Smart Mater. Struct.* **2010**, *19*, 115017. [CrossRef]
302. Lallart, M.; Guyomar, D. Piezoelectric Conversion and Energy Harvesting Enhancement by Initial Energy Injection. *Appl. Phys. Lett.* **2010**, *97*, 014104. [CrossRef]
303. Sankman, J.; Ma, D. A 12-MW to 1.1-MW AIM Piezoelectric Energy Harvester for Time-Varying Vibrations With 450-NA\bm Q. *IEEE Trans. Power Electron.* **2015**, *30*, 632–643. [CrossRef]
304. Wang, S.; Lin, Z.-H.; Niu, S.; Lin, L.; Xie, Y.; Pradel, K.C.; Wang, Z.L. Motion Charged Battery as Sustainable Flexible-Power-Unit. *ACS Nano* **2013**, *7*, 11263–11271. [CrossRef] [PubMed]
305. Zi, Y.; Niu, S.; Wang, J.; Wen, Z.; Tang, W.; Wang, Z.L. Standards and Figure-of-Merits for Quantifying the Performance of Triboelectric Nanogenerators. *Nat. Commun.* **2015**, *6*, 8376. [CrossRef] [PubMed]
306. Hu, Y.; Yue, Q.; Lu, S.; Yang, D.; Shi, S.; Zhang, X.; Yu, H. An Adaptable Interface Conditioning Circuit Based on Triboelectric Nanogenerators for Self-Powered Sensors. *Micromachines* **2018**, *9*, 105. [CrossRef] [PubMed]
307. Cheng, X.; Miao, L.; Song, Y.; Su, Z.; Chen, H.; Chen, X.; Zhang, J.; Zhang, H. High Efficiency Power Management and Charge Boosting Strategy for a Triboelectric Nanogenerator. *Nano Energy* **2017**, *38*, 438–446. [CrossRef]
308. Niu, S.; Wang, X.; Yi, F.; Zhou, Y.S.; Wang, Z.L. A Universal Self-Charging System Driven by Random Biomechanical Energy for Sustainable Operation of Mobile Electronics. *Nat. Commun.* **2015**, *6*, 8975. [CrossRef]
309. Bao, D.; Luo, L.; Zhang, Z.; Ren, T. A Power Management Circuit with 50% Efficiency and Large Load Capacity for Triboelectric Nanogenerator. *J. Semicond.* **2017**, *38*, 095001. [CrossRef]
310. Xi, F.; Pang, Y.; Li, W.; Jiang, T.; Zhang, L.; Guo, T.; Liu, G.; Zhang, C.; Wang, Z.L. Universal Power Management Strategy for Triboelectric Nanogenerator. *Nano Energy* **2017**, *37*, 168–176. [CrossRef]
311. Zhang, C.; LI, W.; Wang, Z.; Fengben, X.I.; Pang, Y. Power Management Circuit and Power Management Method for Triboelectric Nanogenerator, and Energy System. U.S. Patent 20200099316A1, 26 March 2020.
312. Liang, X.; Jiang, T.; Liu, G.; Xiao, T.; Xu, L.; Li, W.; Xi, F.; Zhang, C.; Wang, Z.L. Triboelectric Nanogenerator Networks Integrated with Power Management Module for Water Wave Energy Harvesting. *Adv. Funct. Mater.* **2019**, *29*, 1807241. [CrossRef]

313. Liu, W.; Wang, Z.; Wang, G.; Zeng, Q.; He, W.; Liu, L.; Wang, X.; Xi, Y.; Guo, H.; Hu, C.; et al. Switched-Capacitor-Convertors Based on Fractal Design for Output Power Management of Triboelectric Nanogenerator. *Nat. Commun.* **2020**, *11*, 1883. [CrossRef] [PubMed]

314. Zi, Y.; Guo, H.; Wang, J.; Wen, Z.; Li, S.; Hu, C.; Wang, Z.L. An Inductor-Free Auto-Power-Management Design Built-in Triboelectric Nanogenerators. *Nano Energy* **2017**, *31*, 302–310. [CrossRef]

315. Tang, W.; Zhou, T.; Zhang, C.; Fan, F.R.; Han, C.B.; Wang, Z.L. A Power-Transformed-and-Managed Triboelectric Nanogenerator and Its Applications in a Self-Powered Wireless Sensing Node. *Nanotechnology* **2014**, *25*, 225402. [CrossRef] [PubMed]

316. Xu, L.; Pang, Y.; Zhang, C.; Jiang, T.; Chen, X.; Luo, J.; Tang, W.; Cao, X.; Wang, Z.L. Integrated Triboelectric Nanogenerator Array Based on Air-Driven Membrane Structures for Water Wave Energy Harvesting. *Nano Energy* **2017**, *31*, 351–358. [CrossRef]

317. He, J.; Fan, X.; Zhao, D.; Cui, M.; Han, B.; Hou, X.; Chou, X. A High-Efficient Triboelectric-Electromagnetic Hybrid Nanogenerator for Vibration Energy Harvesting and Wireless Monitoring. *Sci. China Inf. Sci.* **2021**, *65*, 142401. [CrossRef]

318. Li, Z.; Saadatnia, Z.; Yang, Z.; Naguib, H. A Hybrid Piezoelectric-Triboelectric Generator for Low-Frequency and Broad-Bandwidth Energy Harvesting. *Energy Convers. Manag.* **2018**, *174*, 188–197. [CrossRef]

319. Li, X.; Yu, W.; Gao, X.; Liu, H.; Han, N.; Zhang, X. PVDF Microspheres ≅ PLLA Nanofibers-Based Hybrid Tribo/Piezoelectric Nanogenerator with Excellent Electrical Output Properties. *Mater. Adv.* **2021**, *2*, 6011–6019. [CrossRef]

320. Lallart, M.; Lombardi, G. Synchronized Switch Harvesting on ElectroMagnetic System: A Nonlinear Technique for Hybrid Energy Harvesting Based on Active Inductance. *Energy Convers. Manag.* **2020**, *203*, 112135. [CrossRef]

321. Tabesh, A.; Frechette, L.G. A Low-Power Stand-Alone Adaptive Circuit for Harvesting Energy from a Piezoelectric Micropower Generator. *IEEE Trans. Ind. Electron.* **2010**, *57*, 840–849. [CrossRef]

322. Yu, Y.; Li, Z.; Wang, Y.; Gong, S.; Wang, X. Sequential Infiltration Synthesis of Doped Polymer Films with Tunable Electrical Properties for Efficient Triboelectric Nanogenerator Development. *Adv. Mater. Deerfield Beach Fla* **2015**, *27*, 4938–4944. [CrossRef] [PubMed]

323. Jiang, M.; Li, B.; Jia, W.; Zhu, Z. Predicting Output Performance of Triboelectric Nanogenerators Using Deep Learning Model. *Nano Energy* **2022**, *93*, 106830. [CrossRef]

 *nanoenergy advances*

Review

Smart Textile Triboelectric Nanogenerators: Prospective Strategies for Improving Electricity Output Performance

Kai Dong [1,2], **Xiao Peng** [1,2], **Renwei Cheng** [1,2] **and Zhong Lin Wang** [1,2,3,*]

[1] Beijing Institute of Nanoenergy and Nanosystems, Chinese Academy of Sciences, Beijing 101400, China; dongkai@binn.cas.cn (K.D.); pengxiao@binn.cas.cn (X.P.); chengrenwei@binn.cas.cn (R.C.)
[2] College of Nanoscience and Technology, University of Chinese Academy of Sciences, Beijing 100049, China
[3] School of Material Science and Engineering, Georgia Institute of Technology, Atlanta, GA 30332, USA
[*] Correspondence: zhong.wang@mse.gatech.edu

Abstract: By seamlessly integrating the wearing comfortability of textiles with the biomechanical energy harvesting function of a triboelectric nanogenerator (TENG), an emerging and advanced intelligent textile, i.e., smart textile TENG, is developed with remarkable abilities of autonomous power supply and self-powered sensing, which has great development prospects in the next-generation human-oriented wearable electronics. However, due to inadequate interface contact, insufficient electrification of materials, unavoidable air breakdown effect, output capacitance feature, and special textile structure, there are still several bottlenecks in the road towards the practical application of textile TENGs, including low output, high impedance, low integration, poor working durability, and so on. In this review, on the basis of mastering the existing theory of electricity generation mechanism of TENGs, some prospective strategies for improving the mechanical-to-electrical conversion performance of textile TENGs are systematically summarized and comprehensively discussed, including surface/interface physical treatments, atomic-scale chemical modification, structural optimization design, work environmental control, and integrated energy management. The advantages and disadvantages of each approach in output enhancement are further compared at the end of this review. It is hoped that this review can not only provide useful guidance for the research of textile TENGs to select optimization methods but also accelerate their large-scale practical process.

Keywords: smart textiles; triboelectric nanogenerators; electricity generation; output enhancement; air breakdown

Citation: Dong, K.; Peng, X.; Cheng, R.; Wang, Z.L. Smart Textile Triboelectric Nanogenerators: Prospective Strategies for Improving Electricity Output Performance. *Nanoenergy Adv.* **2022**, 2, 133–164. https://doi.org/10.3390/nanoenergyadv2010006

Academic Editor: Joao Oliveira Ventura

Received: 4 January 2022
Accepted: 3 March 2022
Published: 7 March 2022

Publisher's Note: MDPI stays neutral with regard to jurisdictional claims in published maps and institutional affiliations.

Copyright: © 2022 by the authors. Licensee MDPI, Basel, Switzerland. This article is an open access article distributed under the terms and conditions of the Creative Commons Attribution (CC BY) license (https://creativecommons.org/licenses/by/4.0/).

1. Introduction

The rapid consumption of fossil energy and the increasing urgency of environmental security precipitate us to reshape the current energy utilization structures that depend on oil and coal [1]. In addition, ubiquitous wearable electronics and the Internet of Things (IoTs) pose a great challenge to the present energy supply modes in the centralized, fixed, ordered, and high energy density forms, which rely heavily on traditional power plants and cable transmission networks [2]. In general, the power needed to operate millions of wearable sensors is very small, typically at the microwatt to watt level. Although orderly energy supply modes can provide a part of the power for distributed electronic devices, the rest of the power must be provided by random energy sources in our living environment, including solar energy, vibration, motion, wind energy, and other resources [3–9]. What we expect is to make full use of any available resources in the environment where the device is deployed. Therefore, the idea of a self-powered system is proposed, which is one of the most feasible schemes for low power electronic devices by effectively acquiring environmental energy [10–18].

Triboelectrification or contact electrification is a universal phenomenon in which two materials contact each other. A triboelectric nanogenerator (TENG) is a new type of energy collection technology first invented by Wang's team in 2012. By coupling triboelectric charging and electrostatic induction, various forms of irregular, low-frequency, and distributed mechanical energy, which is common in daily life but usually wasted, can be effectively converted into electric energy, including human movement, vibration, wind, mechanical triggering, water waves, and so on [19–24]. With the merits of lightweight, cost-effectiveness, universal availability, abundant materials choice, and especially high conversion efficiency at low frequency, TENGs exhibit a great application prospect in wearable emergency power supply, multifunctional self-powered sensors, healthcare apparatus, and artificial intelligence [25–35]. TENG's fundamental theory can be traced back to Maxwell's equations, which shows that the second term in Maxwell's displacement current has a direct relationship to the output electric current of TENGs [36]. Recently, expanded Maxwell's equations were also derived by assuming that the medium is moving as a rigid translation in space [37]. The expanded Maxwell's equations not only largely expand their applications in various fields but also serve as the fundamental theory of the NGs, including output current and associated electromagnetic radiation.

By combining the traditional flexible and wearable textile materials with emerging and advanced TENG science, a new type of intelligent textile technology, namely textile TENG, is developed, which has two outstanding functions: independent energy collection and active self-powered energy-sensing (Figure 1) [38–42]. With the help of wearable intelligent systems with no burden and self-sufficiency, individuals can easily obtain and make efficient use of electric energy, which will help promote the development of people-centered portable electronics and artificial intelligence in the future [43–47]. However, the low power density and high internal impedance are still the two main factors that hinder the effective commercial utilization of textile TENGs. The maximum energy output per cycle has a quadratic relationship with the charge density of the triboelectric surface and is positively correlated with the average output power and energy conversion efficiency of TENGs. According to Paschen's law, the breakdown effect of high-pressure air has a great influence on the maximum surface charge density [48]. Due to the restriction of high-pressure air breakdown, most of the surface charge densities enhanced by material optimization or external ion implantation are easy to diffuse into the atmosphere and internal triboelectric layer, resulting in charge loss and reduction of surface charge density [49,50]. Breaking through the limitation of air breakdown and prolonging the time of charge decay is especially important for improving the output of TENGs [51]. In order to improve the electromechanical conversion performance of TENGs, people adopted various methods to improve its output performance and expand the applications, such as physical surface modification, chemical surface modification, the embedding of charge trap layer, switching realization, active charge excitation, and so on. Although a large number of reviews summarized these methods to enhance the power output performance of TENGs [52–58], there is a little comprehensive summary about the improvement of the output performance of textile TENGs. Due to the high aspect ratio, complex curved configuration, and surface micro-to-nano structural defects of 1D fiber structure, it is hard and also unreasonable to directly apply these strategies of improving the power output of the common planar membrane structural TENGs to textile TENGs. In addition, because of the limited effective contact area in textile TENGs, their mechanical-to-electrical conversion efficiency is much lower than that of common planar membrane structures. Therefore, it is extremely necessary to make a comprehensive summary and constructive discussion on the potential strategies to improve the electromechanical conversion output performance of textile TENGs, so as to make their power generation meet the actual use demand.

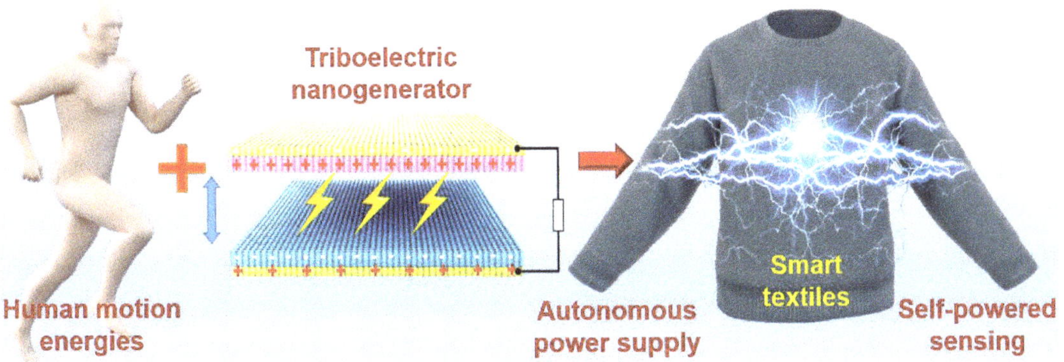

Figure 1. Schematic illustration of smart textile TENGs by converting human motion energies into electric energy through TENG technology.

This review focuses on textile TENGs, aiming at improving their electrical output performance. The electrification mechanism of TENGs is first interpreted with special attention to the atomic-scale electron-cloud/wave function overlap model. Based on the basic understanding of the charge generation mechanism of TENGs, the potential approaches to enhancing the power output of textile TENGs are systematically investigated, which are mainly divided into five categories on the basis of their respective features, including surface/interface physical treatments, atomic-scale chemical modification, structural optimization design, work environmental control, and integrated energy management. Among them, surface/interface physical treatments consist of two schemes, i.e., micro-/nanostructure patterning and soft contacted interface designing. The atomic-scale chemical modification includes surface modification, ion irradiation/implantation, and charge trapping/storage. As for structural optimization design, it can be further classified into intermediate layer adding, charge shuttling/pumping, and direct-current mode designing. In any case, each strategy has its special advantages in improving the triboelectric output performance but meanwhile exists unavoidable issues. The combination of these methods based on the selected structural characteristics and application directions may be a more preferred scheme. This review is not only a timely retrospection of the emerging smart textiles based on advanced TENG technology but also a key process to break through its existing bottlenecks and promote its real application.

2. Electrification Mechanism

Contact electrification (CE) or triboelectrification means that two different materials or materials of the same chemical type will be charged after physical contact. However, the underlying mechanism was debated for a long time, but no conclusion was reached. Recently, some researchers used a variety of experimental methods to explore the atomic-scale contact or friction behaviors as well as their induced electrification phenomena. For example, the combination of in situ high-resolution transmission electron microscope (TEM) and atomic force microscope (AFM) measurements can provide direct real-time observation of atomic-level interface structure in the processes of friction and the formation of a loosely stacked interface layer between two metal asperities can result in low friction under tensile stress (Figure 2a) [59]. In addition, using Kelvin probe force microscope (KPFM) technology and properly functionalized probes, carbon monoxide molecules can be imaged in σ-real space with anisotropic holes and quadrupole charges (Figure 2b) [60]. This method is expected to expand the possibility of characterizing complex molecular systems and surface charge distribution. The atomic-scale motion of nanotubes on a graphene substrate are also investigated based on DFT simulations to explore their atomic-scale rolling friction behavior and induced charge-transfer mechanism [61]. As shown in Figure 2c, a simplified physical model is established to the theoretical basis on atomic-scale rolling and sliding friction behaviors. The typical maximum and minimum energy positions during the rolling and sliding process are selected to characterize its corresponding charge-transfer morphology (Figure 2d). It can be found that the charge interaction is mainly concentrated in the contact area of the moving object, and there is no charge redistribution beyond the contact area. The charge is completely accumulated at the bottom of the carbon nanotube and depleted at the top of the flat graphene substrate, which indicates that selecting the rod rather than the flat structure as a strategy can effectively increase the induced charge density of the triboelectric interface.

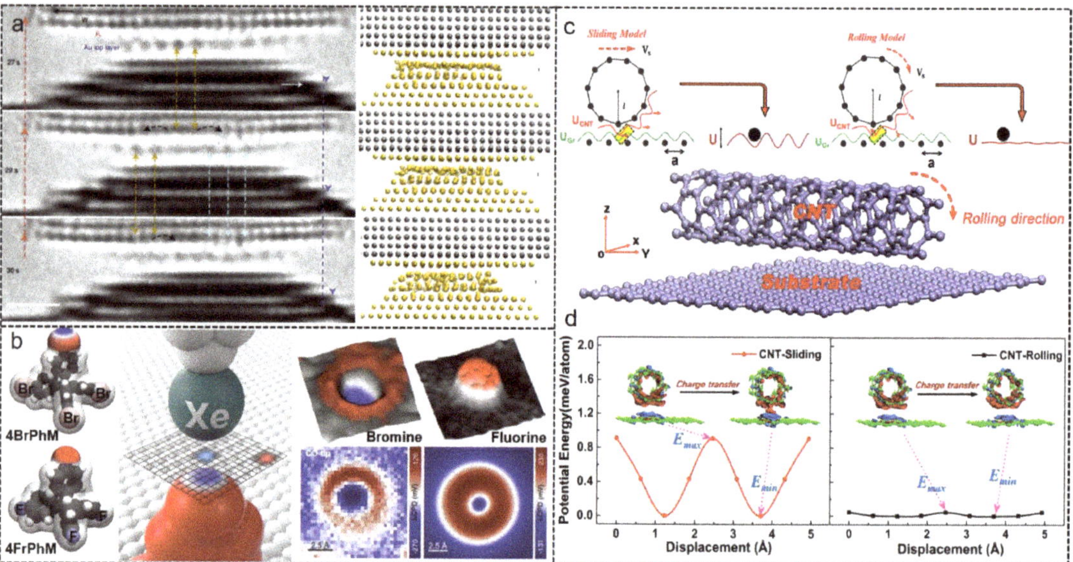

Figure 2. Experimental characterization of atomic-scale contact or friction behaviors as well as their induced electrification phenomena. (**a**) The atomic-scale interface structure in the friction process was observed directly and in real-time. Through in situ high-resolution transmission electron microscopy (TEM) and atomic force microscopy (AFM) measurements, it was found that a loosely stacked interface layer was formed between the two metal micro bumps. Reproduced with the permission of [59], copyright 2021, Springer Nature. (**b**) Real-space images of the anisotropic charge distribution of the σ-hole and the quadrupolar charge of a carbon monoxide molecule obtained by KPFM. Adapted with the permission of [60], copyright 2021, AAAS. (**c**) Atomic-scale rolling and sliding friction behaviors between CNT and graphene substrate. (**d**) The corresponding potential energy distribution during the rolling and sliding process. (**c,d**) Reproduced with the permission of [61], copyright 2020, American Chemical Society.

Contact electrification or triboelectrification mechanism is particularly important for TENGs. As usual, a macro-scale electron or charge transfer model during two friction layers in a complete contact and separation cycle is used to reveal the electrification mechanism (Figure 3a). Taking the typical vertical contact-separation model as an example, when the two friction materials A and B make contact with each other, the same amount of charges are generated on their interface with opposite polarities (Figure 3a(i)). When the two friction materials begin to separate, static charges are induced in the electrodes, generating an instantaneous electrical current (Figure 3a(ii)). When the two friction layers are completely separated, the charges on the friction layers are fully equilibrated by the electrostatic induced charges on their attached electrodes (Figure 3a(iii)). In the reverse case, if the two frictional materials gradually approach each other, the electrons or charges will be transferred in the reverse trend (Figure 3a(iv)). After the whole system returns to the initial state, the charges on the electrodes will be offset by the frictional layers. The contact and separation process in Figure 3a will form an alternative potential or current signal. In addition to the widely used macro-scale charge transfer model, Wang et al. proposed an atomic-scale electron cloud potential well or wave function overlapping model based on the electron-emission-dominated charge transfer mechanism to attempt to describe the CE process between any two materials and even atoms [62,63]. As shown in Figure 3b, once the two atoms approach and make contact with each other, the electron clouds will overlap between the two atoms to form ionic or covalent bonds, resulting in the initial single potential wells becoming an asymmetric double-well potential. Due to the strong overlap of electron clouds, the energy barrier between the two decreases. Then, electrons can then be transferred from one atom to the other, resulting in CE (Figure 3b(ii)). Due to the existence of surface potential barriers that bind the electrons tightly in specific orbits and prevent the charge generated by CE from flowing back, the charges generated in CE can be readily retained by the material as the electrostatic charges for several hours at room temperature (Figure 3b(iii)) [64]. The process presented in Figure 3b is referred to as the Wang transition, which has laid a solid foundation for exploring the meso-scale and macro-scale contact electrification or triboelectrification of textile TENGs.

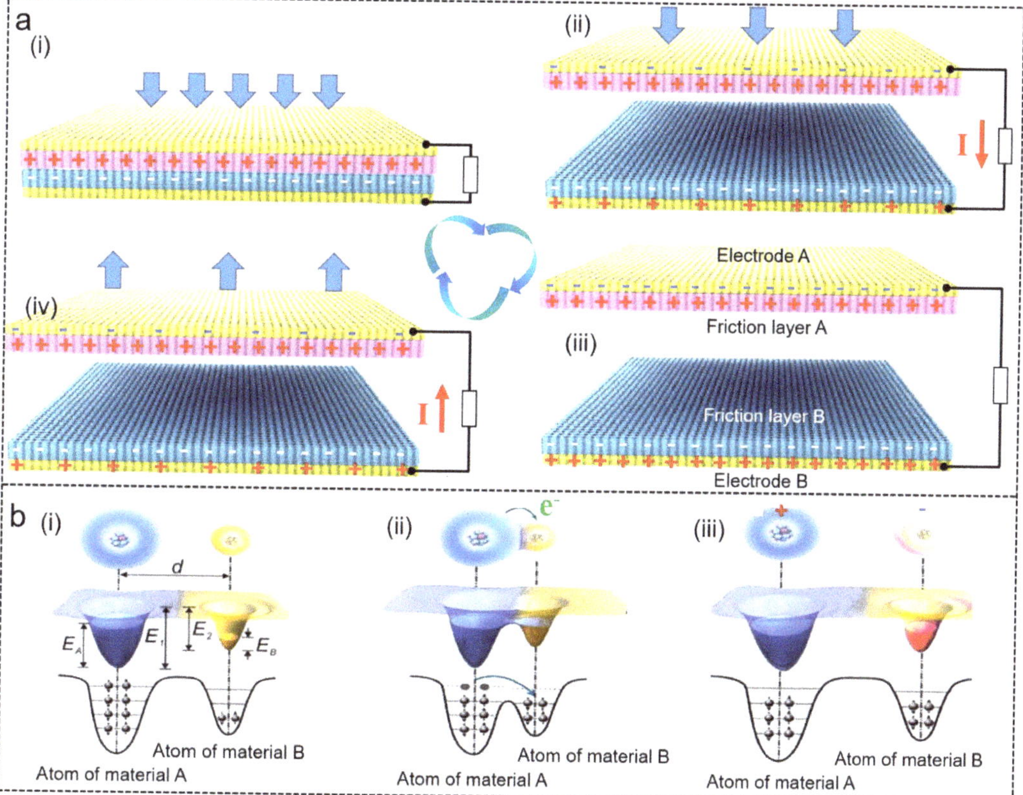

Figure 3. Electrification mechanism of mechanical-to-electrical conversion during contact or friction process. (**a**) Classic electron or charge transfer process in a complete contact and separation cycle. (**b**) An electron-cloud-potential-well model was proposed for explaining CE and Charge transfer between two materials that may not have a well-specified energy band structure. Reproduced with the permission of [65], copyright 2019, Elsevier.

3. Basic Strategies

3.1. Surface/Interface Physical Treatments

3.1.1. Micro-/Nano-Patterned Structures

One of the simplest and most common methods to improve the output performance of TENGs is to increase the effective contact area through fabricating micro-/nano-textures at the contact surfaces or interfaces. Multiple micro-/nano- patterned structures with regular and uniform textures can be fabricated based on various processing techniques, such as lithography, etching, self-assembly, laser patterning, and so on [66–71]. Various types of nanostructures, including nanoparticles, nanorods, and nanowires, are fabricated into TENGs to increase the contact area, thereby obtaining a high charge transfer density. As shown in Figure 4a, three types of regular and uniform polymer patterned arrays, including line, cube, and pyramid, are fabricated to improve the efficiency of TENGs [72]. The obtained results show that the output efficiency of different TENGs follows film < line < cube < pyramid. The correlation between the contact area and electrical performance of TENG with textured surfaces is revealed to make a quantitative analysis of the effect of the effective contact area of micro-/nano-textures on the power output of TENGs [73]. As illustrated in Figure 4b, both the contact area and open-circuit voltage of TENGs with the pyramid texture increase under lighter loading while remaining stable under heavier loading. In addition, for better understanding the impact of interfacial design on the power generation of TENGs, systematical numerical studies on the adhesive contact at the micro-/nano-structured interfaces are conducted, which confirm that the deformation of interfacial structures is related to the pressure-voltage of TENG [74]. The hysteretic behavior in contact force response of TENG is analyzed, which shows that the counterclockwise hysteresis curve of the contact force versus output power originates from the asymmetric time constant between triboelectric charging and natural discharging [75].

The 3D secondary heart-like structure is designed on electrospun PVDF nanofibers to improve the output performance of textile TENGs [76]. The high output results from the combined effect between the heart-shaped structure and porous structure expand the effective contact area through its high surface roughness (Figure 4c). Micro-/nano-patterned structures can also be easily applied to the surface of textiles. For example, a fabric-structured TENG is woven from energy harvesting fibers, which is composed of aluminum wire, vertically arranged nanowires and a PDMS tube with a high aspect ratio nanotextured surface (Figure 4d) [77]. Similarly, by using silver-plated fabric and PDMS nano-pattern based on ZnO nanorod array on silver-plated fabric as two kinds of friction materials, a fully flexible and foldable nano-patterned fabric TENGs with high power generation performance and mechanical strength is designed (Figure 4e) [78]. The dramatic increase in the output performance of TENGs with the micro and nanostructures can be attributed to the following reasons: (1) the frictional layers with micro and nanostructures have better electrification to generate more triboelectric charges during the contact; (2) the triboelectric charges are more easily separated to form a larger dipole moment between the electrodes; (3) the micro and nanostructures increase the effective contact area between triboelectric layers, which can increase the capacitance and dielectric constant. It is also worth noting that the interfacial structures of TENGs can be carefully designed with consideration of a special target application based on numerical simulation methods.

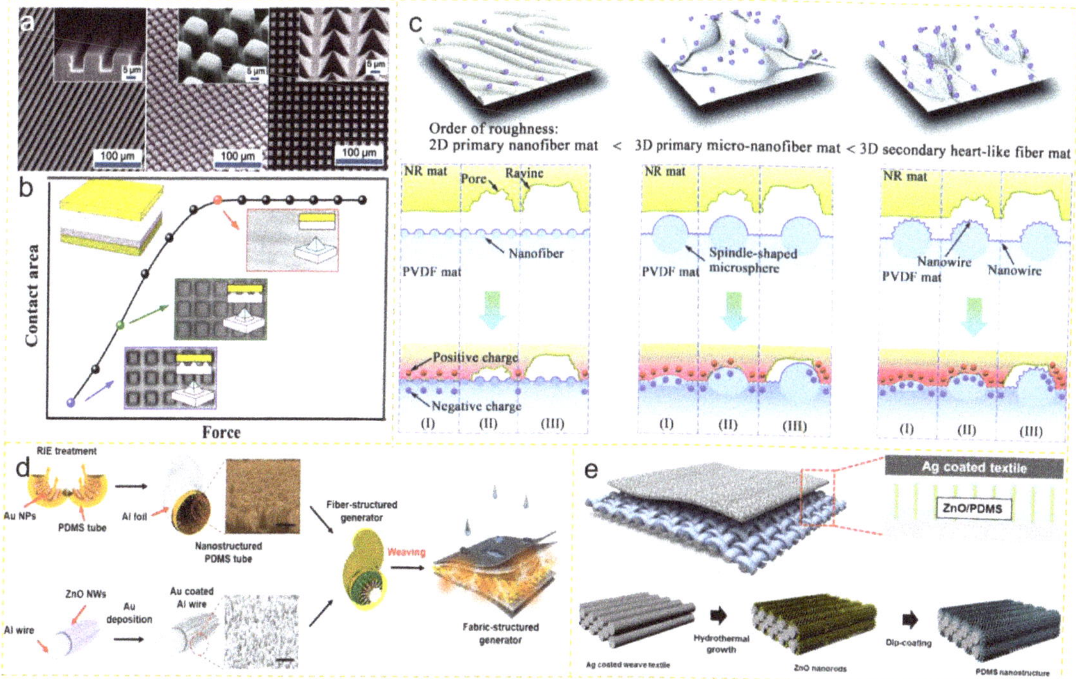

Figure 4. Construction of micro/nanostructures on the contact interface to improve the working performance of textile TENGs. (**a**) SEM images of several patterned micro/nanostructures, such as columns, cubes, and pyramids. Reproduced with the permission of [72], copyright 2012, American Chemical Society. (**b**) Relationship between surface contact area and applied pressure. Reproduced with the permission of [73], copyright 2012, Elsevier. (**c**) Conceptual illustration showing the contact interface states with 2D primary nanofiber mat, 3D primary micro-nanofiber mat, and 3D secondary heart-like fiber mat. Reproduced with the permission of [76], copyright 2019, The Royal Society of Chemistry. (**d**) A highly stretchable 2D fabric TENG with the Al wire grown on ZnO nanowires in the core and nanostructural PDMS tube in the shell. Reproduced with the permission of [77], copyright 2015, American Chemical Society. (**e**) A nanopatterned wearable fabric TENG consisting of an Ag-coated fabric and PDMS nanopatterns based on ZnO nanorod arrays on an Ag-coated textile substrate. Reproduced with the permission of [78], copyright 2015, American Chemical Society.

3.1.2. Soft Contacted Interface

Soft-surface contact is another effective approach that largely improves transferred charges by increasing the contact area. There are several benefits to using the soft contact interface for fabricating TENGs [79,80]. Firstly, more effective energy harvesting and higher sensitivity of soft TENGs can be realized under a tiny external mechanical stimulation. Secondly, the TENGs fabricated with soft materials are easier to adhere to curved surfaces, and more prominently, they can quickly adapt to the deformations caused by complex body movements. Last but not least, compared with the general rigid TENGs, soft-textured TENGs has a wider range of application scopes, especially in implantable medical monitoring. In order to fabricate soft-textured TENGs, supersoft yet tough silicone rubber is often chosen as the elastomeric dielectric, owing to its inherent biocompatibility, superior mechanical properties, excellent flexibility/stretchability, and strong tendency to gain electrons [81,82]. For example, a typical soft-surface-contact TENG is made of an acrylic hollow ball as the shell and a rolling flexible liquid/silicone as the core to effectively collect water wave energy (Figure 5a) [83]. The results show that compared with the traditional PTFE-based hard contact electrode, the maximum output charge of the soft contact spherical TENG is increased by 10 times. In addition to the friction layers, the conductive electrodes can also be made of soft-textured materials. For example, TENG, composed of a conductive liquid electrode and an elastic polymer covering layer, is highly shape-adaptive and stretchable and is designed to effectively collect energy in various working modes [84]. As shown in Figure 5b, TENG can withstand up to 300% strain without lowering the electrical performance due to the unique adaptability of liquid electrodes and the high flexibility of the rubber cover. However, the mechanical stability of TENGs based on liquid electrodes is a great challenge due to the high risk of leakage of liquid electrodes. Therefore, researchers usually take another method by directly mixing with elastic polymer materials with a conductive medium to prepare soft electrodes with better mechanical stability. As exhibited in Figure 5c, a tubular TENG is made of elastic material and spiral inner electrode adhered to the tube with dielectric layer and outer electrode, respectively [85]. These methods to obtain a more fully interface contact state can be used to design high-performance fiber-based TENGs. Among them, the simplest is to directly coat the elastic dielectric material on the surface of the conductive fiber. As shown in Figure 5d, by mixing fiber TENGs and fiber supercapacitor in a fabric, a self-charging electric fabric with biomechanical energy collection capability and energy storage capability based on all yarns is developed [86]. The fiber TENG is obtained by coating silicone rubber on the surface of stainless steel/polyester blended fiber, which can be further knitted into large-scale fabrics. However, the interfacial bonding strength between fiber electrodes and surface coating elastomers is not high, resulting in poor mechanical stability. The poor interface stability can be effectively avoided by directly injecting flowing liquid electrodes into dielectric elastomeric tubes. For example, a triboelectric fiber consists of an elastomeric SEBS hollow fiber filled with metallic EGaIn liquid, fabricated with melt extrusion and injection methods (Figure 5e) [87]. Similarly, a coaxial wet spinning process is used to continuously manufacture inherently stretchable, high conductivity, and stable conductivity liquid metal core microfibers (Figure 5f) [88]. However, the issue of liquid metal-based triboelectric fibers is that liquid electrodes are easy to leak outside. To this end, some researchers use the photo-crosslinking method to solidity the core liquid electrode to obtain triboelectric fibers with more stable mechanical properties. As shown in Figure 5g, a core–shell triboelectric fiber is prepared in silicone hollow fiber by the gel photo-crosslinking method. The fiber has the advantages of softness, flexibility, electrical conductivity, and large-scale production [89].

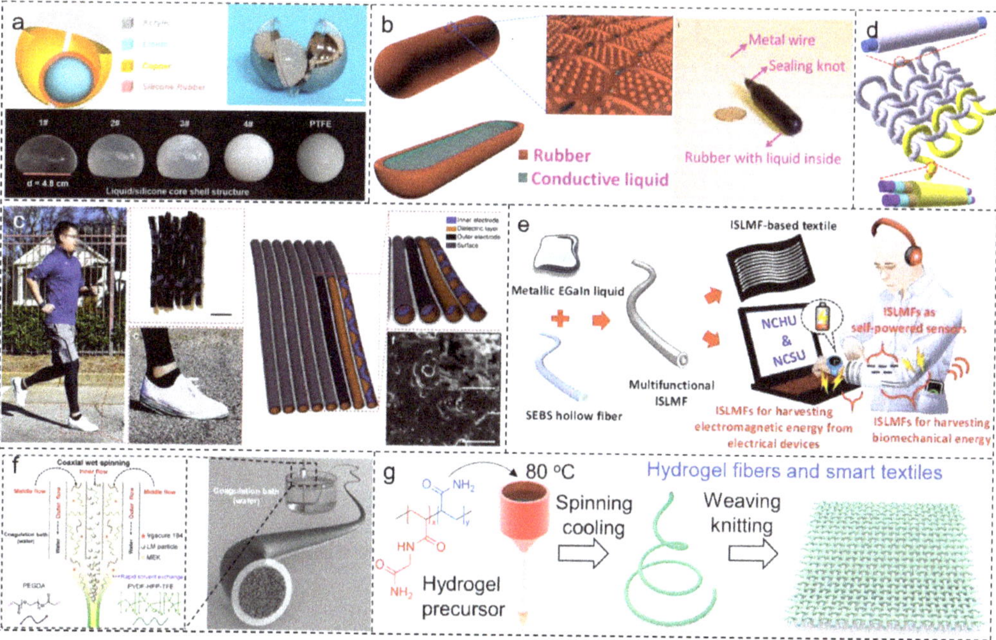

Figure 5. Softness interface to maximize the contact area. (**a**) Soft-contact spherical TENG with a flexible rolling sphere. Reproduced with the permission of [83], copyright 2019, Elsevier. (**b**) A highly shape-adaptive and stretchable conductive liquid-based TENG. Reproduced with the permission of [84], copyright 2016, AAAS. (**c**) Tube-like TENG woven into a coat and assembled under shoes. Reproduced with the permission of [85], copyright 2016, Springer Nature. (**d**) Knitting power textiles fabricated with silicone rubber coated triboelectric fibers. Adapted with the permission of [86], copyright 2017, American Chemical Society. (**e**) Multifunctional liquid-metal triboelectric fibers with metallic EGaIn liquid filled in SEBS hollow fiber. Reproduced with the permission of [87], copyright 2021, Wiley-VCH. (**f**) Liquid-metal sheath-core microfibers consisting of double-network fluoroelastomer as the sheath and that percolated EGaIn alloy nanoparticles as the core. Reproduced with the permission of [88], copyright 2021, AAAS. (**g**) Scalable triboelectric fibers fabricated with an organogel electrode and soft silicone rubber. Reproduced with the permission of [89], copyright 2021, Elsevier.

3.2. Atomic-Level Chemical Modification

3.2.1. Surface Functionalization

The material physicochemical properties of the triboelectric layer are related to its behavior and ability to gain and lose electrons in the process of contact electrification or triboelectrification. In principle, the greater the difference of electron affinity between two triboelectric layers is, the more the triboelectric surface charges will be generated. In addition, the ability of electricity generation of triboelectric layer also depends on their intrinsic material properties, such as dielectric constant, polarity, work function, etc. The triboelectric series for a wide range of polymers is quantified using a universal standard method, which establishes a fundamental materials property of quantitative triboelectrification [90–94]. The normalized triboelectric charge density is defined and deduced to reveal the intrinsic tendency of polymers to gain or lose electrons [62,95]. However, the triboelectric charge density of the material can be changed by adjusting the functional groups on the surface with different electron-withdrawing or electron-donating abilities. Considering that the charge density on polymer surface is closely related to the surface chemical property, surface chemical engineering treatments through appropriate functionalizations, including overall chemical reaction, surface chemical treatment, functional group grafting, and so on, are one of the most fundamental strategies to improve the output performances of TENGs [56,96,97]. For example, the surface charge of polymers can be controlled by carrying their physicochemical properties, such as the strength of macromolecular interactions and surface adhesion. It has been found that polymers with lower modulus show higher surface charge values than those with a higher modulus. In fact, the modulus is directly proportional to the cohesive energy of materials. The effect of polymer cohesion energy on contact electrification is much greater than that of the surface roughness. Polymers showing strong surface adhesion and low cohesion energy in bulk are expected to have higher surface charge [98].

A facile atomic-level chemical functionalization method is proposed to effectively modify the triboelectric property of polymer surfaces by using a series of halogens and amines [99]. As shown in Figure 6a, the negative triboelectric charge on the PET surface is functionalized by arylsilane terminated with electron-accepting elements, halogens, while for the triboelectrically positive side, its surface is functionalized using several aminated molecules. In addition to the PET, the triboelectric properties of other polymers can also be well modulated by chemical modification. The output power of polyimide (PI)-based TENGs is enhanced by introducing electron-withdrawing and electron-donating groups into the main chain [100]. Density functional theory is used to study the relationship between charge retention characteristics and functional groups of the PI films so as to observe the molecular electrostatic potential and the charge distribution (Figure 6b). It is found that the increase of electrical outputs is attributed to the increase of the charge density transferred from the electrode. The chemical functionalization method to improve the electrical output performance of TENGs is also very easy to apply to fibers or fabrics. As shown in Figure 6c, by coating surface interface-engineered PDMS layers on highly conductive Ni-Cu textiles, the electrical performance of the textile TENG is greatly enhanced [101]. The chemical modification of the coated PDMS surface is achieved by one-step Ar-only plasma treatment and a two-step process with consecutive Ar and $CF_4 + O_2$ plasma treatment. In addition, by enriching the fiber surface with hierarchical structures and amide bonds through chemical grafting of CNT and PET via a polyamidation reaction, a fabric-based TENG can easily achieve over 10 times improvement in output voltage and current at a low modifier content of less than 1 wt% (Figure 6d) [102]. A relatively systematic and improved mechanism is proposed to clarify the influence of chemically finishing on contact electrification [103]. The electron cloud of the atomic nucleus with electron-donating ability has a large density and occupies a large area, while that with electron-withdrawing ability occupies a small area. As shown in Figure 6e, for the contact between aminosilane and FEP, the overlap of electron cloud leads to the reduction of the energy barrier between them. Afterwards, the electrons are transferred from the low electronegativity atom (N) to the high electronegativity atom (F). Once they are separated, the electron cloud range of N atoms decreases, while the electron cloud range of F atoms increases. As for the contact between fluorosilane and FEP, the electrons of the H atom are transferred to the F atom, due to that the more F atoms in the FEP polymer, the stronger the ability to obtain electrons (Figure 6f). In summary, creating new functional groups through a variety of chemical methods to obtain strong electron-donating or electron-accepting ability on fabric surface is an effective method to greatly and stably improve the output performance of textile TENGs.

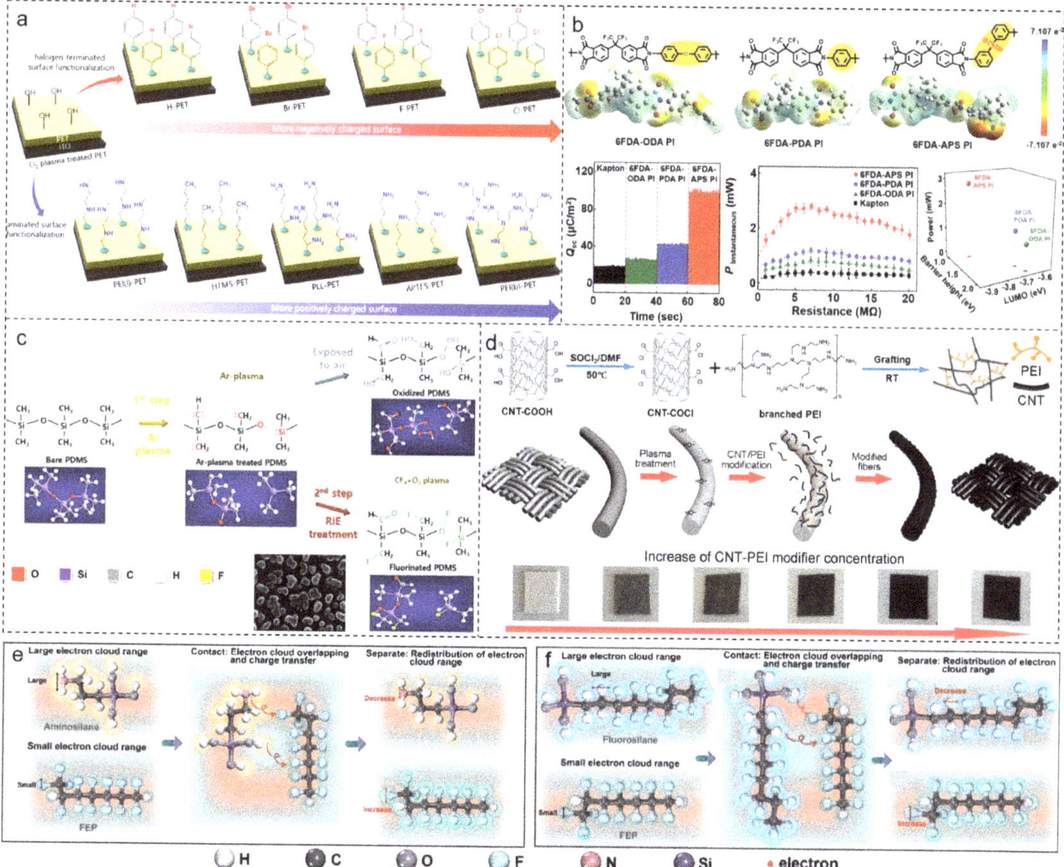

Figure 6. Chemical modification of polymer materials to alter the electron-withdrawing or electron-donating abilities. (**a**) Atomic-level chemical functionalization on PET polymer surface. Reproduced with the permission of [99], copyright 2017, American Chemical Society. (**b**) High-output TENGs based on PI-based polymers by introducing functionalities into the backbone. Adapted with the permission of [100], copyright 2019, Wiley-VCH. (**c**) Improvement of output performance of TENGs by a simple plasma treatment approach. Reproduced with the permission of [101], copyright 2019, Elsevier. (**d**) Enhancing the performance of fabric TENGs by structural and chemical modification. Reproduced with the permission of [102], copyright 2021, American Chemical Society. (**e**) Schematic diagram of the electron cloud overlap model between aminosilane and FEP, in which aminosilane has a large electron cloud range with orange color and FEP has a small electron cloud range with green color. (**e**) Schematic diagram of the electron cloud overlap model between fluorosilane and FEP, where fluorosilane has a large electron cloud range with blue color and FEP has a small electron cloud range with green color. (**e,f**) Reproduced with the permission of [103], copyright 2021, Elsevier.

3.2.2. Ion Injection/Irradiation/Implantation/Decoration

The above surface functionalization methods to alter the original electro-donating and electron-withdrawing ability are limited by their low accuracy and poor stability. Due to this, the surface triboelectric charges only come from two surfaces with the difference in surface potentials; the compensation of surface potential difference by injecting or implanting additional electrons or ions is also an effective method. Therefore, by directly adding ions or single-polarity charged particles on or inside the triboelectric materials, the charge density of triboelectric surfaces can be greatly increased. Ionized-air injection is the simplest method to implant polarized charges into polymers. With the assistance of an air-ionization gun, the ions with negative or positive polarities are generated and subsequently implanted onto the material surfaces [96]. The polarity of the ions injected from the outlet of the gun can be manually controlled by pressing or releasing the trigger bar. As shown in Figure 7a, the negative ions reaching the surface of FEP film transfer the same number of electrons from the bottom electrode to the ground by electrostatic induction so that the bottom electrode has positive charges of the same charge density [104]. As a result, the charge density on the FEP surface from the ion-injection process can reach a much higher level than that on an electrode-free FEP layer. In order to further improve the stability and retention rate of implanted ions, a modification strategy based on ion implantation technology is proposed to precisely control the distribution and concentration of dopant atoms in the polymers. As shown in Figure 7b, under the accelerated fields, polar groups and unsaturated bonds containing N element can not only break the symmetry of the spatial structural of PTFE, but also combine with the free radicals on the chain to form new chemical bonds and chemical groups [105]. The electronegativity and the increase of electron cloud density of the group lead to the improvement of the electron-withdrawing ability. As a result, the ion implantation modified PTFE and FEP films exhibit the most negative triboelectricity in the triboelectric series. Besides nitrogen ion implantation, other ions are also gradually used to change the charging properties of polymer materials. Atomic oxygen irradiation is used to manipulate the surface structure and chemical components to adjust their electrical properties [106]. As seen in the mechanism illustrated in Figure 7c, atomic oxygen infrared radiation can increase the electron-donating groups and enhance electropositivity of PDMS films, which leads to an increase in the work function of PDMS, thus enhancing its surface states to be charged. In addition, a Kapton film modified by low-energy helium ion irradiation also shows several unprecedented characteristics, such as high surface charge density, excellent stability, and ultrahigh electro-donating capability, which makes it lose electrons during the contact electrification process (Figure 7d,e) [107]. The electrification properties of other ion irradiated polymers, such as PET, PTFE, and FEP, are also systematically studied (Figure 7f), which suggests that the same ion irradiation can lead to different performance changes for different polymers. Argon ion implantation is carried out on Kapton to make it serve as a positive triboelectric layer, while pure Kapton behaves as a negative triboelectric layer [108]. The effect of argon plasma treatment (including plasma power and treatment time) on the output performance of TENG is also investigated [109]. Neutral beam is also an advanced plasma-based etching and surface treatment technology of polymers. Kim et al. modified the surfaces of PDMS and TPU using N_2 and O_2 gas-based neutral beam process, which showed the remarkable output performance of TENGs [110].

The approach of ion injection or implantation technology to the study of triboelectric polymers is helpful to further clarify the internal mechanism between molecular composition/structure and contact electrification/triboelectrification ability, which provides useful guidance for developing high-performance triboelectric polymers.

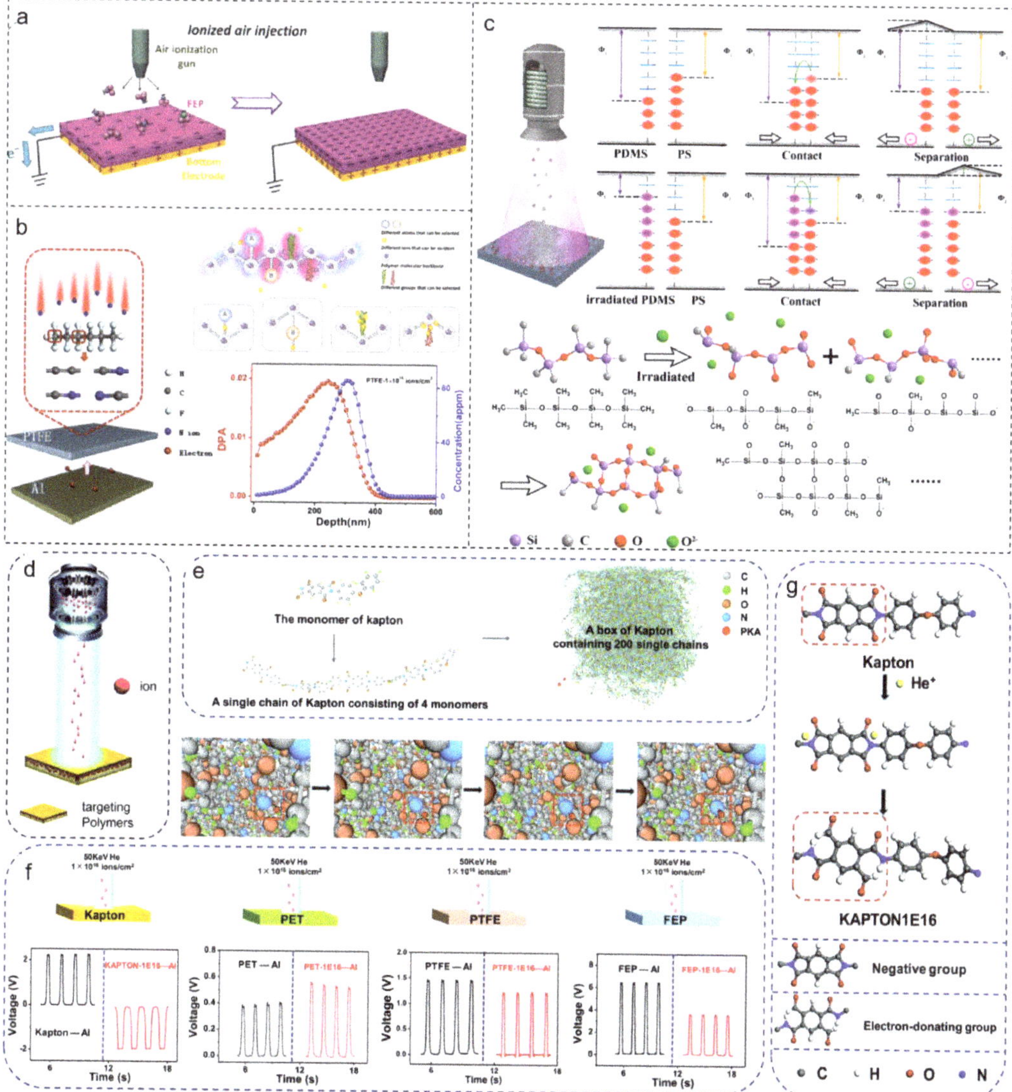

Figure 7. Ion injection or ion irradiation/implantation. (**a**) Ionized-air injection to maximum surface charge density of TENG. Reproduced with the permission of [104], copyright 2014, Wiley-VCH. (**b**) Implanting ions into the polymers to change their chemical structure by ion implanter. Reproduced with the permission of [105], copyright 2021, Elsevier. (**c**) Manipulating electrical properties of silica-based materials via atomic oxygen irradiation. Adapted with the permission of [106], copyright 2021, American Chemical Society. (**d–g**) Manipulating triboelectric surface charge density of polymers by low-energy helium ion irradiation or implantation. Adapted with the permission of [107], copyright 2020, The Royal Society of Chemistry.

3.2.3. Charge Trapping/Storage

Triboelectric charges will delay the surface friction of materials and transfer to the interface between the friction layer and electrode, which will decrease triboelectric charge density [111–114]. Interfacial modifications made by mixing charge trapping or storage elements are considered a suitable method to increase the retention of triboelectric charges for output enhancement [115]. For instance, the ferroelectric polymer-metallic nanowire composite nanofibers are described for use in high-performance TENGs. For example, an improved fiber-based TENG with a breathable-antibacterial Ag NWs electrode and electrostatic induction enhancement electrospun polystyrene (PS) nanofibers as charge storage layer is proposed in order to improve the output performance of TENGs and meanwhile maintain good flexibility and breathability (Figure 8a) [116]. The result shows that the samples with PS film as the intermediate charge storage layer have an obvious higher output than others without one. Similarly, a multilayered nanofiber TENG is proposed to greatly enhance the triboelectric charge density by introducing a charge-transport layer (polystyrene and carbon black) and a charge-storage layer (polystyrene) (Figure 8b) [117]. The short-circuit current and open-circuit voltage of the multilayered TENG are distinctively 3 times and 2.5 times larger compared with the single-layer structured TENG, respectively. Another example is a high-performance TENG which is designed utilizing the electrospun PVDF-Ag NW composite and nylon fibers as the top and bottom triboelectric layers, respectively. The enhanced surface charge potential and the charge trapping capabilities of the PVDF-Ag NW composite nanofibers significantly enhance the output performance of TENG [118].

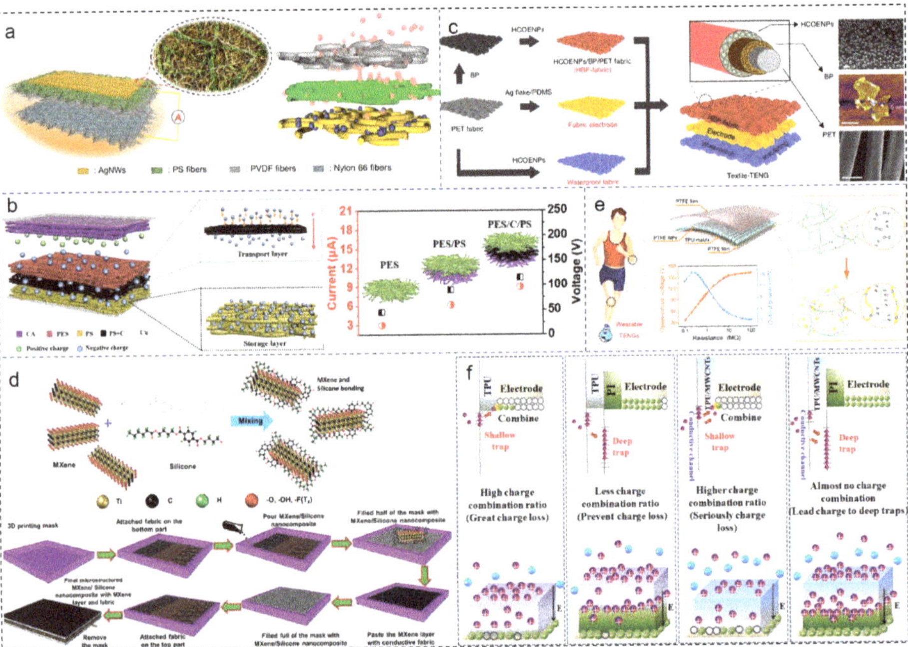

Figure 8. Improving output performance of TENGs by adding charge trapping layers. (**a**) An all-fibrous TENG with enhanced outputs depended on the PS charge storage layer. Adapted with the permission of [116], copyright 2021, Elsevier. (**b**) Enhancement of charge density by introducing a charge-transport layer and a charge-storage layer. Reproduced with the permission of [117], copyright 2018, Elsevier. (**c**) Textile TENGs with black phosphorous as the synergetic electron-trapping coating. Reproduced with the permission of [119], copyright 2018, Springer Nature. (**d**) A double-side-contact fabric-assisted TENG by using 2D MXene as the charge trapping layer. Reproduced with the permission of [120], copyright 2021, Wiley-VCH. (**e**) Boosting the power and lowering the impedance of TENGs through manipulating the permittivity. Reproduced with the permission of [121], copyright 2021, American Chemical Society. (**f**) Schematic illustration of charging trapping mechanisms. Reproduced with the permission of [122], copyright 2021, American Chemical Society.

In addition to adding charge trapping material into nanofibers by electrospinning, it can also be realized on fabrics. As shown in Figure 8c, an all-fabric TENG with excellent durability and high triboelectricity was developed using black phosphorus (BP) encapsulated with hydrophobic cellulose oleoyl ester nanoparticles (HCOENPs) as a synergetic electron-trapping layer [119]. The HCOENPs/BP mixed layer provides a charge storage layer to reduce the dissipation of triboelectric electrons to increase the electricity outputs. With a novel, scalable surface modification method of fabric-assisted micropatterning techniques, a highly negative MXene/silicone nanocomposite surface with 2D MXene as the charge trapping layer was adopted to boost the output performance of double-side-contact fabric-assisted TENG (Figure 8d) [120]. The obtained output voltage and current are appropriately 1.6 and 1.5 times higher than those of the sandpaper-assisted microstructures. In order to decrease the high intrinsic impedance of TENGs, which often range from dozens to hundreds of megohms, a high-performance TENG with boosted electrical output and lower internal impedance is reported, which consisted of a thermoplastic TPU matrix with polyethylene glycol additives and PTFE nanoparticle inclusions (Figure 8e) [121]. The increment of permittivity improves the injected charge density and internal capacitance of the TENG, thus resulting in the enhancement of output power and the reduction of matching impedance, respectively. The charge trapping mechanism combined with increased conductivity for improving the output performance of TENGs is illustrated in Figure 8f [122]. When there is no PI layer and MWCNTs, the triboelectric charges with purple dots generated by TPU are generally trapped in shallow traps, which is easy to combine with the induced charges with green dots on the electrode and charged ions or particles with blue dots from the air, resulting in obvious charge loss and low output. If the PI layer is attached with the TPU, partial triboelectric charges will be trapped in the TPU/PI double layer, resulting in less charge loss. In addition, the volume conductivity of TPU can be improved by adding MWCNTs, which will make the newly generated charge easily move to the PI layer, thus allowing more charges in shallow traps to enter the deep traps.

3.3. Structural Optimization Design

3.3.1. Intermediate Layer Embedding

Embedding a superior intermediate layer into TENGs was also proved as an efficient strategy to enhance the surface charge density and power output performance that are attributed to the enhancement of the dielectric permittivity [6,51,123]. The research shows that the triboelectric charge in the friction dielectric layer will decay with time. The intermediate layer between the triboelectric layer and electrode can capture and accumulate more electrons, which reduces the drift and diffusion of electrons and thus prolongs the charge decay process. When there is no intermediate layer, the electrons on the surface may drift inward under the driving force of the electric field and eventually combine with the opposite charge induced on the electrode, thus reducing the surface charge density. Therefore, it is of great significance to add an intermediate layer between the triboelectric layer and the electrode to reduce the mutual shielding of charges.

The method of adding an intermediate layer to the triboelectric layer is put forward as a novel strategy to prolong the charge decay time and enhance induced charge so as to improve the output performance of TENGs. For example, Cui et al. investigated the relationship between the thickness of the friction layer and the transport process of triboelectric charges in a PVDF film. The results showed that when the thickness of the friction layer was greater than the storage depth, the stored charge could reach a maximum. In addition, with the excess increase of storage depth, the charge did not accumulate further (Figure 9a) [124]. Moreover, by further adding a dielectric layer and a transport layer between the PVDF and the electrode, the output performance of TENG will be more significantly improved. Similarly, by adjusting the content of MXene flakes and the balance between the dielectric property and the percolation effect, the oriented MXene-doped PVDF composite film with highly enhanced triboelectric output performance is developed [125]. MXenes are an emerging class of 2D transition metal carbides, which possess outstanding metallic electronic conductivity and a high electronegative surface [126,127]. It is found that the surface charge density of the MXene-modified TENG obtained 350% enhancement compared to that with the pure PVDF. In order to overcome the limitation of traditional TENG's output performance in the actual environment, an electric double layer made of Al film coated by Au nanoparticles is inserted between a top layer made of Al/PDMS and a bottom layer made of Al to develop sustainable and enhanced output performance of TENG, whose energy conversion efficiency can reach 22.4% (Figure 9b) [128]. When the force is loaded to the top layer and then withdrawn, the positive and negative charges are spatially separated in the middle layer through the sequential contact structure and the direct electrical connection between the middle layer and the ground, forming an electric double layer. It can be found that the power output of the three-layer TENG is much higher than that two-layer TENG [129]. In addition, in order to prevent the problem of poor interface strength caused by simple, intermediate layer stacking, an interlaminar fiber winding structure is adopted. For instance, a 3D braided TENG with shape adaptability and high resilience is developed through a four-step braiding technology, which is easy to implement and can be expanded industrially for power supply and pressure sensing (Figure 9c) [130]. The output of the 3D braided TENG is about twice that of the multilayered 2D TENG fabric, which is due to the special 3D braided structure with more contact and separation areas that can effectively enhance the total power output.

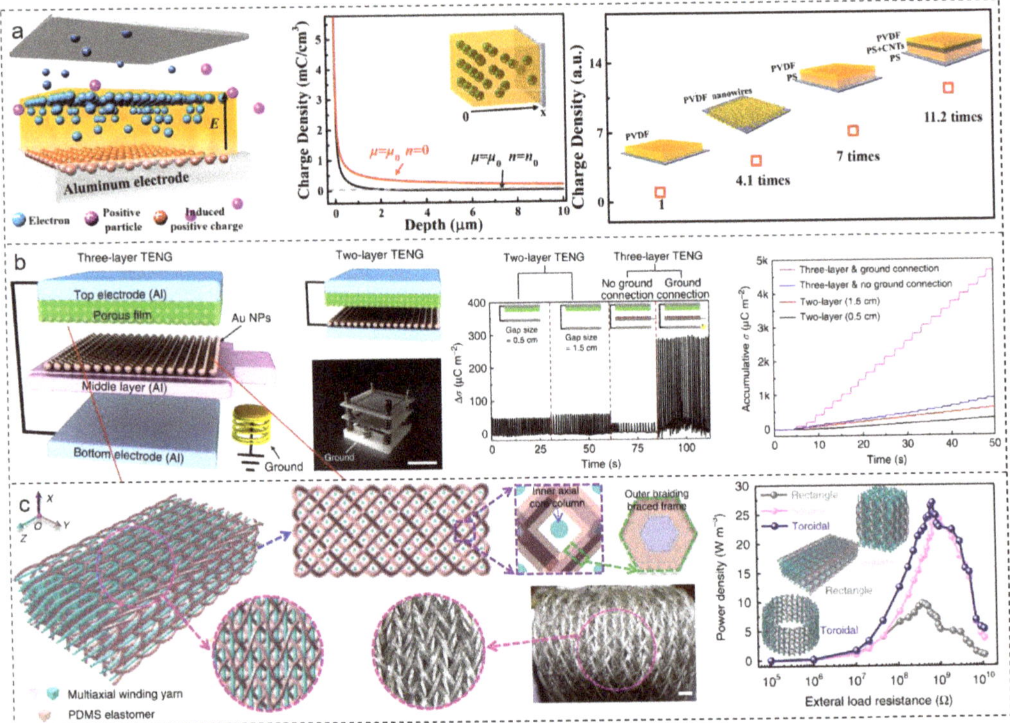

Figure 9. Improvement of output performance of TENGs by embedding superior intermediate layer. (**a**) Improvement of output performance of TENGs by adjusting the depth distribution of the triboelectric charges. Reproduced with the permission of [124], copyright 2016, American Chemical Society. (**b**) Output performance of TENGs enhanced by electric double layer effect. Reproduced with the permission of [128], copyright 2016, Springer Nature. (**c**) Shape adaptable and highly resilient 3D braided TENGs. Reproduced with the permission of [130], copyright 2020, Springer Nature.

3.3.2. Charge Shuttling/Pumping

The surface charge density of tribomaterial, which is highly correlated with the current, voltage and charge transfer of TENG, is considered to be the most important parameter of electrical output. However, the surface charge density of dielectric materials is limited by air breakdown and the number of stored charges in surface energy states. Inspired by the excitation strategy of the electromagnetic generator, In 2018, Cheng et al. [131] and Xu et al. [132] independently discovered the external charge excitation strategy to improve the surface charge density of the tribomaterials of TENG. As shown in Figure 10a, they both choose metal materials that can store high-density charge rather than dielectric materials as the tribomaterials, and charges can be pumped from one of the electrodes to the another one in the main TENG by a pump TENG [132]. In detail, part 1 and part 2 act as the pump and main TENG, respectively. The contact and separation of the pump TENG will drive the positive and negative charges to the two electrodes of the main TENG in one direction through the high-voltage rectifier bridges, which causes the surface charge density of the tribomaterials in the main TENG to increase gradually until it reaches a saturation value (Figure 10b). Therefore, the electrical output of the main TENG can be greatly increased (the surface charge density can reach 1003 μC cm^{-2}). Recently, Bai et al. successfully achieved an ultrahigh average power density of 1.66 kW m^{-3} under a low drive frequency of 2 Hz and high transferred charges of about 4.5 μC with 15 times improvement by charge pumping strategy [133]. Wang et al. added a buffer capacitor based on this strategy (Figure 10c) [134]. The charges pumped into the main TENG will shuttle between it and ceramic capacitors as its capacitance value varies during the contact-separation process, which will introduce two power sources (Figure 10d). The charge density of 1.85 mC m^{-2} proves the success of this strategy. Furthermore, this TENG was successfully applied to efficiently harvest ocean wave energy. Liu et al. also obtained an ultra-high surface charge density of 2.38 mC m^{-2} by a half-wave rectifier bridge [135].

It is well known that self-excitation is also an efficient way to obtain high-output electromagnetic generators. Similarly, the self-charge excitation strategy can also be used to develop a high surface charge density of TENG except for the external charge excitation strategy. Liu et al. realized self-charge excitation through a voltage multiplier circuit, thereby obtaining a high charge density of 1.25 mC m^{-2} [136]. As shown in Figure 10e, three diodes and two capacitors are taken as an example to illustrate the working mechanism of the self-charge excitation strategy. The TENG can be regarded as a tiny capacitor with high voltage when it is in a separated state thus most of the charges in TENG will be pumped into the two ceramic capacitors connected in series. However, when the TENG is in a separated state, the reverse current direction will cause the connection state of the ceramic capacitor to change from parallel to series, which will result in more charges to transfer from the ceramic capacitor to the TENG with almost zero voltage. The surface charge density of the TENG will be saturated to a super high value after several cycles. In addition, a Zener diode is connected to the voltage multiplier circuit to avoid the impact of air breakdown on the output stability. Long et al. further designed the sliding mode TENG using the self-charge excitation strategy [137].

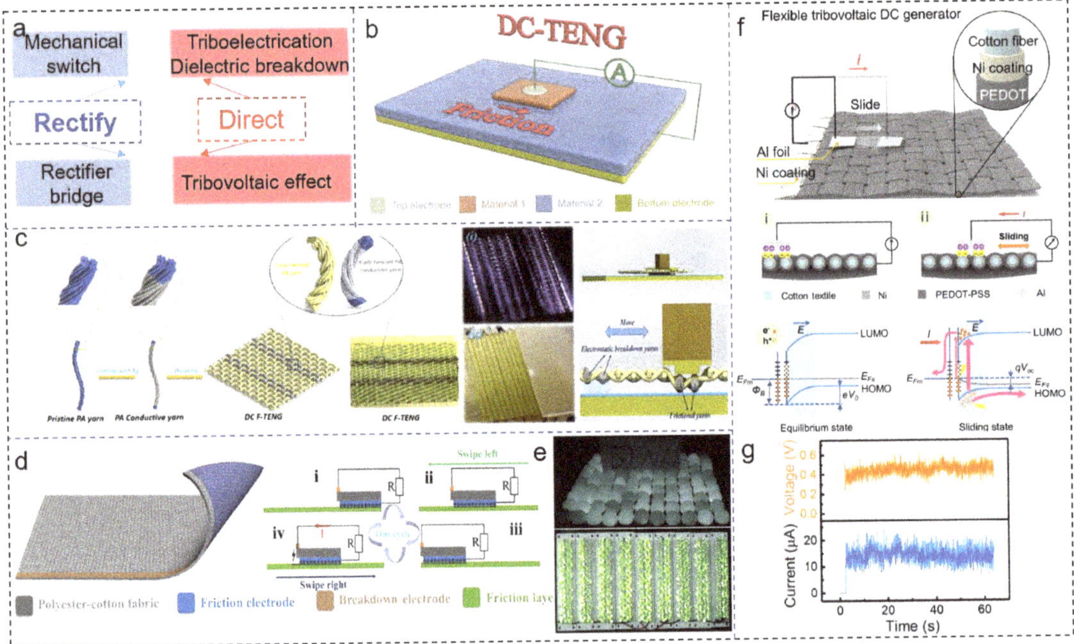

Figure 10. Strategies for direct-current textile TENGs. (**a**) Basic methods to obtain direct-current output from TENGs. (**b**) Schematic illustration of the basic structure of direct-current TENG. Reproduced with the permission of [138], copyright 2020, Wiley-VCH. (**c**) A direct-current fabric TENG by tactfully taking advantage of the electrostatic breakdown phenomenon of clothes. Reproduced with the permission of [139], copyright 2020, American Chemical Society. (**d,e**) A high output direct-current power fabric based on the air breakdown effect by alternately coating two electrodes on the top surface and bottom surface of polyester fabric. Reproduced with the permission of [140], copyright 2021, The Royal Society of Chemistry. (**f,g**) Direct-current textile TENGs based on the tribovoltaic effect at dynamic metal-semiconducting polymer interfaces. Reproduced with the permission of [141], copyright 2021, American Chemical Society.

3.3.3. Direct-Current Mode

It is well known that almost all electronic products need a direct-current (DC) power source. Therefore, the alternating current (AC) output of traditional TENG cannot directly meet their power demands. Researchers developed various methods to obtain DC from TENG (Figure 11a) [138,142–145]. In general, these strategies can be broadly divided into two categories, including direct output DC through other physical effects (tribovoltaic effect and dielectric breakdown effect) and various rectifier devices (rectifier bridge and mechanical switch). Choosing commercial electronic rectifier bridges, including full-wave and half-wave rectification, is the most widely used method of converting AC signals into DC output, which is undoubtedly the most convenient and low-cost method [130,146–148]. However, as a member of electronic devices, the electronic rectifier bridge will inevitably consume extra energy when it is working, which cannot be underestimated for the TENG with low energy output. The high voltage (~kV) of TENGs may also cause fatal damage to the rectifier bridge [144,149,150]. Moreover, the rigid electronic rectifier bridge seems to be unfriendly for fabric-based TENG, which is one of the promising solutions for wearable energy harvesters. A reasonable mechanical rectifier bridge can ensure that the two electrodes in the external circuit are periodically and alternately connected to the two electrodes in the TENG, which is usually realized by a disc structure. However, this periodic contact seems difficult to be guaranteed for the fabric TENG, which requires high flexibility. There is no related report of this method for fabric-based TENG.

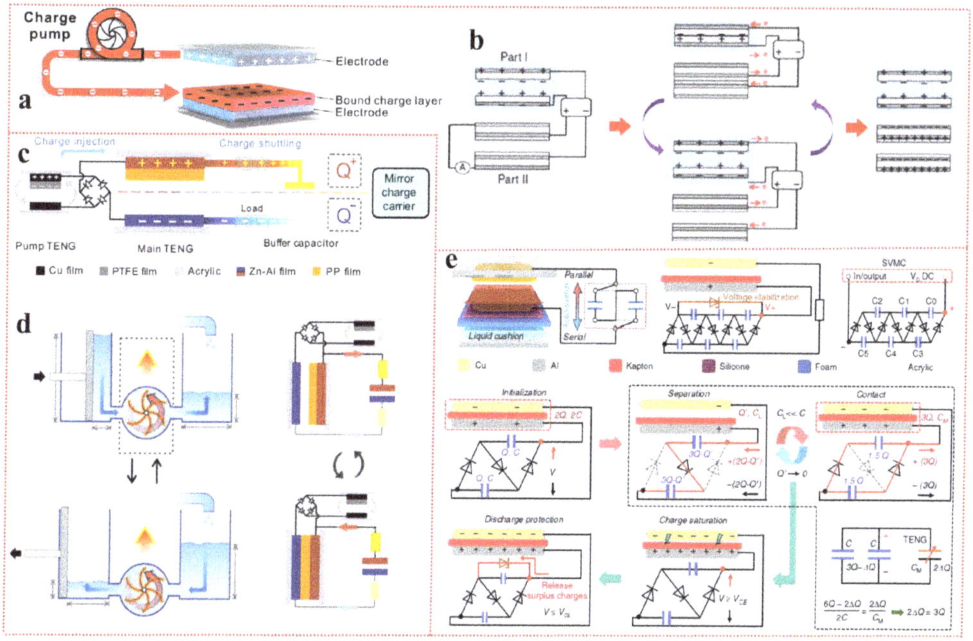

Figure 11. Charging shuttling/pumping approach towards the high and stable output of TENGs. (**a**) Schematic diagram of the charge pumping principle. Reproduced with the permission of [132], copyright 2018, Elsevier. (**b**) Working mechanism of the charge pumping strategy. Adapted with the permission of [131], copyright 2018, Springer Nature. (**c,d**) Increasing the charge density of a TENG by charge-shuttling. Reproduced with the permission of [134], copyright 2020, Springer Nature. (**e**) A self-charge excitation TENG system using part of the energy output from TENG to improve its working charge density. Reproduced with the permission of [136], copyright 2019, Springer Nature.

Dielectric breakdown will reduce the surface charge density of dielectric materials, which was always regarded as an unfavorable factor for TENGs. Yang et al. first designed a DC-TENG based on the coupling effect of dielectric breakdown and triboelectrification in 2014 [151]. Researchers understand that air breakdown can also be beneficial to TENG. Liu et al. simplified the structure in 2019, in which the simplest DC-TENG only needs two kinds of electrodes (Figure 11b) [152]. The friction electrode generates charges on the surface of the friction layer based on the triboelectrification, while the breakdown electrode harvests the surface charges based on the air breakdown effect. However, there is still few works on the flexible DC-TENG for wearable electronics. Chen et al. prepared fabric-based DC-TENG by weaving conductive yarns into the fabric [139]. The electrical output can be scaled up by connecting multiple conductive yarns in series (Figure 11c). Moreover, yarn supercapacitors are integrated into another piece of fabric to form a self-powered system. This is a good attempt. However, to obtain a stable air breakdown gap, the fabric-based DC-TENG must be attached to the hard acrylic board, and almost no flexibility exists. Cheng et al. utilizes the tiny height of the fabric to build an air gap, which shows no obvious effect on the flexibility of the textile (Figure 11d) [140]. This structural design is suitable for any thin and/or flexible material. Benefiting from the high electrical output performance of the fabric-based DC-TENG, various wearable electronics, such as watches, calculators, 1053 LEDs, and 99 bulbs, can be directly powered by a finger-sized fabric-based DC-TENG without rectifier bridge and storage devices (Figure 11e).

Similar to the photovoltaic effect, the friction energy will result in electron-hole pairs between the semiconductor-semiconductor/metal interface, and they will be separated under the built-in electric field, thereby generating a unidirectional current in the external circuit (defined as the tribovoltaic effect). Based on this effect, Meng et al. developed FDC-TENG by utilizing the Schottky contact between poly(3,4-ethylenedioxythiophene) (PEDOT) and nickel (Ni) as well as the ohmic contact between PEDOT and aluminum (Al) [141]. Ni and PEDOT are coated on the cotton fabric successively. DC output can be obtained by sliding Al on the surface of the treated cotton fabric (Figure 11f). Moreover, constant DC can be achieved by sliding the Al slider at a uniform speed (Figure 11g). However, the voltage of DC-TENG based on the tribovoltaic effect is not high owing to the limitation of the built-in electric field.

4. Work Environmental Control

The electricity output performance of TENG largely depends on its external working environments, which mainly includes humidity, temperature, and pressure. Ngugen and Yang studied the effects of humidity and pressure on the output performance of TENGs since 2013, i.e., the second year of its first invention. They found that the generated charge increased more than 20% when the relative humidity decreased from 90% to 10% at the ambient pressure and decreased as the air pressure decreased from atmospheric pressure to 50 Torr at the relative humidity close to 0% [153,154]. In addition, Wen et al. conducted experiments to specifically investigate the influence of temperature on the output performance of TENGs over a wide range of temperatures from 300 K to 500 K in a high-temperature range and from 77 K to 300 K in a low-temperature range [155]. The results found that the performance of TENGs is maximized at around 260 K and degrades at both higher and lower temperatures. Lin et al. studied the contact electrification and triboelectric charging process for a metal-dielectric case under different thermal conditions with the help of AFM and KPFM [156]. It was found that the hotter solids tend to have positive triboelectric charges, while colder solids tend to acquire negative charges. Electrons are excited by heat and transferred from the hot surface to the cold surface. Therefore, a thermionic emission band structure model is put forward to describe the electron transfer phenomenon between two solids at different temperatures.

Electrostatic breakdown is a ubiquitous phenomenon existing in electrostatic discharge for numerous devices. The surface charge density of TENG is limited by triboelectrification charge density, air breakdown, and dielectric breakdown [157,158]. When the voltage exceeds the threshold breakdown voltage, or the electric field exceeds the threshold breakdown electric field, the electrons emitted from the cathode will be accelerated to high kinetic energy, which will continue to collide with the gas molecules and cause collision ionization. Then, during the collision, the secondary electrons will be formed, causing the chain reaction, which is called electron avalanche (Figure 12a) [159]. However, the increase of gas pressure can cause a shorter mean free path and thereby inhibit the gas breakdown. With the increase of temperature, the mean free path of electrons and collide possibility increase. Electrons will gain more energy under an electrostatic field, making the avalanche breakdown more likely to occur (Figure 12b) [160]. Considering the limitation of dielectric breakdown effect in air, Wang et al. reported an optimization method to increase the triboelectric charge density by coupling surface polarization, triboelectrification and hysteretic dielectric polarization of vacuum ferroelectric materials ($\sim 10^{-6}$ torr) [157]. As exhibited in Figure 12c, when the TENG is operated in vacuum ($\sim 10^{-6}$ torr), the air breakdown described by Paschen's law is avoided, and the triboelectric charge density increases from 120 to 660 $\mu C\ m^{-2}$. A high vacuum environment can not only ensure better performance of TENGs but also avoid the performance degradation of TENGs caused by the natural accumulation of dust and moisture. In addition, Yi et al. also proposed a strategy to improve the output performance of TENG in a wide range of atmosphere pressure of oxygen [161]. As shown in Figure 12d, a high vacuum system was built to accurately measure the output performance of TENGs in different environments. The chamber was firstly pumped to a high vacuum environment and then the pure gases was injected into the vacuum chamber to minimize the interference of other environmental factors, such as temperature and humidity. The output charge density of the TENG under different gas environments first increases and then decreases with the increasing of gas pressure (Figure 12e). In addition, the output charge density in oxygen gas is higher than that in air or nitrogen gas, the reason of which can be attributed to the lower threshold voltage and easier breakdown characteristics of oxygen gas compared with air and nitrogen gas.

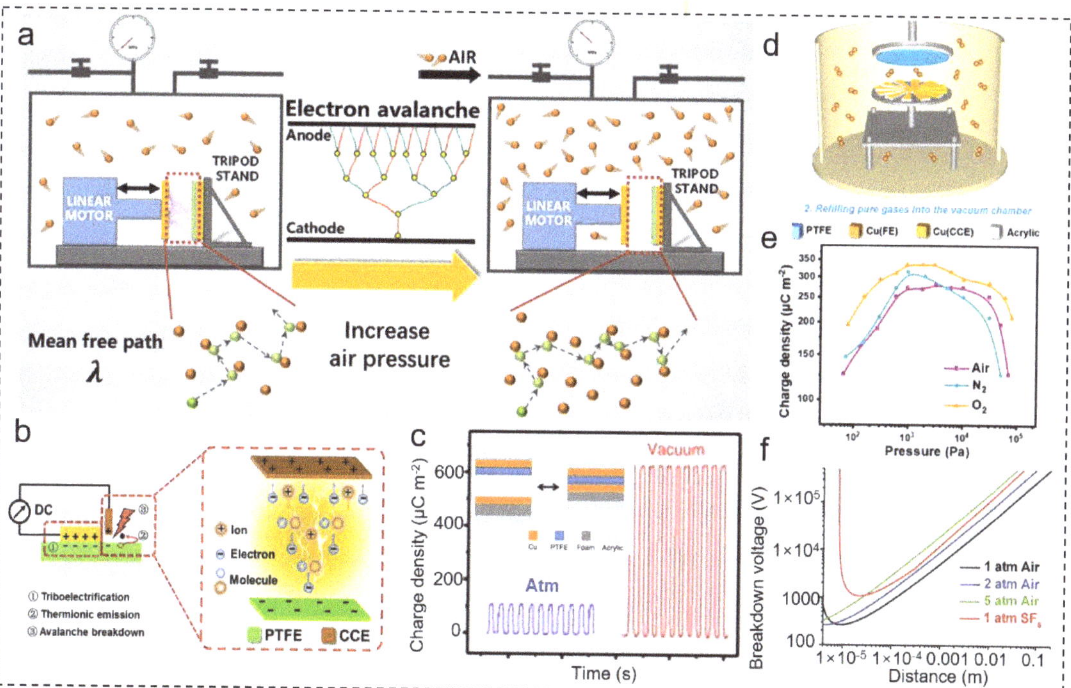

Figure 12. External environmental control to suppress the gas breakdown. (**a**) The threshold output energy density of TENGs is enhanced by suppressing the breakdown effects in a high-pressure environment. Reproduced with the permission of [159], copyright 2020, Wiley-VCH. (**b**) Physical model and the avalanche breakdown of DC-TENG. Reproduced with the permission of [160], copyright 2020, Wiley-VCH. (**c**) Increasing the triboelectric charge density under vacuum working conditions. Reproduced with the permission of [157], copyright 2017, Springer Nature. (**d**) Schematic illustration of the measurement system. (**e**) The charge density of TENG in different gas environments. (**d,e**) Reproduced with the permission of [161], copyright 2021, Elsevier. (**f**) Diagram of Paschen's law and the maximum effective energy output of each cycle in different environments. Reproduced with the permission of [159], copyright 2017, Wiley-VCH.

However, the above performance improvement methods only work at ultrahigh vacuum. In fact, the output performance of TENGs is still limited when the vacuum is lower than the ultrahigh level due to the unavoidable air breakdown effect. By suppressing the breakdown effects under a high-pressure gas environment [159], Zi et al. investigated the air breakdown limits of TENGs in different air pressure and gas atmosphere [48]. As shown in Figure 12f, with the increase of air pressure, the Paschen's curve moves to the left, and the threshold breakdown voltage increases, which is due to lower mean-free-path and less accumulated kinetic energy in electrons. In other words, air breakdown will be easier under a lower vacuum level. In the low-pressure range, the density of air molecules is relatively low, which makes it difficult to cause air breakdown.

5. Integrated Power Management

The capacitive TENG shows the high voltage (~kV) and low current (~μA) output characteristics, which endow it with extremely high internal resistances (~hundreds of MΩ). However, the internal actual resistance of most electronics is several hundred to several thousand ohms. The great mismatched internal impedance results in low energy utilization efficiency. Therefore, many studies focused on decreasing the voltage and increasing the current through the energy management module (EMM). Transformers, the voltage conversion devices widely used in power grids, are successfully applied to increase the current of TENGs [150,162–167]. In general, transformers exhibit high energy conversion efficiency under high-frequency electrical signals thus many researchers utilize transformers to manage the power of turntable mode TENGs. After the management of the transformer, Zhu et al. successfully charge the mobile phone by the TENG at 3000 r min^{-1} [165]. Pu et al. obtained an energy conversion efficiency of 72.4% after optimizing the transformer [167]. Except for the turntable mode TENGs, transient discharge can also be utilized to achieve a high energy conversion efficiency of the transformer for low-frequency TENGs. Niu et al. used an electronic switch to control the charge and discharge of the capacitor charged by TENGs to obtain a rapidly changing electrical signal, and the conversion efficiency reached 60% [164]. Wang et al. combined the spark switch with the transformer and designed an efficient EMM with 78.5% energy conversion efficiency (Figure 13a,b) [150]. The spark switch will be turned on when the voltage of C_{in} reaches its breakdown voltage, and the energy stored in C_{in} will be released to the load or stored in C_{out} through the transformer. Peak current density of 2010 A m^{-2} and a constant power density of 291 mW m^{-2} can be achieved at 3 Hz. After optimizing the spark switch and transformer, 16 series hygrothermographs are continuously powered at 1 Hz (Figure 13c).

The buck circuit, which is widely used in the design of microelectronic circuits, was also successfully developed to reduce the internal impedance of TENGs [168–171]. The advantages of non-magnetic, miniaturization and light weight make buck circuits seem more suitable for flexible electronics. The key to obtaining an effective LC buck circuit is to design a suitable switch. In 2017, Xi et al. achieved an energy conversion efficiency of 80.4% by choosing a MOSFET and a voltage comparator as the switch (Figure 13d,e) [170]. The working process of the circuit can be roughly divided into three stages. Firstly, when the voltage of TENG reaches its optimal value, the switch will be turned on, and the energy of TENG will be transferred to inductor L, capacitor C, and resistor R. Secondly, the energy in the inductor L will move to the capacitor C and the resistor R when the switch is turned off. Finally, the energy stored in capacitor C will power the resistor R. After optimizing the EMM, the human body motion energy can be harvested through the FTENG to drive the thermometer (Figure 13f). However, as an active switch, MOSFET will consume a lot of energy from TENG. Harmon et al. selected a passive switch silicon-controlled rectifier (SCR) in the buck circuit (Figure 13g) [169]. The working cycle of the circuit is similar to the above paper. Therefore, the voltage of C_{out} increases while that of C_{in} decreases (Figure 13h). As for the switch, when the voltage of C_{in} exceeds the regulated value of D_5, the triggered SCR will be turned on. Additionally, the SCR will be turned off when the voltage across the SCR is 0. The whole module can be very small (Figure 13i). The energy conversion efficiency is 89.8%, which is higher than that of MOSFET for the lower loss of passive switch (Figure 13j). Moreover, other switches are also cleverly designed, such as mechanical switches [168] and air breakdown switches [171].

Besides the above energy management strategies, using the switched capacitors is another effective way of decreasing the intrinsic resistance of the TENG. In general, switched capacitors means utilizing switches and capacitors to design a circuit that capacitors are charged in series and discharged in parallel. Tang et al. first designed a switched capacitor EMM with mechanical switches, and the output energy can be enhanced 2200 times after management [172]. However, the mechanical switches will complicate the system. To overcome this problem, Liu et al. used the diodes as electronic switches and successfully designed an EMM (Figure 13k–n) [173]. As shown in Figure 13k, the dotted diodes are turned off when TENG is charging the EMM, and the capacitor is connected in series. When the capacitor in the EMM supplies energy to the external circuit, another part of the turned-off diodes will result in the capacitors connected in parallel, thereby increasing the output current and reducing the output voltage. Furthermore, the filter storage capacitor will provide a constant output for the load. Compared with full-wave rectification, half-wave rectification can provide a higher voltage (Figure 13l) [174]. The photograph of EMM on a printing circuit board demonstrates its high integration (Figure 13m). After optimization, a power density of 954 W m^{-2} was achieved (Figure 13n). Additionally, the matched load resistance can be reduced from 600 MΩ to 0.8 MΩ with a high conversion efficiency of 94.5%.

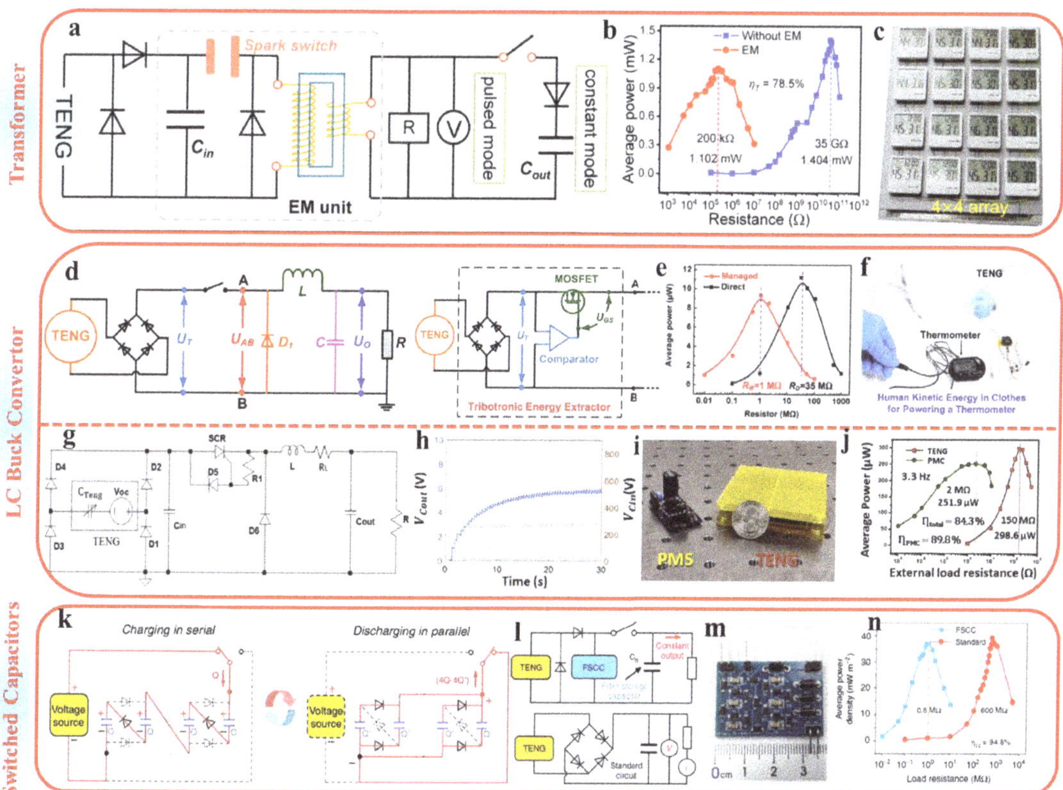

Figure 13. Power management strategies for TENGs. (**a–c**) Circuit diagram (**a**), conversion efficiency (**b**), and application (**c**) of EMM based on transformer and spark switch. Reproduced with the permission of [150], copyright 2020, Elsevier. (**d–f**) Circuit diagram (**d**), conversion efficiency (**e**), and application (**f**) of EMM based on buck circuit and MOSFET. Reproduced with the permission of [170], copyright 2017, Elsevier. (**g–j**) Circuit diagram (**g**), output (**h**), actual photos (**i**), and conversion efficiency (**j**) of EMM based on buck circuit and silicon-controlled rectifier. Reproduced with the permission of [169], copyright 2020, Elsevier. (**k–n**) Working Mechanism (**k**), circuit diagram (**l**), actual photos (**m**), and conversion efficiency (**n**) of EMM based on switched capacitors and silicon-controlled rectifier. Reproduced with the permission of [173], copyright 2020, Springer Nature.

6. Summary and Perspectives

In this review, the current feasible strategies to enhance the mechanical-to-electrical conversion performance of textile TENGs are systematically summarized. The electricity generation mechanism of TENGs is firstly introduced based on an atomic-scale electron-cloud potential well or wave function overlapping model. Afterwards, five major strategies, including surface/interface physical treatments, atomic-scale chemical modification, structural optimization design, external environmental control, and integrated energy management, to enhance the power output of textile TENGs are put forward, each of which was analyzed and discussed in detail. Based on the comprehensive understanding of these approaches, a targeted comparison highlights the superiorities and the potential deficiencies in improving the electricity output of textile TENGs (Figure 14). For example, the intermediate layer structural design can greatly increase the power output by reducing the electron drift and extending charge decay. As for the surface physical modification, the soft contacted interface can enable TENGs to fully contact with the external loading to generate high mechanical deformability, which endows them with high sensitivity and wide application scope. However, the soft texture of the selected materials also leads to their poor mechanical strength. On balance, each method has its advantages and disadvantages. Appropriate material selections and structural designs are required according to the actual requirements. In addition, in order to improve the output performance of textile TENGs to truly meet their large-scale commercial application, the following issues are also worth full consideration.

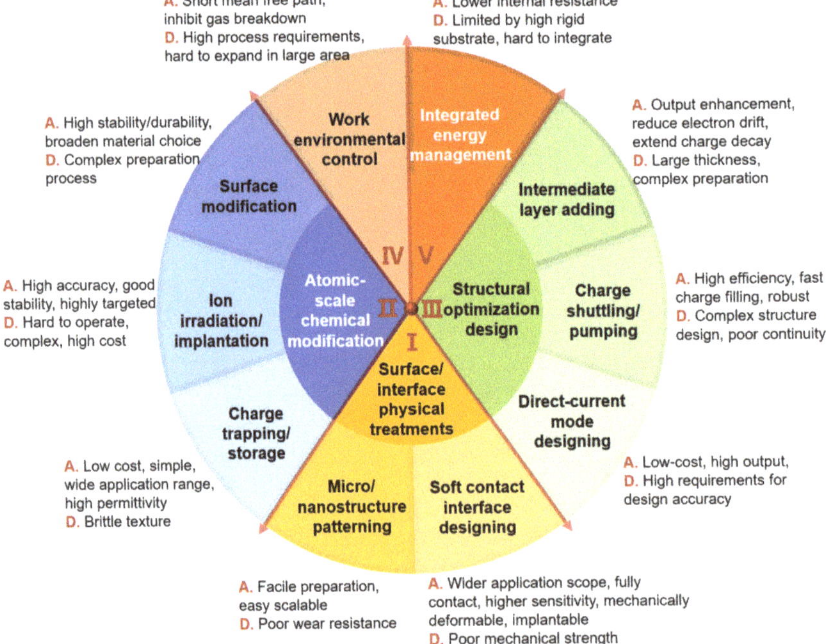

Figure 14. Comparison of the five major strategies to enhance the electricity output performance of textile TENGs. The abbreviations of "A." and "D." represent the advantages and disadvantages, respectively.

(1) The charge transferring and electricity generation mechanism of textile TENGs, particularly for fibers with special structural effects, is still very confusing. Although the electron-cloud potential well or wave function overlapping model can well explain the electron transport behavior between single atoms, there is still a lack of effective exploration on the collision charge generation process between molecules or atomic

clusters with multiple atoms. In addition, although a large number of researches reported the effect of micro/nanostructures on the electrical output of planar membrane structural TENGs, few studies explore the impact of the special structural effect on the surface of high-aspect-ratio curved fiber TENGs on their electrical output performance. Therefore, it is necessary to investigate the physical structure effect and chemical electricity generation mechanism of fiber/textile TENGs by means of macro measurement or micro analysis in more detail.

(2) Most of the above output improvement methods are applicable when the electricity output of textile TENGs is small. However, when the open-circuit voltage reaches the level of air or dielectric breakdown, most of the above methods will lose their function. In this case, direct-current mode structural design and external environment control may be feasible methods. However, the implementation of external environment control is very difficult and can hardly be applied in practice. The direct-current power generation mode depends on the construction of a narrow air breakdown channel and a certain sliding distance, which has high structural and motion mode limitations. Therefore, it is still necessary to explore a better way to improve the triboelectric output performance beyond the air breakdown effect.

(3) It is also a great challenge to effectively and orderly connect the electrodes from different units in textile TENGs and with other energy management or storage devices. Most smart textiles will extend to all parts of the human body, which makes the circuit design across the whole body network more difficult. In fact, most of the current circuit components are rigid and relatively high weight, which will add extra burden to the actual wearing. Therefore, the whole body flexible electrode network design is also urgent for the actual application of textile TENGs.

(4) At present, there is still a lack of comprehensive performance evaluation standards for textile TENGs. Although the structural figure-of-merit and performance figure-of-merit of TENG are proposed to evaluate the effects of structure and material on its output performance, respectively, the impact of other factors is ignored, such as surface condition, working environment, and so on. In addition, the performance evaluation criterion of textile TENGs should still consider the special structural effect and wearing performance of textiles.

(5) The application terminal of textile TENGs is the human body, which not only needs to have a high degree of fitness to meet the diverse and complex human motion and external environment variations but should also bring certain comfort and satisfaction to the daily wearing of the human body. Therefore, for the material selection of textile TENGs, it is best to choose materials with good affinity to human skin to avoid toxic materials or materials that will cause itching to human skin.

(6) During the operation of textile TENGs, frequent, direct and long-term mechanical impacts, as well as negative environmental factors, pose serious challenges to their robustness and reliability [175]. Therefore, improving the robustness and reliability of textile TENGs is an important issue that needs to be addressed urgently. Textile TENGs for wearable use will constantly be subject to abrasion, pressure, and other complex mechanical deformations, which will cause wear and defects. In addition, machine washability is also a great challenge for textile TENGs, which requires appropriate packaging technology to avoid structural or circuit damage.

We hope that this review can not only burst out more inspiration and ideas for the researchers who are focusing on the study of textile TENGs, but also bring a new dawn to industrial designers who are committed to developing commercial products of textile TENGs. We believe that through the persistent efforts of researchers and the continuous follow-up of relevant technical means, textile TENGs will be bound to play an irreplaceable role in the future intelligent wearable market and artificial intelligence field.

Author Contributions: K.D. planned and supervised the whole process. X.P. and R.C. assisted in literature collation and part of chapter writing. Z.L.W. supervised the whole process of this manuscript. All authors revised the manuscript. All authors have read and agreed to the published version of the manuscript.

Funding: The authors are grateful for the support received from the National Key R&D Project from the Minister of Science and Technology (Grant No. 2021YFA1201601), National Natural Science Foundation of China (Grant No. 22109012), Natural Science Foundation of the Beijing Municipality (Grant No. 2212052), and the Fundamental Research Funds for the Central Universities (Grant No. E1E46805).

Conflicts of Interest: The authors declare that they have no known competing financial interests or personal relationships that could have appeared to influence the work reported in this paper.

References

1. Davis, S.J.; Lewis, N.S.; Shaner, M.; Aggarwal, S.; Arent, D.; Azevedo, I.L.; Benson, S.M.; Bradley, T.; Brouwer, J.; Chiang, Y.-M.; et al. Net-zero emissions energy systems. *Science* **2018**, *360*, eaas9793. [CrossRef] [PubMed]
2. Hittinger, E.; Jaramillo, P. Internet of Things: Energy boon or bane? *Science* **2019**, *364*, 326–328. [CrossRef] [PubMed]
3. Liu, L.; Guo, X.; Lee, C. Promoting smart cities into the 5G era with multi-field Internet of Things (IoT) applications powered with advanced mechanical energy harvesters. *Nano Energy* **2021**, *88*, 106304. [CrossRef]
4. Dharmasena, R.D.I.G.; Jayawardena, K.D.G.I.; Mills, C.A.; Deane, J.H.B.; Anguita, J.V.; Dorey, R.A.; Silva, S.R.P. Triboelectric nanogenerators: Providing a fundamental framework. *Energy Environ. Sci.* **2017**, *10*, 1801–1811. [CrossRef]
5. Ryu, H.; Park, H.-M.; Kim, M.-K.; Kim, B.; Myoung, H.S.; Kim, T.Y.; Yoon, H.-J.; Kwak, S.S.; Kim, J.; Hwang, T.H.; et al. Self-rechargeable cardiac pacemaker system with triboelectric nanogenerators. *Nat. Commun.* **2021**, *12*, 4374. [CrossRef]
6. Wang, H.; Han, M.; Song, Y.; Zhang, H. Design, manufacturing and applications of wearable triboelectric nanogenerators. *Nano Energy* **2021**, *81*, 105627. [CrossRef]
7. He, W.; Fu, X.; Zhang, D.; Zhang, Q.; Zhuo, K.; Yuan, Z.; Ma, R. Recent progress of flexible/wearable self-charging power units based on triboelectric nanogenerators. *Nano Energy* **2021**, *84*, 105880. [CrossRef]
8. Feng, H.; Bai, Y.; Qiao, L.; Li, Z.; Wang, E.; Chao, S.; Qu, X.; Cao, Y.; Liu, Z.; Han, X.; et al. An Ultra-Simple Charge Supplementary Strategy for High Performance Rotary Triboelectric Nanogenerators. *Small* **2021**, *17*, 2101430. [CrossRef]
9. Huang, L.-B.; Dai, X.; Sun, Z.; Wong, M.-C.; Pang, S.-Y.; Han, J.; Zheng, Q.; Zhao, C.-H.; Kong, J.; Hao, J. Environment-resisted flexible high performance triboelectric nanogenerators based on ultrafast self-healing non-drying conductive organohydrogel. *Nano Energy* **2021**, *82*, 105724. [CrossRef]
10. Yin, L.; Kim, K.N.; Trifonov, A.; Podhajny, T.; Wang, J. Designing wearable microgrids: Towards autonomous sustainable on-body energy management. *Energy Environ. Sci.* **2022**, *15*, 82–101. [CrossRef]
11. Karan, S.K.; Maiti, S.; Lee, J.H.; Mishra, Y.K.; Khatua, B.B.; Kim, J.K. Recent Advances in Self-Powered Tribo-/Piezoelectric Energy Harvesters: All-In-One Package for Future Smart Technologies. *Adv. Funct. Mater.* **2020**, *30*, 2004446. [CrossRef]
12. Mahapatra, S.D.; Mohapatra, P.C.; Aria, A.I.; Christie, G.; Mishra, Y.K.; Hofmann, S.; Thakur, V.K. Piezoelectric Materials for Energy Harvesting and Sensing Applications: Roadmap for Future Smart Materials. *Adv. Sci.* **2021**, *8*, 2100864. [CrossRef] [PubMed]
13. Sahu, M.; Hajra, S.; Kim, H.-G.; Rubahn, H.-G.; Kumar Mishra, Y.; Kim, H.J. Additive manufacturing-based recycling of laboratory waste into energy harvesting device for self-powered applications. *Nano Energy* **2021**, *88*, 106255. [CrossRef]
14. Parandeh, S.; Etemadi, N.; Kharaziha, M.; Chen, G.; Nashalian, A.; Xiao, X.; Chen, J. Advances in Triboelectric Nanogenerators for Self-Powered Regenerative Medicine. *Adv. Funct. Mater.* **2021**, *31*, 2105169. [CrossRef]
15. Dang, C.; Shao, C.; Liu, H.; Chen, Y.; Qi, H. Cellulose melt processing assisted by small biomass molecule to fabricate recyclable ionogels for versatile stretchable triboelectric nanogenerators. *Nano Energy* **2021**, *90*, 106619. [CrossRef]
16. Zhong, W.; Xu, B.; Gao, Y. Engraved pattern spacer triboelectric nanogenerators for mechanical energy harvesting. *Nano Energy* **2022**, *92*, 106782. [CrossRef]
17. Durukan, M.B.; Cicek, M.O.; Doganay, D.; Gorur, M.C.; Çınar, S.; Unalan, H.E. Multifunctional and Physically Transient Supercapacitors, Triboelectric Nanogenerators, and Capacitive Sensors. *Adv. Funct. Mater.* **2022**, *32*, 2106066. [CrossRef]
18. Jiang, M.; Li, B.; Jia, W.; Zhu, Z. Predicting output performance of triboelectric nanogenerators using deep learning model. *Nano Energy* **2022**, *93*, 106830. [CrossRef]
19. Fan, F.-R.; Tian, Z.-Q.; Wang, Z.L. Flexible triboelectric generator. *Nano Energy* **2012**, *1*, 328–334. [CrossRef]
20. Wu, C.; Wang, A.C.; Ding, W.; Guo, H.; Wang, Z.L. Triboelectric Nanogenerator: A Foundation of the Energy for the New Era. *Adv. Energy Mater.* **2019**, *9*, 1802906. [CrossRef]
21. Ye, C.; Liu, D.; Peng, X.; Jiang, Y.; Cheng, R.; Ning, C.; Sheng, F.; Zhang, Y.; Dong, K.; Wang, Z.L. A Hydrophobic Self-Repairing Power Textile for Effective Water Droplet Energy Harvesting. *ACS Nano* **2021**, *15*, 18172–18181. [CrossRef] [PubMed]
22. Chen, J.; Wang, Z.L. Reviving Vibration Energy Harvesting and Self-Powered Sensing by a Triboelectric Nanogenerator. *Joule* **2017**, *1*, 480–521. [CrossRef]

23. Xu, W.; Zheng, H.; Liu, Y.; Zhou, X.; Zhang, C.; Song, Y.; Deng, X.; Leung, M.; Yang, Z.; Xu, R.X.; et al. A droplet-based electricity generator with high instantaneous power density. *Nature* **2020**, *578*, 392–396. [CrossRef] [PubMed]
24. Rodrigues, C.; Nunes, D.; Clemente, D.; Mathias, N.; Correia, J.M.; Rosa-Santos, P.; Taveira-Pinto, F.; Morais, T.; Pereira, A.; Ventura, J. Emerging triboelectric nanogenerators for ocean wave energy harvesting: State of the art and future perspectives. *Energy Environ. Sci.* **2020**, *13*, 2657–2683. [CrossRef]
25. Dong, K.; Peng, X.; Wang, Z.L. Fiber/Fabric-Based Piezoelectric and Triboelectric Nanogenerators for Flexible/Stretchable and Wearable Electronics and Artificial Intelligence. *Adv. Mater.* **2020**, *32*, 1902549. [CrossRef]
26. Wang, Z.L.; Chen, J.; Lin, L. Progress in triboelectric nanogenerators as a new energy technology and self-powered sensors. *Energy Environ. Sci.* **2015**, *8*, 2250–2282. [CrossRef]
27. Dong, K.; Wang, Z.L. Self-charging power textiles integrating energy harvesting triboelectric nanogenerators with energy storage batteries/supercapacitors. *J. Semicond.* **2021**, *42*, 101601. [CrossRef]
28. Guan, X.; Xu, B.; Wu, M.; Jing, T.; Yang, Y.; Gao, Y. Breathable, washable and wearable woven-structured triboelectric nano-generators utilizing electrospun nanofibers for biomechanical energy harvesting and self-powered sensing. *Nano Energy* **2021**, *80*, 105549. [CrossRef]
29. Gunawardhana, K.R.S.D.; Wanasekara, N.D.; Dharmasena, R.D.I.G. Towards Truly Wearable Systems: Optimizing and Scaling Up Wearable Triboelectric Nanogenerators. *iScience* **2020**, *23*, 101360. [CrossRef]
30. Zou, Y.; Raveendran, V.; Chen, J. Wearable triboelectric nanogenerators for biomechanical energy harvesting. *Nano Energy* **2020**, *77*, 105303. [CrossRef]
31. Shen, S.; Xiao, X.; Xiao, X.; Chen, J. Wearable triboelectric nanogenerators for heart rate monitoring. *Chem. Commun.* **2021**, *57*, 5871–5879. [CrossRef] [PubMed]
32. Zhang, S.; Bick, M.; Xiao, X.; Chen, G.; Nashalian, A.; Chen, J. Leveraging triboelectric nanogenerators for bioengineering. *Matter* **2021**, *4*, 845–887. [CrossRef]
33. Tat, T.; Libanori, A.; Au, C.; Yau, A.; Chen, J. Advances in triboelectric nanogenerators for biomedical sensing. *Biosens. Bioelectron.* **2021**, *171*, 112714. [CrossRef] [PubMed]
34. Pan, M.; Yuan, C.; Liang, X.; Zou, J.; Zhang, Y.; Bowen, C. Triboelectric and Piezoelectric Nanogenerators for Future Soft Robots and Machines. *iScience* **2020**, *23*, 101682. [CrossRef]
35. Elbanna, M.A.; Arafa, M.H.; Bowen, C.R. Experimental and Analytical Investigation of the Response of a Triboelectric Generator Under Different Operating Conditions. *Energy Technol.* **2020**, *8*, 2000576. [CrossRef]
36. Wang, Z.L. On Maxwell's displacement current for energy and sensors: The origin of nanogenerators. *Mater. Today* **2017**, *20*, 74–82. [CrossRef]
37. Wang, Z.L. On the expanded Maxwell's equations for moving charged media system—General theory, mathematical solutions and applications in TENG. *Mater. Today* **2021**. [CrossRef]
38. Dong, K.; Hu, Y.; Yang, J.; Kim, S.W.; Hu, W.; Wang, Z.L. Smart textile triboelectric nanogenerators: Current status and perspectives. *MRS Bull.* **2021**, *46*, 512–521. [CrossRef]
39. Chen, G.; Au, C.; Chen, J. Textile Triboelectric Nanogenerators for Wearable Pulse Wave Monitoring. *Trends Biotechnol.* **2021**, *39*, 1078–1092. [CrossRef]
40. Huang, T.; Zhang, J.; Yu, B.; Yu, H.; Long, H.; Wang, H.; Zhang, Q.; Zhu, M. Fabric texture design for boosting the performance of a knitted washable textile triboelectric nanogenerator as wearable power. *Nano Energy* **2019**, *58*, 375–383. [CrossRef]
41. Liu, J.; Gu, L.; Cui, N.; Xu, Q.; Qin, Y.; Yang, R. Fabric-Based Triboelectric Nanogenerators. *Research* **2019**, *2019*, 1091632. [CrossRef] [PubMed]
42. Tcho, I.-W.; Kim, W.-G.; Choi, Y.-K. A self-powered character recognition device based on a triboelectric nanogenerator. *Nano Energy* **2020**, *70*, 104534. [CrossRef]
43. Zhao, Z.; Yan, C.; Liu, Z.; Fu, X.; Peng, L.-M.; Hu, Y.; Zheng, Z. Machine-Washable Textile Triboelectric Nanogenerators for Effective Human Respiratory Monitoring through Loom Weaving of Metallic Yarns. *Adv. Mater.* **2016**, *28*, 10267–10274. [CrossRef]
44. Paosangthong, W.; Torah, R.; Beeby, S. Recent progress on textile-based triboelectric nanogenerators. *Nano Energy* **2019**, *55*, 401–423. [CrossRef]
45. Hu, Y.; Zheng, Z. Progress in textile-based triboelectric nanogenerators for smart fabrics. *Nano Energy* **2019**, *56*, 16–24. [CrossRef]
46. Kwak, S.S.; Yoon, H.-J.; Kim, S.-W. Textile-Based Triboelectric Nanogenerators for Self-Powered Wearable Electronics. *Adv. Funct. Mater.* **2019**, *29*, 1804533. [CrossRef]
47. De Medeiros, S.M.; Chanci, D.; Moreno, C.; Goswami, D.; Martinez, R.V. Waterproof, Breathable, and Antibacterial Self-Powered e-Textiles Based on Omniphobic Triboelectric Nanogenerators. *Adv. Funct. Mater.* **2019**, *29*, 1904350. [CrossRef]
48. Zi, Y.; Wu, C.; Ding, W.; Wang, Z.L. Maximized Effective Energy Output of Contact-Separation-Triggered Triboelectric Nanogenerators as Limited by Air Breakdown. *Adv. Funct. Mater.* **2017**, *27*, 1700049. [CrossRef]
49. Liu, W.; Wang, Z.; Hu, C. Advanced designs for output improvement of triboelectric nanogenerator system. *Mater. Today* **2021**, *45*, 93–119. [CrossRef]
50. Chen, L.; Shi, Q.; Sun, Y.; Nguyen, T.; Lee, C.; Soh, S. Controlling Surface Charge Generated by Contact Electrification: Strategies and Applications. *Adv. Mater.* **2018**, *30*, 1802405. [CrossRef]
51. Xie, X.; Chen, X.; Zhao, C.; Liu, Y.; Sun, X.; Zhao, C.; Wen, Z. Intermediate layer for enhanced triboelectric nanogenerator. *Nano Energy* **2021**, *79*, 105439. [CrossRef]

52. Kim, W.-G.; Kim, D.-W.; Tcho, I.-W.; Kim, J.-K.; Kim, M.-S.; Choi, Y.-K. Triboelectric Nanogenerator: Structure, Mechanism, and Applications. *ACS Nano* **2021**, *15*, 258–287. [CrossRef]
53. Zhou, Y.; Deng, W.; Xu, J.; Chen, J. Engineering Materials at the Nanoscale for Triboelectric Nanogenerators. *Cell Rep. Phys. Sci.* **2020**, *1*, 100142. [CrossRef]
54. Khandelwal, G.; Maria Joseph Raj, N.P.; Kim, S.-J. Materials Beyond Conventional Triboelectric Series for Fabrication and Applications of Triboelectric Nanogenerators. *Adv. Energy Mater.* **2021**, *11*, 2101170. [CrossRef]
55. Nurmakanov, Y.; Kalimuldina, G.; Nauryzbayev, G.; Adair, D.; Bakenov, Z. Structural and Chemical Modifications Towards High-Performance of Triboelectric Nanogenerators. *Nanoscale Res. Lett.* **2021**, *16*, 122. [CrossRef]
56. Yu, Y.; Wang, X. Chemical modification of polymer surfaces for advanced triboelectric nanogenerator development. *Extreme Mech. Lett.* **2016**, *9*, 514–530. [CrossRef]
57. Yang, H.; Fan, F.R.; Xi, Y.; Wu, W. Design and engineering of high-performance triboelectric nanogenerator for ubiquitous unattended devices. *EcoMat* **2021**, *3*, e12093. [CrossRef]
58. Wang, Z.; Tang, Q.; Shan, C.; Du, Y.; He, W.; Fu, S.; Li, G.; Liu, A.; Liu, W.; Hu, C. Giant performance improvement of triboelectric nanogenerator systems achieved by matched inductor design. *Energy Environ. Sci.* **2021**, *14*, 6627–6637. [CrossRef]
59. He, Y.; She, D.; Liu, Z.; Wang, X.; Zhong, L.; Wang, C.; Wang, G.; Mao, S.X. Atomistic observation on diffusion-mediated friction between single-asperity contacts. *Nat. Mater.* **2021**, *21*, 173–180. [CrossRef]
60. Mallada, B.; Gallardo, A.; Lamanec, M.; de la Torre, B.; Špirko, V.; Hobza, P.; Jelinek, P. Real-space imaging of anisotropic charge of σ-hole by means of Kelvin probe force microscopy. *Science* **2021**, *374*, 863–867. [CrossRef]
61. Zhang, B.; Cheng, Z.; Lu, Z.; Zhang, G.; Ma, F. Atomic-Scale Rolling Friction and Charge-Transfer Mechanism: An Integrated Study of Physical Deductions and DFT Simulations. *J. Phys. Chem. C* **2020**, *124*, 8431–8438. [CrossRef]
62. Wang, Z.L. From contact electrification to triboelectric nanogenerators. *Rep. Prog. Phys.* **2021**, *84*, 096502. [CrossRef] [PubMed]
63. Lin, S.; Chen, X.; Wang, Z.L. Contact Electrification at the Liquid-Solid Interface. *Chem. Rev.* **2021**. [CrossRef]
64. Xu, C.; Zi, Y.; Wang, A.C.; Zou, H.; Dai, Y.; He, X.; Wang, P.; Wang, Y.C.; Feng, P.; Li, D.; et al. On the electron-transfer mechanism in the contact-electrification effect. *Adv. Mater.* **2018**, *30*, 1706790. [CrossRef]
65. Wang, Z.L.; Wang, A.C. On the origin of contact-electrification. *Mater. Today* **2019**, *30*, 34–51. [CrossRef]
66. Ibrahim, M.; Jiang, J.; Wen, Z.; Sun, X. Surface Engineering for Enhanced Triboelectric Nanogenerator. *Nanoenergy Adv.* **2021**, *1*, 58–80. [CrossRef]
67. Wang, H.S.; Jeong, C.K.; Seo, M.-H.; Joe, D.J.; Han, J.H.; Yoon, J.-B.; Lee, K.J. Performance-enhanced triboelectric nanogenerator enabled by wafer-scale nanogrates of multistep pattern downscaling. *Nano Energy* **2017**, *35*, 415–423. [CrossRef]
68. Pradel, K.C.; Fukata, N. Systematic optimization of triboelectric nanogenerator performance through surface micropatterning. *Nano Energy* **2021**, *83*, 105856. [CrossRef]
69. Ahmed, A.; Hassan, I.; Pourrahimi, A.M.; Helal, A.S.; El-Kady, M.F.; Khassaf, H.; Kaner, R.B. Toward High-Performance Triboelectric Nanogenerators by Engineering Interfaces at the Nanoscale: Looking into the Future Research Roadmap. *Adv. Mater. Technol.* **2020**, *5*, 2000520. [CrossRef]
70. Xi, Y.; Zhang, F.; Shi, Y. Effects of surface micro-structures on capacitances of the dielectric layer in triboelectric nanogenerator: A numerical simulation study. *Nano Energy* **2021**, *79*, 105432. [CrossRef]
71. Park, S.; Park, J.; Kim, Y.-g.; Bae, S.; Kim, T.-W.; Park, K.-I.; Hong, B.H.; Jeong, C.K.; Lee, S.-K. Laser-directed synthesis of strain-induced crumpled MoS2 structure for enhanced triboelectrification toward haptic sensors. *Nano Energy* **2020**, *78*, 105266. [CrossRef]
72. Fan, F.-R.; Lin, L.; Zhu, G.; Wu, W.; Zhang, R.; Wang, Z.L. Transparent Triboelectric Nanogenerators and Self-Powered Pressure Sensors Based on Micropatterned Plastic Films. *Nano Lett.* **2012**, *12*, 3109–3114. [CrossRef] [PubMed]
73. Yang, W.; Wang, X.; Li, H.; Wu, J.; Hu, Y.; Li, Z.; Liu, H. Fundamental research on the effective contact area of micro-/nano-textured surface in triboelectric nanogenerator. *Nano Energy* **2019**, *57*, 41–47. [CrossRef]
74. Jin, C.; Kia, D.S.; Jones, M.; Towfighian, S. On the contact behavior of micro-/nano-structured interface used in vertical-contact-mode triboelectric nanogenerators. *Nano Energy* **2016**, *27*, 68–77. [CrossRef]
75. Seol, M.-L.; Han, J.-W.; Moon, D.-I.; Meyyappan, M. Hysteretic behavior of contact force response in triboelectric nanogenerator. *Nano Energy* **2017**, *32*, 408–413. [CrossRef]
76. Zhang, J.-H.; Li, Y.; Du, J.; Hao, X.; Huang, H. A high-power wearable triboelectric nanogenerator prepared from self-assembled electrospun poly(vinylidene fluoride) fibers with a heart-like structure. *J. Mater. Chem. A* **2019**, *7*, 11724–11733. [CrossRef]
77. Kim, K.N.; Chun, J.; Kim, J.W.; Lee, K.Y.; Park, J.-U.; Kim, S.-W.; Wang, Z.L.; Baik, J.M. Highly Stretchable 2D Fabrics for Wearable Triboelectric Nanogenerator under Harsh Environments. *ACS Nano* **2015**, *9*, 6394–6400. [CrossRef]
78. Seung, W.; Gupta, M.K.; Lee, K.Y.; Shin, K.-S.; Lee, J.-H.; Kim, T.Y.; Kim, S.; Lin, J.; Kim, J.H.; Kim, S.-W. Nanopatterned Textile-Based Wearable Triboelectric Nanogenerator. *ACS Nano* **2015**, *9*, 3501–3509. [CrossRef]
79. Wu, M.; Gao, Z.; Yao, K.; Hou, S.; Liu, Y.; Li, D.; He, J.; Huang, X.; Song, E.; Yu, J.; et al. Thin, soft, skin-integrated foam-based triboelectric nanogenerators for tactile sensing and energy harvesting. *Mater. Today Energy* **2021**, *20*, 100657. [CrossRef]
80. Guan, D.; Cong, X.; Li, J.; Shen, H.; Zhang, C.; Gong, J. Quantitative characterization of the energy harvesting performance of soft-contact sphere triboelectric nanogenerator. *Nano Energy* **2021**, *87*, 106186. [CrossRef]

81. Dong, K.; Wu, Z.; Deng, J.; Wang, A.C.; Zou, H.; Chen, C.; Hu, D.; Gu, B.; Sun, B.; Wang, Z.L. A Stretchable Yarn Embedded Triboelectric Nanogenerator as Electronic Skin for Biomechanical Energy Harvesting and Multifunctional Pressure Sensing. *Adv. Mater.* **2018**, *30*, 1804944. [CrossRef] [PubMed]

82. Dong, K.; Deng, J.; Ding, W.; Wang, A.C.; Wang, P.; Cheng, C.; Wang, Y.-C.; Jin, L.; Gu, B.; Sun, B.; et al. Versatile Core–Sheath Yarn for Sustainable Biomechanical Energy Harvesting and Real-Time Human-Interactive Sensing. *Adv. Energy Mater.* **2018**, *8*, 1801114. [CrossRef]

83. Cheng, P.; Guo, H.; Wen, Z.; Zhang, C.; Yin, X.; Li, X.; Liu, D.; Song, W.; Sun, X.; Wang, J.; et al. Largely enhanced triboelectric nanogenerator for efficient harvesting of water wave energy by soft contacted structure. *Nano Energy* **2019**, *57*, 432–439. [CrossRef]

84. Yi, F.; Wang, X.; Niu, S.; Li, S.; Yin, Y.; Dai, K.; Zhang, G.; Lin, L.; Wen, Z.; Guo, H.; et al. A highly shape-adaptive, stretchable design based on conductive liquid for energy harvesting and self-powered biomechanical monitoring. *Sci. Adv.* **2016**, *2*, e1501624. [CrossRef] [PubMed]

85. Wang, J.; Li, S.; Yi, F.; Zi, Y.; Lin, J.; Wang, X.; Xu, Y.; Wang, Z.L. Sustainably powering wearable electronics solely by biomechanical energy. *Nat. Commun.* **2016**, *7*, 12744. [CrossRef]

86. Dong, K.; Wang, Y.-C.; Deng, J.; Dai, Y.; Zhang, S.L.; Zou, H.; Gu, B.; Sun, B.; Wang, Z.L. A Highly Stretchable and Washable All-Yarn-Based Self-Charging Knitting Power Textile Composed of Fiber Triboelectric Nanogenerators and Supercapacitors. *ACS Nano* **2017**, *11*, 9490–9499. [CrossRef]

87. Lai, Y.-C.; Lu, H.-W.; Wu, H.-M.; Zhang, D.; Yang, J.; Ma, J.; Shamsi, M.; Vallem, V.; Dickey, M.D. Elastic Multifunctional Liquid–Metal Fibers for Harvesting Mechanical and Electromagnetic Energy and as Self-Powered Sensors. *Adv. Energy Mater.* **2021**, *11*, 2100411. [CrossRef]

88. Zheng, L.; Zhu, M.; Wu, B.; Li, Z.; Sun, S.; Wu, P. Conductance-stable liquid metal sheath-core microfibers for stretchy smart fabrics and self-powered sensing. *Sci. Adv.* **2021**, *7*, eabg4041. [CrossRef]

89. Jing, T.; Xu, B.; Yang, Y. Organogel electrode based continuous fiber with large-scale production for stretchable triboelectric nanogenerator textiles. *Nano Energy* **2021**, *84*, 105867. [CrossRef]

90. Zou, H.; Zhang, Y.; Guo, L.; Wang, P.; He, X.; Dai, G.; Zheng, H.; Chen, C.; Wang, A.C.; Xu, C.; et al. Quantifying the triboelectric series. *Nat. Commun.* **2019**, *10*, 1427. [CrossRef]

91. Zou, H.; Guo, L.; Xue, H.; Zhang, Y.; Wang, Z.L. Quantifying and understanding the triboelectric series of inorganic non-metallic materials. *Nat. Commun.* **2020**, *11*, 2093. [CrossRef]

92. Kim, M.P.; Lee, Y.; Hur, Y.H.; Park, J.; Kim, J.; Lee, Y.; Ahn, C.W.; Song, S.W.; Jung, Y.S.; Ko, H. Molecular structure engineering of dielectric fluorinated polymers for enhanced performances of triboelectric nanogenerators. *Nano Energy* **2018**, *53*, 37–45. [CrossRef]

93. Lee, B.-Y.; Kim, D.H.; Park, J.; Park, K.-I.; Lee, K.J.; Jeong, C.K. Modulation of surface physics and chemistry in triboelectric energy harvesting technologies. *Sci. Technol. Adv. Mater.* **2019**, *20*, 758–773. [CrossRef] [PubMed]

94. Kim, D.W.; Lee, J.H.; Kim, J.K.; Jeong, U. Material aspects of triboelectric energy generation and sensors. *NPG Asia Mater.* **2020**, *12*, 6. [CrossRef]

95. Dzhardimalieva, G.I.; Yadav, B.C.; Lifintseva, T.Y.V.; Uflyand, I.E. Polymer chemistry underpinning materials for triboelectric nanogenerators (TENGs): Recent trends. *Eur. Polym. J.* **2021**, *142*, 110163. [CrossRef]

96. Xu, J.; Zou, Y.; Nashalian, A.; Chen, J. Leverage Surface Chemistry for High-Performance Triboelectric Nanogenerators. *Front. Chem.* **2020**, *8*, 959. [CrossRef] [PubMed]

97. Xiao, R.; Yu, G.; Xu, B.B.; Wang, N.; Liu, X. Fiber Surface/Interfacial Engineering on Wearable Electronics. *Small* **2021**, *17*, 2102903. [CrossRef] [PubMed]

98. Utka, A.; Mālnieks, K.; Lapčinskis, L.; Kaufelde, P.; Linarts, A.; Bērziņa, A.; Zābels, R.; Jurkāns, V.; Gornevs, I.; Blūms, J.; et al. The role of intermolecular forces in contact electrification on polymer surfaces and triboelectric nanogenerators. *Energy Environ. Sci.* **2019**, *12*, 2417–2421.

99. Shin, S.-H.; Bae, Y.E.; Moon, H.K.; Kim, J.; Choi, S.-H.; Kim, Y.; Yoon, H.J.; Lee, M.H.; Nah, J. Formation of Triboelectric Series via Atomic-Level Surface Functionalization for Triboelectric Energy Harvesting. *ACS Nano* **2017**, *11*, 6131–6138. [CrossRef]

100. Lee, J.W.; Jung, S.; Lee, T.W.; Jo, J.; Chae, H.Y.; Choi, K.; Kim, J.J.; Lee, J.H.; Yang, C.; Baik, J.M. High-Output Triboelectric Nanogenerator Based on Dual Inductive and Resonance Effects-Controlled Highly Transparent Polyimide for Self-Powered Sensor Network Systems. *Adv. Energy Mater.* **2019**, *9*, 1901987. [CrossRef]

101. Lee, C.; Yang, S.; Choi, D.; Kim, W.; Kim, J.; Hong, J. Chemically surface-engineered polydimethylsiloxane layer via plasma treatment for advancing textile-based triboelectric nanogenerators. *Nano Energy* **2019**, *57*, 353–362. [CrossRef]

102. Feng, P.-Y.; Xia, Z.; Sun, B.; Jing, X.; Li, H.; Tao, X.; Mi, H.-Y.; Liu, Y. Enhancing the Performance of Fabric-Based Triboelectric Nanogenerators by Structural and Chemical Modification. *ACS Appl. Mater. Interfaces* **2021**, *13*, 16916–16927. [CrossRef]

103. Liu, Y.; Fu, Q.; Mo, J.; Lu, Y.; Cai, C.; Luo, B.; Nie, S. Chemically tailored molecular surface modification of cellulose nanofibrils for manipulating the charge density of triboelectric nanogenerators. *Nano Energy* **2021**, *89*, 106369. [CrossRef]

104. Wang, S.; Xie, Y.; Niu, S.; Lin, L.; Liu, C.; Zhou, Y.S.; Wang, Z.L. Maximum Surface Charge Density for Triboelectric Nanogenerators Achieved by Ionized-Air Injection: Methodology and Theoretical Understanding. *Adv. Mater.* **2014**, *26*, 6720–6728. [CrossRef] [PubMed]

105. Fan, Y.; Li, S.; Tao, X.; Wang, Y.; Liu, Z.; Chen, H.; Wu, Z.; Zhang, J.; Ren, F.; Chen, X.; et al. Negative triboelectric polymers with ultrahigh charge density induced by ion implantation. *Nano Energy* **2021**, *90*, 106574. [CrossRef]

106. Luo, N.; Feng, Y.; Li, X.; Sun, W.; Wang, D.; Ye, Q.; Sun, X.; Zhou, F.; Liu, W. Manipulating Electrical Properties of Silica-Based Materials via Atomic Oxygen Irradiation. *ACS Appl. Mater. Interfaces* **2021**, *13*, 15344–15352. [CrossRef]

107. Li, S.; Fan, Y.; Chen, H.; Nie, J.; Liang, Y.; Tao, X.; Zhang, J.; Chen, X.; Fu, E.; Wang, Z.L. Manipulating the triboelectric surface charge density of polymers by low-energy helium ion irradiation/implantation. *Energy Environ. Sci.* **2020**, *13*, 896–907. [CrossRef]

108. Sahu, M.; Šafranko, S.; Hajra, S.; Padhan, A.M.; Živković, P.; Jokić, S.; Kim, H.J. Development of triboelectric nanogenerator and mechanical energy harvesting using argon ion-implanted kapton, zinc oxide and kapton. *Mater. Lett.* **2021**, *301*, 130290. [CrossRef]

109. Cheng, G.-G.; Jiang, S.-Y.; Li, K.; Zhang, Z.-Q.; Wang, Y.; Yuan, N.-Y.; Ding, J.-N.; Zhang, W. Effect of argon plasma treatment on the output performance of triboelectric nanogenerator. *Appl. Surf. Sci.* **2017**, *412*, 350–356. [CrossRef]

110. Kim, W.; Okada, T.; Park, H.-W.; Kim, J.; Kim, S.; Kim, S.-W.; Samukawa, S.; Choi, D. Surface modification of triboelectric materials by neutral beams. *J. Mater. Chem. A* **2019**, *7*, 25066–25077. [CrossRef]

111. Park, H.-W.; Huynh, N.D.; Kim, W.; Lee, C.; Nam, Y.; Lee, S.; Chung, K.-B.; Choi, D. Electron blocking layer-based interfacial design for highly-enhanced triboelectric nanogenerators. *Nano Energy* **2018**, *50*, 9–15. [CrossRef]

112. Du, J.; Duan, J.; Yang, X.; Wang, Y.; Duan, Y.; Tang, Q. Charge boosting and storage by tailoring rhombus all-inorganic perovskite nanoarrays for robust triboelectric nanogenerators. *Nano Energy* **2020**, *74*, 104845. [CrossRef]

113. Wang, Y.; Jin, X.; Wang, W.; Niu, J.; Zhu, Z.; Lin, T. Efficient Triboelectric Nanogenerator (TENG) Output Management for Improving Charge Density and Reducing Charge Loss. *ACS Appl. Electron. Mater.* **2021**, *3*, 532–549. [CrossRef]

114. Saadatnia, Z.; Esmailzadeh, E.; Naguib, H.E. High Performance Triboelectric Nanogenerator by Hot Embossing on Self-Assembled Micro-Particles. *Adv. Eng. Mater.* **2019**, *21*, 1700957. [CrossRef]

115. Zhang, J.-H.; Zhang, Y.; Sun, N.; Li, Y.; Du, J.; Zhu, L.; Hao, X. Enhancing output performance of triboelectric nanogenerator via large polarization difference effect. *Nano Energy* **2021**, *84*, 105892. [CrossRef]

116. Wang, H.; Sakamoto, H.; Asai, H.; Zhang, J.-H.; Meboso, T.; Uchiyama, Y.; Kobayashi, E.; Takamura, E.; Suye, S.-I. An all-fibrous triboelectric nanogenerator with enhanced outputs depended on the polystyrene charge storage layer. *Nano Energy* **2021**, *90*, 106515. [CrossRef]

117. Li, Z.; Zhu, M.; Qiu, Q.; Yu, J.; Ding, B. Multilayered fiber-based triboelectric nanogenerator with high performance for biomechanical energy harvesting. *Nano Energy* **2018**, *53*, 726–733. [CrossRef]

118. Cheon, S.; Kang, H.; Kim, H.; Son, Y.; Lee, J.Y.; Shin, H.-J.; Kim, S.-W.; Cho, J.H. High-Performance Triboelectric Nanogenerators Based on Electrospun Polyvinylidene Fluoride–Silver Nanowire Composite Nanofibers. *Adv. Funct. Mater.* **2018**, *28*, 1703778. [CrossRef]

119. Xiong, J.; Cui, P.; Chen, X.; Wang, J.; Parida, K.; Lin, M.-F.; Lee, P.S. Skin-touch-actuated textile-based triboelectric nanogenerator with black phosphorus for durable biomechanical energy harvesting. *Nat. Commun.* **2018**, *9*, 4280. [CrossRef]

120. Salauddin, M.; Rana, S.M.S.; Rahman, M.T.; Sharifuzzaman, M.; Maharjan, P.; Bhatta, T.; Cho, H.; Lee, S.H.; Park, C.; Shrestha, K.; et al. Fabric-Assisted MXene/Silicone Nanocomposite-Based Triboelectric Nanogenerators for Self-Powered Sensors and Wearable Electronics. *Adv. Funct. Mater.* **2021**, *32*, 2107143. [CrossRef]

121. Wang, H.L.; Guo, Z.H.; Zhu, G.; Pu, X.; Wang, Z.L. Boosting the Power and Lowering the Impedance of Triboelectric Nanogenerators through Manipulating the Permittivity for Wearable Energy Harvesting. *ACS Nano* **2021**, *15*, 7513–7521. [CrossRef] [PubMed]

122. Lv, S.; Zhang, X.; Huang, T.; Yu, H.; Zhang, Q.; Zhu, M. Trap Distribution and Conductivity Synergic Optimization of High-Performance Triboelectric Nanogenerators for Self-Powered Devices. *ACS Appl. Mater. Interfaces* **2021**, *13*, 2566–2575. [CrossRef]

123. Liu, Y.; Ping, J.; Ying, Y. Recent Progress in 2D-Nanomaterial-Based Triboelectric Nanogenerators. *Adv. Funct. Mater.* **2021**, *31*, 2009994. [CrossRef]

124. Cui, N.; Gu, L.; Lei, Y.; Liu, J.; Qin, Y.; Ma, X.; Hao, Y.; Wang, Z.L. Dynamic Behavior of the Triboelectric Charges and Structural Optimization of the Friction Layer for a Triboelectric Nanogenerator. *ACS Nano* **2016**, *10*, 6131–6138. [CrossRef]

125. Zhang, B.; Tian, G.; Xiong, D.; Yang, T.; Chun, F.; Zhong, S.; Lin, Z.; Li, W.; Yang, W. Understanding the Percolation Effect in Triboelectric Nanogenerator with Conductive Intermediate Layer. *Research* **2021**, *2021*, 7189376. [CrossRef] [PubMed]

126. Cao, W.-T.; Ouyang, H.; Xin, W.; Chao, S.; Ma, C.; Li, Z.; Chen, F.; Ma, M.-G. A Stretchable Highoutput Triboelectric Nanogenerator Improved by MXene Liquid Electrode with High Electronegativity. *Adv. Funct. Mater.* **2020**, *30*, 2004181. [CrossRef]

127. Gao, L.; Li, C.; Huang, W.; Mei, S.; Lin, H.; Ou, Q.; Zhang, Y.; Guo, J.; Zhang, F.; Xu, S.; et al. MXene/Polymer Membranes: Synthesis, Properties, and Emerging Applications. *Chem. Mater.* **2020**, *32*, 1703–1747. [CrossRef]

128. Chun, J.; Ye, B.U.; Lee, J.W.; Choi, D.; Kang, C.-Y.; Kim, S.-W.; Wang, Z.L.; Baik, J.M. Boosted output performance of triboelectric nanogenerator via electric double layer effect. *Nat. Commun.* **2016**, *7*, 12985. [CrossRef]

129. Dong, K.; Peng, X.; Cheng, R.; Ning, C.; Jiang, Y.; Zhang, Y.; Wang, Z.L. Advances in High-Performance Autonomous Energy and Self-Powered Sensing Textiles with Novel 3D Fabric Structures. *Adv. Mater.* **2022**, 2109355. [CrossRef]

130. Dong, K.; Peng, X.; An, J.; Wang, A.C.; Luo, J.; Sun, B.; Wang, J.; Wang, Z.L. Shape adaptable and highly resilient 3D braided triboelectric nanogenerators as e-textiles for power and sensing. *Nat. Commun.* **2020**, *11*, 2868. [CrossRef]

131. Cheng, L.; Xu, Q.; Zheng, Y.; Jia, X.; Qin, Y. A self-improving triboelectric nanogenerator with improved charge density and increased charge accumulation speed. *Nat. Commun.* **2018**, *9*, 3773. [CrossRef] [PubMed]

132. Xu, L.; Bu, T.Z.; Yang, X.D.; Zhang, C.; Wang, Z.L. Ultrahigh charge density realized by charge pumping at ambient conditions for triboelectric nanogenerators. *Nano Energy* **2018**, *49*, 625–633. [CrossRef]

133. Bai, Y.; Xu, L.; Lin, S.; Luo, J.; Qin, H.; Han, K.; Wang, Z.L. Charge Pumping Strategy for Rotation and Sliding Type Triboelectric Nanogenerators. *Adv. Energy Mater.* **2020**, *10*, 2000605. [CrossRef]
134. Wang, H.; Xu, L.; Bai, Y.; Wang, Z.L. Pumping up the charge density of a triboelectric nanogenerator by charge-shuttling. *Nat. Commun.* **2020**, *11*, 4203. [CrossRef] [PubMed]
135. Liu, Y.; Liu, W.; Wang, Z.; He, W.; Tang, Q.; Xi, Y.; Wang, X.; Guo, H.; Hu, C. Quantifying contact status and the air-breakdown model of charge-excitation triboelectric nanogenerators to maximize charge density. *Nat. Commun.* **2020**, *11*, 1599. [CrossRef]
136. Liu, W.; Wang, Z.; Wang, G.; Liu, G.; Chen, J.; Pu, X.; Xi, Y.; Wang, X.; Guo, H.; Hu, C.; et al. Integrated charge excitation triboelectric nanogenerator. *Nat. Commun.* **2019**, *10*, 1426. [CrossRef]
137. Long, L.; Liu, W.; Wang, Z.; He, W.; Li, G.; Tang, Q.; Guo, H.; Pu, X.; Liu, Y.; Hu, C. High performance floating self-excited sliding triboelectric nanogenerator for micro mechanical energy harvesting. *Nat. Commun.* **2021**, *12*, 4689. [CrossRef]
138. Song, Y.; Wang, N.; Wang, Y.; Zhang, R.; Olin, H.; Yang, Y. Direct Current Triboelectric Nanogenerators. *Adv. Energy Mater.* **2020**, *10*, 2002756. [CrossRef]
139. Chen, C.; Guo, H.; Chen, L.; Wang, Y.-C.; Pu, X.; Yu, W.; Wang, F.; Du, Z.; Wang, Z.L. Direct Current Fabric Triboelectric Nanogenerator for Biomotion Energy Harvesting. *ACS Nano* **2020**, *14*, 4585–4594. [CrossRef]
140. Cheng, R.; Dong, K.; Chen, P.; Ning, C.; Peng, X.; Zhang, Y.; Liu, D.; Wang, Z.L. High output direct-current power fabrics based on the air breakdown effect. *Energy Environ. Sci.* **2021**, *14*, 2460–2471. [CrossRef]
141. Meng, J.; Guo, Z.H.; Pan, C.; Wang, L.; Pan, C.; Li, L.; Pu, X.; Wang, Z.L. Flexible Textile Direct-Current Generator Based on the Tribovoltaic Effect at Dynamic Metal-Semiconducting Polymer Interfaces. *ACS Energy Lett.* **2021**, *6*, 2442–2450. [CrossRef]
142. Liu, D.; Yin, X.; Guo, H.; Zhou, L.; Li, X.; Zhang, C.; Wang, J.; Wang Zhong, L. A constant current triboelectric nanogenerator arising from electrostatic breakdown. *Sci. Adv.* **2019**, *5*, eaav6437. [CrossRef] [PubMed]
143. Dharmasena, R.D.I.G.; Cronin, H.M.; Dorey, R.A.; Silva, S.R.P. Direct current contact-mode triboelectric nanogenerators via systematic phase shifting. *Nano Energy* **2020**, *75*, 104887. [CrossRef]
144. Lei, R.; Shi, Y.; Ding, Y.; Nie, J.; Li, S.; Wang, F.; Zhai, H.; Chen, X.; Wang, Z.L. Sustainable high-voltage source based on triboelectric nanogenerator with a charge accumulation strategy. *Energy Environ. Sci.* **2020**, *13*, 2178–2190. [CrossRef]
145. Luo, J.; Xu, L.; Tang, W.; Jiang, T.; Fan, F.R.; Pang, Y.; Chen, L.; Zhang, Y.; Wang, Z.L. Direct-Current Triboelectric Nanogenerator Realized by Air Breakdown Induced Ionized Air Channel. *Adv. Energy Mater.* **2018**, *8*, 1800889. [CrossRef]
146. Cheng, R.; Dong, K.; Liu, L.; Ning, C.; Chen, P.; Peng, X.; Liu, D.; Wang, Z.L. Flame-Retardant Textile-Based Triboelectric Nanogenerators for Fire Protection Applications. *ACS Nano* **2020**, *14*, 15853–15863. [CrossRef]
147. Kim, T.; Kim, D.Y.; Yun, J.; Kim, B.; Lee, S.H.; Kim, D.; Lee, S. Direct-current triboelectric nanogenerator via water electrification and phase control. *Nano Energy* **2018**, *52*, 95–104. [CrossRef]
148. Ryu, H.; Lee, J.H.; Khan, U.; Kwak, S.S.; Hinchet, R.; Kim, S.-W. Sustainable direct current powering a triboelectric nanogenerator via a novel asymmetrical design. *Energy Environ. Sci.* **2018**, *11*, 2057–2063. [CrossRef]
149. Guo, H.; Chen, J.; Wang, L.; Wang, A.C.; Li, Y.; An, C.; He, J.-H.; Hu, C.; Hsiao, V.K.S.; Wang, Z.L. A highly efficient triboelectric negative air ion generator. *Nat. Sustain.* **2020**, *4*, 147–153. [CrossRef]
150. Wang, Z.; Liu, W.; He, W.; Guo, H.; Long, L.; Xi, Y.; Wang, X.; Liu, A.; Hu, C. Ultrahigh Electricity Generation from Low-Frequency Mechanical Energy by Efficient Energy Management. *Joule* **2021**, *5*, 441–455. [CrossRef]
151. Yang, Y.; Zhang, H.; Wang, Z.L. Direct-Current Triboelectric Generator. *Adv. Funct. Mater.* **2014**, *24*, 3745–3750. [CrossRef]
152. Zhou, Z.; Padgett, S.; Cai, Z.; Conta, G.; Wu, Y.; He, Q.; Zhang, S.; Sun, C.; Liu, J.; Fan, E.; et al. Single-layered ultra-soft washable smart textiles for all-around ballistocardiograph, respiration, and posture monitoring during sleep. *Biosens. Bioelectron.* **2020**, *155*, 112064. [CrossRef]
153. Nguyen, V.; Zhu, R.; Yang, R. Environmental effects on nanogenerators. *Nano Energy* **2015**, *14*, 49–61. [CrossRef]
154. Nguyen, V.; Yang, R. Effect of humidity and pressure on the triboelectric nanogenerator. *Nano Energy* **2013**, *2*, 604–608. [CrossRef]
155. Wen, X.; Su, Y.; Yang, Y.; Zhang, H.; Wang, Z.L. Applicability of triboelectric generator over a wide range of temperature. *Nano Energy* **2014**, *4*, 150–156. [CrossRef]
156. Lin, S.; Xu, L.; Xu, C.; Chen, X.; Wang, A.C.; Zhang, B.; Lin, P.; Yang, Y.; Zhao, H.; Wang, Z.L. Electron transfer in nanoscale contact electrification: Effect of temperature in the metal–dielectric case. *Adv. Mater.* **2019**, *31*, 1808197. [CrossRef]
157. Wang, J.; Wu, C.; Dai, Y.; Zhao, Z.; Wang, A.; Zhang, T.; Wang, Z.L. Achieving ultrahigh triboelectric charge density for efficient energy harvesting. *Nat. Commun.* **2017**, *8*, 88. [CrossRef]
158. Zhao, Z.; Dai, Y.; Liu, D.; Zhou, L.; Li, S.; Wang, Z.L.; Wang, J. Rationally patterned electrode of direct-current triboelectric nanogenerators for ultrahigh effective surface charge density. *Nat. Commun.* **2020**, *11*, 6186. [CrossRef]
159. Fu, J.; Xu, G.; Li, C.; Xia, X.; Guan, D.; Li, J.; Huang, Z.; Zi, Y. Achieving Ultrahigh Output Energy Density of Triboelectric Nanogenerators in High-Pressure Gas Environment. *Adv. Sci.* **2020**, *7*, 2001757. [CrossRef]
160. Liu, D.; Zhou, L.; Li, S.; Zhao, Z.; Yin, X.; Yi, Z.; Zhang, C.; Li, X.; Wang, J.; Wang, Z.L. Hugely Enhanced Output Power of Direct-Current Triboelectric Nanogenerators by Using Electrostatic Breakdown Effect. *Adv. Mater. Technol.* **2020**, *5*, 2000289. [CrossRef]
161. Yi, Z.; Liu, D.; Zhou, L.; Li, S.; Zhao, Z.; Li, X.; Wang, Z.L.; Wang, J. Enhancing output performance of direct-current triboelectric nanogenerator under controlled atmosphere. *Nano Energy* **2021**, *84*, 105864. [CrossRef]
162. Bhatia, D.; Lee, J.; Hwang, H.J.; Baik, J.M.; Kim, S.; Choi, D. Design of Mechanical Frequency Regulator for Predictable Uniform Power from Triboelectric Nanogenerators. *Adv. Energy Mater.* **2018**, *8*, 105864. [CrossRef]

163. Cheng, X.; Miao, L.; Song, Y.; Su, Z.; Chen, H.; Chen, X.; Zhang, J.; Zhang, H. High efficiency power management and charge boosting strategy for a triboelectric nanogenerator. *Nano Energy* **2017**, *38*, 438–446. [CrossRef]
164. Niu, S.; Wang, X.; Yi, F.; Zhou, Y.S.; Wang, Z.L. A universal self-charging system driven by random biomechanical energy for sustainable operation of mobile electronics. *Nat. Commun.* **2015**, *6*, 8975. [CrossRef] [PubMed]
165. Zhu, G.; Chen, J.; Zhang, T.; Jing, Q.; Wang, Z.L. Radial-arrayed rotary electrification for high performance triboelectric generator. *Nat. Commun.* **2014**, *5*, 3426. [CrossRef] [PubMed]
166. Yang, J.; Yang, F.; Zhao, L.; Shang, W.; Qin, H.; Wang, S.; Jiang, X.; Cheng, G.; Du, Z. Managing and optimizing the output performances of a triboelectric nanogenerator by a self-powered electrostatic vibrator switch. *Nano Energy* **2018**, *46*, 220–228. [CrossRef]
167. Pu, X.; Liu, M.; Li, L.; Zhang, C.; Pang, Y.; Jiang, C.; Shao, L.; Hu, W.; Wang, Z.L. Efficient Charging of Li-Ion Batteries with Pulsed Output Current of Triboelectric Nanogenerators. *Adv. Sci.* **2016**, *3*, 1500255. [CrossRef] [PubMed]
168. Cao, Z.; Wang, S.; Bi, M.; Wu, Z.; Ye, X. Largely enhancing the output power and charging efficiency of electret generators using position-based auto-switch and passive power management module. *Nano Energy* **2019**, *66*, 104202. [CrossRef]
169. Harmon, W.; Bamgboje, D.; Guo, H.; Hu, T.; Wang, Z.L. Self-driven power management system for triboelectric nanogenerators. *Nano Energy* **2020**, *71*, 104642. [CrossRef]
170. Yu, A.; Pu, X.; Wen, R.; Liu, M.; Zhou, T.; Zhang, K.; Zhang, Y.; Zhai, J.; Hu, W.; Wang, Z.L. Core-Shell-Yarn-Based Triboelectric Nanogenerator Textiles as Power Cloths. *ACS Nano* **2017**, *11*, 12764–12771. [CrossRef]
171. Zhang, H.; Marty, F.; Xia, X.; Zi, Y.; Bourouina, T.; Galayko, D.; Basset, P. Employing a MEMS plasma switch for conditioning high-voltage kinetic energy harvesters. *Nat. Commun.* **2020**, *11*, 3221. [CrossRef] [PubMed]
172. Tang, W.; Zhou, T.; Zhang, C.; Ru Fan, F.; Bao Han, C.; Lin Wang, Z. A power-transformed-and-managed triboelectric nanogenerator and its applications in a self-powered wireless sensing node. *Nanotechnology* **2014**, *25*, 225402. [CrossRef] [PubMed]
173. Liu, W.; Wang, Z.; Wang, G.; Zeng, Q.; He, W.; Liu, L.; Wang, X.; Xi, Y.; Guo, H.; Hu, C.; et al. Switched-capacitor-convertors based on fractal design for output power management of triboelectric nanogenerator. *Nat. Commun.* **2020**, *11*, 1883. [CrossRef] [PubMed]
174. Wang, Z.; Liu, W.; Hu, J.; He, W.; Yang, H.; Ling, C.; Xi, Y.; Wang, X.; Liu, A.; Hu, C. Two voltages in contact-separation triboelectric nanogenerator: From asymmetry to symmetry for maximum output. *Nano Energy* **2020**, *69*, 104452. [CrossRef]
175. Xu, W.; Wong, M.-C.; Hao, J. Strategies and progress on improving robustness and reliability of triboelectric nanogenerators. *Nano Energy* **2019**, *55*, 203–215. [CrossRef]

nanoenergy advances

Review

Recent Advances in Lubricant-Based Triboelectric Nanogenerators for Enhancing Mechanical Lifespan and Electrical Output

Seh-Hoon Chung [1], Jihoon Chung [2,*] and Sangmin Lee [1,*]

[1] School of Mechanical Engineering, Chung-Ang University, 84 Heukseok-ro, Dongjak-gu, Seoul 06974, Korea; sehhoon1010@cau.ac.kr

[2] Department of Chemical & Biomolecular Engineering, University of California, Berkeley, CA 94720, USA

* Correspondence: jihoon05@berkeley.edu (J.C.); slee98@cau.ac.kr (S.L.)

Abstract: A triboelectric nanogenerator (TENG) is a noteworthy mechanical energy harvester that can convert mechanical energy into electricity by combining triboelectrification and electrostatic induction. However, owing to the nature of its working mechanism, TENGs have critical limitations in mechanical and electrical aspects, which prevent them from being utilized as primary power sources. To overcome these limitations, several studies are turning their attention to utilizing lubricants, which is a traditional method recently applied to TENGs. In this review, we introduce recent advances in lubricant-based TENGs that can effectively enhance their electrical output and mechanical lifespan. In addition, this review provides an overview of lubricant-based TENGs. We hope that, through this review, researchers who are trying to overcome mechanical and electrical limitations to expand the applications of TENGs in industries will be introduced to the use of lubricant materials.

Keywords: triboelectric nanogenerator; energy harvesting; lubricant liquid; output enhancement; mechanical lifespan

Citation: Chung, S.-H.; Chung, J.; Lee, S. Recent Advances in Lubricant-Based Triboelectric Nanogenerators for Enhancing Mechanical Lifespan and Electrical Output. *Nanoenergy Adv.* **2022**, 2, 210–221. https://doi.org/10.3390/nanoenergyadv2020009

Academic Editors: Ya Yang and Zhong Lin Wang

Received: 18 April 2022
Accepted: 13 May 2022
Published: 19 May 2022

Publisher's Note: MDPI stays neutral with regard to jurisdictional claims in published maps and institutional affiliations.

Copyright: © 2022 by the authors. Licensee MDPI, Basel, Switzerland. This article is an open access article distributed under the terms and conditions of the Creative Commons Attribution (CC BY) license (https://creativecommons.org/licenses/by/4.0/).

1. Introduction

As the number of portable electronics and Internet of Things (IoT) devices has increased drastically, the need for on-site energy generation has been in the spotlight to power these devices individually and extend their battery life. Energy-harvesting technologies can convert ambient energy, such as solar [1–3], wind [4–8], wave [9–12], and radio frequency [13–15], into electricity that can provide sufficient energy for small electronic devices. Among these energy-harvesting technologies, harvesting mechanical energy has great potential to effectively power portable and small electronics because it is not affected by external environment such as weather conditions. In order to harvest mechanical motion, energy harvesters that utilizes electromagnetic, piezoelectric effect has been utilized [16,17]. Triboelectric nanogenerators (TENGs) are one of mechanical energy harvesters that can generate electricity by combining triboelectrification and electrostatic induction [18–23]. Owing to their light weight [24], high electrical output [25,26], and availability of raw materials, recent research has focused on developing various TENG designs and structures to effectively collect energy from multiple mechanical sources, such as vibration energy [27–30] and rotation energy [31–34]. Currently, TENGs charge commercial electrical devices and lithium-ion batteries used in building self-powered systems [35–38] and are enhanced for utilization in industrial applications [39].

However, the working mechanisms of TENGs are critically limited in their mechanical and electrical aspects, restricting the use of TENGs as primary power sources. From a mechanical perspective, TENGs have two materials placed in contact to cause triboelectricity, leading to frictional wear. Frictional wear is one of the constant limitations of TENGs because they generate electrical output through mechanical input [40]. Due to the nature of triboelectricity, surface friction from two or more material is necessary to generate surface charge and requires consistent or periodical contact due to degradation of surface charge over time. Particularly, various TENGs utilizing micro-/nano-structures and polymer dielectric materials for surface charge enhancements are more vulnerable to frictional wear owing to their poor mechanical properties [41,42]. Frictional damage on a TENG surface leads to a decrease in electrical output due to surface structure failure as well as device failure when mechanical damage is accumulated.

An electrical limitation is imposed when the surface charge potential of the material is higher than the breakdown potential of air, as the surface charge is lost through field emission and ionization of air [43–45]. Since more and more studies report higher surface charge materials with surface micro and nanostructures, this electrical limitation is becoming restriction factor for developing high power TENG [46,47]. With the introduction of the upper limitation of surface charge due to air breakdown, there have been studies in favor and opposing the use of this phenomenon to enhance the electrical output of TENGs [48–50]. However, the cause of this limitation remains unresolved, as the primary potential difference of TENGs is governed by the surface charge, which is still limited by air breakdown. Therefore, to overcome these limitations, several studies are turning their attention to utilizing lubricants, a traditional method recently applied to TENGs, to overcome both limitations.

In this review, we highlight recent advances in lubricant-based TENGs that can effectively enhance both the electrical output and mechanical lifespan of TENGs. The review will focus on overviewing lubricant-based TENGs, their working mechanisms, their various designs, and an output comparison with conventional TENGs. Especially, the reason for lubricant based TENGs can overcome mechanical and electrical limitation of TENG, and experimental data from various studies will be discussed. Finally, the different perspectives regarding lubricant-based TENGs, and challenges associated with their use are discussed.

2. Lubricant-Based TENGs

When mechanical input is applied to a conventional TENG, two dielectric materials come into contact and cause triboelectrification [51–57]. The surface charge generated by the friction between the two dielectric materials is then transferred to the electrodes through electrostatic induction [58–63]. Contact between the dielectric materials is inevitable during this process, leading to frictional wear. The frictional wear gradually decreases the output of TENG by damaging the surface structure fabricated on the surface and eventually fails after the mechanical damage is accumulated. Therefore, utilizing lubricants between the frictional surfaces is an effective approach to reducing the frictional force. Moreover, lubricating oils such as transformer oil and vegetable oil are also called as dielectric liquid which can be used as insulating material by its high breakdown voltage [64]. As the surface charge of dielectric materials in TENGs can be released to the atmosphere, utilizing the lubricant at TENGs can increase the electrical output. By these effects, recently, various lubricant materials combined with mechanical designs have been introduced to overcome the mechanical limitations of TENGs (Figure 1). As sliding motion-based TENGs are more vulnerable to friction failure, lubricant materials were more actively utilized in these than in other TENGs, such as horizontal contact-separation modes.

As shown in Figure 1a, a lubricating liquid can be applied to a sliding motion-based TENG system [65]. The lubricant liquid on the TENG surface forms a thin layer that can effectively decrease the frictional force between the electrode and polytetrafluoroethylene (PTFE) surface. The main electrical potential difference is generated by friction between the PTFE and polystyrene (PS) surfaces, and the TENG generates an amplified current of over 1 mA because the dielectric liquid acts as a switch during operation. Furthermore, field emission occurs when the PTFE contacts the electrode surface, and electrons can flow directly from the PTFE surface to the electrode. Hence, the lubricant liquid enhanced the electrical output of the TENG by suppressing air breakdown. Various TENGs utilizing other lubricant liquids, such as oleic acid, have been introduced (Figure 1b). In a previous study, oleic acid and PS were dissolved in N,N-dimethylformamide, and then spin-coated

lubricant liquid on the surface. The transferred charge and current of the lubricated TENG were more than twice compared with those of the TENG operated in air. As mentioned in the previous paragraphs, this is result from combination of suppressing air breakdown and increasing the Debye length through using non-polar liquid lubricant.

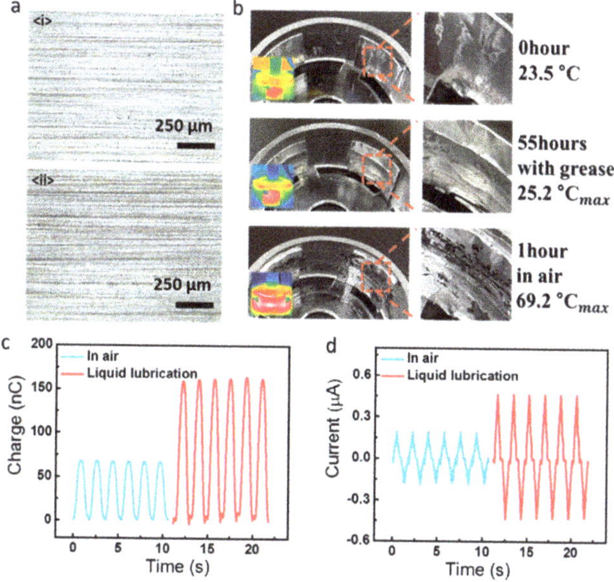

Figure 3. Mechanical and electrical performance of lubricant-based TENGs: (**a**) microscopic photograph of electrode surface <i> before and <ii> after 72 h of continuous operation of the TENG (Reprinted with permission from ref. [69], 2021, Wiley); (**b**) photograph of the electrode surface and the thermal image of TENG during operation (Reprinted with permission from ref. [70]. 2022, Elsevier); (**c**) transferred charge; (**d**) current output of TENG with and without liquid lubrication (Reprinted with permission from ref. [67]. 2020, Wiley).

To further enhance the mechanical lifespan and electrical output, future studies are required to optimize the lubricants specialized for TENG applications. One of the important steps for optimizing lubricant materials is a quantitative study of various liquid lubricants. Many studies are continuing this effort to provide guideline for selecting appropriate lubricant materials to be utilized in various applications. As shown in Figure 4a,b, a recent study showed that liquid lubricants such as squalene, paraffin oil, and PAO 10 have higher electrical outputs than TENG operated under dry conditions [83]. In this work, TENG operating with liquids such as olive oil, rapeseed oil, plurial A 500 PE, PEG 200, water have shown considerably lower electrical output. This study also reported that the relative permittivity and viscosity of lubricant is the key factor to increase output of TENG according to the experimental result. In addition, in other studies, lubricant liquids such as mineral oil and silicone oil show high electrical output, whereas castor oil, water, and ethanol show relatively low output (Figure 4c,d). Overall, various studies have shown that synthetic non-polar liquids such as squalene, mineral oil, silicone oil, and hexadecane have a higher electrical output, whereas polar liquids such as ethylene glycol, water, and ethanol tend to show low electrical output. As shown in Figure 2, the polar liquids screen the surface charge and lead to a decrease in the output. In addition, considering that organic oils such as rapeseed, olive, and castor oils are mixtures of various compounds, they contain polar molecules such as glycerol that would lower the electrical output of TENGs. As shown in Figure 4e–h, a study on the electrical output and friction coefficient depending on PAO 4, perfluoropolyether (PFPE), glycerol, and ethanol, respectively. As shown in the plots, the electrical output increases and friction coefficient decreases when using a non-polar liquid such as PAO 4 and PFPE, whereas the electrical output decreases and friction coefficient increases when using polar liquid such as glycerol and ethanol. Considering that there are vast number of synthetic and natural oils are used for lubrication, there must be further studies on these materials as well as effect of these lubrication materials when lubrication materials are used for a longer extension of time.

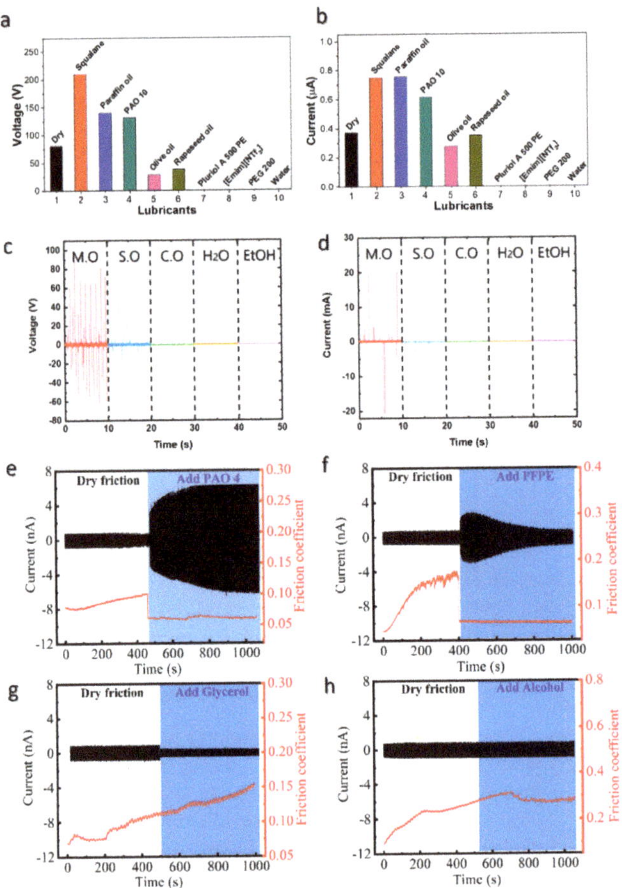

Figure 4. Electrical output depending on various liquids applied on the surface of TENGs: (**a**,**c**) Voltage and (**b**,**d**) current output of TENGs depending on various liquids (Reprinted with permission from ref. [69]. 2021, Wiley. And reprinted with permission from ref. [83]. 2021, Elsevier). Measured current and friction coefficient when (**e**) PAO 4, (**f**) PFPE, (**g**) glycerol, and (**h**) alcohol was applied on the surface of TENGSs.

5. Summary and Perspectives

This review introduces the current strategies and an overview of lubrication-based TENGs. As the air breakdown effect limits the electrical performance and induces frictional wear affecting the mechanical lifespan of TENGs, the use of lubricants has been actively studied to overcome these limitations. These studies have shown that a lubricant liquid applied to the TENG surface can effectively increase the mechanical lifespan and electrical output by lowering the friction coefficient and suppressing air breakdown. Previous studies have shown working mechanism of lubricant-based TENGs can generate high electrical output through suppressing air breakdown and non-polar liquid with high Debye length, which would effectively transfer the surface charge through electrostatic induction. Various mechanical designs and working mechanisms have been developed to further decrease the friction force and enhance the electrical output. For further enhancement of lubricant-based TENGs, optimization of lubricant liquids and mechanical components should be considered at the design level, and additional quantitative studies are required as follows:

(1) Optimization of lubricant-based TENG design to withhold more lubricant on the surface during operation and advanced surface design considering lubrication at the design level;

(2) Further quantitative analysis of the relationship between TENGs and lubricant liquids, especially on the EDL formation of lubricant liquids under high surface charge conditions;

(3) Quantitative analysis of various lubricant liquids affecting mechanical lifespan and electrical output of TENG, including commercial synthetic and organic oil;

(4) Long-term influence of lubricant liquids on TENG surfaces and the effect of long-term operation on polymer surfaces.

We hope that, through this review, researchers who are trying to overcome mechanical and electrical limitations for expanding the applications of TENGs in industries will be introduced to the use of lubricant materials. We believe that constant research efforts and innovations in lubricant-based TENG have great potential for utilizing TENGs as a primary energy source for existing electronics.

Author Contributions: Conceptualization, S.-H.C., J.C. and S.L.; writing—original draft preparation, S.-H.C.; writing—review and editing, J.C. and S.L.; supervision, J.C. and S.L.; project administration, S.L.; funding acquisition, J.C. All authors have read and agreed to the published version of the manuscript.

Funding: This research was supported by Basic Science Research Program through the National Research Foundation of Korea(NRF) funded by the Ministry of Education(2021R1A6A3A03040052) and the National Research Foundation of Korea(NRF) grant funded by the Korea government(MSIT)(No. 2021R1A4A3030268).

Conflicts of Interest: The authors declare no conflict of interest.

References

1. Chalasani, S.; Conrad, J.M. A survey of energy harvesting sources for embedded systems. In Proceedings of the 2008 IEEE Southeastcon, Huntsville, AL, USA, 3–6 April 2008; pp. 442–447. [CrossRef]

2. Raghunathan, V.; Kansal, A.; Hsu, J.; Friedman, J.; Srivastava, M. Design considerations for solar energy harvesting wireless embedded systems. In Proceedings of the IPSN 2005, Fourth International Symposium on Information Processing in Sensor Networks, Boise, ID, USA, 15 April 2005; pp. 457–462.

3. Abdin, Z.; Alim, M.; Saidur, R.; Islam, M.; Rashmi, W.; Mekhilef, S.; Wadi, A. Solar energy harvesting with the application of nanotechnology. *Renew. Sustain. Energy Rev.* **2013**, *26*, 837–852. [CrossRef]

4. Li, S.; Yuan, A.J.; Lipson, H. Ambient wind energy harvesting using cross-flow fluttering. *J. Appl. Phys.* **2011**, *109*, 026104. [CrossRef]

5. Tan, Y.K.; Panda, S.K. Optimized wind energy harvesting system using resistance emulator and active rectifier for wireless sensor nodes. *IEEE Trans. Power Electron.* **2010**, *26*, 38–50.

6. Tan, Y.K.; Panda, S.K. Measurement. Self-autonomous wireless sensor nodes with wind energy harvesting for remote sensing of wind-driven wildfire spread. *IEEE Trans. Instrum. Meas.* **2011**, *60*, 1367–1377. [CrossRef]

7. Lee, J.-S.; Yong, H.; Choi, Y.I.; Ryu, J.; Lee, S. Stackable Disk-Shaped Triboelectric Nanogenerator to Generate Energy from Omnidirectional Wind. *Int. J. Precis. Eng. Manuf. Technol.* **2021**, *9*, 557–565. [CrossRef]

8. Yong, H.; Chung, J.; Choi, D.; Jung, D.; Cho, M.; Lee, S. Highly reliable wind-rolling triboelectric nanogenerator operating in a wide wind speed range. *Sci. Rep.* **2016**, *6*, 33977. [CrossRef]

9. Carrara, M.; Cacan, M.R.; Toussaint, J.; Leamy, M.J.; Ruzzene, M.; Erturk, A. Metamaterial-inspired structures and concepts for elastoacoustic wave energy harvesting. *Smart Mater. Struct.* **2013**, *22*, 065004. [CrossRef]

10. Khan, T.A.; Alkhateeb, A.; Heath, R.W. Millimeter Wave Energy Harvesting. *IEEE Trans. Wirel. Commun.* **2016**, *15*, 6048–6062. [CrossRef]

11. Younesian, D.; Alam, M.-R. Multi-stable mechanisms for high-efficiency and broadband ocean wave energy harvesting. *Appl. Energy* **2017**, *197*, 292–302. [CrossRef]
12. Chung, J.; Heo, D.; Shin, G.; Chung, S.-H.; Hong, J.; Lee, S. Water behavior based electric generation via charge separation. *Nano Energy* **2020**, *82*, 105687. [CrossRef]
13. Le, T.; Mayaram, K.; Fiez, T. Efficient Far-Field Radio Frequency Energy Harvesting for Passively Powered Sensor Networks. *IEEE J. Solid-State Circuits* **2008**, *43*, 1287–1302. [CrossRef]
14. Masotti, D.; Costanzo, A.; Del Prete, M.; Rizzoli, V. Genetic-based design of a tetra-band high-efficiency radio-frequency energy harvesting system. *IET Microw. Antennas Propag.* **2013**, *7*, 1254–1263. [CrossRef]
15. Zungeru, A.M.; Ang, L.-M.; Prabaharan, S.; Seng, K.P. Radio Frequency Energy Harvesting and Management for Wireless Sensor Networks. In *Green Mobile Devices and Networks: Energy Optimization and Scavenging Techniques*; CRC Press: Boca Raton, FL, USA, 2012; pp. 341–368.
16. Beeby, S.P.; Torah, R.N.; Tudor, M.J.; Glynne-Jones, P.; O'Donnell, T.; Saha, C.R.; Roy, S. A micro electromagnetic generator for vibration energy harvesting. *J. Micromech. Microeng.* **2019**, *17*, 1257–1265. [CrossRef]
17. Lefeuvre, E.; Badel, A.; Richard, C.; Petit, L.; Guyomar, D. A comparison between several vibration-powered piezoelectric generators for standalone systems. *Sens. Actuators A Phys.* **2006**, *126*, 405–416. [CrossRef]
18. Zhu, G.; Pan, C.; Guo, W.; Chen, C.-Y.; Zhou, Y.; Yu, R.; Wang, Z.L. Triboelectric-Generator-Driven Pulse Electrodeposition for Micropatterning. *Nano Lett.* **2012**, *12*, 4960–4965. [CrossRef] [PubMed]
19. Wang, S.; Lin, L.; Wang, Z.L. Nanoscale Triboelectric-Effect-Enabled Energy Conversion for Sustainably Powering Portable Electronics. *Nano Lett.* **2012**, *12*, 6339–6346. [CrossRef]
20. Jang, S.; Joung, Y.; Kim, H.; Cho, S.; Ra, Y.; Kim, M.; Ahn, D.; Lin, Z.-H.; Choi, D. Charge transfer accelerating strategy for improving sensitivity of droplet based triboelectric sensors via heterogeneous wettability. *Nano Energy* **2022**, *97*, 107213. [CrossRef]
21. Kim, M.; Ra, Y.; Cho, S.; Jang, S.; Kam, D.; Yun, Y.; Kim, H.; Choi, D. Geometric gradient assisted control of the triboelectric effect in a smart brake system for self-powered mechanical abrasion monitoring. *Nano Energy* **2021**, *89*, 106448. [CrossRef]
22. Chung, J.; Cho, H.; Yong, H.; Heo, D.; Rim, Y.S.; Lee, S. Versatile surface for solid–solid/liquid–solid triboelectric nanogenerator based on fluorocarbon liquid infused surfaces. *Sci. Technol. Adv. Mater.* **2020**, *21*, 139–146. [CrossRef]
23. Wu, C.; Wang, A.; Ding, W.; Guo, H.; Wang, Z.L. Triboelectric Nanogenerator: A Foundation of the Energy for the New Era. *Adv. Energy Mater.* **2018**, *9*, 1802906. [CrossRef]
24. Ra, Y.; Oh, S.; Lee, J.; Yun, Y.; Cho, S.; Choi, J.H.; Jang, S.; Hwang, H.J.; Choi, D.; Kim, J.-G.; et al. Triboelectric signal generation and its versatile utilization during gear-based ordinary power transmission. *Nano Energy* **2020**, *73*, 104745. [CrossRef]
25. Liu, Y.; Liu, W.; Wang, Z.; He, W.; Tang, Q.; Xi, Y.; Wang, X.; Guo, H.; Hu, C. Quantifying contact status and the air-breakdown model of charge-excitation triboelectric nanogenerators to maximize charge density. *Nat. Commun.* **2020**, *11*, 1599. [CrossRef] [PubMed]
26. Chung, J.; Heo, D.; Shin, G.; Choi, D.; Choi, K.; Kim, D.; Lee, S. Ion-Enhanced Field Emission Triboelectric Nanogenerator. *Adv. Energy Mater.* **2019**, *9*, 1901731. [CrossRef]
27. Kim, D.; Chung, J.; Heo, D.; Chung, S.; Lee, G.; Hwang, P.T.J.; Kim, M.; Jung, H.; Jin, Y.; Hong, J.; et al. AC/DC Convertible Pillar-Type Triboelectric Nanogenerator with Output Current Amplified by the Design of the Moving Electrode. *Adv. Energy Mater.* **2022**, *12*, 2103571. [CrossRef]
28. Bhatia, D.; Lee, K.-S.; Niazi, M.U.K.; Park, H.-S. Triboelectric nanogenerator integrated origami gravity support device for shoulder rehabilitation using exercise gaming. *Nano Energy* **2022**, *97*, 107179. [CrossRef]
29. Hwang, H.J.; Kim, J.S.; Kim, W.; Park, H.; Bhatia, D.; Jee, E.; Chung, Y.S.; Kim, D.H.; Choi, D. An ultra-mechanosensitive visco-poroelastic polymer ion pump for continuous self-powering kinematic triboelectric nanogenerators. *Adv. Energy Mater.* **2019**, *9*, 1803786. [CrossRef]
30. Cha, K.; Chung, J.; Heo, D.; Song, M.; Chung, S.-H.; Hwang, P.T.; Kim, D.; Koo, B.; Hong, J.; Lee, S. Lightweight mobile stick-type water-based triboelectric nanogenerator with amplified current for portable safety devices. *Sci. Technol. Adv. Mater.* **2022**, *23*, 161–168. [CrossRef]
31. Son, J.-H.; Heo, D.; Song, Y.; Chung, J.; Kim, B.; Nam, W.; Hwang, P.T.; Kim, D.; Koo, B.; Hong, J.; et al. Highly reliable triboelectric bicycle tire as self-powered bicycle safety light and pressure sensor. *Nano Energy* **2021**, *93*, 106797. [CrossRef]
32. Chung, J.; Heo, D.; Cha, K.; Lin, Z.-H.; Hong, J.; Lee, S. A portable device for water-sloshing-based electricity generation based on charge separation and accumulation. *Iscience* **2021**, *24*, 102442. [CrossRef]
33. Chung, J.; Song, M.; Chung, S.-H.; Choi, W.; Lee, S.; Lin, Z.-H.; Hong, J.; Lee, S. Triangulated Cylinder Origami-Based Piezoelectric/Triboelectric Hybrid Generator to Harvest Coupled Axial and Rotational Motion. *Research* **2021**, *2021*, 1–9. [CrossRef]
34. Chung, S.-H.; Chung, J.; Kim, B.; Kim, S.; Lee, S. Screw Pump-Type Water Triboelectric Nanogenerator for Active Water Flow Control. *Adv. Eng. Mater.* **2020**, *23*, 2000758. [CrossRef]
35. Dharmasena, R.D.I.G.; Silva, S. Towards optimized triboelectric nanogenerators. *Nano Energy* **2019**, *62*, 530–549. [CrossRef]
36. Hwang, H.J.; Hong, H.; Cho, B.G.; Lee, H.K.; Kim, J.S.; Lee, U.J.; Kim, W.; Kim, H.; Chung, K.-B.; Choi, D. Band well structure with localized states for enhanced charge accumulation on Triboelectrification. *Nano Energy* **2021**, *90*, 106647. [CrossRef]
37. Choi, J.H.; Ra, Y.; Cho, S.; La, M.; Park, S.J.; Choi, D. Electrical charge storage effect in carbon based polymer composite for long-term performance enhancement of the triboelectric nanogenerator. *Compos. Sci. Technol.* **2021**, *207*, 108680. [CrossRef]

38. Heo, D.; Chung, J.; Shin, G.; Seok, M.; Lee, C.; Lee, S. Yo-Yo Inspired Triboelectric Nanogenerator. *Energies* **2021**, *14*, 1798. [CrossRef]
39. Pan, M.; Yuan, C.; Liang, X.; Zou, J.; Zhang, Y.; Bowen, C. Triboelectric and Piezoelectric Nanogenerators for Future Soft Robots and Machines. *Iscience* **2020**, *23*, 101682. [CrossRef]
40. Lin, L.; Wang, S.; Xie, Y.; Jing, Q.; Niu, S.; Hu, Y.; Wang, Z.L. Segmentally Structured Disk Triboelectric Nanogenerator for Harvesting Rotational Mechanical Energy. *Nano Lett.* **2013**, *13*, 2916–2923. [CrossRef]
41. Zhao, L.; Zheng, Q.; Ouyang, H.; Li, H.; Yan, L.; Shi, B.; Li, Z. A size-unlimited surface microstructure modification method for achieving high performance triboelectric nanogenerator. *Nano Energy* **2016**, *28*, 172–178. [CrossRef]
42. Xu, W.; Wong, M.C.; Hao, J. Strategies and progress on improving robustness and reliability of triboelectric nanogenerators. *Nano Energy* **2018**, *55*, 203–215. [CrossRef]
43. Yang, B.; Tao, X.-M.; Peng, Z. Upper limits for output performance of contact-mode triboelectric nanogenerator systems. *Nano Energy* **2018**, *57*, 66–73. [CrossRef]
44. Zi, Y.; Wu, C.; Ding, W.; Wang, Z.L. Maximized Effective Energy Output of Contact-Separation-Triggered Triboelectric Nanogenerators as Limited by Air Breakdown. *Adv. Funct. Mater.* **2017**, *27*, 1700049. [CrossRef]
45. Jiang, C.; Dai, K.; Yi, F.; Han, Y.; Wang, X.; You, Z. Optimization of triboelectric nanogenerator load characteristics considering the air breakdown effect. *Nano Energy* **2018**, *53*, 706–715. [CrossRef]
46. Huang, J.; Fu, X.; Liu, G.; Xu, S.; Li, X.; Zhang, C.; Jiang, L. Micro/nano-structures-enhanced triboelectric nanogenerators by femtosecond laser direct writing. *Nano Energy* **2019**, *62*, 638–644. [CrossRef]
47. Xi, Y.; Zhang, F.; Shi, Y. Effects of surface micro-structures on capacitances of the dielectric layer in triboelectric nanogenerator: A numerical simulation study. *Nano Energy* **2020**, *79*, 105432. [CrossRef]
48. Yi, Z.; Liu, D.; Zhou, L.; Li, S.; Zhao, Z.; Li, X.; Wang, Z.L.; Wang, J. Enhancing output performance of direct-current triboelectric nanogenerator under controlled atmosphere. *Nano Energy* **2021**, *84*, 105864. [CrossRef]
49. Wang, J.; Wu, C.; Dai, Y.; Zhao, Z.; Wang, A.; Zhang, T.; Wang, Z.L. Achieving ultrahigh triboelectric charge density for efficient energy harvesting. *Nat. Commun.* **2017**, *8*, 88. [CrossRef]
50. Seol, M.-L.; Han, J.-W.; Moon, D.-I.; Meyyappan, M. Triboelectric nanogenerator for Mars environment. *Nano Energy* **2017**, *39*, 238–244. [CrossRef]
51. Wang, S.; Niu, S.; Yang, J.; Lin, L.; Wang, Z.L. Quantitative Measurements of Vibration Amplitude Using a Contact-Mode Freestanding Triboelectric Nanogenerator. *ACS Nano* **2014**, *8*, 12004–12013. [CrossRef]
52. Wang, Z.L. From contact-electrification to triboelectric nanogenerators. *Rep. Prog. Phys.* **2021**, *84*, 096502. [CrossRef]
53. Niu, S.; Wang, S.; Lin, L.; Liu, Y.; Zhou, Y.S.; Hu, Y.; Wang, Z.L. Theoretical study of contact-mode triboelectric nanogenerators as an effective power source. *Energy Environ. Sci.* **2013**, *6*, 3576–3583. [CrossRef]
54. Cho, H.; Chung, J.; Shin, G.; Sim, J.-Y.; Kim, D.S.; Lee, S.; Hwang, W. Toward sustainable output generation of liquid–solid contact triboelectric nanogenerators: The role of hierarchical structures. *Nano Energy* **2018**, *56*, 56–64. [CrossRef]
55. Chung, J.; Heo, D.; Kim, B.; Lee, S. Superhydrophobic Water-Solid Contact Triboelectric Generator by Simple Spray-On Fabrication Method. *Micromachines* **2018**, *9*, 593. [CrossRef] [PubMed]
56. Kim, B.; Chung, J.; Moon, H.; Kim, D.; Lee, S. Elastic spiral triboelectric nanogenerator as a self-charging case for portable electronics. *Nano Energy* **2018**, *50*, 133–139. [CrossRef]
57. Heo, D.; Kim, T.; Yong, H.; Yoo, K.T.; Lee, S. Sustainable oscillating triboelectric nanogenerator as omnidirectional self-powered impact sensor. *Nano Energy* **2018**, *50*, 1–8. [CrossRef]
58. Niu, S.; Liu, Y.; Wang, S.; Lin, L.; Zhou, Y.S.; Hu, Y.; Wang, Z.L. Theory of Sliding-Mode Triboelectric Nanogenerators. *Adv. Mater.* **2013**, *25*, 6184–6193. [CrossRef]
59. Niu, S.; Liu, Y.; Wang, S.; Lin, L.; Zhou, Y.S.; Hu, Y.; Wang, Z.L. Theoretical Investigation and Structural Optimization of Single-Electrode Triboelectric Nanogenerators. *Adv. Funct. Mater.* **2014**, *24*, 3332–3340. [CrossRef]
60. Shao, J.; Willatzen, M.; Wang, Z.L. Theoretical modeling of triboelectric nanogenerators (TENGs). *J. Appl. Phys.* **2020**, *128*, 111101. [CrossRef]
61. Chung, J.; Yong, H.; Moon, H.; Van Duong, Q.; Choi, S.T.; Kim, D.; Lee, S. Hand-Driven Gyroscopic Hybrid Nanogenerator for Recharging Portable Devices. *Adv. Sci.* **2018**, *5*. [CrossRef]
62. Kim, T.; Kim, D.Y.; Yun, J.; Kim, B.; Lee, S.H.; Kim, D.; Lee, S.J.N.E. Direct-current triboelectric nanogenerator via water electrification and phase control. *Nano Energy* **2018**, *52*, 95–104. [CrossRef]
63. Chung, J.; Yong, H.; Moon, H.; Choi, S.T.; Bhatia, D.; Choi, D.; Kim, D.; Lee, S. Capacitor-Integrated Triboelectric Nanogenerator Based on Metal-Metal Contact for Current Amplification. *Adv. Energy Mater.* **2018**, *8*, 1703024. [CrossRef]
64. Fontes, D.H.; Ribatski, G.; Filho, E.P.B. Experimental evaluation of thermal conductivity, viscosity and breakdown voltage AC of nanofluids of carbon nanotubes and diamond in transformer oil. *Diam. Relat. Mater.* **2015**, *58*, 115–121. [CrossRef]
65. Chung, J.; Chung, S.-H.; Lin, Z.-H.; Jin, Y.; Hong, J.; Lee, S. Dielectric liquid-based self-operating switch triboelectric nanogenerator for current amplification via regulating air breakdown. *Nano Energy* **2021**, *88*, 106292. [CrossRef]
66. Zhang, J.; Zheng, Y.; Xu, L.; Wang, D. Oleic-acid enhanced triboelectric nanogenerator with high output performance and wear resistance. *Nano Energy* **2019**, *69*, 104435. [CrossRef]
67. Zhou, L.; Liu, D.; Zhao, Z.; Li, S.; Liu, Y.; Liu, L.; Gao, Y.; Wang, Z.L.; Wang, J. Simultaneously Enhancing Power Density and Durability of Sliding-Mode Triboelectric Nanogenerator via Interface Liquid Lubrication. *Adv. Energy Mater.* **2020**, *10*. [CrossRef]

68. Wang, K.; Li, J.; Li, J.; Wu, C.; Yi, S.; Liu, Y.; Luo, J. Hexadecane-containing sandwich structure based triboelectric nanogenerator with remarkable performance enhancement. *Nano Energy* **2021**, *87*, 106198. [CrossRef]
69. Chung, S.H.; Chung, J.; Song, M.; Kim, S.; Shin, D.; Lin, Z.H.; Koo, B.; Kim, D.; Hong, J.; Lee, S. Nonpolar Liquid Lubricant Submerged Triboelectric Nanogenerator for Current Amplification via Direct Electron Flow. *Adv. Energy Mater.* **2021**, *11*, 2100936. [CrossRef]
70. Song, M.; Chung, J.; Chung, S.-H.; Cha, K.; Heo, D.; Kim, S.; Hwang, P.T.; Kim, D.; Koo, B.; Hong, J.; et al. Semisolid-lubricant-based ball-bearing triboelectric nanogenerator for current amplification, enhanced mechanical lifespan, and thermal stabilization. *Nano Energy* **2021**, *93*, 106816. [CrossRef]
71. Zhu, D.; Hu, Y.-Z. The study of transition from elastohydrodynamic to mixed and boundary lubrication. The advancing frontier of engineering tribology. In Proceedings of the 1999 STLE/ASME HS Cheng Tribology Surveillance, New York, NY, USA; 1999; pp. 150–156.
72. Jin, Z.; Fisher, J. Tribology in Joint Replacement. In *Joint Replacement Technology*; Elsevier: Amsterdam, The Netherlands, 2014; pp. 31–61.
73. Höglund, E. Influence of lubricant properties on elastohydrodynamic lubrication. *Wear* **1999**, *232*, 176–184. [CrossRef]
74. Prata, A.T.; Barbosa, J.R., Jr. Role of the thermodynamics, heat transfer, and fluid mechanics of lubricant oil in hermetic reciprocating compressors. *Heat Transf. Eng.* **2009**, *30*, 533–548. [CrossRef]
75. Kitamura, T.; Kojima, H.; Hayakawa, N.; Kobayashi, K.; Kato, T.; Rokunohe, T. Influence of space charge by primary and secondary streamers on breakdown mechanism under non-uniform electric field in air. In Proceedings of the 2014 IEEE Conference on Electrical Insulation and Dielectric Phenomena (CEIDP), Des Moines, IA, USA, 19–22 October 2014; pp. 122–125. [CrossRef]
76. Li, Y.; Zhao, Z.; Liu, L.; Zhou, L.; Liu, D.; Li, S.; Chen, S.; Dai, Y.; Wang, J.; Wang, Z.L. Improved Output Performance of Triboelectric Nanogenerator by Fast Accumulation Process of Surface Charges. *Adv. Energy Mater.* **2021**, *11*, 2100050. [CrossRef]
77. Zhang, C.; Zhou, L.; Cheng, P.; Yin, X.; Liu, D.; Li, X.; Guo, H.; Wang, Z.L.; Wang, J. Surface charge density of triboelectric nanogenerators: Theoretical boundary and optimization methodology. *Appl. Mater. Today* **2019**, *18*, 100496. [CrossRef]
78. Wang, S.; Xie, Y.; Niu, S.; Lin, L.; Liu, C.; Zhou, Y.S.; Wang, Z.L. Maximum Surface Charge Density for Triboelectric Nanogenerators Achieved by Ionized-Air Injection: Methodology and Theoretical Understanding. *Adv. Mater.* **2014**, *26*, 6720–6728. [CrossRef] [PubMed]
79. Gonda, A.; Capan, R.; Bechev, D.; Sauer, B. The Influence of Lubricant Conductivity on Bearing Currents in the Case of Rolling Bearing Greases. *Lubricants* **2019**, *7*, 108. [CrossRef]
80. Liu, X.; Zhang, J.; Zhang, L.; Feng, Y.; Feng, M.; Luo, N.; Wang, D. Influence of interface liquid lubrication on triboelectrification of point contact friction pair. *Tribol. Int.* **2021**, *165*, 107323. [CrossRef]
81. Wang, Y.; Wang, T.; Da, P.; Xu, M.; Wu, H.; Zheng, G. Silicon Nanowires for Biosensing, Energy Storage, and Conversion. *Adv. Mater.* **2013**, *25*, 5177–5195. [CrossRef]
82. Wang, Y.; Narayanan, S.R.; Wu, W. Field-Assisted Splitting of Pure Water Based on Deep-Sub-Debye-Length Nanogap Electrochemical Cells. *ACS Nano* **2017**, *11*, 8421–8428. [CrossRef]
83. Wu, J.; Xi, Y.; Shi, Y. Toward wear-resistive, highly durable and high performance triboelectric nanogenerator through interface liquid lubrication. *Nano Energy* **2020**, *72*, 104659. [CrossRef]

Review

Electromechanical Nanogenerators for Cell Modulation

Zhirong Liu [1,2,†], Zhuo Wang [1,2,†] and Linlin Li [1,2,*]

1   Beijing Institute of Nanoenergy and Nanosystems, Chinese Academy of Sciences, Beijing 101400, China; liuzhirong@binn.cas.cn (Z.L.); wangzhuo@binn.cas.cn (Z.W.)
2   School of Nanoscience and Technology, University of Chinese Academy of Sciences, Beijing 100049, China
*   Correspondence: lilinlin@binn.cas.cn
†   These authors contribute equally to this work.

Abstract: Bioelectricity is an indispensable part of organisms and plays a vital role in cell modulation and tissue/organ development. The development of convenient and bio-safe electrical stimulation equipment to simulate endogenous bioelectricity for cell function modulation is of great significance for its clinical transformation. In this review, we introduce the advantages of an electromechanical nanogenerator (EMNG) as a source of electrical stimulation in the biomedical field and systematically overview recent advances in EMNGs for cell modulation, mainly including cell adhesion, migration, proliferation and differentiation. Finally, we emphasize the significance of self-powered and biomimetic electrostimulation in cell modulation and discuss its challenges and future prospects in both basic research and clinical translation.

Keywords: triboelectric nanogenerator; piezoelectric nanogenerator; electromechanical conversion; self-powered; cell modulation

Citation: Liu, Z.; Wang, Z.; Li, L. Electromechanical Nanogenerators for Cell Modulation. *Nanoenergy Adv.* **2022**, *2*, 110–132. https://doi.org/10.3390/nanoenergyadv2010005

Academic Editors: Ya Yang and Zhong Lin Wang

Received: 27 December 2021
Accepted: 3 March 2022
Published: 7 March 2022

Publisher's Note: MDPI stays neutral with regard to jurisdictional claims in published maps and institutional affiliations.

Copyright: © 2022 by the authors. Licensee MDPI, Basel, Switzerland. This article is an open access article distributed under the terms and conditions of the Creative Commons Attribution (CC BY) license (https://creativecommons.org/licenses/by/4.0/).

1. Introduction

Bioelectricity acting as an endogenous biophysical factor can modulate a myriad of cell behaviors, such as cell cycle, adhesion, proliferation, migration and differentiation, and further regulate important biological processes, such as embryogenesis and tissue regeneration [1–3]. Enlightened by endogenous bioelectricity, biomimetic electrical stimulation has been widely employed to regulate cell activities and offers widespread application potential for biomedical therapeutics [4]. At the same time, a series of electrical stimulation devices have been developed for in vitro and in vivo cell electrostimulation [5]. The commercial electrical stimulator is the most widely used device in basic biomedical research because its electrical parameters can be finely tuned from a wide range [6]. However, its bulky size brings a lot of inconvenience and causes poor patient compliance for clinical applications. Additionally, the long-distance wire connection between the stimulator and the target may increase potential safety risks. As another kind of commonly used electrical stimulator, the implantable battery is small and safe enough for in vivo applications; however, due to the limited battery capacity, regular battery replacement is required for long-term electrostimulation, which is expensive and increases the risk of postoperative infection [7].

To meet the requirements of small size, safety and long-term electrical stimulation at the same time, various new types of energy harvesters have been developed, which can collect energy from the surrounding environment or organisms and convert it into electricity [5]. Depending on the energy source, energy harvesters can be divided into three types: (i) environmental energy harvesters, including photovoltaic cells and pyroelectric nanogenerators [8]; (ii) mechanical energy harvesters, including triboelectric nanogenerators (TENG), piezoelectric nanogenerators (PENG) and electromagnetic generator (EMG) [1]; (iii) biochemical energy harvesters, including biofuel cells and non-pulley potential collectors [9,10]. Among them, several kinds of electromechanical nanogenerators (EMNGs) have drawn extensive attention in biomedical fields due to the existence of abundant mechanical energy in organisms from tissue to cell level, such as cell activities, heartbeat, limb movement and respiration. An EMG is a kind of efficient, well-established and versatile mechanical energy harvester, and some high-frequency EMGs have been reported for in vitro cell electroporation [11–13]. However, an EMG only has superior performance at high frequencies and high dimensions [14], so they are less efficient in collecting low-frequency, disordered and weak mechanical energy in living organisms. Therefore, this review mainly focuses on recent advances in TENGs and PENGs for cell modulation (Figure 1). For a TENG, its basic working modes and its working mechanism are introduced. For a PENG, the methods used to generate a surface piezopotential for cell electrostimulation are discussed, including speaker, ultrasonic wave, magnetic field and cell/tissue activities. Then, the applications of EMNGs on cell modulation are elucidated, including cell proliferation, adhesion, migration, differentiation, etc. Finally, we discuss the challenges and scope for development of EMNGs for cell modulation in the future.

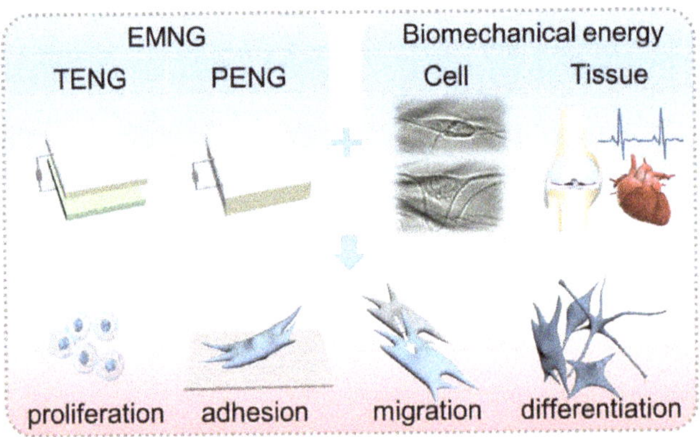

Figure 1. EMNGs driven by biomechanical energy for cell modulation, including cell proliferation, adhesion, migration and differentiation.

2. TENGs for Cell Modulation

2.1. Working Mechanism of TENGs

TENGs can collect abundant biomechanical energy from human bodies and convert it into electricity due to the coupling effect of triboelectrification and electrostatic induction [15–18]. Since its invention in 2012, the TENG mainly presents four basic operating modes, including the vertical contact–separation mode, lateral sliding mode, single electrode mode and freestanding triboelectric layer mode [19] (Figure 2a). Therefore, the structure of a TENG can be flexibly designed according to various applications. As their working principles are similar, herein the contact–separation mode is taken as an example to explain its working principle in detail (Figure 2b). Two films with different triboelectric properties are placed face to face as the friction layers and then metal is attached to their back as the electrode layer. In the original state, there is no induced charge (Figure 2b(i)). When the two friction layers are in contact with each other, an equal amount of opposite triboelectric charge is generated on the contact surface due to the triboelectric effect (Figure 2b(ii)). When the two friction layers are separated, it generates a potential difference between the two electrodes due to electrostatic induction. Thus, the electrons flow from the top electrode to the bottom electrode through the external circuit, thereby generating instantaneous current (Figure 2b(iii)), and then finally, it reaches equilibrium when the two friction layers are completely separated (Figure 2b(iv)). When the two friction layers make contact again, the induced charges flow back through the external circuit to compensate for the potential difference (Figure 2b(v)). The two friction layers continue to make contact and separate in order to generate an alternating current [20,21]. Due to the widespread existence of triboelectricity, TENGs have a wide range of material choices, from natural to synthetic materials, which can meet the requirements for biocompatibility and flexibility in biomedical applications [22–24]. The combinations of friction layer materials that are commonly used in biomedical TENG can be divided into metal–polymer and polymer–polymer combinations [25]. For biomedical applications, especially implantable medical devices, it is necessary to select materials with good biocompatibility, low toxicity and potential biodegradability. Therefore, noble metals (Au, Ag and Pt) and transition metals (Mg, Fe and Zn) are selected as the metal friction or electrode layers. Polytetrafluoroethylene (PTFE), polyethylene terephthalate (PET), polyimide (Kapton), polydimethylsiloxane (PDMS) and polypropylene are common polymer friction layers for TENGs. Specifically, conductive polymers, such as polypyrrole (PPy), poly(3,4-ethylenedioxythiophene) (PEDOT), polythiophene (PTh) and polyaniline (PANi), can act as both polymer friction layers and electrodes. In the process of the contact and separation of two friction layers, their ability to gain and lose electrons depends on the triboelectric properties of the friction layer materials. According to the triboelectric sequence of common friction materials [26,27], the farther the distance between the two materials in the list, the greater the amount of charge transfer during the contact–separation process. In addition, the surface micro–nano structure, chemical modification or electron injection of the friction layer materials can increase its effective contact area and surface charge density, thus greatly improving the output performance of the TENG [28–30]. To meet the needs of short-term electrostimulation and avoid secondary surgery, biodegradable and bioabsorbable TENGs have also been designed using biodegradable or photothermal-tuned degradable materials [31–33], such as Mg, poly (caprolactone) (PCL), polylactic acid (PLA), poly(lactic-co-glycolide acid) (PLGA), poly (vinyl alcohol) (PVA), chitosan, cellulose, chitin and silk fibroin.

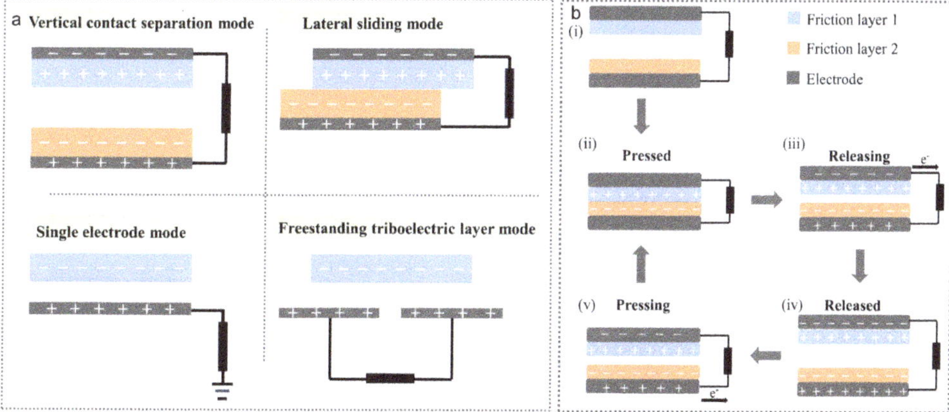

Figure 2. (**a**) The four fundamental working modes of a TENG: vertical contact–separation mode; lateral sliding mode; single electrode mode; and freestanding triboelectric layer mode. (**b**) The working principle of the TENG in contact–separation mode.

2.2. TENGs for Cell Modulation

Due to the above advantages, the TENG has developed rapidly in biomedical applications. Early research mainly focused on biomechanical energy harvesting, self-powered health monitoring and tissue-level electrical stimulation [22,34]. Recently, TENG-based electrostimulation has been used for cell modulation, mainly for promoting neural/osteogenic differentiation and directing cell migration and proliferation for wound healing.

2.2.1. TENGs for Nerve Repair

Neurons are electrical excitable cells from neural tissues, amongst others. It has been proven that electrostimulation can improve the synaptic function of neurons, thereby promoting nerve regeneration and functional recovery [2]. For this application, a biodegradable TENG was designed that can be implanted into an animal for the electrical stimulation of nerve cells and gradually degrade after finishing its work [31]. The employed TENG has a multilayered structure, including the friction layers, the electrode layers and the encapsulation structure. Two biodegradable polymers are assembled together as the friction layers, and there is a 200 nm spacer layer between the two friction layers. The electrode layer is prepared by depositing a 50 nm magnesium (Mg) film on the back of the friction layer. The fabricated TENG is then encapsulated to improve its stability in the surrounding physiological environment. When applying this TENG to stimulate cells seeded on complementary micrograting electrodes (10 V/mm, 1 Hz), the nerve cell growth can achieve directional growth, which is of great significance for neural repair. Guo et al. combined a wearable TENG and electroconductive microfibers to construct a self-powered electrostimulation system for the neural differentiation of stem cells (Figure 3a,b) [35]. The prepared TENG worked based on the continuous contact and separation of the upper aluminum (Al) electrode and the bottom Kapton film attached to a copper (Cu) electrode. A 3 M foam (2 mm) was bound between the poly(methyl methacrylate) (PMMA) substrate and the upper electrode as a stress buffer layer to increase its output stability. The output of the TENG could reach about 300 V and 30 µA when triggered by human walking (Figure 3c). The electroconductive microfibers were composed of poly(3,4-ethylenedioxythiophene) (PEDOT) and reduced graphene oxide (rGO), which can enhance the proliferation ability of mesenchymal stem cells (MSCs) and show a good neural differentiation tendency. Importantly, after 21 days of the continuous stimulation of the stem cells grown on the conductive hybrid microfibers by the TENG, the gene and protein expressions of both Tuj-1 (neuron-specific maker) and GFAP (neurogliocyte-specific maker) were significantly higher than those without electrical stimulation (Figure 3d,e). This work confirmed the feasibility and potential of TENGs in the field of neural differentiation and repair.

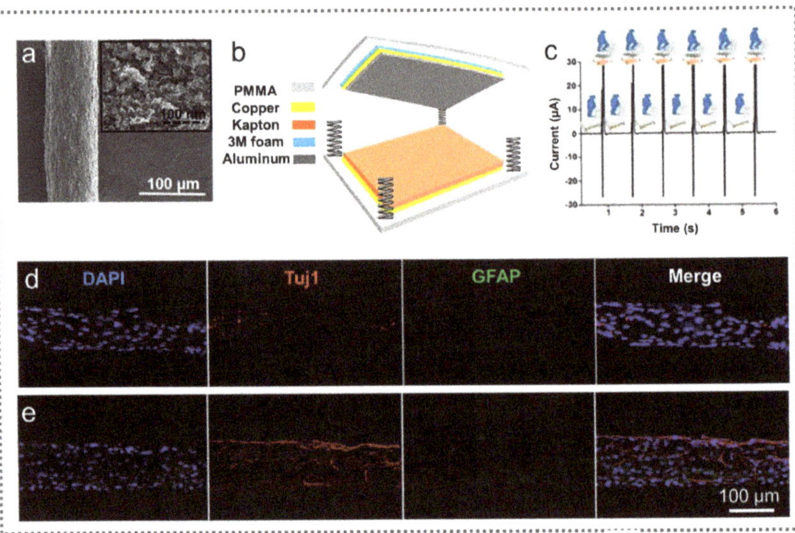

Figure 3. TENGs for nerve repair: (**a**) the micromorphology of the rGO−PEDOT microfiber; (**b**) the structural design of the TENG; (**c**) the output current of the TENG driven by normal human walking. The nucleus (blue), neuron-specific maker Tuj1 (red) and neurogliocyte-specific maker GFAP (green) of the cells (**d**) without electrostimulation and (**e**) with TENG-induced electrostimulation for 21 days. Reprinted with permission from ref. [35], Copyright 2016, American Chemical Society.

2.2.2. TENG for Bone Repair

Bone is a natural composite of piezoelectric materials, which mainly comprises a piezoelectric collagen matrix and hydroxyapatite crystals [36]. In the process of bone repair and regeneration, the electric signal is a vital factor for promoting the osteodifferentiation of stem cells and osteoblast proliferation [37,38]. Due to the abundant mechanical energy during joint movement, TENGs can effectively collect this energy and then convert it into electricity for bone repair. It has been confirmed that the electrostimulation of TENGs can promote osteoblast attachment, proliferation and differentiation. For example, Tian et al. developed a self-powered electrical stimulator for bone repair, consisting of an implantable TENG as the power source, a rectifier and an interdigital electrode [39]. When the flexible TENG was implanted on the SD rat's femur, it could successfully convert the biomechanical energy of the rat's daily exercise into electricity. Applying the electricity generated by the TENG to the cells through the interdigital electrode could effectively promote cell adhesion, spreading and proliferation. Moreover, the osteogenic differentiation level of the cells increased by 28.2% after stimulation for 12 days. Shi et al. utilized a TENG to load and accumulate negative charges on the surface of an anodized titanium implant for promoting the osteogenesis of MC3T3-E1 preosteoblast cells [40]. The prepared TENG worked based on the vertical contact–separation mode (Figure 4a). A PTFE film with a nanorod structure and a titanium (Ti) foil with a nanotube structure were used as the triboelectric layers. A thin cooper (Cu) film was magnetron sputtered on the outer face of the PTFE friction layer as an electrode and a Ti foil was employed as another electrode. Then, PTFE tape and PDMS were used to encapsulate the TENG for waterproofing. The output of the TENG could reach up to 12 V, 0.15 μA and 5.3 nC (Figure 4b) and its output voltage could be well maintained after 1×10^6 cycles. The prepared TENG could convert the biomechanical energy of human daily movement into electricity for constructing a long-term and stable negatively charged implant surface, thereby inhibiting bacterial adhesion and biofilm formation. Moreover, the negatively charged implant could promote preosteoblast adhesion and the osteogenic differentiation of MC3T3-E1 cells (Figure 4c). Yao et al. designed an implantable and bioresorbable TENG that could be attached to living tissue and generate bidirectional electric pulses (Figure 4d) [41]. An island–bridge magnesium (Mg) layer served as both the bottom electrode and a triboelectric layer. PLGA with a micropyramid structure was used as another triboelectric layer and the Mg electrode with the island–bridge structure was coated on the top of the PLGA layer, serving as the top electrode. The micropyramid structure of the PLGA could improve the output of the TENG due to the increase in contact area. The serpentine geometry and island–bridge structure of the electrode layer could effectively improve the robustness of the structure and reduce the modulus of the flexible device. Thus, it could be pasted to irregular tissues and withstand large strains. When applying the electrostimulation on a pair of dressing electrodes, the generated electric field could activate relevant growth factors (Figure 4e,f). The enhanced secretion of the fibroblast growth factor (FGF1) and vascular endothelial growth factor (VEGF) could accelerate vascularization for nutrient supply and metabolic transportation. More transforming growth factor and bone morphogenetic protein could promote cell differentiation and accelerate bone formation and mineralization (Figure 4g). In general, the electrostimulation through the TENG could regulate the bone microenvironment, promote rapid bone regeneration and synergistically increase bone strength and mineral density (Figure 4h).

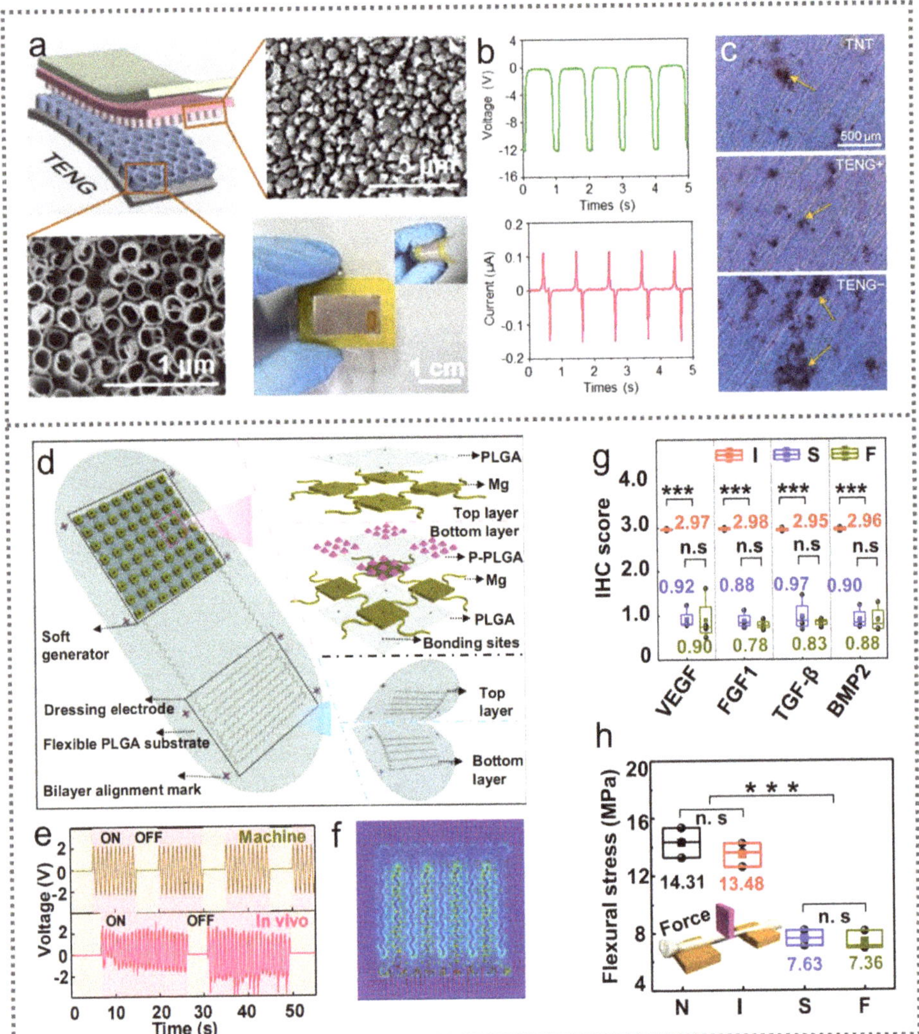

Figure 4. TENGs for bone repair: (**a**) the surface microstructure and photograph of the prepared TENG; (**b**) the output voltage and current of the TENG; (**c**) alizarin red staining showing the calcium deposits of MC3T3-E1 after TENG electrostimulation for 21 days. Reprinted with permission from ref. [40], Copyright 2020, Elsevier. (**d**) The schematics of the implantable self-powered bone fracture electrostimulation device, consisting of a TENG as a power source and a pair of dressing electrodes for applying electrostimulations to the fracture; (**e**) the output voltage of the TENG driven by a mechanical linear motor (top) and on a rat (bottom); (**f**) the simulated electrical field distribution of the dressing electrodes inside a tissue under the voltage of 4 V; (**g**) the immunohistochemistry (IHC) score of the expressed growth factors in different groups; (**h**) the three-point bending flexural stress measurement for the different groups after 6 weeks. N, normal group; I, dressing electrodes with TENG electrostimulation; S, dressing electrodes without TENG electrostimulation; F, no implanted device. n.s. and *** represent non-significant and $p < 0.001$, respectively. Reprinted with permission from ref. [41], Copyright 2021, the Authors. Published by PNAS.

2.2.3. TENGs for Wound Healing

In addition, the endogenous electric field also plays key roles in wound healing. Generally, skin wounds have a transepithelial potential of 15–60 mV, which serves as a directional cue to direct cell proliferation and migration to the wound site for wound healing [42]. Thus, electrostimulation is an attractive auxiliary method for wound repair (Figure 5a). However, it is still largely limited in clinical applications due to the inconvenience of its implementation. To solve this problem, a wearable electrostimulation bandage was designed to accelerate the healing of skin wounds [43,44]. It was composed of a wearable TENG as a power supply and a pair of dressing electrodes, both of which were integrated in a bandage [43]. The TENG was prepared by overlapping a Cu (electropositive material) electrode layer with a Cu/PTFE (electronegative material) film on the PET substrate. After wrapping the bandage on the chest of the rat, the TENG could collect the mechanical energy of the rat's breathing to generate electricity. When the rat was under deep anesthesia, the output voltage was only ~0.2 V at a rate of 30 times per min, which corresponded to a slow and shallow breathing pattern. After the rat recovered from anesthesia, the output voltage reached ~1.3 V at a rate of 40 times per min, corresponding to the calm and stable state of the rat. When the rat moved normally, the output voltage reached the highest level of ~2.2 V at a rate of 110 times per min (Figure 5b). Under the electric field induced by the TENG (2 V cm^{-1}, 1 Hz), fibroblasts exhibited an obvious proliferation and alignment. After 6 h of the electric field stimulation, cells continued to proliferate and differentiate along the direction of the electric field (Figure 5c). Fibroblasts play a crucial role in the healing process of skin wounds. Initially, cells migrate to the wound, proliferate and interact with surrounding cells. After that, fibroblasts differentiate and generate ECM, glycoproteins, adhesive molecules and various cytokines, which then replace the provisional fibrin-based matrix and accelerate wound repair. Under the treatment of the electrostimulation bandage, a full-thickness rectangular skin wound was quickly closed within 3 days, while the normal shrinkage-based healing process needs 12 days. Li et al. also confirmed the promoting effect of TENG electrostimulation on tissue repair. Their prepared TENG was implantable and its degradation rate could be tuned in vivo by near-infrared (NIR) light irradiation [33]. The biodegradable TENG consisted of three parts: an Au deposited, hemisphere array-structured layer served as both the triboelectric layer and the bottom electrode; a biodegradable polymer was another triboelectric layer; and a thin Mg film was deposited on the back as the top electrode. Another biodegradable polymer doped with Au nanorods (AuNRs) was applied as the bottom substrate, which endowed the TENG's ability to respond to NIR irradiation so that the biodegradation process could be rationally controlled (Figure 5d). The open-circuit voltage (V_{oc}) and short-circuit current (I_{sc}) of the TENG reached 28 V and 220 nA, respectively. After implantation in the subdermal region on the back of the SD rats, the V_{oc} was about 2 V. With NIR treatment, the output of the TENG quickly dropped to 0 within 24 h and the device was largely degraded in 14 days (Figure 5e,f). Applying the output of the TENG to stimulate fibroblasts (100 mV mm^{-1}) could significantly accelerate the migration of cells across the scratch, which was essential for the wound healing treatment (Figure 5g,h).

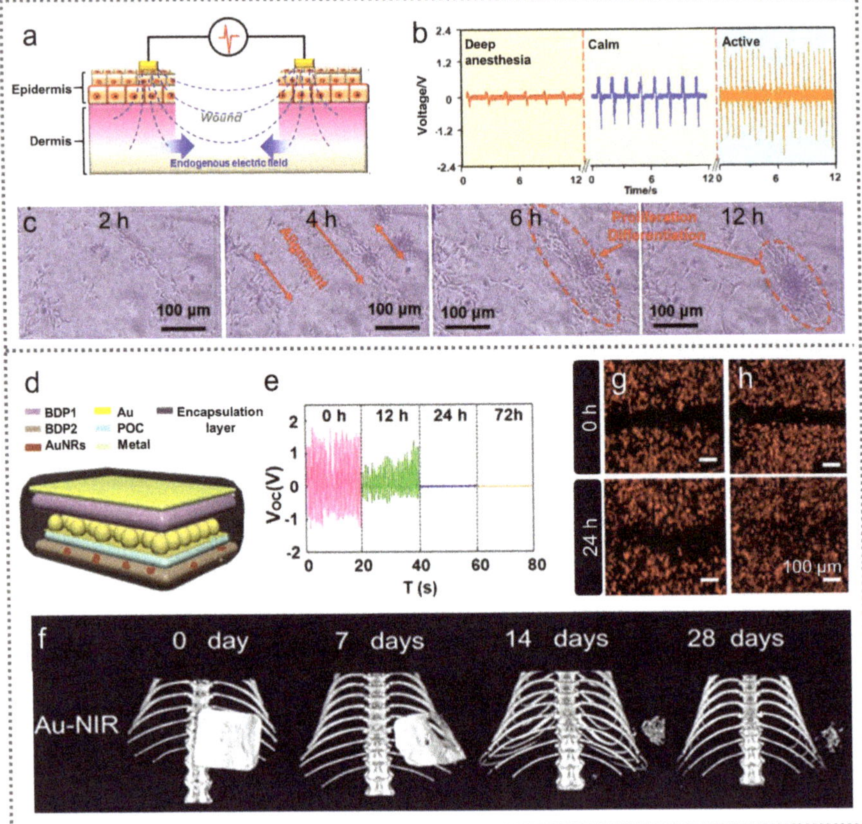

Figure 5. TENGs for wound healing: (**a**) the mechanism of wound healing under the endogenous electric field; (**b**) the output voltage of the TENG driven by different frequencies of breathing in rats; (**c**) the morphology of the NIH 3T3 cells after TENG electrical stimulation of different times. Reprinted with permission from ref. [43], Copyright 2018, American Chemical Society. (**d**) The structure design of the photothermal-controlled biodegradable TENG; (**e**) the output of the TENG with NIR irradiation; (**f**) a micro-CT image of the implanted TENGs over time. Fluorescence microscope images of fibroblast L929 cell migration (**g**) without electrostimulation and (**h**) with TENG electrostimulation. Reprinted with permission from ref. [33], Copyright 2018, Elsevier.

2.2.4. TENGs for Drug Delivery and Cardiac Pacing

In addition to the aforementioned applications, TENGs have also been employed to construct self-powered electroporation systems to improve the permeability of cell membranes for drug delivery or drug release. Electroporation utilizes high-intensity electric fields to create transient nanopores in plasma membranes to introduce biomolecules into cells. Liu et al. integrated a TENG with a silicon nanoneedle array electrode to build a self-powered nanoelectroporation system for facilitating efficient intracellular drug delivery while minimizing cellular damage (Figure 6a) [45]. In this system, the TENG acted as a self-powered and stable electrostimulation source to provide electrical pulses for electroporation, which made the system more convenient in practical applications. On the other hand, a silicon nanoneedle array was employed to replace the traditional planar electrode for nanoelectroporation, which could only generate high local electrical fields in the nanoneedle–adherent cell interface, thereby reducing cell damage (Figure 6b). This system could deliver a variety of exogenous species, such as small molecules, siRNA and biomacromolecules, into different types of cells. Depending on the delivered biomolecule and cell type, the delivery efficiency ranged from 50% to 90% and cell viability was above 94% (Figure 6c). On this basis, the same group further designed a high-throughput electroporation system based on a TENG and silver nanowire-modified foam electrodes [28]. Cell suspension could continuously flow through the foam electrodes in the electroporation channel to achieve electroporation under the action of the TENG's electrical pulses (Figure 6d). The system could achieve high cell processing throughput of up to 10^5 cells/min while ensuring high cell delivery efficiency and viability. Yang et al. also designed a self-powered gene electrotransfection system that consisted of a TENG and a nanowire array electrode [46]. The employed TENG was based on the vertical contact–separation mode, which could harvest the mechanical energy of the simple human tapping motion and convert it into electricity for electrotransfection. Two CuO nanowire array meshes were coaxially placed as electrodes, which could greatly amplify the local electric field strength to enhance the transfection efficiency and reduce cell damage. The system could achieve a high-efficiency siRNA electrotransfection of 95% for MiaPaCa-2 cells and 84% for K562 cells, and then downregulate the expression of targeted mutation genes, thereby significantly inhibiting cell proliferation and anti-apoptosis ability. Zhao et al. loaded doxorubicin (DOX) into red blood cells (D@RBC), which could create transient pores in the cytomembrane through TENG electroporation and release DOX on demand to kill tumor cells for cancer therapy (Figure 6e–g) [47]. These results illustrate that TENG-based self-powered systems have great potential for both fundamental biological research and clinical applications.

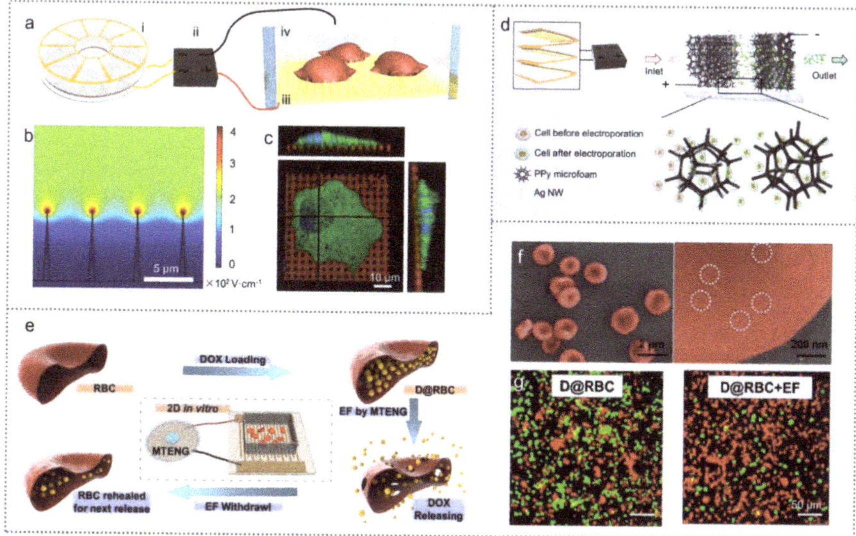

Figure 6. TENGs for drug delivery: (**a**) a schematic illustration of the TENG-driven electroporation system based on a nanoneedle array electrode; (**b**) the simulated electrical field distribution of the nanoneedle array with an applied voltage of 20 V; (**c**) a fluorescence image of MCF-7 after delivering dextran-FITC (green). Reprinted with permission from ref. [45], Copyright 2019, WILEY-VCH. (**d**) A schematic illustration of the TENG-driven high-throughput electroporation system. Reprinted with permission from ref. [28], Copyright 2020, American Chemical Society. (**e**) A schematic illustration to show the controlled release of DOX from RBC driven by the TENG; (**f**) SEM images of D@RBC under the electric field; (**g**) the viabilities of HeLa cells in the D@RBC and D@RBC under the electric field groups (live cells are green and dead cells are red). Reprinted with permission from ref. [47], Copyright 2019, WILEY-VCH.

Cardiomyocytes are also electrically active cells and their action potentials contribute to the spontaneous beating of the cells. Rhythmic action potentials propagate continuously in multicellular cardiac tissue and travel through the myocardium to induce myocardial contraction [1]. Therefore, electrical stimulation has been used in heart failure treatments and myocardial reconstruction. Inspired by this, Ouyang et al. demonstrated a self-powered symbiotic pacemaker based on an implantable TENG [48]. The V_{OC} of the implantable TENG was as high as 65.2 V. The energy collected from each heartbeat was about 0.495 µJ, which was above the endocardial pacing threshold energy (0.377 µJ). Thus, this self-powered symbiotic pacemaker successfully corrected sinus arrhythmia and prevented disease progression on a large animal model.

3. PENGs for Cell Modulation

PENGs can convert mechanical motion into electric energy based on piezoelectric materials and piezoelectric effects. A PENG is composed of piezoelectric materials, positive and negative electrodes and flexible substrates. Among them, the piezoelectric material is the core component of a PENG and is the basis for generating electricity.

3.1. Piezoelectric Materials and Piezoelectricity

Piezoelectric material is a kind of dielectric material with a non-centrosymmetric structure, which can generate and accumulate charges with opposite signs on its surface under the action of mechanical stress or strain [49]. According to compositions, piezoelectric materials can be divided into three categories: (i) inorganic piezoelectric materials; (ii) organic piezoelectric materials; and (iii) composite piezoelectric materials.

Inorganic piezoelectric materials mainly include piezoelectric single crystals, piezoelectric ceramics and piezoelectric semiconductors. Quartz, lithium niobate, lithium gallate, lithium germanate and lithium tantalate are several common piezoelectric single crystals. Piezoelectric ceramics often have strong piezoelectricity and high dielectric constants, such as barium titanate ($BaTiO_3$), lead zirconate titanate (PZT), lead zinc niobium (PZN), lead metaniobate and lead barium lithium niobate. Generally, piezoelectric ceramics containing lead have high piezoelectric coefficients, but the toxicity of lead limits their application in the biomedical field. Currently, most piezoelectric semiconductors have wurtzite structures and have both piezoelectricity and semiconducting properties, such as ZnO, GaN, AlN and BN. The piezoelectricity of inorganic materials originates from the non-centrosymmetric crystal structure. Taking wurtzite-structured ZnO as an example (Figure 7a,b), positive Zn^{2+} and negative O^{2-} have tetrahedral coordination in an unstrained hexagonal ZnO crystal structure. So, the centers of the anions and cations overlap each other and there is no polarization. When the crystal is under a mechanical stress, the displacement of Zn^{2+} and O^{2-} inside the crystal causes the positive and negative charge centers to shift in opposite directions, resulting in a dipole moment in the unit cell [50]. Due to the continuous superposition of the dipole moments, a macroscopic piezoelectric potential (piezopotential) is generated in the ZnO crystal (Figure 6b) [51].

Compared to inorganic piezoelectric materials, although the piezoelectric properties of organic piezoelectric materials are relatively poor, their advantages of good biocompatibility, processability and high flexibility can satisfy the requirements for biomedical applications, especially for wearable and implantable devices. Organic piezoelectric materials can be divided into natural and synthetic piezopolymers. Collagen, chitosan, cellulose and chitin are widely studied natural piezoelectric polymers. PVDF and its trifluoroethylene (TrFE) copolymer (P(VDF–TrFE)), poly-L-lactic acid (PLLA) and polyhydroxyalkanoates (PHAs) are representative synthetic piezopolymers. The piezoelectricity in organic materials arises from their non-centrosymmetric molecular arrangement and orientation. Figure 6c depicts the piezoelectric effect of PVDF ((CH_2–CF_2)n). PVDF has five crystal phases, in which α-phase with trans−gauche−trans−gauche conformation (TGTG′) and a β-phase with all trans conformation (TTTTT) are predominant [52,53]. The dielectric property of PVDF is caused by the difference in electronegativity between F and H atoms that leads to a dipole moment is generated along the F→H direction [54]. For α-phase PVDF, there is no piezoelectricity since the chains are arranged in the unit cell, resulting in a net cancellation of the dipole moments. For β-phase PVDF, since the parallel orientation of the dipole moments leads to the overlay of the electric dipole moments, it has the highest piezoelectricity [55] (Figure 7c). Composite piezoelectric materials are often obtained by dispersing piezoelectric nanoceramics into a polymer matrix, which combines the advantages of inorganic piezoelectric materials and piezoelectric polymers, showing good flexibility, high piezoelectric coefficients and processability. Piezoelectric materials with high mechanical and piezoelectric properties are the basis for the fabrication of high-performance PENGs.

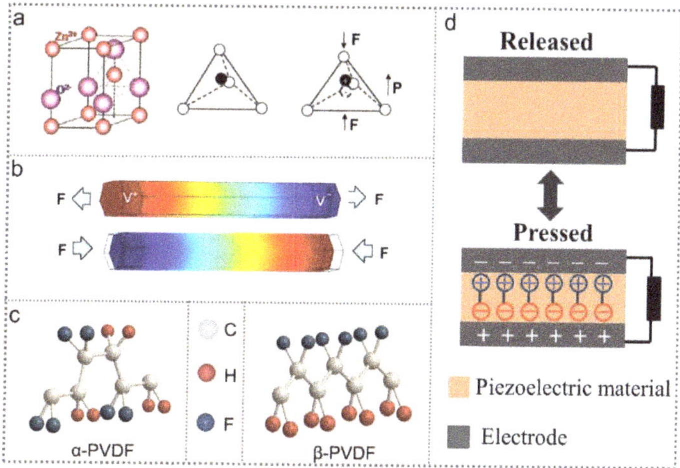

Figure 7. (**a**) An atomic model of a wurtzite structure ZnO and a schematic diagram of a stress-induced electric dipole moment. Reprinted with permission from ref. [50], Copyright 2012, WILEY-VCH. (**b**) Piezopotential distribution along a ZnO nanowire under axial stretch or compress. Reprinted with permission from ref. [51], Copyright 2009, American Institute of Physics. (**c**) The molecular structure of α-phase and β-phase PVDF; (**d**) the working principle of the PENG.

3.2. Working Mechanism of PENGs

In 2006, Wang et al. first developed a zinc oxide (ZnO) nanowire-based PENG to collect tiny vibrational energy [56]. In 2010, a PENG that was based on single ZnO nanowire was successfully implanted in a live rat to harvest mechanical energy from its breath and heartbeat, with an output of around 3 mV and 30 pA [57]. After that, all kinds of piezoelectric materials and structures have been applied to fabricate implantable PENGs, including PVDF, P(VDF–TrFE), ZnO nanowire arrays, PZT, etc. [58–64]. Traditionally, a PENG is composed of an intermediate piezoelectric layer and two metal electrodes in a sandwich structure. Figure 7d demonstrates the electrical generation principles of a typical PENG. At the beginning, without the action of external force, the charge centers of the anions and cations overlap each other, so there is no polarization in the piezoelectric material. When mechanical stress is applied, the piezoelectric material is compressed and deformed, so that the anions and cations inside the crystal are displaced and the positive and negative charge centers are separated to generate a surface piezopotential. The electrons flow through the external circuit to achieve charge balance, thereby generating piezoelectric current output. When the external force is removed, the piezopotential gradually decreases and disappears and the electrons flow back to rebalance the charge. During the cyclic compression–release process, mechanical energy is continuously converted into electrical energy [65,66]. The selection of piezoelectric materials that have excellent performance is the key to improving the output of PENGs. In addition, the crystallinity of piezoelectric materials and the arrangement direction of dipoles can be further improved by means of annealing, stretching and high-voltage polarization, thereby achieving efficient mechanical energy harvesting [66].

3.3. PENGs for Cell Modulation

A PENG is an energy harvesting device that converts mechanical energy into electrical energy using the piezoelectric effect. Therefore, its application is similar to that of a TENG and it too can be used as a self-powered electrical stimulation source. For example, Zhang et al. proposed a biomechanical energy-driven shape–memory PENG for promoting MC3T3-E1 cell proliferation and orientation and osteogenic differentiation [67]. Jin et al. designed a tribo/piezoelectric hybrid nanogenerator to promote the long-term proliferation and migration of Schwann cells and regenerate the myelination of nerve fibers and neuromotor function reconstruction [68]. In addition to being an electrical stimulation source, some piezoelectric materials are also employed as biological scaffolds [54]. Therefore, piezoelectric biomaterials can be designed for in situ cell-scale electrical stimulation, avoiding wire connections and electrode implantation. External mechanical force or deformation is an essential precondition for piezoelectric biomaterials to generate piezoelectric potential (piezopotential) for cell electrostimulation. The methods commonly used to apply mechanical force are acoustic waves, magnetic fields and cell traction.

3.3.1. Acoustic Wave-Driven PENGs for Cell Modulation

An acoustic wave is a mechanical wave that is generated by the vibration of a sound source. Acoustic pressure, i.e., the pressure change caused by sound waves, is the main source of mechanical force. According to the frequency, acoustic waves can be divided into four categories: infrasound (<20 Hz); audible sound (20–20 kHz); ultrasound (20 kHz–1 GHz); and hypersound (>1 GHz). Wang et al. fabricated electrospun poly(vinylidene fluoride-trifluoroethylene) (P(VDF–TrFE)) nanofibers with a d_{31} of 16.17 pC/N [69]. A lab-designed speaker was employed to generate the mechanical vibration to deform the nanofibers. Under the dual effects of the nanofiber morphology and electrical stimulation (0.75 V, 22.5 nA), L929 fibroblast cells aligned perfectly along the direction of the nanofibers and the cell proliferation rate increased by 1.6-fold.

An ultrasonic wave is a commonly used mechanical force source in the field of biomedicine because of its high tissue penetration and good directionality [70] and its ultrasonic power, frequency and period can be adjusted precisely. In the ultrasonic process, acoustic pressure and ultrasonic cavitation effects play major roles in driving the deformation of piezoelectric materials [71]. When the sound pressure gradually reaches a certain value, the tiny bubble cores in the liquid expand rapidly and then suddenly close to generate the shock wave. The ultrasonic cavitation is a series of the dynamic processes, such as expansion, collapse and generation, of microjets. Wan et al. demonstrated an ultrasound-mediated cell sheet harvesting based on a piezoelectric polyvinylidene fluoride (PVDF)/barium titanate (BaTiO$_3$, BTO) composite film (Figure 8a) [72]. Under the ultrasound stimulation (1 MHz, 0.8 W cm^{-2}), the output voltage and current of the PVDF/BTO film could reach ~100 mV and 0.19 nA, respectively. The adsorption and conformation of fibronectin (FN) play an important role in regulating early cell material adhesion. After ultrasound treatment for 20 s, the surface potential of the PVDF/BTO film was more negative, thereby decreasing the FN adsorption to manipulate cell adhesive behavior. On the other hand, the generated piezopotential under ultrasound stimulation could regulate the secondary structure of adhesive protein fibronectin (FN) from β-sheet to α-helix, thereby weakening the interfacial protein interaction and further regulating the cell adhesion behavior (Figure 8b,c). In addition, ultrasound assisted piezoelectric biomaterials have also been employed for inducing neural differentiation [73–76]. Marino et al. prepared piezoelectric tetragonal BTO nanoparticles that could be attached to the cell membrane [76]. Under ultrasound stimulation, the generated piezopotential activated voltage-gated membrane channels, allowed calcium and sodium to flow in and then mediated the enhancement of neurite outgrowth. Liu et al. designed a biohybrid multifunctional micromotor composed of magnetic Fe_3O_4 nanoparticles, piezoelectric BTO nanoparticles and *S. platensis* [74]. The micromotor was actuated and steered magnetically using a low-intensity rotating magnetic field (Figure 8d,e). Due to its diameter ($\approx$0.8 µm) being smaller than the size of neural stem-like cells, this micromotor could reach the desired position at the single cell level. After reaching the targeted neural stem-like cell, the in situ piezopotential of BTO was generated by an ultrasound stimulation, which activated the Ca^{2+} channels and thereby induced the neuronal differentiation of the targeted cells (Figure 8f,g). Additionally, piezoelectric nanoparticles can be endocytosed by the cells to achieve non-destructive intracellular electrical stimulation for promoting osteogenic differentiation. Ma et al. synthesized piezoelectric nylon-11 nanoparticles with uniform morphology that could be easily endocytosed using dental pulp stem cells (DPSCs) [77]. Under ultrasound conditions, piezoelectric stimulation generated by nylon-11 nanoparticles could efficiently promote the osteogenic differentiation of stem cells through electromechanical conversion in a non-invasive way. Ultrasound assisted piezoelectric biomaterials can also regulate proinflammatory macrophage polarization [78], which is critical for antitumor immunity. After ultrasound treatment on β-PVDF, a significant upregulation of the mRNA levels of proinflammatory (M1) markers was observed, including TNF-α, IL-1β and MCP-1 (Figure 8h–j). In addition, the intensity of intracellular M1 marker iNOS was higher than that on tissue culture plates (TCPs), while anti-inflammatory (M2) marker Arg-1 was significantly reduced (Figure 8k,l). These results prove that the piezoelectricity of β-PVDF obviously promotes M1 polarization.

3.3.2. Magnetic Field-Driven PENGs for Cell Modulation

Magnetic fields have also been employed to induce piezopotential by rationally designing magneto-electric (ME) composite materials. ME materials comprise a class of multifunctional nanostructures that are capable of strongly coupling magnetic and electric fields. The ME effect is defined as the electrical polarization of a substance in response to a magnetic field (positive effect) or the magnetization change of a substance under the action of an electric field (converse effect) [79]. ME materials include single phase and composite materials, within which the composites that are composed of a ferromagnetic phase and a piezoelectric phase have lager ME effects at room temperature [80]. Under the action of the magnetic field, the magnetic material is deformed and the strain is further transferred to the closely connected piezoelectric material to generate piezopotential. This phenomenon is the magneto-electric coupling in two-phase ME materials. The properties of ME composite materials can be regulated by changing the phase materials, component sizes and phase volume ratios [81]. Core shell-structured ME materials have attracted considerable attention in the design of drug delivery systems, which are capable of controlling drug release via external magnetic fields. For example, Nair et al. designed $CoFe_2O_4@BaTiO_3$ ME nanoparticles that could generate piezopotential by applying a low alternating current magnetic field to allow controlled on-demand drug release [82]. The application of ME composites for the electrical stimulation of cells and tissues has been intensively studied in tissue engineering. Mushtaq et al. designed a soft hybrid nanorobot that mimicked an electric eel, based on the magneto-electric coupling effect [83]. These bionic nanoeels consisted of three parts: a flexible and elongated piezoelectric PVDF tail connected to a polypyrrole nanowire, which was decorated with nickel rings for magnetic actuation (Figure 9a). Upon the rotating magnetic fields, the nanoeels displayed different swimming modes, including tumbling, wobbling and corkscrewing (Figure 9b). At the same time, the piezoelectric soft tail was driven to deform, causing its electric polarization. As magnetic fields can achieve deep tissue penetration and can control the movement of magnetic materials with high precision, they can be used to achieve targeted local electrical stimulation. Dong et al. introduced a highly integrated multifunctional soft helical microswimmer based on $CoFe_2O_4@BiFeO_3$ (CFO@BFO) core shell ME nanoparticles, which could realize targeted neuronal cell delivery, on-demand localized neuron electrostimulation and post-delivery enzymatic degradation (Figure 9c) [84]. Under a rotating magnetic field, the helical microswimmer rotated around its long axis and moved in translation (Figure 9d). After reaching the target position, an alternating magnetic field was applied to induce the magnetostriction of the CFO core and then the pressure was transferred to the BFO shell to generate piezopotential, thus promoting the neuronal differentiation of the cells (Figure 9e–g). The prepared microswimmer could be degraded by the enzymes in the ECM produced by the cells. Fang et al. prepared a stretchable carbon porous nanocookies@conduit (NC@C) using 3D printing technology, in which the NC was composed of rGO, mesoporous silica and carbon layers with an excellent magneto-electric effect. The prepared NC@C could encapsulate the neuron growth factor (NGF) and achieve on-demand release under the control of a magnetic field. At the same time, it could electrically stimulate cells to effectively induce cell proliferation and neuronal differentiation in vitro and could further improve myelin layers and guide axonal orientation in vivo (Figure 9h–j) [85]. In addition, ME materials have also been developed for deep brain stimulation, bone regeneration, skeletal muscle tissue regeneration and more [85–88]. These materials inspire new approaches to targeted cell therapies for traumatic injuries.

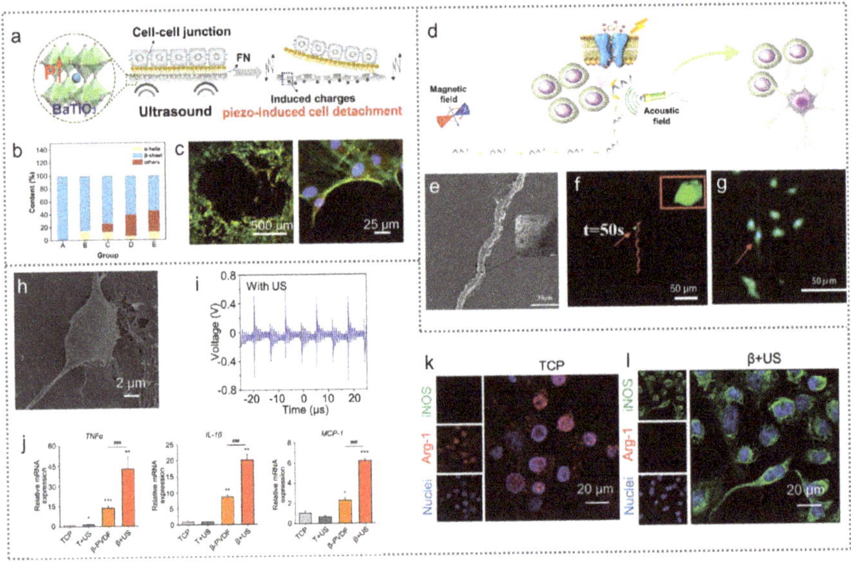

Figure 8. Ultrasound-driven PENGs for cell modulation: (**a**) a schematic diagram of the cell detachment regulated by piezopotential; (**b**) the contents of fibronectin α-helix and β-sheet under different conditions (A, FN in PBS; B, FN on the PVDF/BTO film; C, FN under ultrasound; D, E, FN on the PVDF/BTO film under ultrasound); (**c**) fluorescence microscopy images of cells remaining on the PVDF/BTO composite film after ultrasound. Reprinted with permission from ref. [72], Copyright 2021, Elsevier. (**d**) An illustration of a highly controllable micromotor for inducing the neuronal differentiation of targeted cells; (**e**) SEM images of *S.platensis*@Fe₃O₄@tBaTiO₃; (**f**) representative time-lapse Ca²⁺ imaging of PC12 cell stimulated by *S.platensis*@Fe₃O₄@tBaTiO₃ with ultrasound; (**g**) a fluorescent image of differentiated PC12 cells stimulated by *S.platensis*@Fe₃O₄@tBaTiO₃ with ultrasound (targeted cell indicated with a red arrow). βIII-tubulin is stained in green and the nuclei are in blue. Reprinted with permission from ref. [74], Copyright 2020, WILEY-VCH. (**h**) A representative SEM image of THP-1 cells on β-PVDF for 24 h; (**i**) the output voltage of β-PVDF with ultrasound stimulation; (**j**) the relative mRNA expression of the M1 markers of TNF-α, IL-1β and MCP-1 in different groups; (**k**,**l**) immunofluorescence staining images of M1 marker iNOS and M2 marker Arg-1 after culturing for 3 days on TCP and β-PVDF with ultrasound. * $p < 0.05$, ** $p < 0.01$, *** $p < 0.001$ vs. the TCP control group, ### $p < 0.001$ vs. the β-PVDF group. Reprinted with permission from ref. [78], Copyright 2021, The Authors. Published by WILEY-VCH.

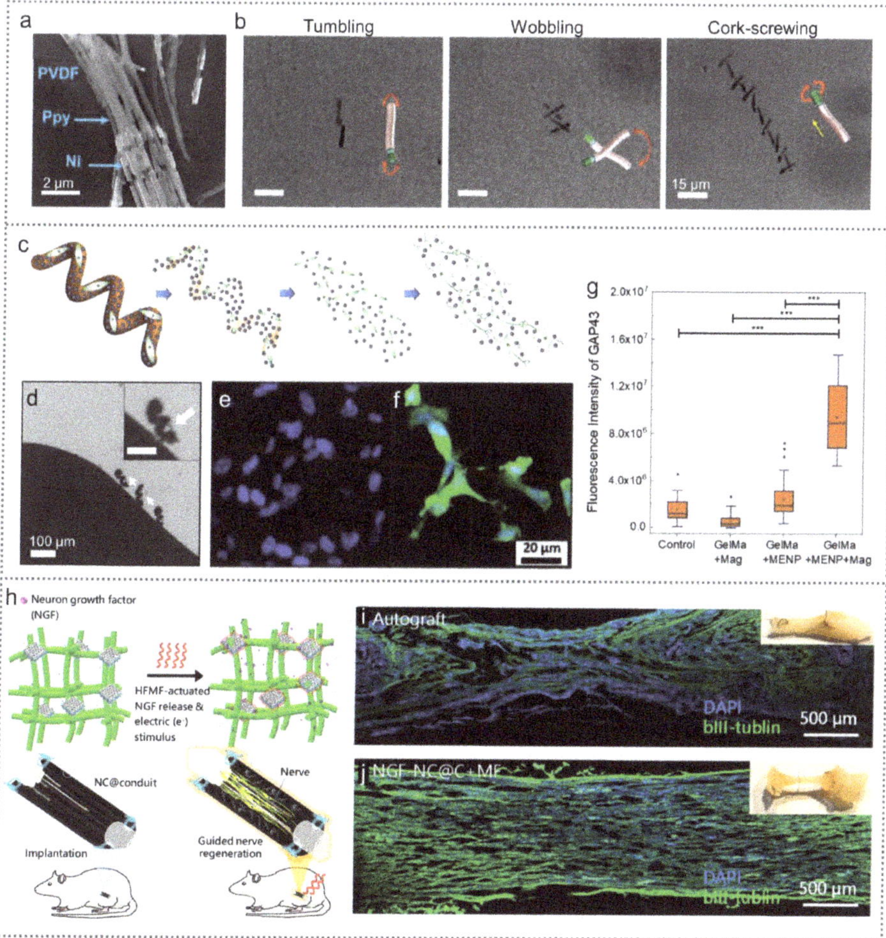

Figure 9. Magnetic field-driven PENGs for cell modulation: (**a**) a SEM image showing the hybrid nanoeels; (**b**) a time-lapse image showing the swimming behavior of the hybrid nanoeels, including tumbling, wobbling and corkscrewing. Reprinted with permission from ref. [83], Copyright 2019, WILEY-VCH. (**c**) A schematic diagram of the degradation process of the soft microswimmers and the neuronal differentiation of SH-SY5Y cells under a magnetic field; (**d**) a microswimmer loaded with cells driven by a rotating magnetic field. Fluorescent images of the cells on the (**e**) control and (**f**) soft microswimmers under magnetic stimulation. The nuclei are stained in blue and the neuronal specific protein GAP43 in green. (**g**) Box-and-whisker plots of the fluorescent intensities representing the level of GAP 43 expressed in SH-SY5Y cells (*** $p < 0.001$). Reprinted with permission from ref. [84], Copyright 2020, WILEY-VCH. (**h**) NC@C under magnetic field treatment promoting magneto-electric conversion into release growth factor and inducing neuron cell differentiation. Sciatic nerve defects harvested from (**i**) autograft and (**j**) NGF-NC@C+ magnetic field. Reprinted with permission from ref. [85], Copyright 2020, Springer Nature.

3.3.3. Cell Traction-Driven PENGs for Cell Modulation

There are abundant mechanical forces in cell activities, such as cell spreading, migration, contraction and cardiomyocyte beating. Thus, within the cellular microenvironment, cells continuously exert mechanical forces on the extracellular matrix (ECM). Similarly, the cells can also exert mechanical forces on the piezoelectric biomaterials. Murillo et al. demonstrated that the electromechanical interaction between living cells and a ZnO nanosheet-based PENG induced a local electric field that could modulate cell activity, such as stimulating the cell motility and activating the calcium channels to induce intracellular calcium transients, depending on the cell type (Figure 10a,d) [89]. The effective piezoelectric coefficient of the ZnO nanosheets was 4–6 pm V^{-1}, according to a piezoresponse atomic force microscope. Under the weak force of cells (0.1–10 nN), ZnO nanosheets could generate a piezopotential in the range of 0.5–50 mV. The number of macrophages grown on ZnO nanosheets with a traveling distance exceeding 150 μm was twice of that grown on the control substrate, indicating that the electromechanical interaction of the ZnO nanosheet-based PENG increased the motility of macrophages and facilitated long-distance displacement (Figure 10c). In addition, 64% of the SaOS-2 cells grown on the ZnO nanosheets presented increases in $[Ca^{2+}]_i$. By contrast, only 6% of the cells grown on glass coverslips showed increases in $[Ca^{2+}]_i$, with low amplitudes of Ca^{2+} transients (Figure 10d). These results indicate that the cell adhesion forces could bend the ZnO nanosheets and induce a local electric field to stimulate the cells and alter their activities. Zhang et al. designed a piezoelectric PVDF with a nanostripe array structure for the neuron-like differentiation of mesenchymal stem cells (MSCs) [90]. The traction force of the living cells on the surface of the nanostriped PVDF could cause the deformation of the PVDF stripes, thereby generating a local piezopotential to provide continuous electrical stimulation to the living cells. According to the simulation, there was a piezopotential from 34 μV to 3.4 mV when cell traction forces increased from 0.1 to 10 nN (Figure 10e,f). However, the piezopotential induced by the cell force (10 nN) on flat PVDF film was only 960 nV, which was insignificant for the physiological activities of the cells. Thus, the nanotopography of the PVDF film could increase the generated piezopotential in response to cell migration and traction and it provided a stronger signal to stimulate the differentiation of the stem cells. Unlike the spindle-shaped and flat cell morphology in the control group, the MSCs on the nanostriped PVDF formed a neuron-like morphology, including highly refracted cell bodies and elongated nanostriped PVDF. Moreover, the mRNA and protein expression levels of neuronal marker Tuj-1 were significantly increased, indicating that the generated piezopotential of the nanostriped PVDF had a positive effect on the neuron-like differentiation of MSCs (Figure 10g). Inspired by the biophysical cues of ECM, Li et al. developed an electromechanical coupling 3D bio-nanogenerator composed of $GO/PEDOT/Fe_3O_4/PAN$ fibers (GO, graphene oxide; PEDOT, poly(3,4-ethylenedioxythiophene); PAN, polyacrylonitrile), in which piezoelectric PAN was used as the electromechanical conversion unit [91]. The unique 3D structure of the bio-nanogenerator provided an ECM-like microenvironment for cell growth. It could also generate piezopotential up to millivolts through cell inherent force, thereby providing in situ electrostimulation for the adherent cells. Liu et al. prepared nanofibers with a suitable stiffness that was analogous to that of collagen using electrospinning technology [92]. Interestingly, the obvious mechanical deformation of the nanofibers was only observed after cell adhesion and mature focal adhesion formation (Figure 10h,i). Based on these dynamic mechanical forces in the cell microenvironment, a smart piezoelectric PVDF scaffold was designed that could generate piezopotential by cell traction force after cell adhesion, activate the transmembrane calcium channels for extracellular Ca^{2+} influx and promote the neuron-like differentiation of stem cells (Figure 10j,k). Since the deformation of the piezoelectric PVDF scaffold only occurred after cell adhesion, the on-demand electrical stimulation was only realized in the differentiation stage, thereby avoiding the inhibitory effect of early electrical stimulation on cell adhesion and spreading. In addition, Liu et al. designed a biodegradable piezoelectric PLLA that could generate piezopotential under joint load. This batteryless electrostimulation could promote protein adsorption and cell migration or recruitment, as well as induce endogenous TGF-β, thereby improving cartilage formation and cartilage regeneration. Rabbits with severe osteochondral defects regenerated hyaline cartilage and achieved complete cartilage healing after 1 to 2 months of self-driven electrical stimulation therapy [93]. This in situ electrical stimulation based on PENGs paves the way for smart scaffold design and future bioelectronic therapies.

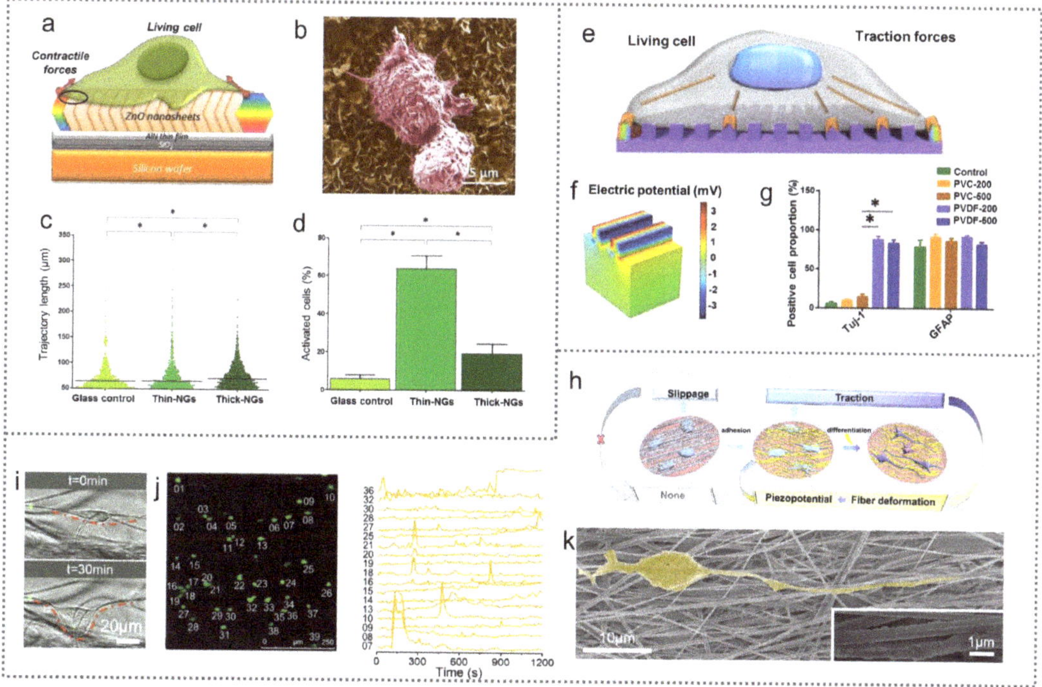

Figure 10. Cell traction-driven PENGs for cell modulation: (**a**) cell forces could bend the ZnO nanosheets of the PENG; (**b**) the micromorphology of the cells on the nanosheets; (**c**) the length of the trajectory represented the macrophage movement; (**d**) the quantification of activated Saos-2 cells with intracellular Ca^{2+} concentration change (* $p < 0.05$). Reprinted with permission from ref. [89], Copyright 2017, WILEY-VCH. (**e**) Cells grown on the nanostriped PVDF; (**f**) a simulation of the nanostriped PVDF generating a piezopotential of 3.4 mV when strained by a tangential force of 10 nN; (**g**) the percentage of Tuj-1 positive cells and GFAP positive cells (* $p < 0.05$). Reprinted with permission from ref. [90], Copyright 2019, WILEY-VCH. (**h**) A schematic diagram of the cell traction triggered, on-demand electrostimulation for neuron-like differentiation; (**i**) cell traction caused deformation along the nanofiber; (**j**) the cells grown on PVDF nanofibers with obvious transient calcium activity; (**k**) the morphology of the cells grown on the PVDF after differentiation. Reprinted with permission from ref. [92], Copyright 2021, WILEY-VCH.

4. Summary and Perspectives

In summary, EMNGs have shown promising applications in self-powered cell modulation, with impressive progress ranging from promoting cell migration, orientation and proliferation to regulating cell adhesion and differentiation (Table 1). Due to the advantages of convenience, good biosafety and patient compliance, EMNGs provide a promising approach for the clinical transformation of electrical stimulation. Generally, TENGs as an electrostimulation source need to cooperate with bioelectrodes to exert electrical stimulation to cells. By contrast, PENGs can achieve in situ wireless electrical stimulation while simultaneously acting as a biological scaffold. However, there are still great challenges and broad spaces for both basic research and clinic applications. Since the mechanical energy in the human body is weak and disordered, the structure of the nanogenerators needs to be reasonably designed and optimized to efficiently harvest the surrounding mechanical energy. Additionally, TENGs and PENGs can be hybridized with EMGs to combine the high voltage of the TENGs/PENGs and the high current of EMGs to realize multimodal electrostimulation, which is of great significance for broadening the application of EMNGs in the biomedical field. At the same time, this combination also poses some challenges for EMG technology, such as miniaturization, flexibility and system integration, etc. [14]. The long-term stability and safety of EMNGs in vivo still need to be verified. The current research is mainly focused on the effect of the open circuit voltage of the EMNG on the cells. In fact, the cell microenvironment is a complex electrophysiological environment. The voltage actually sensed by the cell depends not only on the output of the EMNG, but also on the electrical property of the culture medium and cell membrane [89]. Clarifying the actual voltage sensed by cells is more instructive for future research. In addition, the output of the EMNG is often a pulse wave, which is different from the square wave, sine wave and other waveforms produced by traditional electric stimulators. Thus, the electrostimulation conditions of the EMNG need to be further optimized and its underlying mechanism also needs to be explored. Additionally, the application of EMNGs can be extended to other electroactive cells, such as cardiomyocytes. The contractile force (20–140 nN) of cardiomyocytes is much higher than the cell traction force (1–10 nN) [90,94], which can be utilized to drive EMNGs more efficiently. Research at the cellular level is more conducive to revealing the regulatory mechanism of EMNG-based electrostimulation, with the ultimate goal of tissue function regulation and disease treatment. Both TENGs and PENGs have the advantages of small size, flexibility and good biocompatibility. To date, many wearable and implantable EMNGs have been reported that can efficiently harvest mechanical energy from living organisms, such as from joint motion, heartbeat and respiration, which have laid the foundation for further bioelectronic implants, especially for bones, hearts and muscles with abundant mechanical activities [93,95]. With the current progress and huge development space, EMNGs show tremendous potential for cell modulation and biomedical therapeutics.

Table 1. Electromechanical nanogenerators for cell modulation.

EMNG	Mode/Piezoelectric Scaffolds	Electrostimulation	Cells	Cell Modulation	Ref.
TENG	Implantable, contact–separation mode	10 V mm^{-1}, 1 Hz, 20 min/day, 5 days	Primary neurons	Cell alignment	[31]
	Contact–separation mode	30 µA, 3000 pulses/day, 21 days	MSCs	Neural differentiation	[35]
	Implantable, contact–separation mode	150 V cm^{-1}, 2 Hz, 1 h/day, 18 days	Preosteoblasts MC3T3-E1	Cell proliferation and osteogenic differentiation	[39]
	Contact–separation mode	12 V, 5–21 days	MC3T3-E1	Osteogenic differentiation	[40]
	Wearable, contact–separation mode	2 V cm^{-1}, 1 Hz, 6 h	Fibroblasts	Proliferation and differentiation into myofibroblasts	[43]
	Wearable, contact–separation mode	8 V, 1 Hz, 24 h	Fibroblasts	Cell alignment	[44]
	Implantable, contact–separation mode	100 mV mm^{-1}, 24 h	Fibroblasts	Cell migration	[33]
	Disk, freestanding mode	20 V, 20 Hz, 200 pluses	MCF-7, Hela, MSCs	Cell membrane permeability and drug delivery	[45]
	Contact–separation mode	20 V, 5 Hz, 200 pulses	MCF-7, MSCs	Cell membrane permeability and drug delivery	[28]
	Contact–separation mode	70 V, 4 kV cm^{-1}	RBCs	Cell membrane permeability and drug release	[47]

Table 1. *Cont.*

EMNG	Mode/Piezoelectric Scaffolds	Electrostimulation	Cells	Cell Modulation	Ref.
PENG	Wearable, PVDF	20 µA, 3 Hz, 2 h/day	MC3T3-E1	Cell proliferation and orientation and osteogenic differentiation	[67]
	PVDF–TrFE	d_{31} = 16.17 pC N^{-1} 6 mV, 6 nA,	Fibroblasts	Cell alignment and proliferation	[69]
	PVDF/BaTiO$_3$	d_{33} = 15.7 pC N^{-1}, −89.1 mV	MSCs	Regulate cell adhesion	[72]
	PVDF	d_{33} = 16.22 pC N^{-1}	Macrophages	Proinflammatory macrophage polarization	[78]
	FeOOH/PVDF	d_{33} = 27.2 pC N^{-1}	MSCs	Neural differentiation	[73]
	S. platensis@Fe$_3$O$_4$@ BaTiO$_3$	/	PC12 cells	Neural differentiation	[74]
	PVDF-TrFE/ CoFe$_2$O$_4$	/	PC12 cells	Neural differentiation	[75]
	Nylon-11	/	Dental pulp stem cells	Osteogenic differentiation	[77]
	CFO@BFO/GelMA	/	SHSY5Y cells	Neural differentiation	[84]
	NC@C	28–38 µA	PC12	Neural differentiation	[85]
	ZnO nanosheets	300 µV–45 mV	Macrophages	Stimulate the motility of macrophages	[89]
	PVDF	34 µV–3.4 mV	MSCs	Neural differentiation	[90]
	GO/PEDOT/Fe$_3$O$_4$/PAN	d_{33} = 4.5 pm V^{-1}, 14.1 µV–1.41 mV	Hepatocytes and MSCs	Motility of primary hepatocytes and osteogenic differentiation of MSCs	[91]
	PVDF	d_{33} = 24 pC N^{-1}, 0.73–133 mV	MSCs	Neural differentiation	[92]
PENG + TENG	Contact–separation mode PVDF/ZnO	1.5–2.7 V	Schwann cells	Cell proliferation and migration	[68]

Author Contributions: Conceptualization, L.L. and Z.L.; writing—original draft preparation, Z.L. and Z.W.; writing—review and editing, L.L.; supervision, L.L.; project administration, L.L.; funding acquisition, L.L. and Z.L. All authors have read and agreed to the published version of the manuscript.

Funding: The work was supported by the Strategic Priority Research Program of the Chinese Academy of Sciences (No. XDA16021100), the National Nature Science Foundation (No. 82072065, 81471784), the National Key R&D project from Minister of Science and Technology, China (2016YFA0202703), the National Youth Talent Support Program and the China Postdoctoral Science Foundation (No. BX2021299, 2021M703166).

Conflicts of Interest: The authors declare no conflict of interest.

References

1. Liu, Z.R.; Wan, X.Y.; Wang, Z.L.; Li, L.L. Electroactive Biomaterials and Systems for Cell Fate Determination and Tissue Regeneration: Design and Applications. *Adv. Mater.* **2021**, *33*, 2007429. [CrossRef] [PubMed]
2. da Silva, L.P.; Kundu, S.C.; Reis, R.L.; Correlo, V.M. Electric Phenomenon: A Disregarded Tool in Tissue Engineering and Regenerative Medicine. *Trends Biotechnol.* **2020**, *38*, 24–49. [CrossRef] [PubMed]
3. Balint, R.; Cassidy, N.J.; Cartmell, S.H. Electrical Stimulation: A Novel Tool for Tissue Engineering. *Tissue Eng. Part B—Rev.* **2013**, *19*, 48–57. [CrossRef] [PubMed]
4. Huang, G.; Li, F.; Zhao, X.; Ma, Y.; Li, Y.; Lin, M.; Jin, G.; Lu, T.J.; Genin, G.M.; Xu, F. Functional and Biomimetic Materials for Engineering of the Three-Dimensional Cell Microenvironment. *Chem. Rev.* **2017**, *117*, 12764–12850. [CrossRef]
5. Jiang, D.; Shi, B.; Ouyang, H.; Fan, Y.; Wang, Z.L.; Li, Z. Emerging Implantable Energy Harvesters and Self-Powered Implantable Medical Electronics. *ACS Nano* **2020**, *14*, 6436–6448. [CrossRef]
6. Zhu, B.; Luo, S.-C.; Zhao, H.; Lin, H.-A.; Sekine, J.; Nakao, A.; Chen, C.; Yamashita, Y.; Yu, H.-h. Large enhancement in neurite outgrowth on a cell membrane-mimicking conducting polymer. *Nat. Commun.* **2014**, *5*, 4523. [CrossRef]
7. Zheng, Q.; Shi, B.; Li, Z.; Wang, Z.L. Recent Progress on Piezoelectric and Triboelectric Energy Harvesters in Biomedical Systems. *Adv. Sci.* **2017**, *4*, 1700029. [CrossRef]
8. Jakešová, M.; Sjöström, T.A.; Đerek, V.; Poxson, D.; Berggren, M.; Głowacki, E.D.; Simon, D.T. Wireless organic electronic ion pumps driven by photovoltaics. *NPJ Flex. Electron.* **2019**, *3*, 14997. [CrossRef]

9.   Ogawa, Y.; Kato, K.; Miyake, T.; Nagamine, K.; Ofuji, T.; Yoshino, S.; Nishizawa, M. Organic Transdermal Iontophoresis Patch with Built-in Biofuel Cell. *Adv. Health. Mater.* **2015**, *4*, 506–510. [CrossRef]

10. Xiao, X.X.; McGourty, K.D.; Magner, E. Enzymatic Biofuel Cells for Self-Powered, Controlled Drug Release. *J. Am. Chem. Soc.* **2020**, *142*, 11602–11609. [CrossRef]

11. Sato, H.; Minamitani, Y.; Ohnishi, N.; Fujiwara, Y.; Katsukie, S. Development of a High-Frequency Burst Pulse Generator for Cancer Cell Treatment and Comparison Between a High-Power Burst Pulse and a Single Pulse. *IEEE Trans. Plasma Sci.* **2020**, *48*, 1051–1059. [CrossRef]

12. Brezar, S.K.; Kranjc, M.; Cemazar, M.; Bucek, S.; Sersa, G.; Miklavcic, D. Electrotransfer of siRNA to Silence Enhanced Green Fluorescent Protein in Tumor Mediated by a High Intensity Pulsed Electromagnetic Field. *Vaccines* **2020**, *8*, 49. [CrossRef]

13. Novickij, V.; Kranjc, M.; Staigvila, G.; Dermol-Cerne, J.; Melesko, J.; Novickij, J.; Miklaveic, D. High-Pulsed Electromagnetic Field Generator for Contactless Permeabilization of Cells In Vitro. *IEEE Trans. Magn.* **2020**, *56*, 5000106. [CrossRef]

14. Vidal, J.V.; Slabov, V.; Kholkin, A.L.; dos Santos, M.P.S. Hybrid Triboelectric-Electromagnetic Nanogenerators for Mechanical Energy Harvesting: A Review. *Nano-Micro Lett.* **2021**, *13*, 199. [CrossRef]

15. Fan, F.R.; Tian, Z.Q.; Wang, Z.L. Flexible triboelectric generator! *Nano Energy* **2012**, *1*, 328–334. [CrossRef]

16. Liang, X.; Liu, Z.R.; Feng, Y.W.; Han, J.J.; Li, L.L.; An, J.; Chen, P.F.; Jiang, T.; Wang, Z.L. Spherical triboelectric nanogenerator based on spring-assisted swing structure for effective water wave energy harvesting. *Nano Energy* **2021**, *83*, 105836. [CrossRef]

17. Ren, Z.W.; Wang, Z.M.; Liu, Z.R.; Wang, L.F.; Guo, H.Y.; Li, L.L.; Li, S.T.; Chen, X.Y.; Tang, W.; Wang, Z.L. Energy Harvesting from Breeze Wind (0.7-6 m s(-1)) Using Ultra-Stretchable Triboelectric Nanogenerator. *Adv. Energy Mater.* **2020**, *10*, 2001770. [CrossRef]

18. Yang, Y.; Wang, Z.L. Emerging nanogenerators: Powering the Internet of Things by high entropy energy. *iScience* **2021**, *24*, 102358. [CrossRef]

19. Luo, J.J.; Gao, W.C.; Wang, Z.L. The Triboelectric Nanogenerator as an Innovative Technology toward Intelligent Sports. *Adv. Mater.* **2021**, *33*, 2004178. [CrossRef]

20. Zhao, G.R.; Zhang, Y.W.; Shi, N.; Liu, Z.R.; Zhang, X.D.; Wu, M.Q.; Pan, C.F.; Liu, H.L.; Li, L.L.; Wang, Z.L. Transparent and stretchable triboelectric nanogenerator for self-powered tactile sensing. *Nano Energy* **2019**, *59*, 302–310. [CrossRef]

21. Nie, J.H.; Chen, X.Y.; Wang, Z.L. Electrically Responsive Materials and Devices Directly Driven by the High Voltage of Triboelectric Nanogenerators. *Adv. Funct. Mater.* **2019**, *29*, 1806351. [CrossRef]

22. Long, Y.; Li, J.; Yang, F.; Wang, J.Y.; Wang, X.D. Wearable and Implantable Electroceuticals for Therapeutic Electrostimulations. *Adv. Sci.* **2021**, *8*, 2004023. [CrossRef]

23. Wang, Y.; Wu, H.T.; Xu, L.; Zhang, H.N.; Yang, Y.; Wang, Z.L. Hierarchically patterned self-powered sensors for multifunctional tactile sensing. *Sci. Adv.* **2020**, *6*, eabb9083. [CrossRef]

24. Peng, X.; Dong, K.; Ye, C.Y.; Jiang, Y.; Zhai, S.Y.; Cheng, R.W.; Liu, D.; Gao, X.P.; Wang, J.; Wang, Z.L. A breathable, biodegradable, antibacterial, and self-powered electronic skin based on all-nanofiber triboelectric nanogenerators. *Sci. Adv.* **2020**, *6*, eaba9624. [CrossRef]

25. Sun, M.J.; Li, Z.; Yang, C.Y.; Lv, Y.J.; Yuan, L.; Shang, C.X.; Liang, S.Y.; Guo, B.W.; Liu, Y.; Li, Z.; et al. Nanogenerator-based devices for biomedical applications. *Nano Energy* **2021**, *89*, 106461. [CrossRef]

26. Zou, H.Y.; Zhang, Y.; Guo, L.T.; Wang, P.H.; He, X.; Dai, G.Z.; Zheng, H.W.; Chen, C.Y.; Wang, A.C.; Xu, C.; et al. Quantifying the triboelectric series. *Nat. Commun.* **2019**, *10*, 1427. [CrossRef]

27. Zou, H.Y.; Guo, L.T.; Xue, H.; Zhang, Y.; Shen, X.F.; Liu, X.T.; Wang, P.H.; He, X.; Dai, G.Z.; Jiang, P.; et al. Quantifying and understanding the triboelectric series of inorganic non-metallic materials. *Nat. Commun.* **2020**, *11*, 2093. [CrossRef]

28. Liu, Z.R.; Liang, X.; Liu, H.H.; Wang, Z.; Jiang, T.; Cheng, Y.Y.; Wu, M.Q.; Xiang, D.L.; Li, Z.; Wang, Z.L.; et al. High-Throughput and Self-Powered Electroporation System for Drug Delivery Assisted by Microfoam Electrode. *ACS Nano* **2020**, *14*, 15458–15467. [CrossRef]

29. Liang, X.; Jiang, T.; Liu, G.X.; Feng, Y.W.; Zhang, C.; Wang, Z.L. Spherical triboelectric nanogenerator integrated with power management module for harvesting multidirectional water wave energy. *Energy Environ. Sci.* **2020**, *13*, 277–285. [CrossRef]

30. Zhang, Z.C.; Yan, Q.Y.; Liu, Z.R.; Zhao, X.Y.; Wang, Z.; Sun, J.; Wang, Z.L.; Wang, R.R.; Li, L.L. Flexible MXene composed triboelectric nanogenerator via facile vacuum-assistant filtration method for self-powered biomechanical sensing. *Nano Energy* **2021**, *88*, 106257. [CrossRef]

31. Zheng, Q.; Zou, Y.; Zhang, Y.L.; Liu, Z.; Shi, B.J.; Wang, X.X.; Jin, Y.M.; Ouyang, H.; Li, Z.; Wang, Z.L. Biodegradable triboelectric nanogenerator as a life-time designed implantable power source. *Sci. Adv.* **2016**, *2*, e1501478. [CrossRef] [PubMed]

32. Jiang, W.; Li, H.; Liu, Z.; Li, Z.; Tian, J.; Shi, B.; Zou, Y.; Ouyang, H.; Zhao, C.; Zhao, L.; et al. Fully Bioabsorbable Natural-Materials-Based Triboelectric Nanogenerators. *Adv. Mater.* **2018**, *30*, 1801895. [CrossRef] [PubMed]

33. Li, Z.; Feng, H.Q.; Zheng, Q.; Li, H.; Zhao, C.C.; Ouyang, H.; Noreen, S.; Yu, M.; Su, F.; Liu, R.P.; et al. Photothermally tunable biodegradation of implantable triboelectric nanogenerators for tissue repairing. *Nano Energy* **2018**, *54*, 390–399. [CrossRef]

34. Wang, Z.L. Triboelectric Nanogenerator (TENG)-Sparking an Energy and Sensor Revolution. *Adv. Energy Mater.* **2020**, *10*, 2000137. [CrossRef]

35. Guo, W.B.; Zhang, X.D.; Yu, X.; Wang, S.; Qiu, J.C.; Tang, W.; Li, L.L.; Liu, H.; Wang, Z.L. Self-Powered Electrical Stimulation for Enhancing Neural Differentiation of Mesenchymal Stem Cells on Graphene-Poly(3,4-ethylenedioxythiophene) Hybrid Microfibers. *ACS Nano* **2016**, *10*, 5086–5095. [CrossRef]

36. Pate, F.D. Bone chemistry and paleodiet. *J. Archaeol. Method Theory* **1994**, *1*, 161–209. [CrossRef]

37. Kumar, A.; Nune, K.C.; Misra, R.D.K. Electric field-mediated growth of osteoblasts—The significant impact of dynamic flow of medium. *Biomater. Sci.* **2016**, *4*, 136–144. [CrossRef]

38. Creecy, C.M.; O'Neill, C.F.; Arulanandam, B.P.; Sylvia, V.L.; Navara, C.S.; Bizios, R. Mesenchymal Stem Cell Osteodifferentiation in Response to Alternating Electric Current. *Tissue Eng. Part A* **2013**, *19*, 467–474. [CrossRef]

39. Tian, J.J.; Shi, R.; Liu, Z.; Ouyang, H.; Yu, M.; Zhao, C.C.; Zou, Y.; Jiang, D.J.; Zhang, J.S.; Li, Z. Self-powered implantable electrical stimulator for osteoblasts' proliferation and differentiation. *Nano Energy* **2019**, *59*, 705–714. [CrossRef]

40. Shi, R.; Zhang, J.S.; Tian, J.J.; Zhao, C.C.; Li, Z.; Zhang, Y.Z.; Li, Y.S.; Wu, C.G.; Tian, W.; Li, Z. An effective self-powered strategy to endow titanium implant surface with associated activity of anti-biofilm and osteogenesis. *Nano Energy* **2020**, *77*, 105201. [CrossRef]

41. Yao, G.; Kang, L.; Li, C.C.; Chen, S.H.; Wang, Q.; Yang, J.Z.; Long, Y.; Li, J.; Zhao, K.N.; Xu, W.N.; et al. A self-powered implantable and bioresorbable electrostimulation device for biofeedback bone fracture healing. *Proc. Natl. Acad. Sci. USA* **2021**, *118*, e2100772118. [CrossRef]

42. Nuccitelli, R. A role for endogenous electric fields in wound healing. *Curr. Top. Dev. Biol.* **2003**, *58*, 1–26.

43. Long, Y.; Wei, H.; Li, J.; Yao, G.; Yu, B.; Ni, D.L.; Gibson, A.L.F.; Lan, X.L.; Jiang, Y.D.; Cai, W.B.; et al. Effective Wound Healing Enabled by Discrete Alternative Electric Fields from Wearable Nanogenerators. *ACS Nano* **2018**, *12*, 12533–12540. [CrossRef]

44. Liu, A.P.; Long, Y.; Li, J.; Gu, L.; Karim, A.; Wang, X.D.; Gibson, A.L.F. Accelerated complete human skin architecture restoration after wounding by nanogenerator-driven electrostimulation. *J. Nanobiotechnol.* **2021**, *19*, 280. [CrossRef]

45. Liu, Z.R.; Nie, J.H.; Miao, B.; Li, J.D.; Cui, Y.B.; Wang, S.; Zhang, X.D.; Zhao, G.R.; Deng, Y.B.; Wu, Y.H.; et al. Self-Powered Intracellular Drug Delivery by a Biomechanical Energy-Driven Triboelectric Nanogenerator. *Adv. Mater.* **2019**, *31*, 1807795. [CrossRef]

46. Yang, C.B.; Yang, G.; Ouyang, Q.L.; Kuang, S.Y.; Song, P.Y.; Xu, G.X.; Poenar, D.P.; Zhu, G.; Yong, K.T.; Wang, Z.L. Nanowire-array-based gene electro-transfection system driven by human-motion operated triboelectric nanogenerator. *Nano Energy* **2019**, *64*, 103901. [CrossRef]

47. Zhao, C.C.; Feng, H.Q.; Zhang, L.J.; Li, Z.; Zou, Y.; Tan, P.C.; Ouyang, H.; Jiang, D.J.; Yu, M.; Wang, C.; et al. Highly Efficient In Vivo Cancer Therapy by an Implantable Magnet Triboelectric Nanogenerator. *Adv. Funct. Mater.* **2019**, *29*, 1808640. [CrossRef]

48. Ouyang, H.; Liu, Z.; Li, N.; Shi, B.J.; Zou, Y.; Xie, F.; Ma, Y.; Li, Z.; Li, H.; Zheng, Q.; et al. Symbiotic cardiac pacemaker. *Nat. Commun.* **2019**, *10*, 1821. [CrossRef]

49. Liu, Z.R.; Yu, X.; Li, L.L. Piezopotential augmented photo- and photoelectro-catalysis with a built-in electric field. *Chin. J. Catal.* **2020**, *41*, 534–549. [CrossRef]

50. Wang, Z.L. Progress in Piezotronics and Piezo-Phototronics. *Adv. Mater.* **2012**, *24*, 4632–4646. [CrossRef]

51. Gao, Z.; Zhou, J.; Gu, Y.; Fei, P.; Hao, Y.; Bao, G.; Wang, Z.L. Effects of piezoelectric potential on the transport characteristics of metal-ZnO nanowire-metal field effect transistor. *J. Appl. Phys.* **2009**, *105*, 113707. [CrossRef]

52. Shao, H.; Fang, J.; Wang, H.; Lang, C.; Lin, T. Robust Mechanical-to-Electrical Energy Conversion from Short-Distance Electrospun Poly(vinylidene fluoride) Fiber Webs. *ACS Appl. Mater. Interfaces* **2015**, *7*, 22551–22557. [CrossRef]

53. Yu, H.; Huang, T.; Lu, M.X.; Mao, M.Y.; Zhang, Q.H.; Wang, H.Z. Enhanced power output of an electrospun PVDF/MWCNTs-based nanogenerator by tuning its conductivity. *Nanotechnology* **2013**, *24*, 405401. [CrossRef]

54. Kapat, K.; Shubhra, Q.T.H.; Zhou, M.; Leeuwenburgh, S. Piezoelectric Nano-Biomaterials for Biomedicine and Tissue Regeneration. *Adv. Funct. Mater.* **2020**, *30*, 1909045. [CrossRef]

55. Chorsi, M.T.; Curry, E.J.; Chorsi, H.T.; Das, R.; Baroody, J.; Purohit, P.K.; Ilies, H.; Nguyen, T.D. Piezoelectric Biomaterials for Sensors and Actuators. *Adv. Mater.* **2019**, *31*, 1802084. [CrossRef]

56. Wang, Z.L.; Song, J.H. Piezoelectric nanogenerators based on zinc oxide nanowire arrays. *Science* **2006**, *312*, 242–246. [CrossRef]

57. Li, Z.; Zhu, G.; Yang, R.; Wang, A.C.; Wang, Z.L. Muscle-Driven In Vivo Nanogenerator. *Adv. Mater.* **2010**, *22*, 2534–2537. [CrossRef]

58. Dagdeviren, C.; Yang, B.D.; Su, Y.; Tran, P.L.; Joe, P.; Anderson, E.; Xia, J.; Doraiswamy, V.; Dehdashti, B.; Feng, X.; et al. Conformal piezoelectric energy harvesting and storage from motions of the heart, lung, and diaphragm. *Proc. Natl. Acad. Sci. USA* **2014**, *111*, 1927–1932. [CrossRef]

59. Cheng, X.; Xue, X.; Ma, Y.; Han, M.; Zhang, W.; Xu, Z.; Zhang, H.; Zhang, H. Implantable and self-powered blood pressure monitoring based on a piezoelectric thinfilm: Simulated, in vitro and in vivo studies. *Nano Energy* **2016**, *22*, 453–460. [CrossRef]

60. Jeong, C.K.; Han, J.H.; Palneedi, H.; Park, H.; Hwang, G.-T.; Joung, B.; Kim, S.-G.; Shin, H.J.; Kang, I.-S.; Ryu, J.; et al. Comprehensive biocompatibility of nontoxic and high-output flexible energy harvester using lead-free piezoceramic thin film. *APL Mater.* **2017**, *5*, 074102. [CrossRef]

61. Kondapalli, H.; Alazzawi, Y.; Malinowski, M.; Timek, T.; Chakrabartty, S. Feasibility of Self-Powering and Energy Harvesting Using Cardiac Valvular Perturbations. *IEEE Trans. Biomed. Circuits Syst.* **2018**, *12*, 1392–1400. [CrossRef] [PubMed]

62. Kim, D.H.; Shin, H.J.; Lee, H.; Jeong, C.K.; Park, H.; Hwang, G.-T.; Lee, H.-Y.; Joe, D.J.; Han, J.H.; Lee, S.H.; et al. In Vivo Self-Powered Wireless Transmission Using Biocompatible Flexible Energy Harvesters. *Adv. Funct. Mater.* **2017**, *27*, 1700341. [CrossRef]

63. Zhai, W.C.; Zhu, L.P.; Berbille, A.; Wang, Z.L. Flexible and wearable piezoelectric nanogenerators based on P(VDF-TrFE)/SnS nanocomposite micropillar array. *J. Appl. Phys.* **2021**, *129*, 095501. [CrossRef]

64. Zhou, L.L.; Zhu, L.P.; Yang, T.; Hou, X.M.; Du, Z.T.; Cao, S.; Wang, H.L.; Chou, K.C.; Wang, Z.L. Ultra-Stable and Durable Piezoelectric Nanogenerator with All-Weather Service Capability Based on N Doped 4H-SiC Nanohole Arrays. *Nano-Micro Lett.* **2022**, *14*, 30. [CrossRef]

65. Dong, K.; Peng, X.; Wang, Z.L. Fiber/Fabric-Based Piezoelectric and Triboelectric Nanogenerators for Flexible/Stretchable and Wearable Electronics and Artificial Intelligence. *Adv. Mater.* **2020**, *32*, 1902549. [CrossRef]

66. Song, Y.H.; Shi, Z.Q.; Hu, G.H.; Xiong, C.X.; Isogai, A.; Yang, Q.L. Recent advances in cellulose-based piezoelectric and triboelectric nanogenerators for energy harvesting: A review. *J. Mater. Chem. A* **2021**, *9*, 1910–1937. [CrossRef]

67. Zhang, Y.Z.; Lingling, X.; Liu, Z.; Cui, X.; Xiang, Z.; Bai, J.Y.; Jiang, D.J.; Xue, J.T.; Wang, C.; Lin, Y.X.; et al. Self-powered pulsed direct current stimulation system for enhancing osteogenesis in MC3T3-E1. *Nano Energy* **2021**, *85*, 106009. [CrossRef]

68. Jin, F.; Li, T.; Yuan, T.; Du, L.J.; Lai, C.T.; Wu, Q.; Zhao, Y.; Sun, F.Y.; Gu, L.; Wang, T.; et al. Physiologically Self-Regulated, Fully Implantable, Battery-Free System for Peripheral Nerve Restoration. *Adv. Mater.* **2021**, *33*, 2104175. [CrossRef]

69. Wang, A.C.; Liu, Z.; Hu, M.; Wang, C.C.; Zhang, X.D.; Shi, B.J.; Fan, Y.B.; Cui, Y.G.; Li, Z.; Ren, K.L. Piezoelectric nanofibrous scaffolds as in vivo energy harvesters for modifying fibroblast alignment and proliferation in wound healing. *Nano Energy* **2018**, *43*, 63–71. [CrossRef]

70. Mitragotri, S. Innovation—Healing sound: The use of ultrasound in drug delivery and other therapeutic applications. *Nat. Rev. Drug Discov.* **2005**, *4*, 255–260. [CrossRef]

71. Pan, X.T.; Bai, L.X.; Wang, H.; Wu, Q.Y.; Wang, H.Y.; Liu, S.; Xu, B.L.; Shi, X.H.; Liu, H.Y. Metal-Organic-Framework-Derived Carbon Nanostructure Augmented Sonodynamic Cancer Therapy. *Adv. Mater.* **2018**, *30*, 1800180. [CrossRef]

72. Wan, X.Y.; Zhang, X.D.; Liu, Z.R.; Zhang, J.M.; Li, Z.; Wang, Z.L.; Li, L.L. Noninvasive manipulation of cell adhesion for cell harvesting with piezoelectric composite film. *Appl. Mater. Today* **2021**, *25*, 101218. [CrossRef]

73. Zhang, R.T.; Han, S.W.; Liang, L.L.; Chen, Y.K.; Sun, B.J.; Liang, N.; Feng, Z.C.; Zhou, H.X.; Sun, C.H.; Liu, H.; et al. Ultrasonic-driven electrical signal-iron ion synergistic stimulation based on piezotronics induced neural differentiation of mesenchymal stem cells on FeOOH/PVDF nanofibrous hybrid membrane. *Nano Energy* **2021**, *87*, 106192. [CrossRef]

74. Liu, L.; Chen, B.; Liu, K.; Gao, J.B.; Ye, Y.C.; Wang, Z.; Qin, N.; Wilson, D.A.; Tu, Y.F.; Peng, F. Wireless Manipulation of Magnetic/Piezoelectric Micromotors for Precise Neural Stem-Like Cell Stimulation. *Adv. Funct. Mater.* **2020**, *30*, 1910108. [CrossRef]

75. Chen, X.Z.; Liu, J.H.; Dong, M.; Muller, L.; Chatzipirpiridis, G.; Hu, C.Z.; Terzopoulou, A.; Torlakcik, H.; Wang, X.P.; Mushtaq, F.; et al. Magnetically driven piezoelectric soft microswimmers for neuron-like cell delivery and neuronal differentiation. *Mater. Horiz.* **2019**, *6*, 1512–1516. [CrossRef]

76. Marino, A.; Arai, S.; Hou, Y.Y.; Sinibaldi, E.; Pellegrino, M.; Chang, Y.T.; Mazzolai, B.; Mattoli, V.; Suzuki, M.; Ciofani, G. Piezoelectric Nanoparticle-Assisted Wireless Neuronal Stimulation. *ACS Nano* **2015**, *9*, 7678–7689. [CrossRef]

77. Ma, B.J.; Liu, F.; Li, Z.; Duan, J.Z.; Kong, Y.; Hao, M.; Ge, S.H.; Jiang, H.D.; Liu, H. Piezoelectric nylon-11 nanoparticles with ultrasound assistance for high-efficiency promotion of stem cell osteogenic differentiation. *J. Mater. Chem. B* **2019**, *7*, 1847–1854. [CrossRef]

78. Kong, Y.; Liu, F.; Ma, B.J.; Duan, J.Z.; Yuan, W.H.; Sang, Y.H.; Han, L.; Wang, S.H.; Liu, H. Wireless Localized Electrical Stimulation Generated by an Ultrasound-Driven Piezoelectric Discharge Regulates Proinflammatory Macrophage Polarization. *Adv. Sci.* **2021**, *8*, 2100962. [CrossRef]

79. Kopyl, S.; Surmenev, R.; Surmeneva, M.; Fetisov, Y.; Kholkin, A. Magnetoelectric effect: Principles and applications in biology and medicine—A review. *Mater. Today Bio* **2021**, *12*, 100149. [CrossRef]

80. Liu, X.L.; Li, D.; Zhao, H.X.; Dong, X.W.; Long, L.S.; Zheng, L.S. Inorganic-Organic Hybrid Molecular Materials: From Multiferroic to Magnetoelectric. *Adv. Mater.* **2021**, *33*, 2004542. [CrossRef]

81. Tokura, Y.; Kanazawa, N. Magnetic Skyrmion Materials. *Chem. Rev.* **2021**, *121*, 2857–2897. [CrossRef]

82. Nair, M.; Guduru, R.; Liang, P.; Hong, J.; Sagar, V.; Khizroev, S. Externally controlled on-demand release of anti-HIV drug using magneto-electric nanoparticles as carriers. *Nat. Commun.* **2013**, *4*, 1707. [CrossRef]

83. Mushtaq, F.; Torlakcik, H.; Hoop, M.; Jang, B.J.; Carlson, F.; Grunow, T.; Laubli, N.; Ferreira, A.; Chen, X.Z.; Nelson, B.J.; et al. Motile Piezoelectric Nanoeels for Targeted Drug Delivery. *Adv. Funct. Mater.* **2019**, *29*, 1808135. [CrossRef]

84. Dong, M.; Wang, X.P.; Chen, X.Z.; Mushtaq, F.; Deng, S.Y.; Zhu, C.H.; Torlakcik, H.; Terzopoulou, A.; Qin, X.H.; Xiao, X.Z.; et al. 3D-Printed Soft Magnetoelectric Microswimmers for Delivery and Differentiation of Neuron-Like Cells. *Adv. Funct. Mater.* **2020**, *30*, 1910323. [CrossRef]

85. Fang, J.H.; Hsu, H.H.; Hsu, R.S.; Peng, C.K.; Lu, Y.J.; Chen, Y.Y.; Chen, S.Y.; Hu, S.H. 4D printing of stretchable nanocookie@conduit material hosting biocues and magnetoelectric stimulation for neurite sprouting. *NPG Asia Mater.* **2020**, *12*, 61. [CrossRef]

86. Shuai, C.J.; Yang, W.J.; He, C.X.; Peng, S.P.; Gao, C.D.; Yang, Y.W.; Qi, F.W.; Feng, P. A magnetic micro-environment in scaffolds for stimulating bone regeneration. *Mater. Des.* **2020**, *185*, 108275. [CrossRef]

87. Kozielski, K.L.; Jahanshahi, A.; Gilbert, H.B.; Yu, Y.; Erin, O.; Francisco, D.; Alosaimi, F.; Temel, Y.; Sitti, M. Nonresonant powering of injectable nanoelectrodes enables wireless deep brain stimulation in freely moving mice. *Sci. Adv.* **2021**, *7*, eabc4189. [CrossRef]

88. Singer, A.; Dutta, S.; Lewis, E.; Chen, Z.Y.; Chen, J.C.; Verma, N.; Avants, B.; Feldman, A.K.; O'Malley, J.; Beierlein, M.; et al. Magnetoelectric Materials for Miniature, Wireless Neural Stimulation at Therapeutic Frequencies. *Neuron* **2020**, *107*, 631–643. [CrossRef]

89. Murillo, G.; Blanquer, A.; Vargas-Estevez, C.; Barrios, L.; Ibanez, E.; Nogues, C.; Esteve, J. Electromechanical Nanogenerator-Cell Interaction Modulates Cell Activity. *Adv. Mater.* **2017**, *29*, 1605048. [CrossRef]
90. Zhang, X.D.; Cui, X.; Wang, D.C.; Wang, S.; Liu, Z.R.; Zhao, G.R.; Zhang, Y.; Li, Z.; Wang, Z.L.; Li, L.L. Piezoelectric Nanotopography Induced Neuron-Like Differentiation of Stem Cells. *Adv. Funct. Mater.* **2019**, *29*, 1900372. [CrossRef]
91. Li, T.; Shi, C.M.; Jin, F.; Yang, F.; Gu, L.; Wang, T.; Dong, W.; Feng, Z.Q. Cell activity modulation and its specific function maintenance by bioinspired electromechanical nanogenerator. *Sci. Adv.* **2021**, *7*, eabh2350. [CrossRef] [PubMed]
92. Liu, Z.R.; Cai, M.J.; Zhang, X.D.; Yu, X.; Wang, S.; Wan, X.Y.; Wang, Z.L.; Li, L.L. Cell-Traction-Triggered On-Demand Electrical Stimulation for Neuron-Like Differentiation. *Adv. Mater.* **2021**, *33*, 2106317. [CrossRef] [PubMed]
93. Liu, Y.; Dzidotor, G.; Le, T.T.; Vinikoor, T.; Morgan, K.; Curry, E.J.; Das, R.; McClinton, A.; Eisenberg, E.; Apuzzo, L.N.; et al. Exercise-induced piezoelectric stimulation for cartilage regeneration in rabbits. *Sci. Transl. Med.* **2022**, *14*, 627. [CrossRef]
94. You, J.; Moon, H.; Lee, B.Y.; Jin, J.Y.; Chang, Z.E.; Kim, S.Y.; Park, J.; Hwang, Y.S.; Kim, J. Cardiomyocyte sensor responsive to changes in physical and chemical environments. *J. Biomech.* **2014**, *47*, 400–409. [CrossRef] [PubMed]
95. Soares dos Santos, M.P.; Coutinho, J.; Marote, A.; Sousa, B.; Ramos, A.; Ferreira, J.A.F.; Bernardo, R.; Rodrigues, A.; Marques, A.T.; da Cruz e Silva, O.A.; et al. Capacitive technologies for highly controlled and personalized electrical stimulation by implantable biomedical systems. *Sci. Rep.* **2019**, *9*, 5001. [CrossRef] [PubMed]

MDPI
St. Alban-Anlage 66
4052 Basel
Switzerland
Tel. +41 61 683 77 34
Fax +41 61 302 89 18
www.mdpi.com

Nanoenergy Advances Editorial Office
E-mail: nanoenergyadv@mdpi.com
www.mdpi.com/journal/nanoenergyadv